T0360654

ELEMENTS OF
STOCHASTIC
DYNAMICS

ELEMENTS OF STOCHASTIC DYNAMICS

Guo-Qiang Cai

Florida Atlantic University, USA

Wei-Qiu Zhu

Zhejiang University, China

World Scientific

NEW JERSEY · LONDON · SINGAPORE · BEIJING · SHANGHAI · HONG KONG · TAIPEI · CHENNAI · TOKYO

Published by

World Scientific Publishing Co. Pte. Ltd.

5 Toh Tuck Link, Singapore 596224

USA office: 27 Warren Street, Suite 401-402, Hackensack, NJ 07601

UK office: 57 Shelton Street, Covent Garden, London WC2H 9HE

Library of Congress Cataloging-in-Publication Data
Names: Cai, G. Q. (Guo-Qiang) | Zhu, Weiqiu.
Title: Elements of stochastic dynamics / Guo-Qiang Cai (Florida Atlantic University, USA),
    Weiqiu Zhu (Zhejiang University, China).
Description: New Jersey : World Scientific, 2016. | Includes bibliographical references and index.
Identifiers: LCCN 2015049522 | ISBN 9789814723329 (hc : alk. paper)
Subjects: LCSH: Stochastic processes. | Stochastic analysis. | Stochastic differential equations. |
    Dynamics.
Classification: LCC QA274 .C345 2016 | DDC 519.2/3--dc23
LC record available at http://lccn.loc.gov/2015049522

British Library Cataloguing-in-Publication Data
A catalogue record for this book is available from the British Library.

Copyright © 2017 by World Scientific Publishing Co. Pte. Ltd.

All rights reserved. This book, or parts thereof, may not be reproduced in any form or by any means, electronic or mechanical, including photocopying, recording or any information storage and retrieval system now known or to be invented, without written permission from the publisher.

For photocopying of material in this volume, please pay a copying fee through the Copyright Clearance Center, Inc., 222 Rosewood Drive, Danvers, MA 01923, USA. In this case permission to photocopy is not required from the publisher.

Desk Editors: Kalpana Bharanikumar/Amanda Yun

Typeset by Stallion Press
Email: enquiries@stallionpress.com

Printed in Singapore

PREFACE

The research field of stochastic dynamics has been evolving for several decades, and becomes a relatively mature scientific field now. There are two necessary elements involved: the investigated object is a dynamical system in which the input and output are changing with time, and another is the existence of uncertainties in system properties, initial state, and/or input and output which can be modeled as random variables or stochastic processes by inference. Topics of stochastic dynamics includes: modeling stochastic excitation processes according to their characteristics, methods and procedures to obtain responses of stochastically excited systems, failure and reliability of stochastic systems, stochastic stability and stochastic bifurcation. In simple words, stochastic dynamics investigates both qualitative and quantitative behaviors of stochastically excited dynamical systems. The knowledge on stochastic dynamics becomes more and more essential in many fields where uncertainties are present and play important roles, and enough data are available to form the probabilistic and/or statistical basis. Applications are found in many branches of engineering, such as civil, mechanical, chemical, aeronautical and astronautical, in physics, chemistry, biology and ecology, in meteorology, in social sciences, in economy and finance, and others.

This book is written for a broader audience with several major purposes. It serves as an introductory book for researchers in various areas who have been doing researches on deterministic dynamical systems and want to introduce stochastic elements in their researches. They may have none or little knowledge on stochastic dynamics. It can also provide researchers already in this field some advanced knowledge, solution procedures, and specific topics such as nonlinear systems, stochastic stability, stochastic bifurcation, first-passage failure, etc. The third major purpose of the book is to serve as a text book or a reference book for graduate students in related courses.

For the above objectives, the book is written as self-contained as possible. The prerequisite to read the book is the undergraduate course on probability and statistics. Deep knowledge on probability and knowledge on stochastic processes help, but are not necessary. Chapters 2–4 give fundamentals on random variables and stochastic processes. For new researchers in this field and graduate students, it is important to study these chapters thoroughly, understand the concepts, and do exercises. More comprehensive knowledge on stochastic processes and applications are given in numerous books, including Stratonovich (1963, 1967), Soong (1973), Karlin and Taylor (1975, 1981), Arnold (1974, 1998), Khasminskii (1980) and Gardiner (1986).

To narrow the very wide-ranging topics of stochastic dynamics, only excitations as input, and hence responses as output, are considered to be stochastic, while the properties of dynamical systems are assumed to be deterministic. The randomness in system properties and initial states are not considered here. This is the situation for most practical problems in which uncertainty in system properties and initial states are negligible, and the uncertainties in excitations are the main concern. Typical examples include structural systems in earthquakes, offshore structures and ships subjected to sea wave forces, airplanes and spaceships disturbed by atmosphere turbulences, vehicles on uneven roads, etc.

As indicated by the book title "Elements of Stochastic Dynamics", the book is by no means a comprehensive book covering all aspects of stochastic dynamics. It just provides basic knowledge necessary for pursuing research on stochastic dynamics, and reports limited research results in some specific topics. An incomplete list of excellent book on the field includes those of Crandall (1958), Crandall and Mark (1963), Lin (1967), Bolotin (1969), Elishakoff (1983), Ibrahim (1985), Dimentberg (1988), Zhu (1992, 2003), Soong and Grigoriu (1993), Lin and Cai (1995), Roberts and Spanos (1999) and Sun (2006).

There are several features of the book. First, some recent research results are incorporated in the book, including (i) modeling of stochastic processes, infinite and bounded, Gaussian and non-Gaussian, narrow- and broad-banded, (ii) solution procedures of nonlinear stochastic systems, (iii) stochastic Hamiltonian systems which represent general dynamic systems, (iv) stochastic ecosystems. Another important feature is that the book provides various simulation procedures of useful stochastic processes based on their probabilistic and statistical characteristics. Since analytical solutions are only obtainable for linear systems and limited nonlinear systems subjected to random excitations, Monte Carlo simulation becomes a general and even "default" means for investigation. With rapid advancement of computational techniques, Monte Carlo simulation serves as a powerful and sometimes only tool in stochastic dynamics field. The simulation procedures provided in the book are expected to be useful for research. Due to the nature and difficulty on the topics in this area, majority of books did not have enough exercise problems. This has been causing difficulty for instructors to teach related courses. This book will improve the situation substantially by providing a quite large number of exercise problems. In this book, basic concepts and theories of random variable and stochastic processes are given in detail; therefore, only some preliminary knowledge of probability is needed, and in-depth mathematics course is not required to take in advance. It will be appealing for students and researchers not with pure mathematics background. Finally, since the projected audience is mainly researchers and graduate students in various areas but mathematics, and also the major emphasis is on applications, the book is not written in rigorous mathematical terms.

The first author of the book contributes to the work partially during his stay at Zhejiang University, Hangzhou, China, as a visiting professor. The financial support from Zhejiang University is greatly acknowledged.

Finally, we express our special gratitude to Professors **S. H. Crandall** and Y. K. Lin, who introduced us into the field of random vibration and stochastic dynamics, and guided us in various research topics in the field.

CONTENTS

CHAPTER 3   STOCHASTIC PROCESSES   57

CHAPTER 6   EXACT STATIONARY SOLUTIONS OF NONLINEAR STOCHASTIC SYSTEMS   233

**CHAPTER 7   APPROXIMATE SOLUTIONS
OF NONLINEAR
STOCHASTIC SYSTEMS**   **283**

CHAPTER 8   STABILITY AND BIFURCATION OF STOCHASTICALLY EXCITED SYSTEMS   393

About the Authors

 **Guo-Qiang Cai** is currently a Professor in the Department of Ocean and Mechanical Engineering at Florida Atlantic University, USA. He is a member of the American Society of Mechanical Engineering; a member of the subcommittee on Stochastic Dynamics in IASSAR (International Association for Structural Safety and Reliability); and an editorial board member of the *International Journal of Nonlinear Mechanics*, *Probabilistic Engineering Mechanics*, and *Advances in Theoretical and Applied Mechanics*. He also reviews submissions for numerous scientific journals. His current research interests include: Structural dynamics — analytical and numerical methods; Linear and nonlinear stochastic systems — response, stability and bifurcation; Stochastic ecosystems; Earthquake engineering; and Reliability analysis.

 **Weiqiu Zhu** is currently a Professor and director of Institute of Applied Mechanics in the Department of Mechanics, at Zhejiang University, China. He is a member of Chinese Academy of Sciences; a member of the International Association of Structural Safety and Reliability; a member of the committee on Stochastic Methods in Structural Engineering; and an editorial board member of the following journals: *International Journal of Nonlinear Mechanics*; *Probabilistic Engineering Mechanics*; *Advances in Structural Engineering*; *International Journal of Computational Methods*; *Science China: Technological Sciences*; *Journal of Zhejiang University Sciences*; *Acta Mechanica Sinica* (Chinese Journal of Mechanics); *Acta Mechanica Solida Sinica* (Chinese Journal of Solid Mechanics); *Dynamics and Control* (in Chinese); and *Advances in Mechanics* (in Chinese). His current research interest is Nonlinear stochastic dynamics and control.

CHAPTER 1

INTRODUCTION

Modeling and analysis of dynamical systems are a critical task in almost all areas, such as physics, chemistry, biology, meteorology, ecology, economy, finance, and many branches of engineering including mechanical, ocean, civil, bio and earthquake engineering. In the modeling process, uncertainties are present inevitably due to various factors, such as possible changes of system parameters, variations in excitations, errors in modeling schemes, etc. To take into consideration of the uncertainties more accurately, observations and measurements are usually carried out to obtain data as much as possible. If the amount of the data for a specific uncertainty is large enough, this uncertainty can be described by means of probability and statistics. Specifically, if the uncertain physical quantity is a time-independent variable, it can be represented by a random variable. On the other hand, if the physical quantity is a time-varying process, it can be modeled as a stochastic process.

The earliest investigation of stochastic dynamics was due to Einstein in 1905 (Einstein, 1956), who developed a stochastic model for Brownian motion, a type of chaotic motion of small particles floating on water. The term of "Random Vibration", widely used in civil and mechanical engineering, was first proposed by Rayleigh (1919) for an acoustic problem. Study of random vibration began in the 50's in three aeronautical problems: buffeting of aircrafts due to atmosphere turbulence, acoustic fatigue of aircrafts caused by jet

1

noise and reliability of payloads in rocket-propelled space vehicles. The common factor in the three problems was the random nature of the excitations. Since then, investigation of systems under random excitations has been promoted to solve problems in aeronautical and astronautical, mechanical and civil engineering. The systems have been extended for linear to nonlinear, and the excitations from external to parametric. A survey of the development in the field of random vibration in the first three decades was given by Crandall and Zhu (1983). With rapid advancement of computer technology, more practical problems in various areas with many degrees of freedom and strong nonlinearity can be solved numerically using simulation techniques.

It is noticed that the term of random vibration has been used widely when responses and reliabilities of stochastic systems are of concern. Thus, development of solution methods is the main objective. If, besides the responses and reliabilities, the qualitative behaviors of stochastic systems, such as stability and bifurcation, are also the objectives of the investigation, the term "Stochastic Dynamics" is usually used, and it covers more topics than those in random vibration.

A stochastic dynamical system may be described by the following stochastic differential equations

$$\frac{d}{dt}X_j(t) = f_j[\mathbf{X}(t), t] + \sum_{l=1}^{m} g_{jl}[\mathbf{X}(t), t]\xi_l(t), \quad j = 1, 2, \ldots, n,$$

(1.0.1)

where $\mathbf{X}(t) = [X_1(t), X_2(t), \ldots, X_n(t)]^T$ is a vector of system response, also known as state variables, the superscript T is a notation for matrix transposition, $\xi_l(t)$ are excitations and at least one of them is stochastic process. Note that the capital letters used in (1.0.1) for the state variables signify that they are random or stochastic quantities. Functions f_j and g_{jl} represent system properties which may or may not depend on time explicitly. An excitation $\xi_l(t)$ is called a parametric (or multiplicative) one if the associated function g_{jl} depends on $\mathbf{X}$; otherwise, it is known as an external (additive) excitation.

If all functions f_j are linear functions of $\mathbf{X}$ and functions g_{jl} are all constants, the system is linear. If all functions f_j and g_{jl} are linear functions of X, it is known as parametrically excited linear system, although it is essentially nonlinear since the superposition principle is no longer applicable. If at least one of the functions f_j and g_{jl} is nonlinear, it is a nonlinear system. For the case of $n = 1$, it is a one-dimensional system. Otherwise, it is called a multi-dimensional system. A continuous system governed by a partial differential equation can be discretized to a multi-dimensional system using schemes such as finite-element procedure.

The stochasticity (randomness) may occur in system properties, in which case some parameters in functions f_j and g_{jl} are random variables. It may occur in excitations, namely, some of excitations $\xi_l(t)$ in Eq. (1.0.1) are stochastic processes. In this book, only the latter case is considered, and systems properties represented by functions f_j and g_{jl} are assumed to be deterministic.

Equations of motion of many mechanical and structural systems are usually established by means of the Newton's second law or the Lagrange equations according to the physical nature. The governing equations often appear in the following form

$$\ddot{Z}_j + h_j(\mathbf{Z}, \dot{\mathbf{Z}}) + u_j(\mathbf{Z}) = \sum_{l=1}^{m} g_{jl}(\mathbf{Z}, \dot{\mathbf{Z}})\xi_l(t), \quad j = 1, 2, \ldots, n,$$

(1.0.2)

where $\mathbf{Z} = [Z_1, Z_2, \ldots, Z_n]^T$ and $\dot{\mathbf{Z}} = [\dot{Z}_1, \dot{Z}_2, \ldots, \dot{Z}_n]^T$ are vectors of displacements and velocities, respectively, and $h_j(\mathbf{Z}, \dot{\mathbf{Z}})$ and $u_j(\mathbf{Z})$ represent damping forces and restoring forces, respectively. Letting $X_{2j-1} = Z_j$, $X_{2j} = \dot{Z}_j$ and $\mathbf{X} = [X_1, X_2, \ldots, X_{2n}]^T$, system (1.0.2) is transformed to

$$\dot{X}_{2j-1} = X_{2j},$$

$$\dot{X}_{2j} = -h_j(\mathbf{X}) - u_j(\mathbf{X}) + \sum_{l=1}^{m} g_{jl}(\mathbf{X})\xi_l(t).$$

(1.0.3)

By comparison of (1.0.3) and (1.0.1), it can be seen that equation set (1.0.3), is a special case of system (1.0.1). Conventionally, system (1.0.2) is known as an n-degree-of-freedom system, which is equivalent to a $2n$-dimensional system (1.0.1). These two technical terms will be followed throughout this book, namely, the term of "degree-of-freedom" is used for second-order system, while the term "dimension" is used for first-order systems. For example, a single-degree-of-freedom system is a two-dimensional system, and an n degrees-of-freedom system is a $2n$-dimensional system.

A stochastic dynamical systems can also be formulated as stochastically excited and dissipated Hamiltonian system, governed by the equations

$$\dot{Q}_j = \frac{\partial H}{\partial P_j},$$

$$\dot{P}_j = -\frac{\partial H}{\partial Q_j} - \sum_{k=1}^{n} c_{jk}(\mathbf{Q}, \mathbf{P}) \frac{\partial H}{\partial P_k} + \sum_{l=1}^{m} g_{jl}(\mathbf{Q}, \mathbf{P}) \xi_l(t),$$

(1.0.4)

where Q_j and P_j are generalized displacements and momenta, respectively, $\mathbf{Q} = [Q_1, Q_2, \ldots, Q_n]^T$, $\mathbf{P} = [P_1, P_2, \ldots, P_n]^T$ and $H = H(\mathbf{Q}, \mathbf{P})$ is a Hamiltonian function. Equation set (1.0.2) can be transformed to the form of (1.0.4) by using the Legendre transform. It can be seen that the stochastically excited and dissipated Hamiltonian system (1.0.4) is also a special case of system (1.0.1).

Mathematically, equations of motion of (1.0.1) are more general than (1.0.2) and (1.0.4) since the latter two equation sets can be transformed to the former equation set. However, for many engineering systems, Eqs. (1.0.2) are usually derived directly from Lagrange equations, and then transformed to (1.0.4). They describe the relationships between different degrees of freedom. The methods and procedures introduced in the book, although applicable to the general system (1.0.1), are especially suitable for systems (1.0.2) and (1.0.4).

The vectors $\mathbf{X}(t) = [X_1(t), X_2(t), \ldots, X_n(t)]^T$ in system (1.0.1), $\mathbf{Z} = [Z_1, Z_2, \ldots, Z_n]^T$ and $\dot{\mathbf{Z}} = [\dot{Z}_1, \dot{Z}_2, \ldots, \dot{Z}_n]^T$ in system (1.0.2), $\mathbf{Q} = [Q_1, Q_2, \ldots, Q_n]^T$ and $\mathbf{P} = [P_1, P_2, \ldots, P_n]^T$ in system (1.0.4) are known as system responses. Moreover, their functions, such as the system Hamiltonian, the amplitude envelope of a single response, and the system total energy, also belong to the category of system responses. Although the systems considered in this book are deterministic, the system responses are stochastic processes due to the stochastic excitations, as illustrated in Fig. 1.0.1.

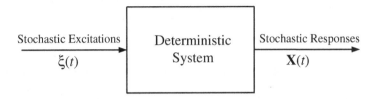

Figure 1.0.1   System excitations and responses.

With system models established deterministically, the most important element is the stochastic excitations, which must be modeled properly based on their characteristics in the involved physical problems. A large amount of data must be acquired from tests, experiments or real-time measurements in order to model a real physical excitation as a stochastic process described by statistical and/or probabilistic characteristics. Among the statistical characteristics, the mean value, mean square or variance, correlation function or spectral density are the most important. The probability distribution may be also important in certain practical problems (Wu and Cai, 2004). If the probability distribution can be inferred from available data, the mean and mean square can be calculated. In general, the power spectral density and probability density are the most desirable to model a stochastic process.

The classification of stochastic processes depends on the criterion chosen. According to their probability distributions, stochastic processes may be called Gaussian processes, Rayleigh processes, Poisson processes, bounded processes, etc. The Gaussian distribution is very popular due to several reasons: (a) the bell-shape of the Gaussian probability density indeed matches the shapes of many practical probability densities, (b) it can be defined completely only by two parameters, the mean and mean square and (c) its mathematical treatment is simple. One of the drawback of the Gaussian distribution is the unbound nature of the process even with a small probability. To overcome it, models of different types of bounded processes are described in the book. If the criterion is the frequency bandwidth of the spectral density, stochastic processes can be classified as narrow-band processes and broad-band processes. A harmonic process is a limiting case of narrow-band process with an infinitely narrow band. On the other hand, the widely used so-called white noise is a limiting case of broad-band process with a constant spectral density over an infinite bandwidth. Although it is an ideation of real broad-band processes, it has been used in many problems since it is easy to treat mathematically and system responses decay very rapidly in the range of higher frequency. Another criterion to classify stochastic processes is their time-evolutionary nature. Conceptually, a process is stationary if its probability density and spectral density do not change with time; otherwise, it is a non-stationary process (accurate definitions of stationarity are given in Chapter 3). Many real excitations lasting for long time duration, such as ocean wave forces, wind forces, forces to vehicles from road roughness, etc., may be considered as stationary processes in certain situations. However, short-time processes, such as ground accelerations during earth-quakes, are generally non-stationary. Other classification schemes may also be possible. Various mathematical models will be described for different types of stochastic processes in the book. The nature of the system responses, as well as solution procedures may be different for different types of stochastic excitation processes.

One of the main objectives in investigating dynamic systems is to acquire the properties of the system responses which may contain or constitute important or critical quantities of the system. In deterministic problems, responses of dynamical systems may be characterized as transient responses and steady-state responses. The transient responses are system responses immediately after the system is exposed to excitations, and initial conditions play an important role. The steady-state responses are those after a sufficiently long period of time since the excitations are applied, and the effects of initial conditions vanish. For stochastic dynamical systems, there are transient responses analogously. The concept corresponding to the steady-state responses in deterministic systems is known as stationary responses. For a given system, there may or may not be stationary responses depending on whether the excitation processes are stationary or non-stationary. In this book, most of solution procedures are for stationary responses by assuming the excitations are stationary.

Since both excitations and system responses are stochastic processes, their properties must be described in statistical and probabilistic terms. A stochastic process can be viewed as a sequence of correlated random variables at different time instants. Therefore, descriptions of random variables, as a prerequisite for stochastic processes, are introduced first in Chapter 2. Characteristics of a single random variable, including the probability distribution, characteristic function, statistical properties such as moments and cumulants, are defined and described. For multiple correlated random variables, the joint probability distribution, conditional probability density, and their joint statistical properties are defined and formulated. The probability density function and statistical moments of a function of random variables are also derived. Finally, simulation schemes are given for a single random variable based on its probability distribution and for two correlated random variables according to their joint probability distribution.

Chapter 3 gives the basic knowledge of stochastic processes. To specify a stochastic process, the first-order properties at one time instant, and the second-order properties involving two different time instants, are the most important and practical. The typical first-order properties include the probability density function and the statistical moments, such as mean and variance, while the most important and useful second-order properties are the correlation function, or covariance function, and the power spectral density. It is well known that the power spectral density, as the Fourier transform of the correlation function, is of practical significance since it describes the energy distribution of the process over the entire frequency domain. In contrast to the deterministic processes, the modes of convergence for stochastic processes, hence, the differentiation and integration, can be defined in different ways. Among them, the most useful and practical definition is the so-called L_2 convergence, which will be adopted throughout the book. Many practical stochastic processes are stationary processes, for which the first-order properties are independent of time and the second-order properties only depend on the time difference. Characteristic of such processes are explored in this chapter. Several commonly used stochastic processes, the Gaussian processes, Poisson and related processes, and evolutionary processes, are described in detail.

Fundamental theories and related topics of a specific stochastic process, Markov diffusion process, are given in Chapter 4. The Markov diffusion processes are especially important in stochastic dynamics since they can be used to generate many practical stochastic processes, and rigorous mathematical theories have been established. Important characteristics of diffusion processes are presented in the chapter. They are (a) the probability density of a diffusion process is governed by the well-known Fokker–Planck–Kolmogorov equation, (b) the simplest diffusion process is the Wiener process (Brownian motion) with the Gaussian white noise as its formal derivative, (c) responses of a system under Gaussian white-noise excitations constitute a diffusion process, and they are governed by the Stratonovich type of stochastic differential equations which can be transformed to Ito type of stochastic differential equations treated mathematically more readily. A diffusion process as the response of a one-dimensional Ito equation is investigated in details. Its dynamical behaviors can be analyzed by identifying the natures of its two boundaries, as shown in the chapter. In terms of Wiener processes, several practical stochastic processes can be generated, including processes filtered from nonlinear systems and randomized harmonic processes. Simulation of stochastic processes is much more complicated than that of random variables. Unlike a random variable, knowledge of the first-order probability distribution is not enough to generate samples of a stochastic process. The power spectral density representing the second-order property is also required to simulate a stochastic process. Schemes to simulate Gaussian and non-Gaussian stochastic processes are presented in this chapter.

Chapters 2–4 are fundamentals of random variables and stochastic processes, which are the basis to investigate the stochastic dynamics. Readers with strong knowledge on these topics may skip these chapters, or review them briefly. Otherwise, it is suggested that the concepts and mathematical treatments described in these chapters be understood and comprehended before getting into the subsequent chapters.

In system dynamics, the term of "linear system" is defined as a system with linear properties and subjected only to external excitations. For linear stochastic systems, exact solutions of certain properties of system responses can be obtained depending on the given knowledge of the excitation processes. Chapter 5 presents some solution procedures for responses of linear stochastic systems. Either time-domain analysis or frequency-domain analysis can be carried out if the stochastic excitations are stationary and also only stationary responses are of concern. If the excitations are Gaussian, the system responses are also Gaussian. On the other hand, the response probability distributions are in general not obtainable if the excitations are not Gaussian even the systems are linear. In the case of Gaussian white-noise excitations, Ito differential rule can be used to derive equations for moments and correlation functions which can be solved exactly.

If systems are nonlinear and/or parametric stochastic excitations are present, exact solutions are available only for special cases. Chapter 6 provides several procedures to obtain such exact stationary solutions: the methods of stationary potential, detailed balance and generalized stationary potential. Among them, the last one, i.e., that method of generalized stationary potential, is the more general one. It is found that quite stringent conditions between the system parameters and excitation intensities are generally required in order to obtain exact stationary solutions. A special type of systems is those in which the system properties are linear and the excitations are parametric on the linear terms of the system state variables. Such type of systems is essentially nonlinear since the superposition principle is no longer applicable; thus, solution methods in Chapter 5 cannot be applied. For such systems excited by Gaussian white noises, exact solutions for moments, correlation functions and spectral densities can be obtained by using the Ito differential rule.

In general, exact stationary solutions are not obtainable for nonlinear systems since the required conditions are not satisfied, especially for practical systems. Therefore, development of approximation procedures is in order. In Chapter 7, several approximation techniques for obtaining probabilistic and/or statistical solutions of nonlinear systems are presented. Equivalent linearization is the most common approximation method, and has been used widely. As extensions, the method of partial linearization and the linearization procedure for parametrically excited nonlinear systems are also introduced. The emphasis of the book is on the stochastic averaging method. Two versions of the stochastic averaging method for mechanical and structural systems are derived to obtain averaged equations of amplitude envelope and energy envelope, respectively. The stochastic averaging method is in fact effective as long as one or more slowly varying response processes are present in a dynamical system. As an application of the method other than in engineering problems, a stochastic nonlinear ecosystem is investigated as an illustration.

In Chapters 6 and 7, focuses are on mechanical and structural systems modeled in term of equations of motion (1.0.2). For more general dynamical systems, stochastically excited and dissipated Hamiltonian systems, governed by (1.0.4), are more suitable. Systematic methods to obtain exact and approximate solutions have been developed (Zhu, 2006) and will be briefly described in the two chapters. The Hamiltonian formulation is especially suitable to deal with multi-degree-of-freedom strongly nonlinear stochastic systems. Due to the complexity of Hamiltonian formulation, however, only basic principles and relatively simple solution procedures are covered in the book.

Chapter 8 deals with the stochastic stability and bifurcation, which are important topics of stochastic dynamics. It is well-known that stability problem arises for dynamical systems when parametric excitations are present or negative damping mechanisms exist. Definitions of stochastic stabilities in different senses are given, and stability analyses are conducted for linear systems and nonlinear systems which can be reduced to one-dimensional systems. For the topic of bifurcations, deterministic bifurcations are introduced briefly, and then two types of stochastic bifurcations, D- and P-bifurcations, are described and illustrated by examples.

Failures of stochastically excited dynamical systems may be classified into different types. One of the most important failure modes is that a critical physical quantity of the system exceeds a prescribed safety boundary for the first time, known as the first-passage failure. The reliability of a stochastically excited system associated with the first-passage failure is investigated in Chapter 9. The systems involved are those excited by Gaussian white noises. For such a system, the governing equation for the reliability function is of the same form as the backward Kolmogorov equation, and the moments of the first-passage time are governed by the generalized Pontryagin equations. While numerical solutions are necessary in general, several analytical solutions are presented. Finally, as a typical first-passage problem, a crack length in a material reaches a critical limit leading to fracture is investigated. It belongs to the general category of fatigue failure. The deterministic fatigue model is randomized, and the stochastic averaging method is applied to solve the problem.

Quite large number of exercise problems is given for each chapter of Chapters 2–9. By practicing these problems, readers will better understand and may have more thoughtful insights about the concepts and procedures given in the book. These exercise problems, although not complex, serve as an introduction and may give some useful hints to deal with more complicated problems in future researches.

CHAPTER 2

RANDOM VARIABLES

2.1 Introduction

Consider a random phenomenon. A single observation of the phenomenon is called a trial. Due to the randomness, the outcome of the trial is unknown in advance. However, we can always identify a set of elements which contains all possible outcomes of the trial. This set is called the sample space of the random phenomenon. For example, the sample space of tossing a die is the set $\{1, 2, 3, 4, 5, 6\}$, and the sample space of measuring the acceleration of a jet plane is all real numbers in a certain range, i.e., $(-A, A)$, where A is the limit of the jet plane acceleration. Every element in the sample space of a random phenomenon is called a sample point, representing a possible outcome. We denote a sample space by Ω and a sample point by ω, $\omega \in \Omega$.

An event is a subset of the sample space. It is a certain event if it is the same as the sample space, containing all sample points, i.e., all possible outcomes. On the contrast, it is an impossible event if it contains no sample point. Take the toss of a die as an example. An event may be that (a) the outcome is 3, (b) the outcome is a number less than 3, (c) the outcome is a number less than or equal to 6, (d) the outcome is larger than 6, etc. Obviously, event (c) is a certain event, while event (d) is an impossible event. Every event has a probability of occurrence associated with it. It is called the probability measure of the event.

For most random phenomenon of physical problems, outcomes are numerical values, such as wind speed to a building, ground acceleration during an earthquake, etc. For outcomes not in the form of numerical values, numbers can be assigned to outcomes by suitable choices. Therefore, we can represent a random phenomenon by a random number, say X. Since the value of X depends on the trial outcome represented by a sample point ω, X is therefore a function defined on the sample space Ω, $X = X(\omega)$, $\omega \in \Omega$. We have the following definition:

A random variable $X(\omega)$, $\omega \in \Omega$, is a function defined on a sample space Ω, such that for every real number x there exists a probability that $X(\omega) \leq x$, denoted by Prob[ω: $X(\omega) \leq x$].

For simplicity, the argument ω in $X(\omega)$ of a random variable is usually omitted, and the probability can be written as Prob[$X \leq x$].

There are two types of random variables: discrete and continuous. A discrete random variable takes only a countable number, finite or infinite, of distinct values. On the other hand, the sample space of a continuous random variable is an uncountable continuous space.

A random variable can be either a single-valued quantity or an n-dimensional vector described by n values. Except specified otherwise, we assume implicitly that a random variable is a single-valued quantity in the book.

2.2 Probability Distribution

Since a random variable is uncertain, we can describe it in terms of probability measure. For a discrete random variable, the simplest and direct way is to specify its probability to take a possible discrete value, written as

$$P_X(x) = \text{Prob}[X = x]. \tag{2.2.1}$$

In the notation $P_X(x)$, X is the random variable, and x is the state variable, i.e., the possible value of X. The convention of using a capital letter to denote a random quantity and a low case letter to represent its corresponding state variable will be followed throughout the book.

Another probability measure to describe a random variable is known as the probability distribution function, denoted as $F_X(x)$. It is the probability that the value of the random variable is smaller than or equal to a state variable, i.e.,

$$F_X(x) = \text{Prob}[X \leq x]. \tag{2.2.2}$$

Equation (2.2.2) implies that $F_X(x)$ is a non-decreasing function. For a discrete random variable,

$$F_X(x) = \text{Prob}[X \leq x] = \sum_{x_i \leq x} P_X(x_i), \tag{2.2.3}$$

where x_i are all possible values of X. Assume that the random variable represents a real physical quantity; thus, all state variables are real numbers, and we have

$$F_X(-\infty) = 0, \quad F_X(\infty) = 1. \tag{2.2.4}$$

As an example, denote X as the outcome of tossing a die. Its probability function is $P_X(x) = 1/6$, $x = 1, 2, 3, 4, 5, 6$ and its probability distribution function $F_X(x)$ is shown in Fig. 2.2.1.

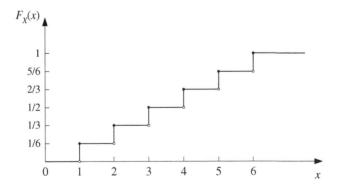

Figure 2.2.1 Probability distribution function of a discrete random variable.

For a continuous random variable associated with an uncountable sample space, the probability for it to be a certain value is generally zero, and the probability function is inappropriate to describe it.

However, the probability distribution function defined in Eq. (2.2.2) is still applicable. More important and useful is the probability density function, defined as the derivative of the probability distribution function if it exists, i.e.,

$$p_X(x) = \frac{dF_X(x)}{dx} = \lim_{\Delta x \to 0} \frac{F_X(x + \Delta x) - F_X(x)}{\Delta x}. \qquad (2.2.5)$$

Equation (2.2.5) indicates

$$p_X(x)dx = F_X(x + dx) - F_X(x) = \text{Prob}[x \leq X \leq x + dx]. \quad (2.2.6)$$

Although the probability for a continuous random variable to be a specified value x is zero, its probability in the vicinity of x can be measured by $p_X(x)$. A large $p_X(x)$ indicates that the probability of random variable X taking the values near x is large.

Given the probability density function $p_X(x)$, the probability for random variable X in a certain sample space, say $[a, b]$, can be found as follows:

$$\text{Prob}[a \leq X \leq b] = \int_a^b p_X(x)dx = F(b) - F(a). \qquad (2.2.7)$$

Also from Eqs. (2.2.3) and (2.2.4), we have

$$F_X(x) = \int_{-\infty}^{x} p_X(x')dx', \qquad (2.2.8)$$

$$F_X(\infty) = \int_{-\infty}^{\infty} p_X(x)dx = 1. \qquad (2.2.9)$$

For a discrete random variable, the probability density function can be represented in terms of Dirac delta function, i.e.,

$$p_X(x) = \sum P_X(x_i)\delta(x - x_i). \qquad (2.2.10)$$

It satisfies the normalization condition in Eq. (2.2.9) since

$$\int_{-\infty}^{\infty} \delta(x - x_i)dx = 1. \tag{2.2.11}$$

Equation (2.2.9) indicates that the total area under the probability density curve is one, and it is referred as the normalization condition of a probability density function. Figure 2.2.2 illustrates the

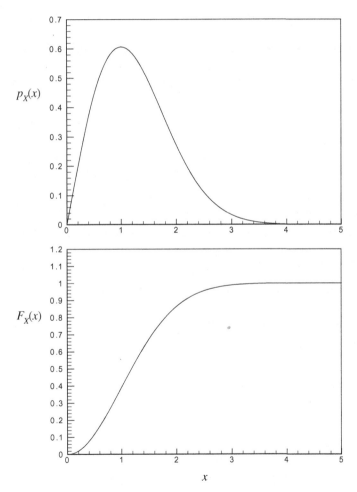

Figure 2.2.2 Probability density and distribution functions of a continuous random variable.

probability density function and probability distribution function of a continuous random variable defined in $[0, \infty)$.

2.3 Statistical Moments

A random variable can also be described by its statistical moments. The nth statistical moments of a random variable X is defined as

$$m_n[X] = E[X^n] = \int\limits_{-\infty}^{\infty} x^n p_X(x)dx. \qquad (2.3.1)$$

where the symbol $E[\cdot]$, denoting the integration with the probability density function as a weighting function, is called ensemble average, statistical average or mathematical expectation. The argument X in $m_n[X]$ may be omitted for simplicity.

The first-order moment $E[X]$ is the mean or average of the random variable X, and the second-order moment $E[X^2]$ is the mean-square value of X. Equation (2.3.1) shows that any order moments of a random variable X can be calculated if its probability density function $p_X(x)$ is known. Based on this, we may say that the probability density function $p_X(x)$ is a complete stochastic description of the random variable X. On the contrary, a finite number of moments may be not a complete description of the random variable. Nevertheless, the mean value and mean-square value are important in characterizing the random variable X.

Denoting the mean value by μ_X, i.e., $\mu_X = m_1 = E[X]$, the nth central moment is defined as $E[(X - \mu_X)^n]$. Among them, the most important one is the second central moment, called variance, denoted by σ_X^2, i.e.,

$$\sigma_X^2 = E[(X - \mu_X)^2] = E[X^2] - \mu_X^2 = m_2 - m_1^2. \qquad (2.3.2)$$

The square root of the variance, i.e., σ_X, is called the standard deviation. It gives a quantitative measure for the diffusion of a random variable from its mean value. A smaller σ_X indicates a larger probability near the mean value μ_X, while the probability far from μ_X is increasing with an increasing σ_X, as shown in Fig. 2.3.1 schematically.

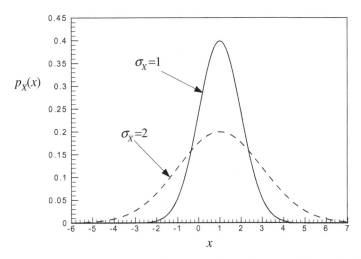

Figure 2.3.1 Probability density functions of random variables with different standard deviations.

Besides the variance, two more parameters, skewness γ_1 and kurtosis γ_2, related to the central moments are introduced to reflect the geometric shape of the probability density function. Their definitions are

$$\gamma_1 = \frac{E[(X - \mu_X)^3]}{\sigma_X^3}, \quad \gamma_2 = \frac{E[(X - \mu_X)^4]}{\sigma_X^4}. \quad (2.3.3)$$

The skewness γ_1 is a measure of asymmetry of the probability density function. If $\gamma_1 = 0$, the probability density is symmetric about its mean value. The kurtosis γ_2 reflects the shape of the peak of the probability density. A larger γ_2 corresponds to a sharp peak. It will be shown in Section 2.5.1 that $\gamma_1 = 0$ and $\gamma_2 = 1$ for a Gaussian distributed random variable. Therefore, the skewness and kurtosis can be used to identify non-Gaussian characteristics of probability distributions.

2.4 Characteristic Function and Cumulants

2.4.1 *Characteristic Function*

The characteristic function of a random variable is defined as the ensemble average of the exponential function of the random variable as follows:

$$M_X(\theta) = E[e^{i\theta X}] = \int_{-\infty}^{\infty} e^{i\theta x} p_X(x) dx, \qquad (2.4.1)$$

where $i = \sqrt{-1}$. Equation (2.4.1) shows that the characteristic function is the Fourier transform of the probability density function. The inversion of (2.4.1) is

$$p_X(x) = \frac{1}{2\pi} \int_{-\infty}^{\infty} M_X(\theta) e^{-i\theta x} d\theta. \qquad (2.4.2)$$

The characteristic function $M_X(\theta)$ is a complex function, and its existence is guaranteed by the non-negative property and normalization condition (2.2.9) of the probability density function.

Differentiating (2.4.1) with respect to θ, the moments of random variable X, if exist, are obtained as follows:

$$m_n = E[X^n] = \frac{1}{i^n} \left[\frac{d^n M_X(\theta)}{d\theta^n} \right]_{\theta=0} \qquad (2.4.3)$$

and the characteristic function in Eq. (2.4.1) can be expanded into a Maclaurin series

$$M_X(\theta) = 1 + \sum_{n=1}^{\infty} \frac{m_n}{n!} (i\theta)^n. \qquad (2.4.4)$$

Equation (2.4.4) indicates that, in general, all moments are required in order to fully describe a random variable.

2.4.2 *Cumulants*

Taking the logarithm of the characteristic function and expand it into the Maclaurin series, we have

$$\ln M_X(\theta) = \sum_{n=1}^{\infty} \frac{\kappa_n[X]}{n!} (i\theta)^n, \qquad (2.4.5)$$

where the coefficient $\kappa_n[X]$ is called the nth cumulant or semi-invariant of X, given by

$$\kappa_n[X] = \frac{1}{i^n}\left[\frac{d^n}{d\theta^n}\ln M_X(\theta)\right]_{\theta=0}. \tag{2.4.6}$$

Similar to the case of moments, the argument X in $\kappa_n[X]$ may be omitted for simplicity.

Take the logarithm of Eq. (2.4.4) and expand the right-hand side into the power series to obtain

$$\begin{aligned}
\ln M_X(\theta) &= \ln\left\{1 + \sum_{n=1}^{\infty}\frac{m_n}{n!}(i\theta)^n\right\} \\
&= \sum_{n=1}^{\infty}\frac{m_n}{n!}(i\theta)^n - \frac{1}{2}\left\{\sum_{n=1}^{\infty}\frac{m_n}{n!}(i\theta)^n\right\}^2 \\
&\quad + \frac{1}{3}\left\{\sum_{n=1}^{\infty}\frac{m_n}{n!}(i\theta)^n\right\}^3 - \cdots . \\
&= m_1(i\theta) + \frac{1}{2!}(m_2 - m_1^2)(i\theta)^2 \\
&\quad + \frac{1}{3!}(m_3 - 3m_1 m_2 + 2m_1^3)(i\theta)^3 - \cdots
\end{aligned} \tag{2.4.7}$$

Comparison of Eqs. (2.4.5) and (2.4.7) leads to the relationship between the moments and cumulants. For the first several orders, we have,

$$\begin{aligned}
\kappa_1 &= m_1 = \mu_X, \\
\kappa_2 &= m_2 - m_1^2 = \sigma_X^2, \\
\kappa_3 &= m_3 - 3m_1 m_2 + 2m_1^3, \\
\kappa_4 &= m_4 - 3m_2^2 - 4m_1 m_3 + 12m_1^2 m_2 - 6m_1^4.
\end{aligned} \tag{2.4.8}$$

In terms of the central moments, (2.4.8) can be converted to

$$\begin{aligned}
\kappa_1 &= m_1 = \mu_X, \\
\kappa_2 &= E[(X - \mu_X)^2] = \sigma_X^2, \\
\kappa_3 &= E[(X - \mu_X)^3], \\
\kappa_4 &= E[(X - \mu_X)^4] - 3\left\{E[(X - \mu_X)^2]\right\}.
\end{aligned} \tag{2.4.9}$$

2.5 Common Probability Distributions

Commonly, random variables are classified according to their probability distributions. In this section, several common probability distributions to be used in this book will be introduced. For simplicity, the subscript to specify the random variable in the probability density function may be omitted if no confusion occurs.

2.5.1 *Gaussian (Normal) Distribution*

The probability density of Gaussian distribution is given by

$$p(x) = \frac{1}{\sqrt{2\pi}\sigma} e^{-\frac{(x-\mu)^2}{2\sigma^2}}, \quad -\infty < x < \infty, \quad \sigma > 0. \tag{2.5.1}$$

It can be shown that the mean and variance are

$$\mu_X = \mu, \quad \sigma_X^2 = \sigma^2. \tag{2.5.2}$$

These two constants completely specify the probability density, and hence all other statistics. The notation $N(\mu, \sigma)$ is used to represents the Gaussian distribution, and the notation $X \sim N(\mu, \sigma)$ stands for a Gaussian random variable X. $N(0, 1)$ is called the standard (or unit) Gaussian distribution, denoted usually as $U \sim N(0, 1)$. Figure 2.5.1 shows two Gaussian probability density functions of different means and standard deviations.

The characteristic function of a $N(\mu, \sigma)$ random variable is

$$M_X(\theta) = e^{i\mu\theta - \frac{1}{2}\sigma^2\theta^2}. \tag{2.5.3}$$

Equations (2.5.1) and (2.5.3) constitute a Fourier transformation pair. Equation (2.5.3) combined with (2.4.5) shows that all cumulants of orders higher than two vanish, which can be considered as an alternative definition for the Gaussian distribution.

One of the reasons that the Gaussian distribution is often assumed for many physical quantities is due to the central limit theorem. Assume that random variables X_i, $i = 1, 2, \ldots, n$ are independent, and with zero mean and unit standard deviation. Then the central limit theorem states that the sum $S_n = \sum_{i=1}^{n} X_i$ tends to the standard Gaussian random variable as n tends to infinity,

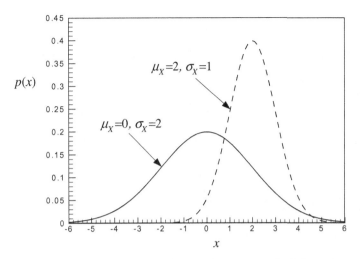

Figure 2.5.1 Gaussian probability density functions with different means and standard deviations.

regardless the individual distribution of X_i. This theorem suggests that, if a random phenomenon is a consequence of a large number of random factors and none of them is predominant, it may be assumed to be Gaussian distributed.

Another reason for the Gaussian distribution to be popular is its simplicity of mathematical properties, one of which is that, it is defined completely only by the first two moments, mean and mean square. Another important property is that the linear functions of Gaussian random variables remain Gaussian distributed. For example, if X and Y are two Gaussian random variables, $aX + bY$ is also a Gaussian random variable. Furthermore, there exists a simple relationship between the moments of various orders. Applying (2.5.1) or (2.4.3), we obtain an iterative equation

$$E[X^n] = \mu E[X^{n-1}] + (n-1)\sigma^2 E[X^{n-2}], \quad n = 2, 3, \ldots. \quad (2.5.4)$$

The derivation of (2.5.4) is left to readers in Exercise Problem 2.4. For a standard Gaussian random variable $U \sim N(0,1)$, (2.5.4) leads to

$$E[U^n] = \begin{cases} 0, & n = \text{odd} \\ 1 \times 3 \times \cdots \cdots \times (n-1), & n = \text{even}. \end{cases} \quad (2.5.5)$$

Using (2.5.4) and (2.5.5), higher-order moments can be readily calculated from the lower-order moments.

A Gaussian random variable $X \sim N(\mu, \sigma)$ can always be expressed in terms of the standard one as

$$X = \sigma U + \mu. \tag{2.5.6}$$

Using (2.5.6), the central moments of X can be calculated from

$$E[(X - \mu)^n] = \sigma^n E[U^n]. \tag{2.5.7}$$

Equations (2.5.5) and (2.5.7) show that the skewness $\gamma_1 = 0$ and the kurtosis $\gamma_2 = 1$.

2.5.2 Uniform Distribution

$$p(x) = \frac{1}{b - a}, \quad a \le x \le b; \quad \mu_X = \frac{1}{2}(a + b), \quad \sigma_X^2 = \frac{1}{12}(b - a)^2. \tag{2.5.8}$$

The uniform distribution is controlled by the lower and upper limits of the interval.

2.5.3 Rayleigh Distribution

$$p(x) = \frac{x}{\sigma^2} e^{-\frac{x^2}{2\sigma^2}}, \quad x \ge 0; \quad \mu_X = \sqrt{\frac{\pi}{2}}\sigma, \quad \sigma_X^2 = \frac{4 - \pi}{2}\sigma^2. \tag{2.5.9}$$

There is only a single parameter σ in the Rayleigh distribution. The probability density functions of Rayleigh distribution (2.5.9) are shown in Fig. 2.5.2 for three different σ values.

2.5.4 Exponential Distribution

$$p(x) = \lambda e^{-\lambda x}, \quad x \ge 0, \quad \lambda > 0; \quad \mu_X = \frac{1}{\lambda}, \quad \sigma_X^2 = \frac{1}{\lambda^2}. \tag{2.5.10}$$

For different values of parameter λ, the probability density functions of exponential distribution (2.5.10) are shown in Fig. 2.5.3.

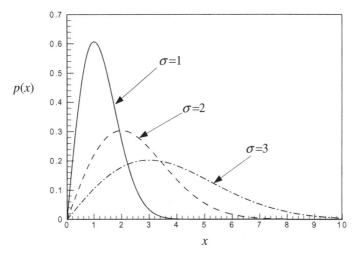

Figure 2.5.2 Probability density functions of Rayleigh distributions with different σ values.

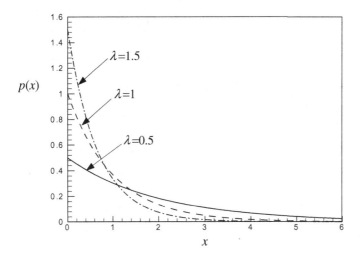

Figure 2.5.3 Probability density functions of exponential distribution with different λ values.

For the distributions given in Sections 2.5.3 and 2.5.4, the random variables are defined in $[0, \infty)$. These distributions can be used to model random variables defined in $[x_0, \infty)$ by simply replacing x by $x - x_0$ in the probability density functions.

2.5.5 λ-*Distribution*

$$p(x) = \left[B\left(\frac{1}{2}, \lambda + \frac{1}{2} \right) \right]^{-1} (1 - x^2)^{\lambda - \frac{1}{2}},$$

$$-1 \le x \le 1 \text{ for } \lambda \ge \frac{1}{2} \quad \text{or} \quad -1 < x < 1 \text{ for } -\frac{1}{2} < \lambda < \frac{1}{2},$$

$$\mu_X = 0, \quad \sigma_X^2 = \frac{1}{2(\lambda + 1)} \tag{2.5.11}$$

where $B(\cdot, \cdot)$ is the Beta function, defined as

$$B(u, v) = \int_0^1 t^{u-1}(1 - t)^{v-1}dt, \quad u, v > 0. \tag{2.5.12}$$

The λ-distribution can be used to model a random variable defined in a finite range. By selecting different values of λ, it is able to describe a variety of different shapes for the probability density function, as shown in Fig. 2.5.4. It is noted that the uniform distribution is a special case of $\lambda = 0.5$.

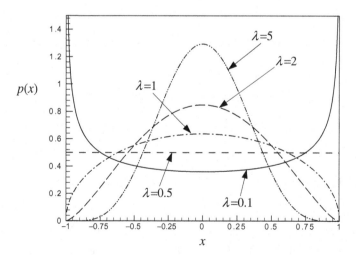

Figure 2.5.4 Probability density functions of λ-distribution with different parameters.

2.5.6 *Poisson Distribution*

The above probability distributions are for continuous random variables. A useful probability distribution known as Poisson distribution for a type of discrete random variables is presented below.

The Poisson distribution is commonly used in problems that deal with the occurrence of some random events. Assume the number of occurrence of a random event is N, which is a discrete random number with all non-negative integers as its sample space. It is said that N obeys Poisson distribution if

$$P_N(n) = \text{Prob}[N = n] = \frac{\mu^n}{n!} e^{-\mu}, \quad n = 0, 1, 2, \ldots \ldots \quad (2.5.13)$$

The only parameter for the Poisson distribution is μ. From (2.5.13), we obtain the moments

$$E[N^k] = \sum_{n=0}^{\infty} n^k P_N(n) = \sum_{n=1}^{\infty} \frac{n^k \mu^n}{n!} e^{-\mu}. \quad (2.5.14)$$

The mean, mean square and variance are then

$$\mu_N = E[N] = \sum_{n=1}^{\infty} \frac{\mu^n}{(n-1)!} e^{-\mu} = \mu e^{-\mu} \sum_{n=1}^{\infty} \frac{\mu^{n-1}}{(n-1)!} = \mu, \quad (2.5.15)$$

$$E[N^2] = \sum_{n=1}^{\infty} \frac{n \mu^n}{(n-1)!} e^{-\mu}$$

$$= \mu^2 e^{-\mu} \sum_{n=2}^{\infty} \frac{\mu^{n-2}}{(n-2)!} + \mu e^{-\mu} \sum_{n=1}^{\infty} \frac{\mu^{n-1}}{(n-1)!} = \mu^2 + \mu, \quad (2.5.16)$$

$$\sigma_N^2 = E[N^2] - \mu_N^2 = \mu. \quad (2.5.17)$$

As the mean value μ tends to infinity, the discrete random variable N tends to a continuous random variable, and the Poisson distribution approaches the Gaussian distribution. The proof is left to the readers in Exercise Problem 2.9.

2.6 Multiple Random Variables

In practical problems, we may deal with two or more random variables which are related to one another in some probabilistic way. They are called jointly distributed random variables. Then it is necessary to extend the above analysis of a single random variable to multiple random variables. In the following, we discuss in detail the case of two random variables. It can be readily generalized to cases of more than two random variables.

2.6.1 *Joint Probability Distribution*

Let X_1 and X_2 be two jointly distributed random variables. Their behavior is described by the joint probability distribution function, defined as

$$F_{X_1X_2}(x_1, x_2) = \text{Prob}[(X_1 \leq x_1) \cap (X_2 \leq x_2)], \qquad (2.6.1)$$

where $\cap$ denotes the intersection of Boolean operation. Similar to the case of single random variable, $F_{X_1X_2}(x_1, x_2)$ satisfies

$$\begin{aligned}
&F_{X_1X_2}(\infty, \infty) = 1, \\
&F_{X_1X_2}(-\infty, x_2) = F_{X_1X_2}(x_1, -\infty) = 0, \\
&F_{X_1X_2}(x_1, \infty) = F_{X_1}(x_1), \quad F_{X_1X_2}(\infty, x_2) = F_{X_2}(x_2).
\end{aligned} \qquad (2.6.2)$$

The joint probability density function is the mixed derivative of the joint distribution function, i.e.,

$$p_{X_1X_2}(x_1, x_2) = \frac{\partial^2}{\partial x_1 \partial x_2} F_{X_1X_2}(x_1, x_2). \qquad (2.6.3)$$

The inverse of (2.6.3) is

$$F_{X_1X_2}(x_1, x_2) = \int_{-\infty}^{x_1} \int_{-\infty}^{x_2} p_{X_1X_2}(x_1', x_2')dx_1'dx_2'. \qquad (2.6.4)$$

It is clear that $F_{X_1X_2}(x_1, x_2)$ is a non-decreasing function with respect to x_1 and x_2, respectively, and $p_{X_1X_2}(x_1, x_2)$ is a non-negative function with the normalization condition as

$$\int_{-\infty}^{\infty} \int_{-\infty}^{\infty} p_{X_1X_2}(x_1, x_2)dx_1dx_2 = 1. \qquad (2.6.5)$$

The respective probability density functions of X_1 and X_2, i.e., $p_{X_1}(x_1)$ and $p_{X_2}(x_2)$, are called the marginal probability density functions. They can be calculated from the joint probability density as follows

$$p_{X_1}(x_1) = \int_{-\infty}^{\infty} p_{X_1 X_2}(x_1, x_2) dx_2, \qquad (2.6.6)$$

$$p_{X_2}(x_2) = \int_{-\infty}^{\infty} p_{X_1 X_2}(x_1, x_2) dx_1. \qquad (2.6.7)$$

Two random variables X_1 and X_2 are said to be mutually independent if

$$p_{X_1 X_2}(x_1, x_2) = p_{X_1}(x_1) p_{X_2}(x_2). \qquad (2.6.8)$$

2.6.2 *Conditional Distribution*

Under the condition of a fixed $X_2 = x_2$, the probability distribution of X_1 is called the conditional probability distribution, denoted by $F_{X_1|X_2}(x_1|x_2)$, given by

$$F_{X_1|X_2}(x_1|x_2) = \frac{F_{X_1 X_2}(x_1, x_2)}{F_{X_2}(x_2)}. \qquad (2.6.9)$$

In more explicit physical meaning, (2.6.9) is identical to

$$\text{Prob}[X_1 \leq x_1 | X_2 \leq x_2] = \frac{\text{Prob}[(X_1 \leq x_1) \cap (X_2 \leq x_2)]}{\text{Prob}[X_2 \leq x_2]}. \qquad (2.6.10)$$

In terms of the probability density functions,

$$p_{X_1|X_2}(x_1|x_2) = \frac{p_{X_1 X_2}(x_1, x_2)}{p_{X_2}(x_2)} = \frac{p_{X_1 X_2}(x_1, x_2)}{\displaystyle\int_{-\infty}^{\infty} p_{X_1 X_2}(x_1, x_2) dx_1}. \qquad (2.6.11)$$

Equation (2.6.11) shows that the conditional probability density $p_{X_1|X_2}(x_1|x_2)$ is the probability density of X_1 if X_2 is fixed at x_2. The independence of two random variables X_1 and X_2 means that the conditional probability density $p_{X_1|X_2}(x_1|x_2)$ is the same as the marginal probability density $p_{X_1}(x_1)$; thus, (2.6.11) leads to (2.6.8).

In the case of multiple random variables, we have

$$p_{X_1 X_2 \cdots X_n}(x_1, x_2, \ldots, x_n)$$
$$= p_{X_1 \cdots X_k | X_{k+1} \cdots X_n}(x_1, \ldots, x_k | x_{k+1}, \ldots, x_n)$$
$$\times p_{X_{k+1} \cdots X_n}(x_{k+1}, \ldots, x_n). \qquad (2.6.12)$$

2.6.3 *Statistical Moments*

The joint moments of two random variables X_1 and X_2 is given by

$$m_{nk}[X_1, X_2] = E[X_1^n X_2^k]$$
$$= \int\limits_{-\infty}^{\infty} \int\limits_{-\infty}^{\infty} x_1^n x_2^k p_{X_1 X_2}(x_1, x_2) dx_1 dx_2,$$
$$n, k = 0, 1, 2, \ldots. \qquad (2.6.13)$$

The variables X_1 and X_2 in $m_{nk}[X_1, X_2]$ may be omitted, and the two subscripts n and k indicate that they are the joint moments of two random variables. If either n or k is zero, they are degenerated to the moments of a single random variable. The joint central moments are $E[(X_1 - \mu_{X_1})^n (X_2 - \mu_{X_2})^k]$. Especially useful is the case of $n = 1$ and $k = 1$, and the joint central moments is called the covariance, denoted by $\kappa_{X_1 X_2}$ i.e.,

$$\kappa_{X_1 X_2} = E[(X_1 - \mu_{X_1})(X_2 - \mu_{X_2})]$$
$$= E[X_1 X_2] - \mu_{X_1} \mu_{X_2} = m_{11} - m_{10} m_{01}. \qquad (2.6.14)$$

The correlation coefficient is the normalized covariance, defined as

$$\rho_{X_1 X_2} = \frac{\kappa_{X_1 X_2}}{\sigma_{X_1} \sigma_{X_2}}. \qquad (2.6.15)$$

Using the well-known Schwarz inequality

$$E[|UV|] \le \{E[U^2] E[V^2]\}^{\frac{1}{2}}, \qquad (2.6.16)$$

it can be shown that $|\rho_{X_1 X_2}| \le 1$ (see Exercise Problem 2.23). If the two random variables are independent, it is known from Eq. (2.6.8) that

$$\kappa_{X_1 X_2} = E[X_1]E[X_2] - \mu_{X_1}\mu_{X_2} = 0, \tag{2.6.17}$$

$$m_{nk} = E[X_1^n X_2^k] = E[X_1^n]E[X_2^k] = m_{n0}m_{0k}. \tag{2.6.18}$$

Two random variables are said to be uncorrelated if their covariance is zero. It is clear from (2.6.17) that two independent random variables are uncorrelated. But the reverse is not generally true, i.e., two uncorrelated random variables are not necessarily independent (see Exercise Problem 2.21).

Consider the case of n random variables, $X_1, X_2, \ldots, X_n$. The vector $\mathbf{X} = [X_1, X_2, \ldots, X_n]^T$ is a random vector. Its mean vector is $\mu_{\mathbf{X}} = [\mu_{X_1}, \mu_{X_2}, \ldots \mu_{X_n}]^T$, and covariance matrix is

$$\mathbf{C_{XX}} = E[(\mathbf{X} - \mu_{\mathbf{X}})(\mathbf{X} - \mu_{\mathbf{X}})^T] = \begin{bmatrix} \sigma_{X_1}^2 & \kappa_{X_1 X_2} & \cdots & \kappa_{X_1 X_n} \\ \kappa_{X_2 X_1} & \sigma_{X_2}^2 & \cdots & \kappa_{X_2 X_n} \\ \cdots & \cdots & \cdots & \cdots \\ \kappa_{X_n X_1} & \kappa_{X_n X_2} & \cdots & \sigma_{X_n}^2 \end{bmatrix}.$$

$$\tag{2.6.19}$$

It can be proved that $\mathbf{C_{XX}}$ is positive definite.

2.6.4 *Characteristic Function and Cumulants*

The characteristic function of two jointly distributed random variables is defined as

$$M_{X_1 X_2}(\theta_1, \theta_2) = E[e^{i(\theta_1 X_1 + \theta_2 X_2)}]$$

$$= \int_{-\infty}^{\infty} \int_{-\infty}^{\infty} e^{i(\theta_1 x_1 + \theta_2 x_2)} p_{X_1 X_2}(x_1, x_2) dx_1 x_2. \tag{2.6.20}$$

Its Maclaurin series is

$$M_{X_1 X_2}(\theta_1, \theta_2) = \sum_{n,k=0}^{\infty} \frac{m_{nk}}{(n+k)!}(i\theta_1)^n(i\theta_2)^k, \tag{2.6.21}$$

where m_{nk} are the joint moments given by (2.6.13), and also can be obtained from the characteristic function as follows

$$m_{nk} = E[X_1^n X_2^k] = \frac{1}{i^{n+k}} \left[\frac{\partial^{n+k}}{\partial \theta_1^n \partial \theta_2^k} M_{X_1 X_2}(\theta_1, \theta_1) \right]_{\theta_1 = \theta_2 = 0}. \quad (2.6.22)$$

The expansion of the logarithmic characteristic function is

$$\ln M_{X_1 X_2}(\theta_1, \theta_2) = \sum_{j=1}^{2} \kappa_1[X_j](i\theta_j) + \frac{1}{2!} \sum_{j,k=1}^{2} \kappa_2[X_j X_k](i\theta_j)(i\theta_k)$$
$$+ \frac{1}{3!} \sum_{j,k,l=1}^{2} \kappa_3[X_j X_k X_l](i\theta_j)(i\theta_k)(i\theta_l) + \cdots \cdots,$$

$$(2.6.23)$$

where κ_n is the nth-order joint cumulant with its expression illustrated for the third-order as follows

$$\kappa_3[X_j X_k X_l]$$
$$= \frac{1}{i^3} \left[\frac{\partial^3}{\partial \theta_j \partial \theta_k \partial \theta_l} \ln M_{X_1 X_2}(\theta_1, \theta_2) \right]_{\theta_1 = \theta_2 = 0}, \quad j, k, l = 1, 2.$$

$$(2.6.24)$$

Similar to the case of single random variable, the cumulants of the first several orders are related to the joint moments as follows

$$E[X_j] = \kappa_1[X_j],$$
$$E[X_j X_k] = \kappa_2[X_j X_k] + \kappa_1[X_j]\kappa_1[X_k],$$
$$E[X_j X_k X_l] = \kappa_3[X_j X_k X_l] + 3\{\kappa_1[X_j]\kappa_2[X_k X_l]\}_s$$
$$+ \kappa_1[X_j]\kappa_1[X_k]\kappa_1[X_l],$$
$$E[X_j X_k X_l X_m] = \kappa_4[X_j X_k X_l X_m] + 3\{\kappa_2[X_j X_k]\kappa_2[X_l X_m]\}_s$$
$$+ 4\{\kappa_1[X_j]\kappa_3[X_k X_l X_m]\}_s$$
$$+ 6\{\kappa_1[X_j]\kappa_1[X_k]\kappa_2[X_l X_m]\}_s$$
$$+ \kappa_1[X_j]\kappa_1[X_k]\kappa_1[X_l]\kappa_1[X_m],$$

$$(2.6.25)$$

where symbol $\{\cdot\}_s$ denotes a symmetrizing operation with respect to all its arguments, i.e., an operation taking the arithmetic mean of different permuted terms similar to the one within the braces. For example,

$$\{\kappa_1[X_j]\kappa_2[X_kX_l]\}_s = \frac{1}{3}\{\kappa_1[X_j]\kappa_2[X_kX_l] + \kappa_1[X_k]\kappa_2[X_lX_j]$$
$$+ \kappa_1[X_l]\kappa_2[X_jX_k]\}. \tag{2.6.26}$$

If the mean values of X_1 and X_2 are zeros, (2.6.25) is simplified to

$$E[X_jX_k] = \kappa_2[X_jX_k],$$

$$E[X_jX_kX_l] = \kappa_3[X_jX_kX_l],$$

$$E[X_jX_kX_lX_m] = \kappa_4[X_jX_kX_lX_m] + 3\{\kappa_2[X_jX_k]\kappa_2[X_lX_m]\}_s,$$

$$E[X_jX_kX_lX_mX_p] = \kappa_5[X_jX_kX_lX_mX_p]$$
$$+ 10\{\kappa_2[X_jX_k]\kappa_3[X_lX_mX_p]\}_s, \tag{2.6.27}$$

$$E[X_jX_kX_lX_mX_pX_r] = \kappa_6[X_jX_kX_lX_mX_pX_r]$$
$$+ 15\{\kappa_2[X_jX_k]\kappa_4[X_lX_mX_pX_r]\}_s$$
$$+ 15\{\kappa_2[X_jX_k]\kappa_2[X_lX_m]\kappa_2[X_pX_r]\}_s$$
$$+ 10\{\kappa_3[X_jX_kX_l]\kappa_3[X_mX_pX_r]\}_s.$$

Although Eqs. (2.6.25) and (2.6.27) are presented here for the case of two random variables, i.e., $j, k, l, m, p, r = 1, 2$; they can be also applied to cases of more random variables.

2.6.5 *Multiple Gaussian Random Variables*

Since Gaussian distribution is of particular importance, properties of jointly Gaussian distributed random variables are described below.

The probability density function of two jointly Gaussian distributed random variables X_1 and X_2 is

$$p(x_1, x_2) = \frac{1}{2\pi\sigma_{X_1}\sigma_{X_2}\sqrt{1 - \rho_{X_1X_2}^2}}$$

$$\times \exp\left[-\frac{\begin{array}{c}\sigma_{X_2}^2(x_1 - \mu_{X_1})^2 - 2\sigma_{X_1}\sigma_{X_2}\rho_{X_1X_2}(x_1 - \mu_{X_1})\\ \times (x_2 - \mu_{X_2}) + \sigma_{X_1}^2(x_2 - \mu_{X_2})^2\end{array}}{2\sigma_{X_1}^2\sigma_{X_2}^2(1 - \rho_{X_1X_2}^2)}\right].$$

$$\tag{2.6.28}$$

Equation (2.6.28) shows that there are five parameters to define the joint Gaussian probability distribution of two random variables, i.e., their mean values μ_{X_1} and μ_{X_2}, standard deviations σ_{X_1} and σ_{X_2}, and correlation coefficient $\rho_{X_1 X_2}$ defined in (2.6.15). If the correlation coefficient $\rho_{X_1 X_2}$ is zero, Eq. (2.6.28) leads to $p_{X_1 X_2}(x_1, x_2) = p_{X_1}(x_1) p_{X_2}(x_2)$, indicating that the two random variables X_1 and X_2 are independent. Therefore, the independence and uncorrelation are equivalent for Gaussian random variables.

For n jointly Gaussian distributed variables $X_1, X_2, \ldots, X_n$, their joint probability density is

$$p(x_1, x_2, \ldots, x_n)$$
$$= \frac{1}{(2\pi)^{\frac{n}{2}} |\mathbf{C_{XX}}|^{\frac{1}{2}}}$$
$$\times \exp\left[-\frac{1}{2|\mathbf{C_{XX}}|} \sum_{j,k=1}^{n} |\mathbf{C_{XX}}|_{jk} (x_j - \mu_{X_j})(x_k - \mu_{X_k})\right],$$
$$(2.6.29)$$

where $\mathbf{C_{XX}}$ is the covariance matrix given by Eq. (2.6.19), $|\mathbf{C_{XX}}|$ is its determinant and $|\mathbf{C_{XX}}|_{jk}$ is the co-factor of its (j, k) element. Equation (2.6.29) can be written as the matrix form

$$p(\mathbf{x}) = \frac{1}{(2\pi)^{\frac{n}{2}} |\mathbf{C_{XX}}|^{\frac{1}{2}}} \exp\left[-\frac{1}{2}(\mathbf{x} - \mu_\mathbf{X})^{\mathrm{T}} \mathbf{C_{XX}^{-1}} (\mathbf{x} - \mu_\mathbf{X})\right]. \tag{2.6.30}$$

Corresponding to Eq. (2.6.30), the characteristic function is

$$M_\mathbf{X}(\theta) = \exp\left[i\mu_\mathbf{X}^{\mathrm{T}}\theta - \frac{1}{2}\theta^{\mathrm{T}} \mathbf{C_{XX}^{-1}} \theta)\right], \tag{2.6.31}$$

which has the scalar form

$$M_{X_1 X_1 \cdots X_n}(\theta_1, \theta_2, \ldots, \theta_n) = \exp\left[i \sum_{j=1}^{n} \mu_{X_j} \theta_j - \frac{1}{2} \sum_{j,k=1}^{n} \kappa_{X_j X_k} \theta_j \theta_k\right]. \tag{2.6.32}$$

As shown in Eqs. (2.6.30) through (2.6.32), one of the most important characteristics of Gaussian distributed variables is that they are completely specified by the first and second moments. Therefore, moments of orders higher than two can be calculated from the first and second moments. Considering a set of jointly Gaussian distributed variables $X_1, X_2, \ldots$, and assuming all mean values are zero, we have (Lin, 1967):

$$E[X_1 X_2 \cdots X_{2m+1}] = 0, \tag{2.6.33}$$

$$E[X_1 X_2 \cdots X_{2m}] = \sum E[X_n X_l] E[X_j X_k], \tag{2.6.34}$$

where the summation is carried out over all different ways to group $2m$ elements into m pairs. The number of terms in the summation in Eq. (2.6.34) is $N = (2m)!/(m!2^m)$. The case of $m = 2$ is shown below,

$$E[X_1 X_2 X_3 X_4] = E[X_1 X_2] E[X_3 X_4] + E[X_1 X_3] E[X_2 X_4]$$
$$+ E[X_1 X_4] E[X_2 X_3]. \tag{2.6.35}$$

$$E[X_1^2 X_2 X_3] = E[X_1^2] E[X_2 X_3] + 2E[X_1 X_2] E[X_1 X_3]. \tag{2.6.36}$$

It is also known from Eq. (2.6.32) that all cumulants of order higher than two are zero, which is another important characteristic of jointly Gaussian distributed random variables.

It can be proved that any linear operations of Gaussian random variables, including algebraic operations, differentiations and integrations, result in Gaussian random variables (Lin, 1967).

2.7 Function of Random Variables

A function of random variable X, say $Y = f(X)$, is also a random variable. Its probabilistic and statistical properties can be obtained from those of X.

2.7.1 *Moments*

The moments of Y can be calculated from

$$E[Y^n] = E[f^n(X)] = \int_{-\infty}^{\infty} f^n(x) p_X(x) dx. \tag{2.7.1}$$

If Y is a function of multiple random variables, i.e., $Y = f(X_1, X_2, \ldots, X_n)$, we have

$$E[Y^n] = \int\limits_{-\infty}^{\infty} \int\limits_{-\infty}^{\infty} \cdots \int\limits_{-\infty}^{\infty} f^n(x_1, x_2, \ldots, x_n)$$

$$\times p_{X_1 X_2 \cdots X_n}(x_1, x_2, \ldots, x_n) dx_1 dx_2, \ldots, dx_n. \quad (2.7.2)$$

2.7.2 *Probability Distribution*

If Y is a monotone function of X, i.e., Y and X constitute a one-to-one mapping and single-valued relationship, the probability distribution function of Y can be derived as

$$F_Y(y) = \text{Prob}[Y \leq y]$$
$$= \text{Prob}[f(X) \leq y] = \text{Prob}[X \leq g(y)] = F_X[g(y)], \quad (2.7.3)$$

where $X = g(Y)$ is the inverse function of $Y = f(X)$. Then the probability density of Y is calculated from

$$p_Y(y) = \frac{d}{dy} F_Y(y) = p_X[g(y)] \left| \frac{dx}{dy} \right|. \quad (2.7.4)$$

If the function relationship is not monotonic, the problem is more complicated. The entire domain may be partitioned into regions so that the function is monotonic in each region.

Now let X_1 and X_2 be jointly distributed random variables. Consider two new random variables Y_1 and Y_2 which have functional relationships with X_1 and X_2, i.e., $Y_1 = f_1(X_1, X_2)$ and $Y_2 = f_2(X_1, X_2)$. Assume that f_1 and f_2 define a one-to-one mapping. We derive the probability distribution function of Y_1 and Y_2 as follows

$$F_{Y_1 Y_2}(y_1, y_2) = \text{Prob}[(Y_1 \leq y_1) \cap (Y_2 \leq y_2)]$$
$$= \text{Prob}[\{f_1(X_1, X_2) \leq y_1\} \cap \{f_2(X_1, X_2) \leq y_2\}]$$
$$= \text{Prob}[\{X_1 \leq g_1(y_1, y_2)\} \cap \{X_2 \leq g_2(y_1, y_2)\}]$$
$$= F_{X_1 X_2}[g_1(y_1, y_2), g_2(y_1, y_2)],$$
$$(2.7.5)$$

where $x_1 = g_1(y_1, y_2)$ and $x_2 = g_2(y_1, y_2)$ are the inverse functions of $y_1 = f_1(x_1, x_2)$ and $y_2 = f_2(x_1, x_2)$. Differentiation of $F_{Y_1 Y_2}(y_1, y_2)$ with respect to y_1 and y_2 results in

$$p_{Y_1 Y_2}(y_1, y_2) = p_{X_1 X_2}[g_1(y_1, y_2), g_2(y_1, y_2)]|J_2|, \qquad (2.7.6)$$

where J_2 is the second-order Jacobian of transformation, given by

$$J_2 = \begin{vmatrix} \dfrac{\partial x_1}{\partial y_1} & \dfrac{\partial x_1}{\partial y_2} \\[2ex] \dfrac{\partial x_2}{\partial y_1} & \dfrac{\partial x_2}{\partial y_2} \end{vmatrix}. \qquad (2.7.7)$$

Consider a random variable as a function of two random variables, i.e., $Y = f(X_1, X_2)$. We may write

$$F_Y(y) = \text{Prob}[Y \le y]$$

$$= \text{Prob}[f(X_1, X_2) \le y] = \iint\limits_{f(x_1, x_2) \le y} p_{X_1 X_2}(x_1, x_2) dx_1 x_2$$

$$= \int_{a_1}^{a_2} \left[\int_{b_1}^{g_1(y, x_2)} p_{X_1 X_2}(x_1, x_2) dx_1 \right] dx_2,$$

$$(2.7.8)$$

where the integration limits a_1, a_2 and b_1 are determined by the area $f(x_1, x_2) \le y$, and $x_1 = g_1(y, x_2)$ is the inverse function of $y = f(x_1, x_2)$. By differentiating (2.7.8) with respect to y, we obtain the probability density of Y as

$$p_Y(y) = \int_{a_1}^{a_2} p_{X_1 X_2}[g_1(y, x_2), x_2]|J_1| dx_2, \qquad (2.7.9)$$

where

$$J_1 = \left| \frac{\partial g_1(y, x_2)}{\partial y} \right| = \left| \frac{\partial x_1}{\partial y} \right|. \qquad (2.7.10)$$

The above analysis can be generalized to cases of multiple random variables. Let $X_i, i = 1, 2, \ldots, n$, be random variables, and

$$Y_k = f_k(X_1, X_2, \ldots, X_n), \quad k = 1, 2, \ldots, m; \quad m \leq n. \quad (2.7.11)$$

For the case of $m = n$, we have

$$p_{Y_1 Y_2 \cdots Y_n}(y_1, y_2, \ldots, y_n) = p_{X_1 X_2 \cdots X_n}(x_1, x_2, \ldots, x_n)|J_n|. \quad (2.7.12)$$

Here,

$$J_n = \begin{vmatrix} \dfrac{\partial x_1}{\partial y_1} & \dfrac{\partial x_1}{\partial y_2} & \cdots & \dfrac{\partial x_1}{\partial y_n} \\ \dfrac{\partial x_2}{\partial y_1} & \dfrac{\partial x_2}{\partial y_2} & \cdots & \dfrac{\partial x_2}{\partial y_n} \\ \cdots & \cdots & \cdots & \cdots \\ \dfrac{\partial x_n}{\partial y_1} & \dfrac{\partial x_n}{\partial y_2} & \cdots & \dfrac{\partial x_n}{\partial y_n} \end{vmatrix}. \quad (2.7.13)$$

In (2.7.12), state variables $x_1, x_2, \ldots, x_n$ on the right-hand side should be expressed in terms of $y_1, y_2, \ldots, y_n$ by using relations (2.7.11). In calculating (2.7.12) and (2.7.13), certain conditions must be satisfied, which we assume for most practical problems.

For the case of $m < n$, in addition to (2.7.11), let

$$y_{m+1} = x_{m+1}, \ y_{m+2} = x_{m+2}, \ldots, y_n = x_n. \quad (2.7.14)$$

Since $|J_m| = |J_n|$, we have

$$p_{Y_1 Y_2 \cdots Y_m}(y_1, y_2, \ldots, y_m)$$

$$= \int \cdots \int p_{Y_1 Y_2 \cdots Y_n}(y_1, y_2, \ldots, y_n) dy_{m+1} dy_{m+2} \ldots dy_n$$

$$= \int \cdots \int p_{X_1 X_2 \cdots X_n}(x_1, x_2, \ldots, x_n)|J_m| dx_{m+1} dx_{m+2} \ldots dx_n.$$

$$(2.7.15)$$

On the right-hand side of (2.7.15), $x_1, x_2, \ldots, x_m$ should be expressed in terms of $y_1, y_2, \ldots, y_m, x_{m+1}, x_{m+2}, \ldots,$ and x_n by using relations (2.7.11).

For the case of $m = 1$, i.e., $Y = f(X_1, X_2, \ldots, X_n)$, we have specifically,

$$p_Y(y) = \int \cdots \int p_{X_1 X_2 \cdots X_n}[x_1 = g(y, x_2, x_3, \ldots, x_n), x_2, \ldots, x_n]$$

$$\times |J_1| dx_2 dx_3, \ldots dx_n, \tag{2.7.16}$$

where $x_1 = g(y, x_2, x_3, \ldots, x_n)$ is obtained from $y = f(x_1, x_2, \ldots, x_n)$, and

$$J_1 = \left| \frac{\partial x_1}{\partial y} \right| = \left| \frac{\partial g(y, x_2, x_3, \ldots, x_n)}{\partial y} \right|. \tag{2.7.17}$$

Example 2.7.1 Given the probability density of X and the functional relationship $Y = X^2$, we have

$$\begin{aligned}
F_Y(y) &= \text{Prob}[Y \le y] \\
&= \text{Prob}[X^2 \le y] \\
&= \text{Prob}[-\sqrt{y} \le X \le \sqrt{y}] = F_X[\sqrt{y}] - F_X[-\sqrt{y}]. \quad (2.7.18)
\end{aligned}$$

$$p_Y(y) = \frac{d}{dy} F_Y(y) = \frac{1}{2\sqrt{y}}[p_X(\sqrt{y}) + p_X(-\sqrt{y})]. \tag{2.7.19}$$

Or we may use another method as follows

$$\int_{-\infty}^{\infty} p_X(x) dx = \int_{-\infty}^{0} p_X(x) dx + \int_{0}^{\infty} p_X(x) dx$$

$$= \int_{\infty}^{0} p_X(-\sqrt{y}) \left(-\frac{1}{2\sqrt{y}} \right) dy + \int_{0}^{\infty} p_X(\sqrt{y}) \left(\frac{1}{2\sqrt{y}} \right) dy,$$

$$= \int_{0}^{\infty} \frac{1}{2\sqrt{y}} [p_X(\sqrt{y}) + p_X(-\sqrt{y})] \, dy$$

$$\tag{2.7.20}$$

which indicates that the probability density of Y is given in (2.7.19).

Example 2.7.2 Consider $Y = X_1 + X_2$ and the joint probability density of X_1 and X_2 is known. Application of (2.7.15) leads to

$$p_Y(y) = \int\limits_{-\infty}^{\infty} p_{X_1 X_2}(x_1, y - x_1)dx_1 = \int\limits_{-\infty}^{\infty} p_{X_1 X_2}(y - x_2, x_2)dx_2.$$

$$(2.7.21)$$

2.8 Simulation of Random Variables

Simulation is a numerical procedure to solve stochastic problems for which analytical procedures are not obtainable. The main idea is to transfer the stochastic problems into deterministic problems. The first step is to generate samples for random variables and stochastic processes. For each sample, the problem becomes deterministic and may be solved using existing techniques. Thus, the solution of the problem can be obtained statistically or probabilistically if enough number of solution samples is calculated. The simulation involving stochasticity is known as the Monte Carlo simulation. Figure 2.8.1 illustrates the general procedure of the Monte Carlo simulation.

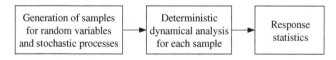

Figure 2.8.1 Monte Carlo simulation method.

The simulation method is the most general method to conduct stochastic dynamical analysis. It can be applied to any stochastic systems provided the deterministic analysis for each sample can be carried out by either analytical or numerical methods. It is obvious that the accuracy of the statistical results depends on the number of samples. Thus, the major drawback of simulation method is that the computational time may be excessive for large and complex systems. However, it becomes less and less serious with the rapid advance of computational technology.

2.8.1 *Random Numbers*

To simulate a random variable X with a specified probability distribution, i.e., to generate samples for X, it is of great importance to generate samples first for a random variable uniformly distributed in interval $[0, 1]$. Let ξ be a such random variable, i.e.,

$$p(\xi) = 1, \quad 0 \leq \xi \leq 1. \tag{2.8.1}$$

The samples of ξ are called random numbers, denoted by $\xi_k, k = 1, 2, \ldots$. Mathematical algorithms have been developed to generate random numbers ξ_k. Although they obey the uniform probability distribution (2.8.1), they are not truly random since they can be repeatedly generated; thus, they are called pseudo-random. Almost all programming languages and engineering software supply functions or procedures for generating random numbers uniformly distributed in interval $[0, 1]$.

2.8.2 *Discrete Random Variables*

Assume that the discrete random variable X has a sample space of $x_i (i = 1, 2, \ldots, n)$ associated with probabilities P_i, respectively, i.e.,

$$P_i = P_X(x_i) = \text{Prob}[X = x_i], \quad \sum_{i=1}^{n} P_i = 1. \tag{2.8.2}$$

Let ξ be the random variable uniformly distributed in $[0, 1]$, and $0 \leq a \leq 1$. We have

$$\text{Prob}[\xi \leq a] = \int_0^a p(\xi) d\xi = a. \tag{2.8.3}$$

Thus,

$$P_i = \text{Prob}[X = x_i] = \sum_{j=1}^{i} P_j - \sum_{j=1}^{i-1} P_j = \text{Prob}\left[\sum_{j=1}^{i-1} P_j \leq \xi < \sum_{j=1}^{i} P_j\right]. \tag{2.8.4}$$

The procedure to generate a sample of X, denoted as X_k, is then (i) generating random numbers ξ_k, (ii) determining i according to $\sum_{j=1}^{i-1} p_j \leq \xi_k < \sum_{j=1}^{i} p_j$ and (iii) taking $X_k = x_i$.

For the special case of equal probability, i.e.,

$$P_1 = P_2 = \cdots = P_n = \frac{1}{n}, \qquad (2.8.5)$$

the samples of X can be generated from

$$X_k = x_i, \quad i = [n\xi_k] + 1, \qquad (2.8.6)$$

where the bracket $[\cdot]$ represents the integer part of the inside real number.

2.8.3 *Single Continuous Random Variable*

Let ξ be the random variable uniformly distributed in $[0, 1]$ and X be a continuous random variable with a probability distribution function $F_X(x)$. Since $0 \leq F_X(x) \leq 1$, we have from (2.8.3)

$$F_X(x) = \text{Prob}[\xi \leq F_X(x)] = \text{Prob}[F_X^{-1}(\xi) \leq x]. \qquad (2.8.7)$$

Comparison of (2.2.2) and (2.8.7) leads to $X = F_X^{-1}(\xi)$, and samples of X can be generated from

$$X_k = F_X^{-1}(\xi_k), \quad k = 1, 2, \ldots, \qquad (2.8.8)$$

where ξ_k are random numbers, i.e., the samples of random variable ξ uniformly distributed in $[0, 1]$.

Consider a random variable X uniformly distributed in $[a, b]$. We have

$$p(x) = \begin{cases} \dfrac{1}{b-a}, & a \leq x \leq b \\ 0, & \text{elsewhere}, \end{cases} \qquad F_X(x) = \frac{x-a}{b-a}, \quad X_k = a + (b-a)\xi_k.$$

$$(2.8.9)$$

For the exponential distribution,

$$p(x) = \begin{cases} \lambda e^{-\lambda x}, & x \geq 0 \\ 0, & x < 0, \end{cases} \qquad F_X(x) = 1 - e^{-\lambda x}, \qquad X_k = -\frac{1}{\lambda} \ln(1 - \xi_k).$$

$$(2.8.10)$$

For the Rayleigh distribution,

$$p(x) = \begin{cases} \dfrac{x}{\sigma^2} e^{-\frac{x^2}{2\sigma^2}}, & x \geq 0 \\ 0, & x < 0, \end{cases}$$

$$F_X(x) = 1 - e^{-\frac{x^2}{2\sigma^2}}, \qquad X_k = \sigma\sqrt{-2\ln(1 - \xi_k)}. \qquad (2.8.11)$$

For the standard normal distribution $N(0,1)$,

$$p(x) = \frac{1}{\sqrt{2\pi}} e^{-\frac{1}{2}x^2}, \qquad F_X(x) = \frac{1}{\sqrt{2\pi}} \int_{-\infty}^{x} e^{-\frac{1}{2}x^2}\, dx. \qquad (2.8.12)$$

In this case, $X_k = F_X^{-1}(\xi_k)$ cannot be expressed analytically, and must be calculated numerically. In fact, all programming languages and engineering software supply functions or procedures for generating samples of a $N(0,1)$ distributed random variable.

2.8.4 *Multiple Continuous Random Variables*

Consider two random variables X_1 and X_2 with a joint probability density $p_{X_1 X_2}(x_1, x_2)$. From (2.6.11),

$$p_{X_1 X_2}(x_1, x_2) = p_{X_1|X_2}(x_1|x_2)p_{X_2}(x_2). \qquad (2.8.13)$$

The procedure to generate a sample point (X_{1k}, X_{2k}) is then (a) to calculate the marginal probability density $p_{X_2}(x_2)$ according to (2.6.7), (b) to generate X_{2k} from $p_{X_2}(x_2)$ and (c) to generate X_{1k} from $p_{X_1|X_2}(x_1|X_{2k})$ calculated from (2.8.13).

Figure 2.8.2 depicts sample points of two-dimensional random variables (X, Y) for different sample numbers. X and Y are assumed to be independent, and uniformly distributed in $[0, 1]$. It is clear that, more the sample number, more is the accuracy of simulation.

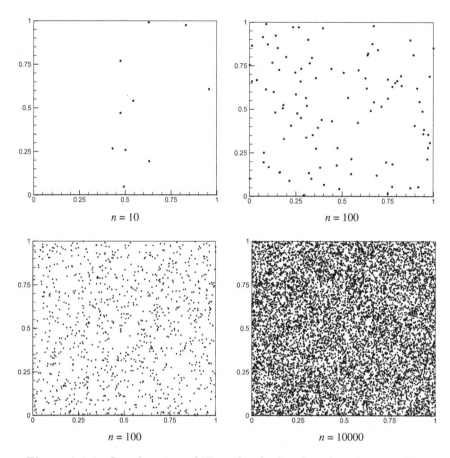

Figure 2.8.2   Sample points of 2D uniformly distributed random variables.

The above procedure can be readily extended to more random variables if their joint probability distribution is provided.

2.8.5 *Two Correlated Gaussian Random Variables*

If X and Y are correlated standard Gaussian random variables, i.e., each obeys $N(0,1)$ distribution, their joint distribution is obtained from (2.6.28) as

$$p(x,y) = \frac{1}{2\pi\sqrt{1-\rho^2}} \exp\left[-\frac{x^2 - 2\rho xy + y^2}{2(1-\rho^2)}\right], \qquad (2.8.14)$$

where ρ is the correlation coefficient, i.e., $E[XY] = \rho$. Following the procedure given in Section 2.8.4, samples of random variables X and Y can be generated. However, a simpler procedure is presented here.

Assume X_1 is another standard Gaussian random variable, independent of X. Letting

$$Y = \rho X + \sqrt{1 - \rho^2} X_1, \qquad (2.8.15)$$

we have

$$E[Y] = 0, \quad E[Y^2] = 1, \quad E[XY] = \rho, \qquad (2.8.16)$$

indicating that Y is also $N(0, 1)$ random variable, and X and Y have a correlation coefficient ρ. Since every program or computer language supplies a command to generate $N(0, 1)$ random numbers, it is simple and straightforward to generate two sequences of correlated $N(0, 1)$ random numbers.

Using (2.8.15), sample points of two-dimensional $N(0, 1)$ random variables (X, Y) are generated and depicted for the cases of uncorrelated $(\rho = 0)$ and correlated $(\rho = 0.8)$. The effect of the correlation is clearly revealed in Fig. 2.8.3.

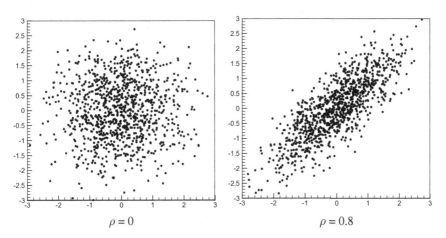

$\rho = 0$ $\rho = 0.8$

Figure 2.8.3 Sample points of 2D standard Gaussian random variables, uncorrelated $(\rho = 0)$ and correlated $(\rho = 0.8)$.

2.8.6 *Transformation Method*

Assume that X is a random variable with a probability density $p_X(x)$ and Y is another random variable transformed from X according to $Y = f(X)$. If samples of X, X_k, can be generated from $p_X(x)$, samples of Y can be calculated from $Y_k = f(X_k)$. For example, if a function or procedure in a programming language or a software is available to generate samples U_k with standard normal distribution N(0, 1), samples of random variable $Y \sim N(\mu, \sigma)$ can be calculated from transformation $Y_k = \mu + \sigma U_k$.

The transformation method can be employed to generate samples of two independent random variables X_1 and X_2, both with standard normal distribution N(0, 1). The joint probability density of X_1 and X_2 is

$$p_{X_1 X_2}(x_1, x_2) = p_{X_1}(x_1) p_{X_2}(x_2) = \frac{1}{2\pi} e^{-\frac{1}{2}(x_1^2 + x_2^2)}. \qquad (2.8.17)$$

Define two new random variables R and Φ through the transformation

$$\begin{cases} X_1 = R\cos\Phi \\ X_2 = R\sin\Phi. \end{cases} \qquad (2.8.18)$$

Using (2.7.12), the joint probability density of R and Φ can be obtained as

$$p_{R\Phi}(r, \phi) = \frac{1}{2\pi} r e^{-\frac{1}{2}r^2}. \qquad (2.8.19)$$

Equation (2.8.19) shows that R and Φ are independent with their marginal probability density functions as, respectively,

$$p_R(r) = r e^{-\frac{1}{2}r^2}, \quad 0 \le r < \infty,$$

$$p_\Phi(\phi) = \frac{1}{2\pi}, \quad 0 \le \phi < 2\pi. \qquad (2.8.20)$$

Equation (2.8.20) indicates that Φ is uniformly distributed in $[0, 2\pi)$ and R is Rayleigh distributed. Samples of Φ and R can be generated from (2.8.9) and (2.8.11) respectively. Consequently, sample of X_1

and X_2 are calculated from (2.8.18), i.e.,

$$
\begin{aligned}
X_{1k} &= \sqrt{-2\ln(\xi_{1k})}\cos(2\pi\xi_{2k}), \\
X_{2k} &= \sqrt{-2\ln(\xi_{1k})}\sin(2\pi\xi_{2k}),
\end{aligned}
\tag{2.8.21}
$$

where ξ_{1k} and ξ_{2k} are independent uniformly distributed random numbers.

Assume that samples of n independent standard Gaussian random variables, $\mathbf{X} = [X_1, X_2, \ldots, X_n]^{\mathrm{T}}$, are generated, then samples of another set of n Gaussian random variables, $\mathbf{Y} = [Y_1, Y_2, \ldots, Y_n]^{\mathrm{T}}$, of mean vector $\mu_{\mathbf{Y}}$ and covariance matrix $\mathbf{C_{YY}}$, can be obtained. Assume the transformation

$$
\mathbf{Y} = \mathbf{A} + \mathbf{B}\mathbf{X}, \tag{2.8.22}
$$

where $\mathbf{A}$ is a vector and $\mathbf{B}$ is a matrix, and both are to be determined. Taking the ensemble average, we have

$$
\mu_{\mathbf{Y}} = \mathbf{A} + \mathbf{B}\mu_{\mathbf{X}} = \mathbf{A}. \tag{2.8.23}
$$

The covariance matrix of $\mathbf{Y}$ is given by

$$
\begin{aligned}
\mathbf{C_{YY}} &= \mathrm{E}[(\mathbf{Y} - \mu_{\mathbf{Y}})(\mathbf{Y} - \mu_{\mathbf{Y}})^{\mathrm{T}}] \\
&= \mathrm{E}[\mathbf{B}\mathbf{X}\mathbf{X}^{\mathrm{T}}\mathbf{B}^{\mathrm{T}}] = \mathbf{B}\mathbf{C_{XX}}\mathbf{B}^{\mathrm{T}} = \mathbf{B}\mathbf{B}^{\mathrm{T}}.
\end{aligned}
\tag{2.8.24}
$$

In deriving (2.8.23) and (2.8.24), use is made of $\mu_{\mathbf{X}} = \mathbf{0}$ and $\mathbf{C_{XX}} = \mathbf{I}$. The covariance matrix $\mathbf{C_{YY}}$ can be expressed as

$$
\mathbf{C_{YY}} = \mathbf{\Phi}
\begin{bmatrix}
\lambda_1 & & & \\
& \lambda_2 & & \\
& & \ddots & \\
& & & \lambda_n
\end{bmatrix}
\mathbf{\Phi}^{\mathrm{T}},
\tag{2.8.25}
$$

where $\lambda_1, \lambda_2, \ldots, \lambda_n$ are eigenvalues of $\mathbf{C_{YY}}$ and $\mathbf{\Phi}$ is the eigenmatrix. Since $\mathbf{C_{YY}}$ is positive definite, all eigenvalues are non-negative, and

(2.8.25) can be written as:

$$\mathbf{C_{YY}} = \boldsymbol{\Phi} \begin{bmatrix} \sqrt{\lambda_1} & & & \\ & \sqrt{\lambda_2} & & \\ & & \ddots & \\ & & & \sqrt{\lambda_n} \end{bmatrix} \begin{bmatrix} \sqrt{\lambda_1} & & & \\ & \sqrt{\lambda_2} & & \\ & & \ddots & \\ & & & \sqrt{\lambda_n} \end{bmatrix} \boldsymbol{\Phi}^{\mathrm{T}}.$$

$$(2.8.26)$$

Comparison of (2.8.24) and (2.8.26) leads to

$$\mathbf{B} = \boldsymbol{\Phi} \begin{bmatrix} \sqrt{\lambda_1} & & & \\ & \sqrt{\lambda_2} & & \\ & & \ddots & \\ & & & \sqrt{\lambda_n} \end{bmatrix}. \qquad (2.8.27)$$

Given $\mathbf{A}$ and $\mathbf{B}$, samples of $\mathbf{Y}$ can be calculated from samples of $\mathbf{X}$ from (2.8.22).

Example 2.8.1 Use the above procedure for two-dimensional case in which two $N(0,1)$ random variables Y_1 and Y_2 are to be simulated with a correlation coefficient ρ. It is found that

$$\mathbf{C_{YY}} = \begin{bmatrix} 1 & \rho \\ \rho & 1 \end{bmatrix} \qquad (2.8.28)$$

and its eigenvalues and eigenmatrix are

$$\lambda_1 = 1 + \rho, \quad \lambda_2 = 1 - \rho, \quad \boldsymbol{\Phi} = \frac{1}{\sqrt{2}} \begin{bmatrix} 1 & 1 \\ 1 & -1 \end{bmatrix}. \qquad (2.8.29)$$

Matrix $\mathbf{B}$ is then found from (2.8.27) as

$$\mathbf{B} = \frac{1}{\sqrt{2}} \begin{bmatrix} \sqrt{1+\rho} & \sqrt{1-\rho} \\ \sqrt{1+\rho} & -\sqrt{1-\rho} \end{bmatrix}, \qquad (2.8.30)$$

and random variables Y_1 and Y_2 are obtained as

$$Y_1 = \frac{1}{\sqrt{2}} \left(\sqrt{1+\rho}X_1 + \sqrt{1-\rho}X_2 \right),$$

$$Y_2 = \frac{1}{\sqrt{2}} \left(\sqrt{1+\rho}X_1 - \sqrt{1-\rho}X_2 \right),$$

(2.8.31)

where X_1 and X_2 are two independent $N(0,1)$ random variables. Either (2.8.31) or (2.8.15) can be used to generate random numbers for two correlated Gaussian random variables.

Exercise Problems

2.1 The Pareto distributed random variable is defined by the probability density function as follows

$$p(x) = \begin{cases} rA^r x^{-(r+1)}, & x \geq A > 0 \\ 0, & x < A, \end{cases}$$

where $r > 0$. Find the nth moment of X, and determine the condition for the nth moment to exist.

2.2 Random variable X has Gamma distribution, i.e.,

$$p_X(x) = \frac{\beta^\alpha}{\Gamma(\alpha)} x^{\alpha-1} e^{-\beta x}, \quad x \geq 0, \quad \beta > 0.$$

Show that its mean and variance are

$$\mu_X = \frac{\alpha}{\beta}, \quad \sigma_X^2 = \frac{\alpha}{\beta^2}.$$

2.3 Determine the characteristic functions for the random variables with the following probability density functions:

(a) Delta function distribution $p(x) = \delta(x - a)$.

(b) Uniform distribution $p(x) = \dfrac{1}{2a}$, $-a \leq x \leq a$, $a > 0$.

(c) Exponential distribution $p(x) = \lambda e^{-\lambda x}$, $x \geq 0$, $\lambda > 0$.

(d) Cauchy distribution $p(x) = \dfrac{a}{\pi(a^2 + x^2)}$, $a > 0$.

Find the mean and mean squares from the characteristic functions if possible.

2.4 Letting $X \sim \mathrm{N}(\mu, \sigma)$, derive Eq. (2.5.4), i.e.,

$$E[X^n] = \mu E[X^{n-1}] + (n-1)\sigma^2 E[X^{n-2}], \quad n = 2, 3, \ldots$$

2.5 Let $X \sim \mathrm{N}(\mu_X, \sigma_X)$, show that

$$E[(X - \mu_X)^n] = \begin{cases} 1 \cdot 3 \cdot 5 \cdots (n-1)\sigma_X^n, & n = \text{even} \\ 0, & n = \text{odd}. \end{cases}$$

2.6 Calculate skewness γ_1 and kurtosis γ_2 for random variable X with the probability density

$$p(x) = Ce^{-x^2 - ax^4}, \quad -\infty < x < \infty,$$

where C is a normalization constant. Draw the two curves versus changing parameter a.

2.7 Find the characteristic function of X defined in Problem 2.2, and then calculate the nth moment by using the characteristic function.

2.8 The probability density function of the Poisson distribution given by Eq. (2.5.13) can be written as

$$p_X(x) = e^{-\mu} \sum_{n=0}^{\infty} \frac{\mu^n}{n!} \delta(x - n).$$

Find its characteristic function, and calculate the moments from the characteristic function.

2.9 Prove the Poisson distribution given in Eq. (2.5.13) approaches the Gaussian distribution as parameter μ tends to infinity.

2.10 The characteristic function of random variable X is given by

$$M_X(\theta) = \frac{1}{1 + \theta^2}.$$

Find the probability density function of X, and calculate its nth moment.

2.11 Derive expressions for the fifth- and sixth-order cumulants in terms of moments and central moments.

2.12 Let X and Y be two random variables. Define the following concepts:

Inner product of X and Y: $\langle X, Y \rangle = E[XY]$.

Norm of X: $\|X\| = \sqrt{\langle X, X \rangle} = \sqrt{E[X^2]}$.

Distance between X and Y: $d(X, Y) = \|X - Y\|$.

Show that (a) $d(X, Y) \geq 0$, (b) $d(X, Y) = 0$ if and only if $X = Y$, (c) $d(X, Y) = d(Y, X)$, and (d) $d(X, Y) \leq d(X, Z) + d(Z, Y)$.

2.13 The joint probability density function of two random variables X and Y is

(a) $p(x, y) = \dfrac{1}{x^2 y^2}$, $1 \leq x, y < \infty$,

(b) $p(x, y) = \dfrac{1}{\pi}$, $x^2 + y^2 \leq 1$,

(c) $p(x, y) = y^2 e^{-y(1+x)}$, $x, y > 0$.

For each case, determine the marginal probability density functions of X and Y. Are X and Y uncorrelated and independent?

2.14 X and Y are two Gaussian random variables with the same probability distribution. Show that $X + Y$ and $X - Y$ are independent.

2.15 X_1 and X_2 are two Gaussian random variables with zero mean and following moments

$$E[X_1^2] = 4, \quad E[X_1 X_2] = 2, \quad E[X_2^2] = 9.$$

(a) Find the joint probability density function of X_1 and X_2, the marginal probability density functions of X_1 and X_2, respectively, and the conditional probability density function of $X_1|X_2$.

(b) Define

$$Y_1 = 2X_1 - X_2, \quad Y_2 = 3X_1 - aX_2.$$

Determine the value of a so that Y_1 and Y_2 are independent.

2.16 Let the joint probability density function of random variables X and Y be

$$p_{XY}(x,y) = Ce^{-(x+3y)}, \quad x, y \geq 0.$$

Are X and Y independent? Determine the probability of $X > Y$.

2.17 The random variables X and Y have the joint probability density function

$$p_{XY}(x,y) = Ce^{-\frac{1}{2}x^2+\sqrt{3}xy-4y^2}, \quad -\infty < x, y < \infty.$$

Find the marginal probability density functions of X and Y, respectively, and calculate their covariance and correlation coefficient.

2.18 Let the joint probability density of random variables X_1 and X_2 be

$$p_{X_1X_2}(x_1, x_2)$$

$$= C(\Delta^2 - x_1^2 - kx_2^2)^{\delta-\frac{1}{2}}, \quad x_1^2 + kx_2^2 < \Delta^2, \quad \delta > -\frac{1}{2}.$$

Find the marginal probability density of X_1.

2.19 The joint probability density function of random variables X and Y is given by

$$p_{XY}(x,y) = Ce^{-(x+3y)}, \quad 0 \leq x < y < \infty.$$

Are X and Y independent?

2.20 The joint probability density function of random variables X and Y is

$$p_{XY}(x,y) = \frac{x}{\pi a^2}, \quad 0 \leq x \leq a, \quad 0 \leq y \leq 2\pi.$$

Find $p_{X|Y}(x|y)$ and $p_{Y|X}(y|x)$. Are X and Y independent?

2.21 Let the joint probability density function of random variables X and Y be

$$p_{XY}(x,y) = C(1 - x^{2n} + y^{2n}), \quad -1 \leq x, y \leq 1,$$

where n is a positive integer. Show that X and Y are uncorrelated, but not independent.

2.22 Let the joint probability density function of random variables X_1 and X_2 be

$$p(x_1, x_2) = C(a\lambda + b)^\delta, \quad \lambda = \frac{1}{2}\omega_0^2 x_1^2 + \frac{1}{2}x_2^2,$$

$$-\infty < x_1, x_2 < \infty, \quad a, b > 0.$$

(a) Find the normalization constant C, and determine the condition for $p(x_1, x_2)$ to be a valid probability density.

(b) Calculate $E[X_1^2]$ and determine the condition for $E[X_1^2]$ to exist.

2.23 Use the following non-negative expectation

$$E[(aX - Y)^2] = a^2 E[X^2] - 2aE[XY] + E[Y^2] \geq 0,$$

to prove (a) $\sqrt{E[X^2]E[Y^2]} \geq E[XY]$ and (b) correlation coefficient $|\rho_{XY}| \leq 1$.

2.24 Random variable X is Gaussian distributed with a mean μ_X and standard deviation σ_X, i.e., $X \sim N(\mu_X, \sigma_X)$. Find the probability density functions of (a) $Y = a + bX$, $b \neq 0$ and (b) $Y = e^{aX}$.

2.25 Given $p_X(x) = \frac{1}{\pi}$, $0 \leq x \leq \pi$ and $Y = \sin X$, show that

$$p_Y(y) = \begin{cases} \dfrac{2}{\pi\sqrt{1 - y^2}}, & 0 \leq y < 1 \\ 0, & \text{otherwise.} \end{cases}$$

2.26 Let X be zero-mean Gaussian random variable and $Y = X^2$. Find the probability density function of Y, and the joint probability density function of X and Y. Are X and Y uncorrelated and independent?

2.27 X is a standard Gaussian random variable, i.e., $X \sim N(0, 1)$. Find the probability density functions of (a) $Y = |X|$, and (b) $Y = aX + bX^2 (b > 0)$.

2.28 Let random variable N be Poisson-distributed, i.e.,

$$P_N(n) = \text{Prob}[N = n] = \frac{\mu^n}{n!}e^{-\mu}, \quad n = 0, 1, 2, \ldots.$$

Find the mean and variance of random variable $X = (N - \mu)\mu^{-1/2}$.

2.29 Let X and Y be two random variables with a relationship of $Y = X - \mu_X$, where μ_X is the mean value of X. Show that the cumulants of X can be expressed in terms of the cumulants of Y as follows

$$\kappa_1[X] = \kappa_1[Y] + \mu_X; \quad \kappa_n[X] = \kappa_n[Y], \quad n = 2, 3, \ldots$$

where $\kappa_n[X]$ and $\kappa_n[Y]$ are the nth cumulants of X and Y respectively.

2.30 Let X and Y be independent random variables, and

$$p_X(x) = \frac{1}{\pi\sqrt{1 - x^2}}, \quad -1 < x < 1; \quad p_Y(y) = ye^{-y^2/2},$$

$$0 \le y < \infty.$$

Show that the probability density of $Z = XY$ is

$$p_Z(z) = \frac{1}{\sqrt{2\pi}} e^{-\frac{z^2}{2}}, \quad -\infty < z < \infty.$$

2.31 Let the joint probability density of random variables X and Y be

$$p_{XY}(x, y) = 1, \quad 0 \le x, y \le 1.$$

Find the probability density of $Z = X + Y$.

2.32 Random variables X and Y are uniformly distributed in $[0, 2]$ and $[0, 1]$, respectively. Calculate the probability density function $p_Z(z)$ and probability distribution functions $F_Z(z)$ of $Z = X + Y$.

2.33 The joint probability density of random variables X and Y is

$$p_{XY}(x, y) = \frac{1}{x^2 y^2}, \quad 1 \le x, y < \infty.$$

Define $U = XY$ and $V = X/Y$. Show that

$$p_{UV}(u, v) = \frac{1}{2u^2 v}, \quad 1 \le u < \infty, \quad \frac{1}{u} \le v \le u.$$

2.34 Random variables X and Y have a joint probability density $p_{XY}(x, y)$. Let $U = X + Y$ and $V = X - Y$. Show that

$$p_{UV}(u, v) = \frac{1}{2} p_{XY}\left(\frac{u+v}{2}, \frac{u-v}{2}\right).$$

Using the above result, derive expressions for the marginal probability density functions $p_U(u)$ and $p_V(v)$.

2.35 Consider the transformation $X = R\cos\Phi$ and $Y = R\sin\Phi$.

(a) If R and Φ have Rayleigh and uniform distributions respectively, i.e.,

$$p_R(r) = re^{-\frac{1}{2}r^2}, \quad 0 \le r < \infty,$$

$$p_\Phi(\phi) = \frac{1}{2\pi}, \quad 0 \le \phi < 2\pi,$$

Show that X and Y are mutually independent standard Gaussian random variables.

(b) If $X \sim N(0, 1)$, $Y \sim N(0, 1)$ and X and Y are independent, show that R and Φ obey Rayleigh and uniform distributions, as given in (a).

2.36 U and V are two independent uniformly distributed random variables in $(0, 1]$. Show that X and Y are two independent standard Gaussian random variables if

$$X = \sqrt{-2\ln(U)}\cos(2\pi V), \quad Y = \sqrt{-2\ln(U)}\sin(2\pi V).$$

2.37 Let $X_1, X_2, \ldots, X_n$ be independent Standard Gaussian random variables of an identical probability distribution $N(0, 1)$. Find the probability density of Y defined as

$$Y = \sum_{i=1}^{n} X_i.$$

2.38 Let random variables X and Y be jointly Gaussian distributed, i.e., their joint probability density is given by Eq. (2.6.28). Find two new random variables U and V by a linear transformation of X and Y, such that U and V are independent. Determine the means and variances of U and V.

2.39 Let $X_1, X_2, \ldots, X_n$ be independent Standard Gaussian random variables of an identical probability distribution $N(0, 1)$. Random variable Y is defined as

$$Y = \sum_{i=1}^{n} X_i^2.$$

Show that the probability density of Y is

$$p(y) = \frac{1}{2^{\frac{n}{2}} \Gamma(\frac{n}{2})} y^{\frac{n}{2}-1} e^{-\frac{y}{2}}, \quad 0 \leq y < \infty.$$

2.40 Use Monte Carlo simulation to generate sample points and calculate the mean, standard deviation, and probability density for each of the following cases, and compare the simulation results with the exact analytical results.

(a) An angle Θ uniformly distributed in $[0, 2\pi)$,

(b) $X = \sin \Theta$, where Θ is defined in (a),

(c) X of an exponential distribution, i.e.,

$$p(x) = \begin{cases} \lambda e^{-\lambda x}, & x \geq 0 \\ 0, & x < 0. \end{cases}$$

Consider three cases of $\lambda = 0.5$, 1.0, 1.5, respectively.

2.41 Use Monte Carlo simulation to generate sample points of two correlated random variables $X \sim N(2, 0.5)$, $Y \sim N(2, 0.5)$, with correlation coefficient $\rho_{XY} = -1$, -0.5, 0, 0.5 and 1, respectively.

CHAPTER 3

STOCHASTIC PROCESSES

3.1 Introduction

Consider a physical phenomenon that evolves with respect to time in random fashion, such as vibration of a building during an earthquake, the motion of a ship in a sea, etc. Denoting $X(t)$ as a physical quantity investigated in the random phenomenon, then at different time instants, $X(t_1), X(t_2), \ldots$ are random variables. Such a random phenomenon can be investigated by introducing a concept called stochastic process, or random process, which is defined below:

A stochastic process $X(t)$ is a family of random variables with t as a parameter belonging to an index set T, denoted by $\{X(t), t \in T\}$.

Although the index set can be of various types, conventionally and also in this book, $X(t)$ is referred as stochastic process only for the case in which the index set is time. Another common situation, although not involved in this book, is that the parameter t is a spatial variable, in which case $X(t)$ is called random field.

A stochastic process is, in strict sense, a function of two arguments, $\{X(t, \omega); t \in T, \omega \in \Omega\}$, where Ω is the sample space. For a fixed time t, $X(t, \omega)$ is a random variable defined in the sample space Ω; while each possible $X(t, \omega)$ function with respect to time t is called a sample function.

In general, the index set T can be discrete or continuous, and the sample space Ω can also be discrete or continuous. When the index set is time which is continuous, an adjective of "discrete" or "continuous" is referred to the sample space. For example, a continuous stochastic process means a stochastic process with a continuous sample space. Except specified otherwise, only stochastic processes with a continuous index set T and continuous sample space Ω are considered.

Figure 3.1.1 illustrates three sample functions of a stochastic process schematically. At a specific time instant, say t_1 or t_2, the values of the sample functions constitute a random variable.

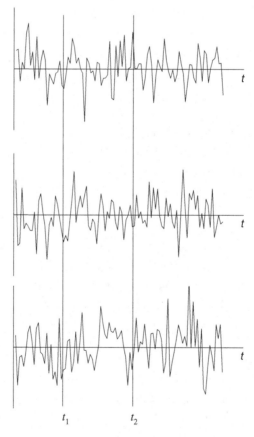

Figure 3.1.1 Sample functions of a stochastic process.

3.2 Specification of Stochastic Processes

As defined, a stochastic process is a family of random variables; thus, its specification is the same as that of jointly distributed random variables, except the number of random variables may be infinite or even uncountable. Therefore, a complete description of a stochastic process may be only possible for very limited cases. Practically, the most important features, although incomplete, may be sufficient for analysis in various applications, such as in engineering, biology, ecology, finance, etc.

3.2.1 *Probability Distributions*

As the same situation for random variables, the probability functions are only meaningful for discrete stochastic processes; whereas the probability distribution functions are applicable for both discrete and continuous stochastic processes. More important are the probability density functions which can be used to describe both the discrete and continuous stochastic processes if Dirac delta functions are included.

Consider a stochastic process $X(t)$. Its first-order, second-order, and up to nth-order probability density functions are denoted respectively as

$$p(x,t), \quad p(x_1,t_1;x_2,t_2), \ldots, p(x_1,t_1;x_2,t_2;\ldots;x_n,t_n). \quad (3.2.1)$$

Since a lower-order probability density can be obtained from a higher-order one by integration, the higher-order one contains more information than the lower-order ones.

Besides the probability distributions, a stochastic process can also be specified by the corresponding characteristic functions

$$M_X(\theta,t) = E[e^{i\theta\,X(t)}],$$

$$M_X(\theta_1,t_1;\theta_2,t_2) = E\{e^{i\,[\theta_1\,X(t_1)+\theta_2\,X(t_2)]}\},$$

$$\vdots$$

$$M_X(\theta_1,t_1;\theta_2,t_2;\ldots;\theta_n,t_n) = E\{e^{i\,[\theta_1\,X(t_1)+\theta_2\,X(t_2)+\cdots+\theta_n\,X(t_n)]}\}.$$

$$(3.2.2)$$

3.2.2 *Moment Functions*

A stochastic process can also be described by its moment functions

$$E[X(t)] = \int xp(x,t)dx,$$

$$E[X(t_1)X(t_2)] = \iint x_1x_2p(x_1,t_1;x_2,t_2)dx_1dx_2,$$

$$\vdots$$

$$E[X(t_1)X(t_2)\cdots X(t_n)] = \int \cdots \int x_1x_2\cdots x_n$$
$$\times p(x_1,t_1;x_2,t_2;\ldots;x_n,t_n)$$
$$\times dx_1dx_2\cdots dx_n. \qquad (3.2.3)$$

The moment functions are related to the characteristic function as follows

$$M_X(\theta_1,t_1;\theta_2,t_2;\ldots;\theta_n,t_n)$$
$$= 1 + \sum_{j=1}^{n} i\theta_j E[X(t_j)] + \frac{1}{2!} \sum_{j,k=1}^{n} (i\theta_j)(i\theta_k)E[X(t_j)X(t_k)] + \cdots .$$
$$(3.2.4)$$

The first and second moment functions, called the mean function and autocorrelation function respectively, are

$$\mu_X(t) = E[X(t)], \quad R_{XX}(t_1,t_2) = E[X(t_1)X(t_2)]. \qquad (3.2.5)$$

The autocorrelation function is non-negative definite, i.e., it satisfies

$$R_{XX}(t_1,t_2)h(t_1)h^*(t_2) \geq 0 \quad \text{for any } t_1 \text{ and } t_2, \qquad (3.2.6)$$

where $h(t)$ is an arbitrary function, and the asterisk denotes the complex conjugate. The proof is given in (Lin, 1967).

The autocovariance function is defined as the second central moment functions, i.e.,

$$\kappa_{XX}(t_1, t_2) = E\{[X(t_1) - \mu_X(t_1)][X(t_2) - \mu_X(t_2)]\}$$
$$= R_{XX}(t_1, t_2) - \mu_X(t_1)\mu_X(t_2) \qquad (3.2.7)$$

The variance function is the special case of the autocovariance function if $t_1 = t_2$, i.e.,

$$\sigma_X^2(t) = E\{[X(t) - \mu_X(t)]^2\}. \qquad (3.2.8)$$

The autocorrelation coefficient function is defined as

$$\rho_{XX}(t_1, t_2) = \frac{\kappa_{XX}(t_1, t_2)}{\sigma_X(t_1)\sigma_X(t_2)}. \qquad (3.2.9)$$

Since $\rho_{XX}(t_1, t_2)$ is in fact the correlation coefficient of two random variables $X(t_1)$ and $X(t_2)$, it is known from Section 2.6.3 that $\rho_{XX}(t_1, t_2) \leq 1$ for any t_1 and t_2.

The autocorrelation or autocovariance function is a measure of correlation of a stochastic process at two different time instants. A larger value of the autocorrelation function indicates a closer relationship of the process at two different time instants.

The mean and variance functions are the first-order statistical properties of the stochastic process since it is related to the first-order probability distribution involving only one time instant t. The autocorrelation, autocovariance and autocorrelation coefficient functions are the second-order statistical properties involving two time instants t_1 and t_2. Although they may be incomplete description for a stochastic process, they are among the most important properties in practical applications.

3.2.3 *Cumulant Functions*

Equivalent to the moment functions, the cumulant functions can also be used to describe a stochastic process. The cumulant functions of various orders are denoted as

$$\kappa_1[X(t)], \kappa_2[X(t_1)X(t_2)], \ldots, \kappa_n[X(t_1)X(t_2)\cdots X(t_n)]. \quad (3.2.10)$$

The cumulant functions are the coefficients of the Maclaurin series expansion of the log-characteristic function shown as follows

$$\ln M_X(\theta_1, t_1; \theta_2, t_2; \ldots; \theta_n, t_n)$$

$$= \sum_{j=1}^{n} (i\theta_j)\kappa_1[X(t_j)] + \frac{1}{2!} \sum_{j,k=1}^{n} (i\theta_j)(i\theta_k)\kappa_2[X(t_j)X(t_k)] + \cdots.$$

$$(3.2.11)$$

As shown in Sections 2.4.2 and 2.6.4, the first cumulant function is the same as the first moment function, i.e., the mean function, and the second and third cumulant functions are identical to the second and third central moment functions, i.e.,

$$\kappa_1[X(t)] = E[X(t)] = \mu_X(t),$$
$$\kappa_2[X(t_1)X(t_2)] = E\{[X(t_1) - \mu_X(t_1)][X(t_2) - \mu_X(t_2)]\},$$
$$\kappa_3[X(t_1)X(t_2)X(t_3)] = E\{[X(t_1) - \mu_X(t_1)][X(t_2) - \mu_X(t_2)]$$
$$[X(t_3) - \mu_X(t_3)]\}. \quad (3.2.12)$$

The equations in Section 2.6.4 for multiple random variables can be used to calculate cumulant functions of various orders of a stochastic process.

3.2.4 *Two Jointly Distributed Stochastic Processes*

Consider two stochastic processes $X_1(t)$ and $X_2(t)$. Analogous to the case of a single stochastic process, the cross-correlation, cross-covariance and cross-correlation coefficient functions are defined to describe the relationship of the two processes at two time instants. They are

$$R_{X_1X_2}(t_1, t_2) = E[X_1(t_1)X_2(t_2)], \quad (3.2.13)$$

$$\kappa_{X_1 X_2}(t_1, t_2) = E\{[X_1(t_1) - \mu_{X_1}(t_1)][X_2(t_2) - \mu_{X_2}(t_2)]\}$$
$$= R_{X_1 X_2}(t_1, t_2) - \mu_{X_1}(t_1)\mu_{X_2}(t_2), \qquad (3.2.14)$$

$$\rho_{X_1 X_2}(t_1, t_2) = \frac{\kappa_{X_1 X_2}(t_1, t_2)}{\sigma_{X_1}(t_1)\sigma_{X_2}(t_2)}. \qquad (3.2.15)$$

It is clear that the autocorrelation and the cross-correlation functions are both symmetric in the following sense,

$$R_{XX}(t_1, t_2) = R_{XX}(t_2, t_1), \quad R_{X_1 X_2}(t_1, t_2) = R_{X_2 X_1}(t_2, t_1).$$
$$(3.2.16)$$

3.3 Stationary Processes

Stochastic processes can be classified according to different criteria. One of the criteria is whether the probabilistic and statistical properties are independent of a shift of the time, i.e., when $t \to t + \tau$. A stochastic process is said to be strongly stationary, or stationary in strict sense, if its complete probability structure is invariant under the shift, i.e., the following equations hold for any τ,

$$p(x_1, t_1; x_2, t_2; \ldots; x_n, t_n) = p(x_1, t_1 + \tau; x_2, t_2 + \tau; \ldots; x_n, t_n + \tau),$$
$$n = 1, 2, \ldots. \qquad (3.3.1)$$

Equation (3.3.1) indicates that the first-order probability density is independent of the time, i.e., $p(x, t + \tau) = p(x, t) = p(x)$, and the higher-order probability densities only depend on the time difference τ. When Eq. (3.3.1) only holds for $n = 1$ and 2, the stochastic process is said to be weakly stationary, or stationary in wide (or weak) sense. In most practical problems, only weakly stationary processes are involved; therefore, we may omit the term "weakly" for simplicity.

For a weakly stationary process, the first-order properties are independent of the time; thus $E[X^n(t)] = E[X^n]$, $\mu_X(t) = \mu_X$, and $\sigma_X^2(t) = \sigma_X^2$. The second-order properties only depend on the time difference, i.e.,

$$R_{XX}(t_1, t_2) = R_{XX}(\tau), \quad \kappa_{XX}(t_1, t_2) = \kappa_{XX}(\tau),$$

$$\rho_{XX}(\tau) = \frac{\kappa_{XX}(\tau)}{\sigma_X^2}; \quad \tau = t_2 - t_1, \tag{3.3.2}$$

$$R_{XX}(0) = E[X^2(t)] = E[X^2], \quad \kappa_{XX}(0) = \sigma_X^2. \tag{3.3.3}$$

The physical meaning of the autocorrelation or autocovariance function is better revealed for the stationary processes. For a given time difference, a larger value of the autocorrelation function indicates a closer relationship of the process at two different time instants. On the other hand, a smaller autocorrelation indicates that the process changes very rapidly in stochastic fashion. With an increasing time difference, the correlation function generally becomes smaller.

Two stochastic processes are jointly stationary if each is stationary, and the following holds for any t_1 and t_2,

$$R_{X_1X_2}(t_1, t_2) = R_{X_1X_2}(t_2 - t_1) = R_{X_1X_2}(\tau). \tag{3.3.4}$$

From the definitions of autocorrelation and cross-correlation functions, we have

$$R_{XX}(\tau) = R_{XX}(-\tau), \quad R_{X_1X_2}(\tau) = R_{X_2X_1}(-\tau). \tag{3.3.5}$$

Using the following inequalities

$$E\left\{\left[\frac{X_1(t_1)}{\sqrt{R_{X_1X_1}(0)}} \pm \frac{X_2(t_2)}{\sqrt{R_{X_2X_2}(0)}}\right]^2\right\}$$

$$= 2 \pm 2\frac{R_{X_1X_2}(\tau)}{\sqrt{R_{X_1X_1}(0)R_{X_2 \geq X_2}(0)}} \geq 0, \tag{3.3.6}$$

we obtain

$$|R_{X_1X_2}(\tau)| \leq \sqrt{R_{X_1X_1}(0)R_{X_2X_2}(0)} = \sqrt{E[X_1^2]E[X_2^2]}, \tag{3.3.7}$$

$$|R_{XX}(\tau)| \leq R_{XX}(0) = E[X^2]. \tag{3.3.8}$$

Inequality (3.3.8) shows that the autocorrelation function reaches the maximum at $\tau = 0$. It makes sense intuitively since the correlation of a random variable and itself is of course the strongest. Similar properties are present for the autocovariance and cross-covariance functions

$$|\kappa_{X_1 X_2}(\tau)| \le \sqrt{\kappa_{X_1 X_1}(0) \kappa_{X_2 X_2}(0)} = \sigma_{X_1} \sigma_{X_2}, \quad |\kappa_{XX}(\tau)| \le \sigma_X^2.$$
(3.3.9)

A parameter called correlation time is defined as a quantitative measure for the correlation of a stationary process

$$\tau_0 = \int_0^\infty |\rho_{XX}(\tau)| \, d\tau, \tag{3.3.10}$$

where $\rho_{XX}(\tau)$ is the autocorrelation coefficient given in (3.3.2). When the process is totally uncorrelated, $\rho_{XX}(0) = 1$ and $\rho_{XX}(\tau) = 0$ for $\tau \ne 0$, the correlation time $\tau_0 = 0$. In this case, the process at two different time instants is not correlated no matter how the two time instants are closed with each other. When $\rho_{XX}(\tau) \ne 0$ even for $\tau \to \infty$, τ_0 is infinite, indicating a very long correlation time.

3.4 Ergodic Processes

For a stochastic process to be modeled and analyzed in practical applications, its characteristics need to be found; at least, its mean and correlation functions need to be evaluated. Assume that N sample functions $x_i(t)$ ($i = 1, 2, \ldots, N$) are obtained from measurements, and then the mean and correlation functions can be calculated from the ensemble averaging as follows

$$\mu_X(t) = E[X(t)] \approx \frac{1}{N} \sum_{i=1}^N x_i(t), \tag{3.4.1}$$

$$R_{XX}(t_1, t_2) = E[X(t_1)X(t_2)] \approx \frac{1}{N} \sum_{i=1}^N x_i(t_1) x_i(t_2). \tag{3.4.2}$$

The accuracy of the estimations depends on the number of sample functions. The more the sample functions are, the more reliable the estimations are. However, for many physical stochastic processes, the number of sample functions obtained from measurements is inadequate to provide reliable estimations.

The difficulty may be overcome for stationary processes for which the first-order properties are independent of the time and the higher-order properties only depend on the time shift. For such a case, a single sample function in a sufficient long time period may be used to obtain the characteristics of the stochastic process. Consider a record (sample function) $x(t)$ of a stochastic process $X(t)$ in the time period $[0, T]$ where T is sufficiently large. The time average of $X(t)$ is defined as

$$\langle X(t) \rangle_t = \lim_{T \to \infty} \frac{1}{T} \int_0^T x(t) dt. \tag{3.4.3}$$

If the time average is the same for all sample functions, and it is the same as the ensemble average, i.e., $\langle X(t) \rangle_t = E[X(t)] = \mu_X$, we say that the process is ergodic in the mean.

Ergodicity of a stochastic process can be defined on different levels. A process is ergodic in the mean square if the following is satisfied

$$\langle X^2(t) \rangle_t = \lim_{T \to \infty} \frac{1}{T} \int_0^T x^2(t) dt = E[X^2(t)]. \tag{3.4.4}$$

Ergodicity in the correlation is defined as

$$\langle X(t)X(t+\tau) \rangle_t = \lim_{T \to \infty} \frac{1}{T} \int_0^T x(t)x(t+\tau) dt$$

$$= E[X(t)X(t+\tau)] = R_{XX}(\tau), \tag{3.4.5}$$

which is equivalent to ergodicity in the covariance defined as

$$\langle [X(t) - \mu_X][X(t+\tau) - \mu_X] \rangle_t$$

$$= \lim_{T \to \infty} \frac{1}{T} \int_0^T [x(t)x(t+\tau) dt - [\langle X(t) \rangle_t]^2$$

$$= E\left\{ [X(t) - \mu_X][X(t+\tau) - \mu_X] \right\} = \kappa_{XX}(\tau). \tag{3.4.6}$$

It is obvious that ergodicity in a higher-order statistics implies ergodicity in the lower orders, and that ergodicity in correlation or covariance implies a weakly stationary process. However, the reverse of either of the above conclusions is not necessarily true.

Ergodicity in the correlation is often assumed for a physical stationary process, in which case the mean value, the mean-square value and correlation function can be estimated from the time averaging shown in (3.4.3)–(3.4.5) from one sample function in a very long time period. It greatly reduces the time and effort in analytical and numerical investigations.

3.5 Stochastic Calculus

In contrast to the case of deterministic functions, the differentiation and integration of stochastic processes can be defined in different senses.

3.5.1 *Modes of Convergence*

In calculus of a deterministic function, we need to deal with the convergence of a function, i.e., $\lim_{t \to t_0} X(t)$. When t approached t_0, a sequence of function values is calculated, and the limit exists if the sequence converges to a definite value. Now $X(t)$ is a stochastic process, and the sequence consists of random variables. The limit is said to exist if the sequence converges to a single random variable in some sense. Denote X_n as a sequence of random variables, and X as a random variable. The following are the several different modes of convergence:

Convergence with probability one:

$$\text{Prob}\left[\lim_{n \to \infty} X_n = X \right] = 1, \tag{3.5.1}$$

Convergence in probability:

$$\lim_{n \to \infty} \text{Prob}[|X_n - X| \geq \varepsilon] = 0 \quad \text{for every } \varepsilon > 0, \tag{3.5.2}$$

Convergence in distribution:

$$\lim_{n \to \infty} F_{X_n}(x) = F_X(x), \tag{3.5.3}$$

Convergence in mean square:

$$\lim_{n \to \infty} E[(X_n - X)^2] = 0. \tag{3.5.4}$$

The convergence with probability one in (3.5.1) is also known as the almost sure convergence. As shown in Fig. 3.5.1, the convergence in distribution is the weakest, while either the convergence with probability one or the convergence in mean square implies the other two modes (see Lin, 1967). However, the convergence with probability one and the convergence in mean square do not imply each other.

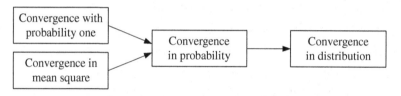

Figure 3.5.1   Relationships between different convergence modes.

The mode of convergence in mean square, also called in L_2 sense, is the most useful for practical problems, and will be used thereafter in the book except specified otherwise.

3.5.2 *Second-Order Stochastic Processes*

The convergence in L_2 sense is conventionally expressed as

$$\underset{n \to \infty}{\text{l.i.m.}}\, X_n = X, \tag{3.5.5}$$

where the symbol l.i.m. reads the limit in mean, and the limit X is a random variable. For the L_2 convergence (3.5.4) to be meaningful, the mean squares of the sequence X_n and the limit X must be finite, i.e.,

$$E[X_n^2] < \infty \quad \text{and} \quad E[X^2] < \infty. \tag{3.5.6}$$

Such random variables are called the second-order random variables. A random process $X(t)$ is said to be a second-order random process if the following is satisfied

$$E[X^2(t)] < \infty \quad \text{for any time instant } t. \tag{3.5.7}$$

Using (3.5.7) and the Schwarz inequality (2.6.16), we obtain

$$|E[X(t)]| \leq E[|X(t) \times 1|]$$
$$\leq \{E[X^2(t)]E[1^2]\}^{1/2} < \infty. \tag{3.5.8}$$
$$|E[X(t_1)X(t_2)]| \leq E[|X(t_1)X(t_2)|]$$
$$\leq \{E[X^2(t_1)]E[X^2(t_2)]\}^{1/2} < \infty. \tag{3.5.9}$$

Inequalities (3.5.8) and (3.5.9) shows that the second-order stochastic process also has the properties

$$-\infty < E[X(t)] < \infty. \tag{3.5.10}$$
$$-\infty < R_{XX}(t_1, t_2) < \infty. \tag{3.5.11}$$

Since $R_{XX}(t,t) = E[X^2(t)]$, (3.5.11) and (3.5.7) are equivalent.

Assume $X(t)$ and $Y(s)$ are two second-order processes, and $\underset{t \to t_0}{\text{l.i.m.}} X(t) = X$, $\underset{s \to s_0}{\text{l.i.m.}} Y(s) = Y$. We write

$$E[X(t)Y(s) - XY] = E\{[X(t) - X][Y(s) - Y]\} + E\{Y[X(t) - X]\}$$
$$+ E\{X[Y(s) - Y]\}. \tag{3.5.12}$$

According to the definition of the convergence in mean square and the Schwarz inequality, the three terms on the right-hand side of (3.5.12) vanish as $t \to t_0$ and $s \to s_0$, resulting in

$$\lim_{t \to t_0, \, s \to s_0} E[X(t)Y(s)] = E[XY]. \tag{3.5.13}$$

Taking $Y(s) = 1$, (3.5.13) leads to

$$\lim_{t \to t_0} E[X(t)] = E[X] = E\left[\underset{t \to t_0}{\text{l.i.m.}} X(t)\right]. \tag{3.5.14}$$

Equation (3.5.14) indicates that the operations of expectation and l.i.m. are commutative.

An important theorem known as the L_2 convergence criterion states: a second-order stochastic process $X(t)$ converges in mean square to a random variable X as $t \to t_0$ if and only if the limit of the functions $E[X(t_1)X(t_2)] = R_{XX}(t_1, t_2)$ exists and is finite as

$t_1, t_2 \to t_0$, no matter how t_1 and t_2 approach t_0. The proof is given in the book by Lin (1967).

3.5.3 *Differentiation of Stochastic Processes*

Similar to deterministic functions, differentiability of a stochastic process in L_2 sense requires the continuity in the same sense, i.e.,

$$\underset{\Delta t \to 0}{\text{l.i.m.}} X(t + \Delta t) = X(t). \qquad (3.5.15)$$

According to the L_2 convergence criterion, the necessary and sufficient condition for the L_2 continuity is that its autocorrelation function $R_{XX}(t_1, t_2) = E[X(t_1)X(t_2)]$ is finite and continuous for any $t_1 = t_2$, i.e., $R_{XX}(t_1, t_2)$ is continuous on the diagonal $t_1 = t_2$.

The L_2 derivative of a stochastic process is also a stochastic process, defined as

$$\frac{d}{dt}X(t) = \dot{X}(t) = \underset{\Delta t \to 0}{\text{l.i.m.}} \frac{X(t + \Delta t) - X(t)}{\Delta t}. \qquad (3.5.16)$$

The necessary and sufficient condition for (3.5.16) to be valid is that the correlation function $R_{XX}(t_1, t_2)$ has continuous mixed second derivative at $t_1 = t_2$, i.e., $\frac{\partial^2 R_{XX}(t_1,t_2)}{\partial t_1 \partial t_2}$ exists at $t_1 = t_2$ and is finite. If $X(t)$ is stationary, the condition is simplified as $\frac{d^2 R_{XX}(\tau)}{d\tau^2}$ exists at $\tau = 0$.

3.5.4 *Statistical Properties of Derivative Process*

The statistical properties of the derivative process $\dot{X}(t)$ defined in (3.5.16) can be obtained from those of the original process $X(t)$. For the mean function $\mu_{\dot{X}}(t)$,

$$\mu_{\dot{X}}(t) = E[\dot{X}(t)] = E\left[\underset{\Delta t \to 0}{\text{l.i.m.}} \frac{X(t + \Delta t) - X(t)}{\Delta t}\right]$$

$$= \lim_{\Delta t \to 0} \frac{E[X(t + \Delta t)] - E[X(t)]}{\Delta t} = \frac{d\mu_X(t)}{dt} = \dot{\mu}_X(t).$$

$$(3.5.17)$$

In deriving (3.5.17), use has been made of the commutability between operations of the expectation and limit in mean, as shown in (3.5.14).

Similarly, we have

$$R_{\dot{X}X}(t_1, t_2) = E[\dot{X}(t_1)X(t_2)] = \frac{\partial}{\partial t_1} R_{XX}(t_1, t_2). \tag{3.5.18}$$

$$R_{\dot{X}\dot{X}}(t_1, t_2) = E[\dot{X}(t_1)\dot{X}(t_2)] = \frac{\partial^2}{\partial t_1 \partial t_2} R_{XX}(t_1, t_2). \tag{3.5.19}$$

More generally,

$$\mu_{X^{(n)}}(t) = E[X^{(n)}(t)] = \frac{d^n \mu_X(t)}{dt^n}. \tag{3.5.20}$$

$$R_{X^{(n)}X^{(m)}}(t_1, t_2) = E[X^{(n)}(t_1)X^{(m)}(t_2)]$$

$$= \frac{\partial^{n+m}}{\partial t_1^n \partial t_2^m} R_{XX}(t_1, t_2). \tag{3.5.21}$$

If process $X(t)$ is stationary, (3.5.17), (3.5.18) and (3.5.19) lead to

$$\mu_{\dot{X}}(t) = \frac{d\mu_X(t)}{dt} = 0, \tag{3.5.22}$$

$$R_{\dot{X}X}(t_1, t_2) = R_{\dot{X}X}(\tau) = -\frac{d}{d\tau} R_{XX}(\tau), \quad \tau = t_2 - t_1, \tag{3.5.23}$$

$$R_{\dot{X}\dot{X}}(t_1, t_2) = R_{\dot{X}\dot{X}}(\tau) = -\frac{d^2}{d\tau^2} R_{XX}(\tau), \quad \tau = t_2 - t_1. \tag{3.5.24}$$

Since $R_{XX}(\tau)$ is an even function, (3.5.23) results in

$$R_{\dot{X}X}(0) = E[X(t)\dot{X}(t)] = -\frac{d}{d\tau} R_{XX}(\tau)\Big|_{\tau=0} = 0. \tag{3.5.25}$$

Equation (3.5.25) indicates that a stationary process $X(t)$ and its derivative process $\dot{X}(t)$ are uncorrelated.

3.5.5 *Integration of Stochastic Processes in L_2-Riemann Form*

Consider the integral of stochastic process $X(t)$

$$Y = \int_a^b X(t)dt. \qquad (3.5.26)$$

It is clear that Y is a random variable, which can be approximated by a Riemann sum

$$Y_n = \sum_{j=1}^n X(t'_j)(t_{j+1} - t_j); \quad t_j \le t'_j \le t_{j+1},$$

$$a = t_1 < t_2 < \cdots < t_{n+1} = b. \qquad (3.5.27)$$

Equation (3.5.27) indicates that Y_n is also a random variable. The integral (3.5.26) is said to exist in L_2 sense if the sequence Y_n converges to Y in L_2 sense, i.e.,

$$\underset{\substack{n \to \infty \\ \Delta_n \to 0}}{\text{l.i.m.}} Y_n = Y, \qquad (3.5.28)$$

where $\Delta_n = \max(\tau_{j+1} - \tau_j)$. It can be proved that the necessary and sufficient condition for (3.5.28) is the integrability of the autocorrelation function, i.e.,

$$\left| \int_a^b \int_a^b R_{XX}(t_1, t_2)dt_1 dt_2 \right| < \infty. \qquad (3.5.29)$$

Under this condition, the operations of expectation and L_2 integration commute, and the statistical properties of the integral Y in (3.5.26) can be evaluated from

$$\mu_Y = E[Y] = E\left[\int_a^b X(t)dt\right] = \int_a^b E[X(t)]dt = \int_a^b \mu_X(t)dt.$$

$$(3.5.30)$$

$$\sigma_Y^2 = E[Y^2] = E\left\{\left[\int_a^b X(t)dt\right]^2\right\}$$

$$= \int_a^b\int_a^b E[X(t_1)X(t_2)]dt_1 dt_2$$

$$= \int_a^b\int_a^b R_{XX}(t_1, t_2)dt_1 dt_2. \tag{3.5.31}$$

More useful is the integration of a stochastic process $X(t)$ including a weighting function $h(t,\tau)$,

$$Y(t) = \int_a^b X(\tau)h(t,\tau)d\tau, \tag{3.5.32}$$

where h is a bounded deterministic function. The integral (3.5.32) is the L_2 limit of the following random sequence in Riemann form

$$Y_n(t) = \sum_{j=1}^n X(\tau_j')h(t,\tau_j')(\tau_{j+1} - \tau_j);$$

$$\tau_j \le \tau_j' \le \tau_{j+1}, \quad a = \tau_1 < \tau_2 < \cdots < \tau_{n+1} = b. \tag{3.5.33}$$

That is

$$\underset{\substack{n\to\infty\\ \Delta_n\to 0}}{\text{l.i.m.}} Y_n(t) = Y(t). \tag{3.5.34}$$

The integral (3.5.32) exists in L_2 sense if and only if

$$\left|\int_a^b\int_a^b R_{XX}(\tau_1, \tau_2)h(t,\tau_1)h(t,\tau_2)d\tau_1 d\tau_2\right| < \infty \quad \text{for all } t. \tag{3.5.35}$$

In this case, we can obtain the mean and autocorrelation functions of $Y(t)$

$$\mu_Y(t) = E\left[\int_a^b X(\tau)h(t,\tau)d\tau\right] = \int_a^b \mu_X(\tau)h(t,\tau)d\tau.$$

(3.5.36)

$$R_{YY}(t_1,t_2) = \int_a^b\int_a^b R_{XX}(\tau_1,\tau_2)h(t_1,\tau_1)h(t_2,\tau_2)d\tau_1 d\tau_2. \qquad (3.5.37)$$

3.5.6 *Integration of Stochastic Processes in L_2-Stieltjes Form*

A more general L_2 integral in the Stieltjes form is introduced as

$$Y(t) = \int_a^b h(t,\tau)dZ(\tau) = \underset{\substack{n\to\infty \\ \Delta_n\to 0}}{\text{l.i.m.}} \sum_{j=1}^n h(t,\tau_j')[Z(\tau_{j+1}) - Z(\tau_j)],$$

(3.5.38)

where $Z(\tau)$ is a random process. Equation (3.5.38) reduces to (3.5.32) if $Z(t)$ is differentiable so that $dZ(\tau) = X(\tau)d\tau$. However, (3.5.38) is meaningful even if $Z(t)$ is not differentiable. The L_2-Siteltjes integral (3.5.38) exists if and only if

$$\left|\int_a^b\int_a^b h(t,\tau_1)h(t,\tau_2)E[dZ(\tau_1)dZ(\tau_2)]\right| < \infty \quad \text{for all } t. \qquad (3.5.39)$$

3.6 Spectral Analysis

The autocorrelation function is a second-order statistical property of a stochastic process since it involves two different time instants, and is related to the second-order probability density function. Equivalent to the autocorrelation function is another second-order statistical property, power spectral density, which is one of the most important characteristic of a stochastic process in practice.

In this section, we only consider stationary stochastic processes with zero mean such that the autocorrelation function is the same as the autocovariance function, and both depend only on the time difference.

3.6.1 *Spectral Density Functions of Stationary Processes*

Consider a stationary stochastic processes $X(t)$ with zero mean. The power spectral density function of $X(t)$ is defined as the Fourier transform of its autocorrelation function, i.e.,

$$\Phi_{XX}(\omega) = \frac{1}{2\pi} \int_{-\infty}^{\infty} R_{XX}(\tau)e^{-i\omega\tau}d\tau. \tag{3.6.1}$$

Since $R_{XX}(\tau)$ is non-negative definite, as shown in (3.2.6), its Fourier transform $\Phi_{XX}(\omega)$ is also non-negative according to Bochner (1959). The inverse of (3.6.1) is

$$R_{XX}(\tau) = \int_{-\infty}^{\infty} \Phi_{XX}(\omega)e^{i\omega\tau}d\omega. \tag{3.6.2}$$

The Fourier transformation pair of (3.6.1) and (3.6.2) is referred to as the well-known Wiener–Khintchine theorem.

To reveal the physical meaning of the power spectral density, we write

$$R_{XX}(0) = E[X^2(t)] = \int_{-\infty}^{\infty} \Phi_{XX}(\omega)d\omega. \tag{3.6.3}$$

Equation (3.6.3) shows that $\Phi_{XX}(\omega)$ describes the distribution of the mean-square value over the entire frequency domain. In many cases, the mean-square value is a measure of energy, such as the case where $X(t)$ is the displacement of a mechanical system and $X^2(t)$ is proportional to the potential energy. In such cases, the power spectral density $\Phi_{XX}(\omega)$ represents the energy distribution of $X(t)$ in the frequency domain. It is also called the mean-square spectral density or power spectral density.

For two joint stationary processes $X_1(t)$ and $X_2(t)$, the cross spectral density is defined as the Fourier transform of the cross-correlation function,

$$\Phi_{X_1 X_2}(\omega) = \frac{1}{2\pi} \int_{-\infty}^{\infty} R_{X_1 X_2}(\tau) e^{-i\omega\tau} d\tau, \qquad (3.6.4)$$

$$R_{X_1 X_2}(\tau) = \int_{-\infty}^{\infty} \Phi_{X_1 X_2}(\omega) e^{i\omega\tau} d\omega. \qquad (3.6.5)$$

Using the symmetric properties of the autocorrelation and cross-correlation functions, as shown in (3.3.4), we obtain

$$\Phi_{XX}(\omega) = \Phi_{XX}(-\omega), \quad \Phi_{X_1 X_2}(\omega) = \Phi_{X_2 X_1}^{*}(\omega), \qquad (3.6.6)$$

where an asterisk denotes the complex conjugate. Equation (3.6.6) indicates that the power spectral density is an even function, and the cross spectral density is a Hermitian function.

3.6.2 *Spectral Density Functions of Derivative Processes*

From (3.6.5), we obtain

$$\frac{d^n}{d\tau^n} R_{XX}(\tau) = i^n \int_{-\infty}^{\infty} \omega^n \Phi_{XX}(\omega) e^{i\omega\tau} d\omega. \qquad (3.6.7)$$

Using (3.6.7), (3.5.18) and (3.5.19) lead to

$$R_{\dot{X}X}(\tau) = -\frac{d}{d\tau} R_{XX}(\tau) = \int_{-\infty}^{\infty} (-i\omega) \Phi_{XX}(\omega) e^{i\omega\tau} d\omega. \qquad (3.6.8)$$

$$R_{\dot{X}\dot{X}}(\tau) = -\frac{d^2}{d\tau^2} R_{XX}(\tau) = \int_{-\infty}^{\infty} \omega^2 \Phi_{XX}(\omega) e^{i\omega\tau} d\omega. \qquad (3.6.9)$$

Equations (3.6.8) and (3.6.9) indicate that

$$\Phi_{\dot{X}X}(\omega) = (-i\omega)\Phi_{XX}(\omega), \quad \Phi_{\dot{X}\dot{X}}(\omega) = \omega^2\Phi_{XX}(\omega). \qquad (3.6.10)$$

More generally,

$$\Phi_{X^{(m)}X^{(n)}}(\omega) = (-1)^m i^{m+n} \omega^{m+n} \Phi_{XX}(\omega). \qquad (3.6.11)$$

Thus, the spectral density functions of a stationary process and its derivative processes can be calculated with one another. It is quite useful for a displacement process, its velocity process and its acceleration processes, if the spectral density of one process is known.

3.6.3 *Spectral Moments*

Energy distribution over the entire frequency domain is an important characteristic of a stochastic process. For some processes, the energy is concentrated in a small frequency band, and they are referred as narrowband processes. On the contrary, a process is called broadband or wideband process if its power spectrum has significant values over a large frequency band. Figure 3.6.1 illustrates schematically the correlation functions and spectral density functions of narrowband and broadband processes, Fig. 3.6.2 depicts three sample functions for a narrowband process, and Fig. 3.6.3 depicts three sample functions for a broadband process.

To quantify the bandwidth of a stochastic process, it is meaningful to introduce certain parameters. Note that the power spectral density function in the domain $[0, \infty)$ is analogous to the probability density of a random variable due to their non-negativity and integrability. From the probability density, we can calculate moments of the random variable, including the mean value and variance. The latter represents the dispersion of the process about its mean value. The smaller the variance is, the larger values the probability has near the mean value. Similarly, we define the spectral moments as

$$\lambda_n = \int_{-\infty}^{\infty} |\omega|^n \Phi_{XX}(\omega)d\omega = 2\int_{0}^{\infty} \omega^n \Phi_{XX}(\omega)d\omega. \qquad (3.6.12)$$

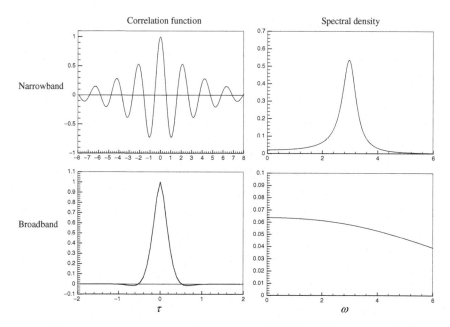

Figure 3.6.1   Schematics of narrowband and broadband processes.

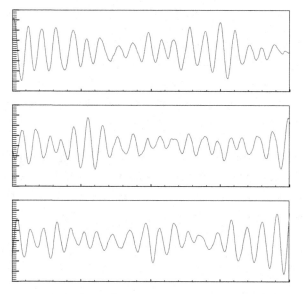

Figure 3.6.2   Sample functions of a narrowband process.

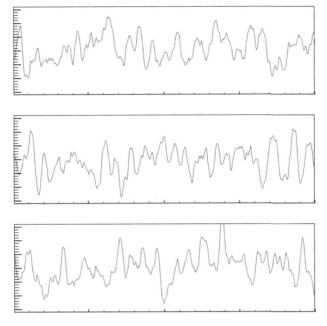

Figure 3.6.3   Sample functions of a broadband process.

We reinstate again that only stochastic processes with zero mean are considered, and then $\lambda_0 = \sigma_X^2$ according to (3.6.3). Follow the same practice of normalizing a probability density function, the spectral moments defined in (3.6.12) need to be normalized for more explicit physical meanings. Specifically, we define the central frequency in terms of the first-order spectral moment,

$$\gamma_1 = \frac{\lambda_1}{\lambda_0} = \frac{2}{\sigma_X^2} \int\limits_0^\infty \omega \Phi_{XX}(\omega) d\omega, \qquad (3.6.13)$$

which is analogous to the mean value. To specifying the dispersion of $\Phi_{XX}(\omega)$ about the central frequency, the spectrum variance parameter δ is defined as

$$\delta = \frac{\sqrt{(\lambda_2/\lambda_0) - (\lambda_1/\lambda_0)^2}}{(\lambda_1/\lambda_0)} = \sqrt{\frac{\lambda_0 \lambda_2}{\lambda_1^2} - 1}. \qquad (3.6.14)$$

Equation (3.6.14) shows that δ is analogous to the coefficient of variance. It can be shown using the Schwarz inequality that $\lambda_0\lambda_2 \geq \lambda_1^2$ and $0 \leq \delta < \infty$. A larger δ corresponds to a broader bandwidth. Another parameter ε, called the bandwidth parameter, is defined as

$$\varepsilon = \sqrt{1 - \frac{\lambda_2^2}{\lambda_0\lambda_4}}. \tag{3.6.15}$$

Also according to the Schwarz inequality, $\lambda_0\lambda_4 \geq \lambda_2^2$ and $0 \leq \varepsilon \leq 1$. A larger ε corresponds to a broader bandwidth.

Consider a harmonic process with a random initial phase, $X(t) = A\sin(\omega_0 t + U)$, where A and ω_0 are constants, and U is a random variable uniformly distributed in $[0, 2\pi)$. The incursion of the random initial phase U renders $X(t)$ as a stationary process. It can be shown that $R_{XX}(\tau) = \frac{1}{2}A^2\cos\omega_0\tau$ and $\Phi_{XX}(\omega) = \frac{1}{4}A^2[\delta(\omega - \omega_0) + \delta(\omega + \omega_0)]$ (see Exercise Problem 3.8). From Equations (3.6.12) through (3.6.15), we have $\gamma_1 = \omega_0$, $\delta = \varepsilon = 0$ (see Exercise Problem 3.23), indicating a zero band width.

Example 3.6.1 Gaussian white noise

A stochastic process is called a Gaussian white noise if it is Gaussian distributed, with zero mean and a constant power spectral density over the entire frequency range $(-\infty, \infty)$. Denoting it as $W(t)$, we have its power spectral density and autocorrelation function as follows

$$\Phi_{WW}(\omega) = K, \quad R_{WW}(\tau) = 2\pi K\delta(\tau). \tag{3.6.16}$$

The second equation in (3.6.16) shows that $\sigma_W^2 = R_{WW}(0) = \infty$ and $R_{WW}(\tau) = 0$ for any $\tau \neq 0$. That $\sigma_W^2 = \infty$ indicates that the process has infinite amount of energy, and $R_{WW}(\tau) = 0$ for $\tau \neq 0$ tells that the process changes infinitely fast. Obviously, such a process does not exist in reality, and all spectral moments do not exist. Nevertheless, it has been used frequently to approximate many physical processes with a broad spectrum due to (a) its simplicity in mathematical treatment and (b) rapid decay of system responses to the white noise excitation in the high frequency range.

Example 3.6.2 Banded white noise

To make the white-noise process more practical, the banded white noise is introduced with the power spectral density in positive ω domain as

$$
\Phi_{XX}(\omega) = \begin{cases} S_0, & \omega_0 - \dfrac{B}{2} \le |\omega| \le \omega_0 + \dfrac{B}{2} \\ 0, & \text{otherwise} \end{cases} \tag{3.6.17}
$$

and the associated correlation function as

$$
R_{XX}(\tau) = \frac{2S_0}{\tau} \cos(\omega_0 \tau) \sin\left(\frac{1}{2} B\tau\right), \tag{3.6.18}
$$

where parameter ω_0 is the center frequency, B specifies the band width, and $B \le 2\omega_0$. The spectral moments are

$$
\lambda_n = 2S_0 \int_{\omega_0 - \frac{B}{2}}^{\omega_0 + \frac{B}{2}} \omega^n d\omega = \frac{2S_0}{n+1} \left[\left(\omega_0 + \frac{B}{2}\right)^{n+1} - \left(\omega_0 - \frac{B}{2}\right)^{n+1} \right].
$$

$$\tag{3.6.19}$$

The variance parameter δ is calculated as

$$
\delta = \frac{B}{2\sqrt{3}\omega_0} \tag{3.6.20}
$$

and the bandwidth parameter ε as

$$
\varepsilon = \frac{B^2(15\omega_0^2 + B^2/4)}{9(5\omega_0^4 + 5\omega_0^2 B^2/2 + B^4/16)}. \tag{3.6.21}
$$

Figure 3.6.4 shows the spectral density and correlation function of the banded white noise for the cases of $\omega_0 = 1$ and three different values, 0.05, 0.1 and 0.5, for the band-width parameter B. It is seen that, for the case of a broader band, $B = 0.5$, the correlation function has higher values in the vicinity of $\tau = 0$, and decays rapidly.

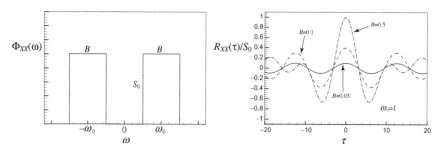

Figure 3.6.4 Spectral density and correlation functions of the banded white noise.

Example 3.6.3 Consider a weakly stationary process $X(t)$ with a power spectral density and an autocorrelation as follows

$$\Phi_{XX}(\omega) = \frac{\sigma_X^2 \alpha}{\pi(\omega^2 + \alpha^2)}, \quad R_{XX}(\tau) = \sigma_X^2 e^{-\alpha|\tau|}. \qquad (3.6.22)$$

Figure 3.6.5 shows the spectral density and correlation functions for several different α values. A larger α value corresponds to a broader band spectrum. Since the central frequency is zero, the processes are called the low-pass processes.

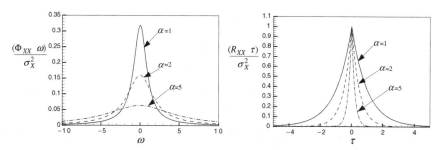

Figure 3.6.5 Spectral density and correlation functions of the low-pass processes.

It is noted that the spectral moments λ_n do not exist for the low-pass processes for $n \geq 1$. This is because the spectral density does not decrease fast enough with an increasing ω value. In reality, however, there always exists a cut-off frequency above which the spectral density is negligible.

3.6.4 *Spectral Density Functions of Non-Stationary Processes*

For non-stationary processes, the correlation functions depend on two time instants, not just their difference, as shown in Eqs. (3.2.5) and (3.2.13). There are two different ways to carry out the spectral analysis. One way is to define the power spectral density and cross spectral density as follows

$$\Phi_{XX}(\omega_1,\omega_2) = \frac{1}{(2\pi)^2} \int_{-\infty}^{\infty}\int_{-\infty}^{\infty} R_{XX}(t_1,t_2)e^{-i(\omega_1 t_1 - \omega_2 t_2)}dt_1 dt_2.$$

(3.6.23)

$$\Phi_{X_1 X_2}(\omega_1,\omega_2) = \frac{1}{(2\pi)^2} \int_{-\infty}^{\infty}\int_{-\infty}^{\infty} R_{X_1 X_2}(t_1,t_2)e^{-i(\omega_1 t_1 - \omega_2 t_2)}dt_1 dt_2.$$

(3.6.24)

Corresponding to the two time instants t_1 and t_2, two frequencies ω_1 and ω_2 are used to specify the frequency domain property of the stochastic processes.

The power spectral density and cross spectral density can also be defined in another way as

$$\Phi_{XX}(\omega,t) = \frac{1}{2\pi} \int_{-\infty}^{\infty} R_{XX}(t,t+\tau)e^{-i\omega\tau}d\tau \qquad (3.6.25)$$

$$\Phi_{X_1 X_2}(\omega,t) = \frac{1}{2\pi} \int_{-\infty}^{\infty} R_{X_1 X_2}(t,t+\tau)e^{-i\omega\tau}d\tau. \qquad (3.6.26)$$

Intuitively, the definition (3.6.25) has clearer physical meaning in the sense that it describes the time-varying energy distribution over the frequency domain. The spectral densities defined in (3.6.23) through (3.6.26) are called the general spectral densities (Lin, 1967).

3.7 Gaussian Stochastic Processes

A stochastic process $X(t)$ is a family of random variables with t as a parameter. A stochastic process $X(t)$ is said to be Gaussian distributed if all random variables at different time instants are jointly Gaussian distributed. As described in Section 2.5.1, a Gaussian distributed random variable can be defined in terms of either the probability density function or characteristic function. The definition for a Gaussian distributed stochastic process can also be given in terms of either the probability density function or the characteristic function. One definition for a Gaussian stochastic process $X(t)$, $t \in T$, is that its characteristic function is of the form

$$M_X[\theta(t)]$$

$$= \exp \left[i \int_T \mu_X(t)\theta(t)dt - \frac{1}{2} \iint_T \kappa_{XX}(t_1, t_2)\theta(t_1)\theta(t_2)dt_1 dt_2 \right],$$

(3.7.1)

where $\mu_X(t)$ and $\kappa_{XX}(t_1,t_2)$ are mean and covariance functions of $X(t)$, respectively. Equation (3.7.1) is an equivalent form of (2.6.28) for jointly Gaussian distributed random variables.

The first-order probability density function of a Gaussian stochastic process is

$$p(x, t) = \frac{1}{\sqrt{2\pi}\sigma_X(t)} \exp \left\{ -\frac{[x - \mu_X(t)]^2}{2\sigma_X^2(t)} \right\}.$$

(3.7.2)

The probabilistic and statistical properties of orders higher than two can be obtained using the equations in Section 2.6.5 for jointly Gaussian distributed random variables. For example, the second-order probability density is

$$p(x_1, t_1; x_2, t_2)$$

$$= \frac{1}{2\pi\sigma_1\sigma_2\sqrt{1 - \rho^2}}$$

$$\times \exp \left[-\frac{\sigma_2^2(x_1 - \mu_1)^2 - 2\sigma_1\sigma_2\rho(x_1 - \mu_1)(x_2 - \mu_2) + \sigma_1^2(x_2 - \mu_2)^2}{2\sigma_1^2\sigma_2^2(1 - \rho^2)} \right],$$

(3.7.3)

where $\mu_1 = \mu_X(t_1)$, $\mu_2 = \mu_X(t_2)$, $\sigma_1 = \sigma_X(t_1)$, $\sigma_2 = \sigma_X(t_2)$ and $\rho = \rho_{XX}(t_1, t_2)$.

If a Gaussian stochastic process is weakly stationary, its mean function $\mu_X(t)$ is a constant and covariance function $\kappa_{XX}(t_1, t_2)$ only depends on the time difference $\tau = t_2 - t_1$. Since a Gaussian stochastic process is completely specified by its mean function and covariance function, the weakly stationary property implies strongly stationary property.

Similar to the case of Gaussian random variables, a linear algebraic operation of a Gaussian stochastic process results in another Gaussian stochastic process. This conclusion can be extended to non-algebraic linear operations of differentiation and integration (Lin, 1967).

3.8 Poisson Processes and Related Stochastic Processes

3.8.1 *Homogeneous Poisson Process*

In Section 2.5.6, we introduced the Poisson distribution for a type of discrete random variables representing the occurrence of some random events. The Poisson distribution is

$$P_X(n) = \text{Prob}[N = n] = \frac{\mu^n}{n!} e^{-\mu}, \quad n = 0, 1, 2, \ldots, \tag{3.8.1}$$

where random variable N is defined in the domain of non-negative integer. In practical problems, a counting process is often required to record the number of events which occur repeatedly and randomly, such as the number of passengers arriving at an airport or a train station, or a sequence of impact forces from sea waves. The Poisson process is defined for such purpose.

Let $N(t)$ denote the number of events arriving (occurring) in the time interval $[0, t)$ and $P_N(n, t)$ denote the probability of the event $N(t) = n$. The counting process is said to be a homogeneous Poisson process or Poisson process with stationary increments, if the following conditions are satisfied:

(a) Independent arrival. Each arrival of the event is independent of the arrivals in the past.

(b) Stationary arrival rate. The probability of one arrival in an interval $(t, t + dt]$ equals to λdt where λ is a positive constant. It depends only on the interval length dt, analogous to the stationary case.

(c) Isolated arrival. In an infinitesimal interval $(t, t+dt]$, only one arrival with a probability density λdt is considered, and the probability of two or more arrivals is negligible.

It is shown that the Poisson process has the probability (Lin, 1967)

$$P_N(n, t) = \frac{(\lambda t)^n}{n!} e^{-\lambda t}. \tag{3.8.2}$$

Comparing (3.8.2) and (3.8.1), we find that (3.8.2) is the Poisson distribution with mean value $\mu_N(t) = E[N(t)] = \lambda t$ and variance $\sigma_N^2(t) = E[N^2(t)] = \lambda t$. The constant λ has the physical meaning of mean arrival rate, indicating that the arrival rate is a stationary stochastic process. The correlation function and the covariance function are (Sun, 2006)

$$R_N(t_1, t_2) = \lambda \min(t_1, t_2) + \lambda^2 t_1 t_2, \quad \kappa_N(t_1, t_2) = \lambda \min(t_1, t_2). \tag{3.8.3}$$

The mean value of the Poisson process increases with time, and the correlation function depends on both time instants; therefore, it is a non-stationary process although the process is called the Poisson process with stationary increments.

3.8.2 *Inhomogeneous Poisson Process*

As stated in the above condition 2, the probability of one arrival in an interval $(t, t + dt]$ is λdt and λ is a constant for a homogeneous Poisson process. If λ is not a constant, the counting process is said to be an inhomogeneous Poisson process, or Poisson process with non-stationary increments. In this case, the probability can be obtained by replacing λt by $\int_0^t \lambda(\tau)d\tau$ as follows

$$P_N(n,t) = \frac{1}{n!}\left[\int_0^t \lambda(\tau)d\tau\right]^n \exp\left[-\int_0^t \lambda(\tau)d\tau\right], \quad (3.8.4)$$

where the mean arrival rate $\lambda(t)$ is a function of t.

3.8.3 *Compound Poisson Process*

The Poisson process is a counting process, specifying the number of events arriving randomly. A physical quantity related to the random events may be of interest, such as the impact forces from random waves. To represent such a random process, the compound Poisson process is defined as

$$X(t) = \sum_{j=1}^{N(t)} Y_j, \quad (3.8.5)$$

where $N(t)$ is a homogeneous Poisson process, Y_j, $j = 1, 2, \ldots$, are independent and identically distributed random variables, and $N(t)$ and Y_j are independent. The mean function, correlation function and covariance function are respectively

$$\mu_X(t) = E[X(t)] = E\{E[X(t)|N(t)]\}$$
$$= E[N(t)\mu_Y] = \lambda t \mu_Y, \quad (3.8.6)$$

$$R_{XX}(t_1, t_2) = E[X(t_1)X(t_2)] = E\{E[X(t_1)X(t_2)|N(t_1)N(t_2)]\}$$
$$= E\{\min[N(t_1), N(t_2)]E[Y^2]\}$$
$$= \lambda \min(t_1, t_2)E[Y^2], \quad (3.8.7)$$

$$\kappa_{XX}(t_1, t_2) = R_{XX}(t_1, t_2) - \mu_X(t_1)\mu_X(t_2)$$
$$= \lambda \min(t_1, t_2)E[Y^2] - \lambda^2 t_1 t_2 \mu_Y^2. \quad (3.8.8)$$

3.8.4 *Impulsive Noise Process*

More general than the compound Poisson process is the random pulse train defined as

$$X(t) = \sum_{j=1}^{N(T)} Y_j w(t - \tau_j), \quad 0 < t \le T, \tag{3.8.9}$$

where $N(T)$ is a counting process specifying the total number of pulses that arrive in the time interval $(0, T]$, τ_j is the random time at which the jth pulse is initiated (to be referred to hereafter as the pulse-arrival time), $w(t - \tau)$ represents a deterministic pulse shape with $w(t - \tau) = 0$ for $\tau > t$ and Y_j is the random magnitude of the jth pulse arriving at τ_j. Moreover, we assume that $N(T)$ is a homogeneous Poisson process with a constant average pulse arrival rate λ, Y_j for different j are independent and identically distributed random variables with zero mean, and Y_j are independent of the pulse arrival time τ_j.

The mth cumulant function of $X(t)$ is (Lin, 1967)

$$\kappa_m[X(t_1), X(t_2), \ldots, X(t_m)]$$

$$= \lambda E[Y^m] \int_0^{\min(t_1, t_2, \ldots, t_m)} w(t_1 - \tau) w(t_2 - \tau) \cdots w(t_m - \tau) d\tau, \tag{3.8.10}$$

where $E[Y^m]$ is the mth moment of any Y_j. The mean, variance and the covariance functions are obtained from (3.8.10) as

$$\mu_X(t) = \kappa_1[X(t)] = \lambda \mu_Y \int_0^t w(t - \tau) d\tau, \tag{3.8.11}$$

$$\sigma_X^2(t) = \kappa_2[X(t), X(t)] = \lambda E[Y^2] \int_0^t w^2(t - \tau) d\tau, \tag{3.8.12}$$

$$\kappa_{XX}(t_1, t_2) = \kappa_2[X(t_1), X(t_2)]$$

$$= \lambda E[Y^2] \int\limits_0^{\min(t_1, t_2)} w(t_1 - \tau)w(t_2 - \tau)d\tau. \qquad (3.8.13)$$

It can be shown (Lin, 1967) that the cumulant function (3.8.10) remains unchanged upon changing the time origin; therefore, $X(t)$ is a stationary process. Assuming that $X(t)$ is a second-order process, i.e., the second cumulant is bounded, then $\lambda E[Y^2]$ is finite. Consider the case in which the average arrival rate λ approaches to infinite. Thus, $E[Y^2]$ is of order of λ^{-1} and $E[Y^m]$ is at most of order $\lambda^{-m/2}$. In this case, the cumulants of order higher than two vanish, and process $X(t)$ is Gaussian distributed.

The above random pulse train process (3.8.9) is in fact one particular type of more general random pulse train processes. The counting process $N(t)$ may be an inhomogeneous Poisson process or a more general counting process. The pulse shape function may be different for different pulses, i.e., $w(t, \tau)$ instead of $w(t - \tau)$.

The impulsive noise process is a special type of random pulse train processes with the pulse shape function being a unit impulse, i.e.,

$$X(t) = \sum_{j=1}^{N(T)} Y_j \delta(t - \tau_j). \qquad (3.8.14)$$

The mth cumulant equation (3.8.10) becomes

$$\kappa_m[X(t_1), X(t_2), \ldots, X(t_m)]$$

$$= \lambda E[Y^m]\delta(t_2 - t_1)\delta(t_3 - t_1) \cdots \delta(t_m - t_1). \qquad (3.8.15)$$

Equation (3.8.15) shows that the impulsive noise is a Poisson white noise. If the average arrival rate λ of the counting Poisson process

$N(t)$ is low, it is a non-Gaussian Poisson white noise, With λ approaches infinity, the impulsive noise becomes a Gaussian white noise.

The impulsive noise process (3.8.14) is a mathematical idealization for real stochastic processes in which the duration of the pulse shape function $w(t - \tau)$ is much shorter than a characteristic time of the investigated dynamical system, known as the relaxation time (see Section 7.4).

3.9 Evolutionary Stochastic Processes

3.9.1 *Stochastic Processes with Orthogonal Increments*

Let $Z(\omega)$ be a complex-valued stochastic process with its sample space in the frequency domain, and satisfy the condition

$$E[|Z(\omega_2) - Z(\omega_1)|^2] < \infty. \tag{3.9.1}$$

$Z(\omega)$ is said to be a stochastic process with orthogonal increments if

$$E\{[Z(\omega_2) - Z(\omega_1)][Z^*(\omega_4) - Z^*(\omega_3)]\}$$
$$= E[Z(\omega_2) - Z(\omega_1)]E[Z^*(\omega_4) - Z^*(\omega_3)] = 0, \tag{3.9.2}$$

for any non-overlapping intervals $(\omega_1, \omega_2]$ and $(\omega_3, \omega_4]$ where $\omega_1 < \omega_2 \le \omega_3 < \omega_4$. It can be proved that the necessary and sufficient condition for $Z(\omega)$ to be an orthogonal-increment stochastic process is (Lin and Cai, 1995)

$$E[dZ(\omega_1)dZ^*(\omega_2)] = \begin{cases} d\Psi(\omega), & \omega_1 = \omega_2 \\ 0, & \omega_1 \ne \omega_2, \end{cases} \tag{3.9.3}$$

where $\Psi(\omega)$ is a deterministic function. Equation (3.9.3) can also be used as an alternative definition for a stochastic process with orthogonal increments.

The orthogonal-increment stochastic processes can be used to construct certain types of stochastic processes.

3.9.2 *Weakly Stationary Process in Terms*
of Orthogonal-Increment Process

Let $X(t)$ be a real-valued stochastic process, $E[X(t)] = 0$, and has a Fourier–Stieltjes representation

$$X(t) = \int_{-\infty}^{\infty} e^{i\omega t} dZ(\omega), \tag{3.9.4}$$

where $Z(\omega)$ is an orthogonal-increment process. Since $X(t)$ is real, it remains unchanged after conjugation, and we have

$$X(t - \tau) = \int_{-\infty}^{\infty} e^{i\omega(t-\tau)} dZ(\omega) = \int_{-\infty}^{\infty} e^{-i\omega'(t-\tau)} dZ^*(\omega'). \tag{3.9.5}$$

The correlation function of $X(t)$ is then calculated as

$$E[X(t)X(t - \tau)] = \int_{-\infty}^{\infty} \int_{-\infty}^{\infty} e^{i(\omega-\omega')t+i\omega'\tau} E[dZ(\omega)dZ^*(\omega')]$$

$$= \int_{-\infty}^{\infty} e^{i\omega\tau} d\Psi(\omega). \tag{3.9.6}$$

In deriving (3.9.6), use has been made of (3.9.3). Equation (3.9.6) shows that the correlation function of $X(t)$ depends only on the time difference τ; therefore, $X(t)$ is a weakly stationary process.

Function $\Psi(\omega)$ may or may not be differentiable. If it is, denoting $\Phi_{XX}(\omega) = \frac{d\Psi(\omega)}{d\omega}$, Eq. (3.9.6) can be written as

$$R_{XX}(\tau) = E[X(t)X(t - \tau)] = \int_{-\infty}^{\infty} e^{i\omega\tau} \Phi_{XX}(\omega) d\omega, \tag{3.9.7}$$

which is the same as (3.6.2), and $\Phi_{XX}(\omega)$ is the power spectral density. The expression (3.9.6) is more general than (3.9.7) by including the cases in which $\Psi(\omega)$ is not differentiable. Since the power spectral density $\Phi_{XX}(\omega)$ is non-negative, $\Psi(\omega)$ is a non-decreasing function, and called the spectral distribution function and denoted by $\Psi_{XX}(\omega)$. If we include Dirac delta functions in $\Phi_{XX}(\omega)$, we may write

$$\Psi_{XX}(\omega) = \int_{-\infty}^{\omega} \Phi_{XX}(\omega')d\omega', \tag{3.9.8}$$

and

$$E[X^2(t)] = \Psi_{XX}(\infty) = \int_{-\infty}^{\infty} \Phi_{XX}(\omega)d\omega. \tag{3.9.9}$$

Figures 3.9.1 and 3.9.2 illustrate the power spectral density and spectral distribution function for two cases. In Fig. 3.9.1, $\Psi_{XX}(\omega)$ is differentiable and $\Phi_{XX}(\omega)$ is continuous, while in Fig. 3.9.2, $\Psi_{XX}(\omega)$ is not differentiable and $\Phi_{XX}(\omega)$ includes Dirac delta functions. It is noted that a Dirac delta function at a certain frequency in the power spectral density represents a harmonic component at the frequency.

3.9.3 *Evolutionary Stochastic Processes*

The orthogonal-increment process $Z(\omega)$ can be used to construct a type of non-stationary stochastic processes, named as evolutionary stochastic processes. Let a stochastic process $X(t)$ with zero mean be expressed as

$$X(t) = \int_{-\infty}^{\infty} a(t,\omega)e^{i\omega t}dZ(\omega), \tag{3.9.10}$$

where $a(t,\omega)$ is a deterministic function. The correlation function of $X(t)$ is calculated as

$$\begin{aligned}
E[X(t_1)X(t_2)] &= \int_{-\infty}^{\infty}\int_{-\infty}^{\infty} a(t_1,\omega_1)a^*(t_2,\omega_2)e^{i(\omega_1 t_1 - \omega_2 t_2)} \\
&\quad \times E[dZ(\omega_1)dZ^*(\omega_2)] \\
&= \int_{-\infty}^{\infty} a(t_1,\omega)a^*(t_2,\omega)e^{i\omega(t_1-t_2)}d\Psi(\omega). \tag{3.9.11}
\end{aligned}$$

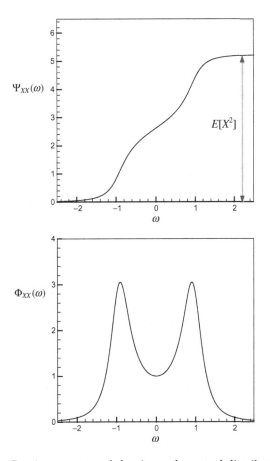

Figure 3.9.1 Continuous spectral density and spectral distribution functions.

Letting $t_1 = t_2$, we have the mean-square value

$$E[X^2(t)] = \int\limits_{-\infty}^{\infty} |a(t,\omega)|^2 d\Psi(\omega). \qquad (3.9.12)$$

Equations (3.9.11) and (3.9.12) show that the mean-square value is a function of time, and the correlation function depends on two time instants, not only the time difference. Therefore, $X(t)$ is a non-stationary stochastic process. If $\Psi(\omega)$ is differentiable with

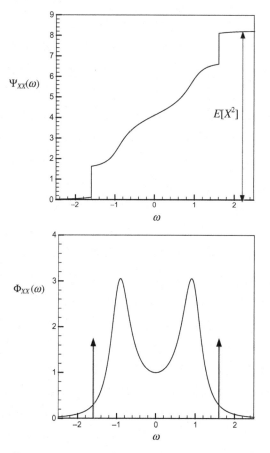

Figure 3.9.2 Discontinuous spectral density and spectral distribution functions.

$d\Psi(\omega) = \Phi(\omega)d\omega$, we have

$$E[X(t_1)X(t_2)] = \int\limits_{-\infty}^{\infty} a(t_1,\omega)a^*(t_2,\omega)e^{i\omega(t_1-t_2)}\Phi(\omega)d\omega. \qquad (3.9.13)$$

$$E[X^2(t)] = \int\limits_{-\infty}^{\infty} |a(t,\omega)|^2 \, \Phi(\omega)d\omega. \qquad (3.9.14)$$

The evolutionary spectral density is defined as (Priestly, 1965)

$$\Phi_{XX}(\omega, t) = |a(t, \omega)|^2 \, \Phi(\omega). \tag{3.9.15}$$

Functions $\Psi(\omega)$ and $\Phi(\omega)$ in (3.9.12) and (3.9.13) can be the spectral distribution function and spectral density function of a suitable stationary process. Function $a(t, \omega)$ modifies the spectrum $\Phi(\omega)$ both in time and frequency domains.

If $a(t, \omega) = a(t)$, it modifies the spectral density $\Phi(\omega)$ uniformly, i.e., changing the spectrum magnitude proportionally and maintaining the same distribution over the frequency. In this case, the representation is

$$X(t) = a(t) \int\limits_{-\infty}^{\infty} e^{i\omega t} dZ(\omega). \tag{3.9.16}$$

$X(t)$ is called the modulated stationary process and $a(t)$ is called the envelope function.

3.9.4 *Random Pulse Train — A Type of Evolutionary Process*

In Section 3.8.4, the random pulse train is introduced as

$$X(t) = \sum_{j=1}^{N(T)} Y_j w(t - \tau_j), \quad 0 < t \leq T, \tag{3.9.17}$$

where $N(T)$ is a counting process specifying the total number of pulses that arrive in the time interval $(0, T]$, τ_j is the random time at which the jth pulse is initiated known as the pulse-arrival time, $w(t - \tau)$ represents a deterministic pulse shape with $w(t - \tau) = 0$ for $\tau > t$, and Y_j is the random magnitude of the jth pulse arriving at τ_j. Without loss of generality, it is assumed that $E[Y_j] = 0$. In Section 3.8.4, the counting process $N(t)$ is assumed to be a homogeneous Poisson process with a stationary arrival rate. Thus, $X(t)$ is a stationary process. For more general application, the pulse arrival rate is now assumed to be a non-stationary stochastic process, denoted by $\Lambda(t)$. The autocovariance function of the random pulse train $X(t)$ can be derived as (Lin and Cai, 1995)

$$\kappa_{XX}(t_1, t_2) = E[Y^2] \int_{-\infty}^{\infty} \mu_\Lambda(t) w(t_1 - \tau) w(t_2 - \tau) d\tau$$

$$= \frac{1}{2\pi} E[Y^2] \int_{-\infty}^{\infty} b(t_1, \omega) b^*(t_2, \omega) e^{i\omega(t_1 - t_2)} d\omega, \qquad (3.9.18)$$

where $\mu_\Lambda(t)$ is the mean function of the pulse arrival rate, and

$$b(t, \omega) = \int_{-\infty}^{\infty} \sqrt{\mu_\Lambda(t - u)} w(u) e^{-i\omega u} du. \qquad (3.9.19)$$

Comparing (3.9.18) with (3.9.13), we find that $b(t, \omega)$ and $\frac{1}{2\pi} E[Y^2]$ in (3.9.18) play the roles of $a(t, \omega)$ and $\Phi(\omega)$ in (3.9.13), respectively. Therefore, the random pulse train $X(t)$ is an evolutionary process, and its evolutionary spectral density is obtained from (3.9.15) as

$$\Phi_{XX}(\omega, t) = \frac{1}{2\pi} E[Y^2] |b(t, \omega)|^2. \qquad (3.9.20)$$

If the Poisson process $N(t)$ in (3.9.17) is homogeneous, then the pulse arrival rate $\Lambda(t)$ is a stationary process, $\mu_\Lambda(t) = \mu_\Lambda$, $b(t, \omega) = b(\omega)$, $\Phi(\omega, t) = \Phi(\omega)$, and the random pulse train $X(t)$ becomes a stationary process, as described in Section 3.8.4.

Exercise Problems

3.1 Let R be a random variable of Rayleigh distribution given in Eq. (2.5.9), Θ be a random variable uniformly distributed in $[0, 2\pi)$, and R and Θ are independent. Define random process $X(t)$ as $X(t) = R \cos(\omega t + \Theta)$

(a) Find the mean and variance of $X(t)$.

(b) Find the probability density and characteristic functions of $X(t)$.

(c) Determine whether $X(t)$ is stationary.

3.2 A stochastic process $\dot{X}(t)$ is defined as

$$X(t) = R\cos(\Omega t + \Theta),$$

where R, Θ and Ω are independent random variables, R and Θ are the same as those in Problem 3.1, and Ω has a probability density function $p_\Omega(w)$, $w > 0$. Determine whether $X(t)$ is stationary.

3.3 Let $Y_1(t), Y_2(t), \ldots, Y_n(t)$ be independent stationary random processes. Process $X(t)$ is defined as

$$X(t) = \prod_{i=1}^{N} Y_i(t).$$

Determine whether $X(t)$ is stationary.

3.4 $X_1(t), X_2(t), \ldots, X_n(t)$ are jointly stationary random processes. Find the general expression for the correlation function of random process $X(t)$ defined as

$$X(t) = \sum_{i=1}^{N} X_i(t).$$

Reduce the expression for the special case in which all $X_i(t)$ are independent.

3.5 Specify the admissible parameter values in the following auto-correlation functions so that the associated stationary random processes are physically realizable.

(a) $|\tau|^{-a}$ (b) $\dfrac{\cosh(b\tau)}{\cosh(\pi\tau)}$ (c) $\dfrac{\sinh(c\tau)}{\sinh(\pi\tau)}$.

3.6 Are the following autocorrelation functions permissible? If not, give the reasons.

(a) $\dfrac{A^2}{1+\tau^4}$ (b) $A^2(\cos\tau^2 \pm \sin\tau^2)$

(c) $\begin{cases} A^2(1-\tau^2)^p, & -1 < \tau < 1, p > -\dfrac{3}{2} \\ 0, & \text{otherwise} \end{cases}$

(d) $a(\cos\tau + \sin\tau)$, $a > 0$ (e) $(1+|\tau|)e^{-|\tau|}$ (f) $|\tau|e^{-|\tau|}$.

3.7 Random process $X(t)$ is defined as a sum of random phase processes

$$X(t) = \sum_{j=1}^{N} \sqrt{2}\sigma_j \cos(\omega_j t + \theta_j).$$

where θ_j are independent random variables uniformly distributed in $[0, 2\pi)$. Determine the autocorrelation function of $X(t)$.

3.8 Random process $X(t)$ is defined as $X(t) = A\sin(\omega_0 t + Y)$ where A and ω_0 are constants and Y is a random number uniformly distributed in $[0, 2\pi)$. Calculate its correlation and power spectral density functions, and show that $X(t)$ is ergodic in mean as well as in correlation.

3.9 A stochastic process is defined as

$$X(t) = A\sin \omega t + B\cos \omega t,$$

where ω is a constant and A and B are independent and identically distributed random variables with zero mean and variance σ^2. Determine (a) whether $X(t)$ is stationary and (b) whether $X(t)$ is ergodic.

3.10 Let $X(t)$ be a stationary Gaussian process with zero mean and correlation function $R_{XX}(\tau) = e^{-\alpha|\tau|}$. Determine whether the following processes are stationary processes, (a) $Y(t) = X^2(t)$ and (b) $Z(t) = X^3(t)$. If they are, find their correlation functions.

3.11 If $\text{l.i.m.}_{t\to t_0} X(t) = X$ and $\text{l.i.m.}_{t\to t_0} X(t) = Y$, prove that random variables X and Y are equivalent in the sense that $\text{Prob}[X \neq Y] = 0$.

3.12 Prove that if the autocorrelation function $R_{XX}(t_1, t_2), t_1, t_2 \in T$, is continuous on the diagonal $t_1 = t_2$, then it is continuous over the entire space $T \times T$.

3.13 For the autocorrelation function given below, determine in each case whether the derivative of the random process $\dot{X}(t)$, exists in the sense of convergence in mean square. If it does, find $R_{\dot{X}\dot{X}}(\tau)$.

(a) $R_{XX}(\tau) = (1 - \frac{a}{2}|\tau|)e^{-a|\tau|}$, $a > 0$.

(b) $R_{XX}(\tau) = e^{-a\tau^2}$, $a > 0$.

(c) $R_{XX}(\tau) = \frac{\sin b\tau}{\tau}$, $b > 0$.

(d) $R_{XX}(\tau) = \frac{a^2}{a^2 + \tau^2}$.

(e) $R_{XX}(\tau) = e^{-\zeta\omega_0\tau}(\cos\omega_d\tau + \frac{\zeta\omega_0}{\omega_d}\sin\omega_d\tau)$, $\omega_d = \omega_0\sqrt{1 - \zeta^2}$.

3.14 Let $X(t)$ be a stationary process with zero mean and a correlation function

$$R_{XX}(\tau) = e^{-a|\tau|}, \quad a > 0.$$

Determine the correlation function of process $Y(t)$, defined as

$$Y(t) = \frac{1}{t}\int_0^t X(\tau)d\tau.$$

3.15 Determine whether the stochastic integral

$$Y(s) = \int_{-\infty}^{\infty} X(t)h(t,s)dt,$$

exists in the sense of mean square for the following cases:

(a) $R_{XX}(\tau) = \frac{a^2}{a^2 + \tau^2}$, $h(t,s) = \begin{cases} s - t, & s > t \\ 0 & s \le t. \end{cases}$

(b) $R_{XX}(\tau) = e^{-a|\tau|}$, $a > 0$, $h(t,s) = \begin{cases} \sqrt{s - t}, & s > t \\ 0 & s \le t. \end{cases}$

3.16 Let $X(t)$ be a random process and $Y(t) = \frac{dX(t)}{dt}$ be the mean-square derivative. Give the general condition for $E[Y(t_1) Y(t_2)\ldots Y(t_n)]$ to exist. Show that if $X(t)$ is Gaussian, the condition reduces to the existence of the mixed second derivative of $E[X(t_j)X(t_k)]$.

3.17 Calculate the power spectral density functions of the two processes $Y(t)$ and $Z(t)$ defined in Problem 3.10.

3.18 Find the power spectral density functions corresponding to the following auto correlation functions

$$\text{(a) } R_{XX}(\tau) = \begin{cases} \sigma_X^2 \left(1 - \dfrac{|\tau|}{T}\right), & |\tau| \leq T \\ 0, & |\tau| > T, \end{cases}$$

$$\text{(b) } R_{XX}(\tau) = \dfrac{\sigma_X^2 \sin \omega_0 \tau}{\pi \tau}.$$

3.19 Let random process $Z(t)$ be

$$Z(t) = aX(t) + bY(t),$$

where a and b are constants and $X(t)$ and $Y(t)$ are stationary random processes. Obtain the power spectral density of $Z(t)$ in terms of the spectral density functions of $X(t)$ and $Y(t)$.

3.20 Random processes $X(t)$ and $Y(t)$ are independent and weakly stationary.

(a) Find the general expression for the spectral density of $Z(t) = X(t)\ Y(t)$ in terms of the spectral densities of $X(t)$ and $Y(t)$.

(b) Apply the obtained expression to the special case where

$$R_{XX}(\tau) = A^2 e^{-a|\tau|}, \quad a > 0, \quad R_{YY}(\tau) = B^2 \cos(b\tau).$$

3.21 $X(t)$ is a band-limited white noise with a power spectral density as follows:

$$\Phi_{XX}(\omega) = \begin{cases} S_0, & |\omega| \leq \omega_0 \\ 0, & \text{otherwise.} \end{cases}$$

Determine its autocorrelation function and mean square, and calculate the central frequency γ_1, spectrum variance parameter δ and bandwidth parameter ε.

3.22 Let the power spectral density of random process $X(t)$ be

$$\Phi_{XX}(\omega) = \begin{cases} S_0, & 0 \le \omega_a \le |\omega| \le \omega_b \\ 0, & \text{otherwise.} \end{cases}$$

Find its autocorrelation function and mean square, and calculate the central frequency γ_1, spectrum variance parameter δ and bandwidth parameter ε.

3.23 As shown in Problem 3.8, stochastic process $X(t) = \sin(\omega_0 t + Y)$ is stationary, where ω_0 is a constant and Y is a random number uniformly distributed in $[0, 2\pi)$. Calculate the central frequency γ_1, spectrum variance parameter δ and bandwidth parameter ε.

3.24 The power spectral density of random process $X(t)$ is given by

$$\Phi_{XX}(\omega) = \frac{\alpha \sigma_X^2}{\pi(\omega^2 + \alpha^2)}.$$

Use the residue theory to find the correlation function of $X(t)$.

3.25 Let the power spectral density of random process $X(t)$ be

$$\Phi_{XX}(\omega) = \frac{K}{(\omega^2 - \omega_0^2)^2 + (2\zeta\omega\omega_0)^2}.$$

Use the residue theory to calculate the mean-square value of $X(t)$ given by

$$E[X^2(t)] = \int_{-\infty}^{\infty} \Phi_{XX}(\omega)d\omega.$$

3.26 Let the power spectral density of a stationary stochastic process $X(t)$ be

$$\Phi_{XX}(\omega) = \begin{cases} K, & |\omega| \le \omega_b \\ 0, & |\omega| > \omega_b. \end{cases}$$

Determine the correlation function. Show that, as $\omega_b \to \infty$, $X(t)$ approaches the white noise.

3.27 Let the power spectral density of random process $X(t)$ be

(a) $\Phi(\omega) = \frac{K}{(\omega^2 - \omega_0^2)^2 + (2\zeta\omega\omega_0)^2}.$

(b) $\Phi(\omega) = \frac{K\omega^2}{(\omega^2 - \omega_0^2)^2 + (2\zeta\omega\omega_0)^2}.$

Use the residue theory to calculate the correlation function of $X(t)$.

3.28 Denote T as the inter-arrival time of a Poisson process with stationary increments, i.e., the time between two adjacent events. Clearly, T is a random variable. Find the probability distribution of T.

3.29 A random pulse train process is defined by (3.8.9), in which the shape function $w(t - \tau_j) = H(t - \tau_j)$ is the Heaviside's step function. Determine its autocorrelation function.

3.30 Determine the autocorrelation function for the random pulse train process defined by Eq. (3.8.9) with the following shape function

$$w(t - \tau_j) = \begin{cases} \dfrac{1}{\Delta}, & t \geq \tau_j \geq t - \Delta \\ 0, & \text{otherwise.} \end{cases}$$

3.31 Determine the covariance function and evolutionary spectral density of the random pulse train

$$X(t) = \sum_{j=1}^{N(T)} Y_j w(t - \tau_j),$$

where Y_j are independent and identically distributed random variables with the probability density

$$p_Y(y) = \begin{cases} \dfrac{1}{2a}, & -a \leq y \leq a \\ 0, & \text{otherwise,} \end{cases}$$

τ_j are random pulse arrival times, with an average arrival rate

$$\mu_\Lambda(\tau) = e^{-\alpha\tau}, \quad \alpha > 0$$

and the pulse shape function is given as

$$w(t - \tau) = \begin{cases} e^{-\beta(t-\tau)}, & \beta > 0, \quad t > \tau \\ 0, & t < \tau. \end{cases}$$

3.32 A random pulse train is defined as

$$X(t) = \sum_{j=1}^{N(T)} Y_j h(t - \tau_j), \quad 0 < t \le T,$$

where Y_j are independent, identically distributed random variables with a zero mean and a finite mean-square value $E[Y^2]$, the shape function is given by

$$h(t) = \begin{cases} 0, & t < 0 \\ Ae^{-\zeta\omega_0 t} \sin \omega_d t, & t \ge 0, \end{cases}$$

where $\omega_d = \sqrt{1 - \zeta^2}\omega_0$, $\zeta < 1$. Determine the covariance function of $X(t)$, assuming that the average pulse arrival rate is

$$\mu_\Lambda(\tau) = e^{-\alpha\tau}, \quad \beta > \alpha > 0.$$

CHAPTER 4

MARKOV AND RELATED
STOCHASTIC PROCESSES

4.1 Introduction

A type of stochastic processes is known as Markov process, which is especially important in stochastic dynamics due to (a) its capability of modeling many practical stochastic processes, (b) the possibility of applying existing mathematical theories of Markov processes to various stochastic problems in different fields and (c) the convenience of generating and simulating Markov processes.

Denote a Markov Process by $X(t)$. If both the value of the process X and the parameter t are discrete, it is called Markov chain. If the value of X is continuous, but the parameter t is discrete, it is called the Markov series. In many applications, both the X value and the parameter t are continuous, and it is conventionally called the Markov process, which we will deal with in this book exclusively.

4.2 Markov Diffusion Processes

4.2.1 *Markov Processes*

When a stochastic process is related to a system with short memory, the process at the current instant may be affected only by the most recent history. Such a process belongs to a general category of Markov process. A stochastic process $X(t)$ is said to be Markovian if its conditional probability satisfies

$$\text{Prob}[X(t_n) \le x_n | X(t_{n-1}) \le x_{n-1}, \ldots, X(t_1) \le x_1]$$
$$= \text{Prob}[X(t_n) \le x_n | X(t_{n-1}) \le x_{n-1}], \tag{4.2.1}$$

where $t_1 < t_2 < \cdots < t_n$. A sufficient condition for a stochastic process $X(t)$ to be Markovian is that its increments in any two non-overlapping time intervals are independent, i.e., $X(t_2) - X(t_1)$ and $X(t_4) - X(t_3)$ are independent if $t_1 < t_2 \le t_3 < t_4$. If $X(t)$ is a Gaussian process, then the sufficient condition is

$$E\{[X(t_2) - X(t_1)][X(t_4) - X(t_3)]\} = 0, \quad t_1 < t_2 \le t_3 < t_4. \tag{4.2.2}$$

It is obvious that a Markov process is a mathematical idealization of a real stochastic process. Nevertheless, a great number of stochastic processes can be represented by Markov process. The Brownian motion in Physics is a Markov process. Many noise and signal processes in various fields, such as engineering, communication, ecology, biology, etc., are frequently modeled as Markov processes or in terms of Markov processes.

For a Markov process, the definition, Eq. (4.2.1), can be expressed in terms of the probability density function, i.e.,

$$p(x_n, t_n | x_{n-1}, t_{n-1}; \ldots; x_1, t_1) = p(x_n, t_n | x_{n-1}, t_{n-1}). \tag{4.2.3}$$

Using (4.2.3) and the property of the conditional probability density, Eq. (2.6.12), we have

$$p(x_1, t_1; x_2, t_2; \ldots; x_n, t_n)$$
$$= p(x_n, t_n | x_{n-1}, t_{n-1}) p(x_{n-1}, t_{n-1} | x_{n-2}, t_{n-2}) \cdots p(x_1, t_1). \tag{4.2.4}$$

It suggests that a higher order probability density function can be obtained from the initial probability density and the conditional probability density. In other words, a Markov process is completely characterized by its conditional probability density and the initial probability density. The latter includes the special case in which the initial sate is fixed, i.e., the initial probability density is a Dirac delta function. Therefore, the conditional probability density

$p(x_k, t_k | x_j, t_j)$ is the most important for quantifying a Markov process $X(t)$. It is also called the transition probability density in literature. A Markov process is said to be stationary if its transition probability density is invariant with a time shift, i.e., for any Δt,

$$p(x_k, t_k | x_j, t_j) = p(x_k, t_k + \Delta t | x_j, t_j + \Delta t) = p(x_\tau, \tau | x_0, 0), \quad (4.2.5)$$

where $\tau = t_k - t_j$ and x_τ and x_0 are the state variables of $X(\tau)$ and $X(0)$, respectively. In this case, the stationary probability density may be obtained by letting the transition time interval approach infinity, i.e.,

$$p(x) = \lim_{\tau \to \infty} p(x, \tau | x_0). \quad (4.2.6)$$

The above concept of a scalar Markov process is readily generalized to a vector Markov process. Let $\mathbf{X}(t) = [X_1(t), X_2(t), \ldots, X_n(t)]^T$ be a n-dimensional Markov vector. It satisfies

$$p(\mathbf{x}_n, t_n | \mathbf{x}_{n-1}, t_{n-1}; \ldots; \mathbf{x}_1, t_1) = p(\mathbf{x}_n, t_n | \mathbf{x}_{n-1}, t_{n-1}). \quad (4.2.7)$$

It is important to note that a component of Markov vector process may or may not be a scalar Markov process.

4.2.2 *Fokker–Planck–Kolmogorov (FPK) Equation*

Consider three time instants $t_1 < t < t_2$. Equation (4.2.7) implies

$$p(\mathbf{x}_2, t_2; \mathbf{y}, t | \mathbf{x}_1, t_1) = p(\mathbf{x}_2, t_2 | \mathbf{y}, t) p(\mathbf{y}, t | \mathbf{x}_1, t_1). \quad (4.2.8)$$

Integrating (4.2.8) with respect to $\mathbf{y}$, we obtain

$$p(\mathbf{x}_2, t_2 | \mathbf{x}_1, t_1) = \int p(\mathbf{x}_2, t_2 | \mathbf{y}, t) p(\mathbf{y}, t | \mathbf{x}_1, t_1) d\mathbf{y}. \quad (4.2.9)$$

Equation (4.2.9) is known as the Chapman–Kolmogorov–Smoluwski equation, which can be considered as a governing equation for the transition probability density. For more convenient analysis, the integral equation (4.2.9) can be converted to an equivalent form of differential equation, known as Fokker–Planck–Kolmogorov (FPK) equation shown below:

$$\frac{\partial}{\partial t}p + \sum_{j=1}^{n}\frac{\partial}{\partial x_j}(a_j p) - \frac{1}{2}\sum_{j,k=1}^{n}\frac{\partial^2}{\partial x_j \partial x_k}(b_{jk}p)$$

$$+ \frac{1}{3!}\sum_{j,k,l=1}^{n}\frac{\partial^3}{\partial x_j \partial x_k \partial x_l}(c_{jkl}p) - \cdots = 0, \qquad (4.2.10)$$

where $p = p(\mathbf{x}, t|\mathbf{x}_0, t_0)$ is the transition probability density, and

$$a_j = a_j(\mathbf{x}, t) = \lim_{\Delta t \to 0}\frac{1}{\Delta t}E[X_j(t + \Delta t) - X_j(t)|\mathbf{X}(t) = \mathbf{x}],$$

$$b_{jk} = b_{jk}(\mathbf{x}, t) = \lim_{\Delta t \to 0}\frac{1}{\Delta t}E\{[X_j(t + \Delta t) - X_j(t)]$$

$$\times [X_k(t + \Delta t) - X_k(t)]|\mathbf{X}(t) = \mathbf{x}\},$$

$$c_{jkl} = c_{jkl}(\mathbf{x}, t) = \lim_{\Delta t \to 0}\frac{1}{\Delta t}E\{[X_j(t + \Delta t) - X_j(t)]$$

$$\times [X_k(t + \Delta t) - X_k(t)][X_l(t + \Delta t) - X_l(t)]|\mathbf{X}(t) = \mathbf{x}\},$$

$$\vdots$$

$$(4.2.11)$$

The functions $a_j, b_{jk}, c_{jkl}, \ldots$, are called the derivate moments, giving the time rate of moments of various increments in $\mathbf{X}(t)$ at time t, on the condition that $\mathbf{X}(t) = \mathbf{x}$. The derivation of the FPK equation (4.2.10) from the Chapman–Kolmogorov–Smoluwski equation (4.2.9) is given in detail in the book (Lin and Cai, 1995).

In many practical problems, the derivate moments of orders higher than two vanish or can be neglected, then Eq. (4.2.10) reduces to

$$\frac{\partial}{\partial t}p + \sum_{j=1}^{n}\frac{\partial}{\partial x_j}(a_j p) - \frac{1}{2}\sum_{j,k=1}^{n}\frac{\partial^2}{\partial x_j \partial x_k}(b_{jk}p) = 0. \qquad (4.2.12)$$

Conventionally, the name of FPK equation refers to Eq. (4.2.12). In this case, the Markov process $\mathbf{X}(t)$ is called a Markov diffusion process, or simply a diffusion process.

The FPK equation (4.2.12) can be rewritten as

$$\frac{\partial}{\partial t}p + \sum_{j=1}^{n} \frac{\partial}{\partial x_j} G_j = 0, \qquad (4.2.13)$$

where

$$G_j = a_j p - \frac{1}{2} \sum_{k=1}^{n} \frac{\partial}{\partial x_k}(b_{jk}p). \qquad (4.2.14)$$

Equation (4.2.13) is analogous to the continuity equation in fluid mechanics for the conservation of mass in a fluid flow. Thus, Eq. (4.2.13) may be interpreted as conservation of probability, and G_j as the jth component of the probability flow vector $\mathbf{G}(\mathbf{x}, t|\mathbf{x}_0, t_0)$.

Equation (4.2.12) is a partial differential equation with first-order derivative in time t and second-order derivatives in the state variable $\mathbf{x}$. For a practical problem, the derivate moments $a_j(\mathbf{x}, t)$ and $b_{jk}(\mathbf{x}, t)$ are determined from the system equations of motion. Besides, proper initial condition and boundary conditions required to solve the FPK equation also need to be established based on the underlined physical problem. In many practical problems, the initial state is fixed, and we have the following initial condition

$$p(\mathbf{x}, t_0 \,|\, \mathbf{x}_0, t_0) = \delta(\mathbf{x} - \mathbf{x}_0) = \Pi_{j=1}^{n}\delta(x_j - x_{j0}). \qquad (4.2.15)$$

The boundary conditions depend on the sample behavior of the system. For boundaries not at infinity, there are several typical cases: reflective boundary, absorbing boundary and periodic boundary. For many engineering problems, a boundary at infinity is of importance. At infinity boundary, the probability flow must vanish, i.e.,

$$\lim_{x_j \to \pm\infty} \mathbf{G}(\mathbf{x}, t|\mathbf{x}_0, t_0) = 0. \qquad (4.2.16)$$

Moreover, since the total probability is finite, we also have

$$\lim_{x_j \to \pm\infty} p(\mathbf{x}, t|\mathbf{x}_0, t_0) = 0 \qquad (4.2.17)$$

and it approaches zero at least as fast as $|x_j|^{-\alpha}$, where $\alpha > 1$ and is determined from the specific system.

If a Markov diffusion process reaches the stationary state, its stationary probability density is the limit of the transition probability density, as shown in Eq. (4.2.6). Then the time derivative term in Eq. (4.2.12) vanishes, resulting in the so-called reduced FPK equation,

$$\sum_{j=1}^{n} \frac{\partial}{\partial x_j}(a_j p) - \frac{1}{2} \sum_{j,k=1}^{n} \frac{\partial^2}{\partial x_j \partial x_k}(b_{jk} p) = 0, \qquad (4.2.18)$$

where $p = p(\mathbf{x})$, the stationary probability density and a_j and b_{jk} are independent of t. Equation (4.2.18) can also be written as

$$\sum_{j=1}^{n} \frac{\partial}{\partial x_j} G_j = 0, \quad G_j = a_j p - \frac{1}{2} \sum_{k=1}^{n} \frac{\partial}{\partial x_k}(b_{jk} p). \qquad (4.2.19)$$

4.2.3 *Kolmogorov Backward Equation*

In the FPK equation, the unknown $p(\mathbf{x}, t | \mathbf{x}_0, t_0)$ is treated as a function of t and $\mathbf{x}$, and the resulting solution contains t_0 and $\mathbf{x}_0$ as parameters. It is also known as the Kolmogorov forward equation since the term $\frac{\partial p}{\partial t}$ is a derivative with respect to the later time. The state variable $\mathbf{x}$ associated with the later time t is known as the forward variable. Conversely, $p(\mathbf{x}, t | \mathbf{x}_0, t_0)$ can also be treated as a function of t_0 and $\mathbf{x}_0$ with t and $\mathbf{x}$ as parameters. The alternative equation corresponding to (4.2.10) is given by

$$\frac{\partial p}{\partial t_0} + \sum_{j=1}^{n} a_j \frac{\partial p}{\partial x_{j0}} + \frac{1}{2} \sum_{j,k=1}^{n} b_{jk} \frac{\partial^2 p}{\partial x_{j0} \partial x_{k0}}$$

$$+ \frac{1}{3!} \sum_{j,k,l=1}^{n} c_{jkl} \frac{\partial^3 p}{\partial x_{j0} \partial x_{k0} \partial x_{l0}} + \cdots = 0, \qquad (4.2.20)$$

where $a_j, b_{jk}, c_{jkl}, \ldots$, are the same derivate moments, except expressed as functions of $\mathbf{x}_0$ and t_0. The derivation of the backward equation (4.2.20) is given in the book by Lin and Cai (1995).

In the case of a Markov diffusion process, (4.2.20) reduces to

$$\frac{\partial p}{\partial t_0} + \sum_{j=1}^{n} a_j(\mathbf{x}_0, t_0) \frac{\partial p}{\partial x_{j0}} + \frac{1}{2} \sum_{j,k=1}^{n} b_{jk}(\mathbf{x}_0, t_0) \frac{\partial^2 p}{\partial x_{j0} \partial x_{k0}} = 0.$$

$$(4.2.21)$$

Equations (4.2.20) and (4.2.21) are known as the Kolmogorov backward equations, and $\mathbf{x}_0$ is the backward variable. The FPK equation (forward equation) is commonly used to obtain the probability density function, while the backward equation may be employed to investigate the first-passage problem (see Chapter 9).

The initial condition for the backward equation is the same as that for the forward equation, for example, Eq. (4.2.15) for a fixed state. It is obvious that the initial state variables x_{j0}, $j = 1, 2, \ldots, n$, cannot be at infinity. For finite boundaries, they can also be classified as reflective, absorbing and periodic, similar to the cases for the forward equation.

4.2.4 *Wiener Process*

The simplest Markov diffusion process is the Wiener process, also known as the Brownian motion process, denoted by $B(t)$. A stochastic process $B(t)$ is said to be a Wiener process if the following are satisfied: (a) $B(t)$ is a Gaussian process, (b) $B(0) = 0$, (c) $E[B(t)] = 0$, and (d)

$$E[B(t_1)B(t_2)] = \sigma^2 \min(t_1, t_2), \qquad (4.2.22)$$

where σ^2 is known as the intensity of the Wiener process. Equation (4.2.22) shows that the Wiener process is not a stationary process. Letting $t_1 < t_2 \le t_3 < t_4$, it follows from (4.2.22) that

$$E\{[B(t_2) - B(t_1)][B(t_4) - B(t_3)]\}$$
$$= E[B(t_2)B(t_4) - B(t_1)B(t_4) - B(t_2)B(t_3) + B(t_1)B(t_3)]$$
$$= \sigma^2(t_2 - t_1 - t_2 + t_1) = 0. \qquad (4.2.23)$$

According to the sufficient condition (4.2.2), $B(t)$ is a Markov process.

Additional properties can be derived for Wiener process. First, it is continuous in L_2 sense since its correlation function is continuous along the diagonal $t_1 = t_2$, as shown from (4.2.22). Furthermore, use Eq. (3.5.19) to find the correlation function of the derivative $\dot{B}(t)$ as

$$E[\dot{B}(t_1)\dot{B}(t_2)] = \frac{\partial^2}{\partial t_1 \partial t_2} E[B(t_1)B(t_2)]$$

$$= \sigma^2 \frac{\partial^2}{\partial t_1 \partial t_2} \min(t_1, t_2) = \sigma^2 \frac{\partial H(t_2 - t_1)}{\partial t_2}$$

$$= \sigma^2 \delta(t_2 - t_1), \tag{4.2.24}$$

where $H(t)$ is a Heaviside unit step function

$$H(t) = \begin{cases} 1, & t > 0 \\ 0, & t < 0 \end{cases}. \tag{4.2.25}$$

In deriving (4.2.24), use has been made of

$$\frac{\partial}{\partial t_1} \min(t_1, t_2) = \begin{cases} 1, & t_1 < t_2 \\ 0, & t_1 > t_2 \end{cases} = H(t_2 - t_1). \tag{4.2.26}$$

Equation (4.2.24) shows that $B(t)$ is not differentiable in L_2 sense since the mixed second derivative of its correlation function is unbounded along $t_1 = t_2$.

Denote the differential increment of $B(t)$ as

$$dB(t) = B(t + dt) - B(t). \tag{4.2.27}$$

Using (4.2.22), we have

$$E[dB(t_1)dB(t_2)] = \begin{cases} \sigma^2 dt, & t_1 = t_2 \\ 0, & t_1 \neq t_2 \end{cases} \tag{4.2.28}$$

and hence,

$$E[B(t + dt) - B(t)] = 0, \quad E\{[B(t + dt) - B(t)]^2\} = \sigma^2 dt. \tag{4.2.29}$$

According to (4.2.11), the first and second derivate moments are

$$a = 0, \quad b = \sigma^2. \tag{4.2.30}$$

Since the Wiener process $B(t)$ is Gaussian, it can be proved that all derivate moments of orders higher than two vanish. Thus, it is a diffusion process, and the FPK equation is

$$\frac{\partial}{\partial t}p - \frac{1}{2}\sigma^2\frac{\partial^2 p}{\partial z^2} = 0, \tag{4.2.31}$$

where z is the state variable of $B(t)$ and $p = p(z,t|z_0,t_0)$ is the transition probability density. With the initial condition and boundary conditions given as

$$\lim_{t\to t_0} p(z,t|z_0,t_0) = \delta(z - z_0), \quad \lim_{z\to\pm\infty} \frac{\partial}{\partial z}p(z,t|z_0,t_0) = 0, \tag{4.2.32}$$

the solution of (4.2.31) is

$$p(z,t|z_0,t_0) = \frac{1}{\sqrt{2\pi(t - t_0)}\sigma} \exp\left[-\frac{(z - z_0)^2}{2\sigma^2(t - t_0)}\right]. \tag{4.2.33}$$

As expected, $B(t)$ is a Gaussian process with mean z_0 and standard deviation $\sigma\sqrt{t - t_0}$.

Another important property of the Wiener process is the well-known Levy's oscillatory property (Levy, 1948), described next. Let $B(t)$ be a Wiener process defined in a finite time interval $[a, b]$. Divide $[a, b]$ into n subintervals $a = t_0 < t_1 < \cdots < t_{n-1} < t_n = b$, and denote $\Delta t_j = t_j - t_{j-1}$ and $\Delta_n = \max_{1\leq j\leq n}\Delta t_j$. The Levy's oscillatory property states

$$\lim_{\substack{n\to\infty \\ \Delta_n\to 0}} \sum_{j=1}^{n} [B(t_j) - B(t_{j-1})]^2 = \sigma^2(b - a). \tag{4.2.34}$$

The proof is given in the book (Lin and Cai, 1995). It is noted that the convergence in (4.2.34) is in the mean square (L_2) sense. It is also known that the Levy's oscillatory property is valid almost surely, i.e., with probability one (see. e.g., Karlin and Taylor, 1975). This suggests that

$$dB(t_1)dB(t_2) = \begin{cases} \sigma^2 dt, & t_1 = t_2 \\ 0, & t_1 \neq t_2, \end{cases} \tag{4.2.35}$$

which is a much stronger statement than (4.2.28). Equation (4.2.35) indicates that $dB(t)$ is of the order of $(dt)^{1/2}$. Thus, it is proved again that $dB(t)/dt$ is unbounded as $dt \to 0$, and process $B(t)$ is not differentiable in L_2 sense.

Besides that $B(t)$ is not differentiable in L_2 sense, it can also be proved that $B(t)$ has unbounded variation within any finite time interval. Therefore, the Wiener process is a mathematical idealization of a type of physical processes.

4.2.5 *Relationship between Wiener Process and Gaussian White Noise*

Consider the following equation

$$\frac{dX(t)}{dt} = W(t), \quad X(0) = 0, \tag{4.2.36}$$

where $W(t)$ is a Gaussian white noise with a spectral density K, i.e.,

$$E[W(t)] = 0, \quad E[W(t)W(t+\tau)] = 2\pi K\delta(\tau). \tag{4.2.37}$$

In terms of (4.2.36), $X(t)$ can be expressed as

$$X(t) = \int_0^t W(u)du. \tag{4.2.38}$$

Thus, we have

$$E[X(t)] = \int_0^t E[W(u)]du = 0. \tag{4.2.39}$$

$$E[X(t_1)X(t_2)] = \int_0^{t_1} \int_0^{t_2} E[W(u)W(v)]dudv$$

$$= 2\pi K \int_0^{t_1} \int_0^{t_2} \delta(u-v)dudv. \tag{4.2.40}$$

Without loss of generality, assume $t_1 < t_2$, and the last integral in (4.2.40) can be calculated as

$$\int_0^{t_1}\int_0^{t_2} \delta(u - v)dudv = \int_0^{t_1}\int_0^{t_1} \delta(u - v)dudv + \int_0^{t_1}\int_{t_1}^{t_2} \delta(u - v)dudv$$

$$= \int_0^{t_1} dv = t_1. \tag{4.2.41}$$

Thus, we have

$$E[X(t_1)X(t_2)] = 2\pi K \min(t_1, t_2). \tag{4.2.42}$$

According to the definition of the Wiener process, $X(t)$ is a Wiener process, and we may write

$$\frac{dB(t)}{dt} = W(t), \tag{4.2.43}$$

with an intensity σ^2 of the Wiener process and the spectral density K of the white noise related as

$$\sigma^2 = 2\pi K. \tag{4.2.44}$$

It is noted that the relationship (4.2.43) is only a formal one since the Wiener process $B(t)$ is not differentiable in L_2 sense.

The validity of (4.2.43) can also be substantiated by comparing Eqs. (4.2.24) and (3.6.16), which shows that the correlation functions of $\dot{B}(t)$ and the white noise $W(t)$ are both the delta function.

4.2.6 *Ito Stochastic Differential Equations*

As the simplest Markov diffusion process, the Wiener process $B(t)$ can be used to construct other Markov diffusion processes through stochastic differential equations. According to Ito (1951a), a scalar

Markov diffusion process $X(t)$ may be generated from

$$dX(t) = m(X,t)dt + \sigma(X,t)dB(t), \qquad (4.2.45)$$

where $B(t)$ is a unit Wiener process, i.e.,

$$E[B(t_1)B(t_2)] = \min(t_1, t_2),$$

$$E[dB(t_1)dB(t_2)] = \begin{cases} 0, & t_1 \neq t_2 \\ dt, & t_1 = t_2 = t. \end{cases} \qquad (4.2.46)$$

In (4.2.45), functions m and σ may depend on $X(t)$, but may also depend explicitly on t. Equation (4.2.45) has the formal solution

$$X(t) = X(0) + \int_0^t m(X, \tau)d\tau + \int_0^t \sigma(X, \tau)dB(\tau). \qquad (4.2.47)$$

The last term in (4.2.47) is a Stieltjes integral described in Section 3.5.6, i.e.,

$$\int_0^t \sigma[X(\tau), \tau]dB(\tau) = \lim_{\substack{n \to \infty \\ \Delta_n \to 0}} \sum_{j=1}^n \sigma[X(\tau_j'), \tau_j'][B(\tau_{j+1}) - B(\tau_j)],$$

$$(4.2.48)$$

where $0 = \tau_1 < \tau_2 < \cdots < \tau_n < \tau_{n+1} = t$, $\Delta_n = \max(\tau_{j+1} - \tau_j)$ and $\tau_j \leq \tau_j' \leq \tau_{j+1}$. Since $B(t)$ is a very unusual stochastic process, being not differentiable and of unbounded variation within any finite time interval, this Stieltjes integral must be interpreted properly. The key issue is the choice of τ_j'. Two different choices were proposed, one is the Ito type and another is the Stratonovich type.

The Ito integral selects $\tau_j' = \tau_j$, and the Stieltjes integral (4.2.48) becomes

$$\int_0^t \sigma[X(\tau), \tau]dB(\tau) = \lim_{\substack{n \to \infty \\ \Delta_n \to 0}} \sum_{j=1}^n \sigma[X(\tau_j), \tau_j][B(\tau_{j+1}) - B(\tau_j)].$$

$$(4.2.49)$$

In terms of Ito integral, the differential equation (4.2.45) is known as the Ito stochastic differential equation (Ito, 1951a). The stochastic process $X(t)$ is a diffusion process, and functions m and σ in (4.2.45) are called the drift and diffusion coefficients, respectively. In the Ito integral (4.2.49), the difference $B(\tau_{j+1}) - B(\tau_j)$ is taken in a forward time interval following the time instant τ_j at which function $\sigma[X(\tau), \tau]$ is evaluated; thus, ensuring that $dB(t)$ is independent of $X(t)$ in (4.2.45).

To obtain the first and second derivate moments, consider a very small Δt and write

$$X(t + \Delta t) - X(t) = \int\limits_{t}^{t+\Delta t} m[X(\tau), \tau]d\tau + \int\limits_{t}^{t+\Delta t} \sigma[X(\tau), \tau]dB(\tau)$$

$$\approx m[X(t), t]\Delta t + \sigma[X(t), t][B(t + \Delta t) - B(t)].$$
(4.2.50)

From (4.2.11), we have

$$a(x, t) = \lim_{\Delta t \to 0} \frac{1}{\Delta t} E[X(t + \Delta t) - X(t) | X(t) = x] = m(x, t).$$
(4.2.51)

$$b(x, t) = \lim_{\Delta t \to 0} \frac{1}{\Delta t} E\{[X(t + \Delta t) - X(t)]^2 | X(t) = x\} = \sigma^2(x, t).$$
(4.2.52)

Therefore, the first and second derivate moments in the FPK equation can be obtained directly from the drift and diffusion coefficients, respectively. However, note that the first and second derivate moments are functions of the state variable x, while the drift and diffusion coefficients are functions of the stochastic process $X(t)$.

The above analysis can be extended to n-dimensional case. An n-dimensional diffusion vector process may be generated from the following set of Ito stochastic differential equations

$$dX_j(t) = m_j(\mathbf{X}, t)dt + \sum_{l=1}^{m} \sigma_{jl}(\mathbf{X}, t)dB_l(t), \quad j = 1, 2, \ldots, n,$$
(4.2.53)

where $B_l(t)$, $l = 1, 2, \ldots, m$, are assumed to be independent unit Wiener process. The first and second derivate moments for the corresponding FPK equation are obtained as

$$a_j(\mathbf{x}, t) = m_j(\mathbf{x}, t), \quad b_{jk}(\mathbf{x}, t) = \sum_{l=1}^{m} \sigma_{jl}(\mathbf{x}, t)\sigma_{kl}(\mathbf{x}, t), \qquad (4.2.54)$$

which can be written as a matrix form

$$\mathbf{a}(\mathbf{x}, t) = \mathbf{m}(\mathbf{x}, t), \mathbf{b}(\mathbf{x}, t) = \boldsymbol{\sigma}(\mathbf{x}, t)\boldsymbol{\sigma}^{\mathrm{T}}(\mathbf{x}, t). \qquad (4.2.55)$$

Consider a function of the diffusion vector process $\mathbf{X}(t)$, say $F(\mathbf{X}, t)$, differentiable with respect to t and twice differentiable with respect to the components of $\mathbf{X}(t)$. The differential of $F(\mathbf{X}, t)$ is obtained as

$$dF(\mathbf{X}, t) = \frac{\partial F}{\partial t}dt + \sum_{j=1}^{n} \frac{\partial F}{\partial X_j}dX_j + \frac{1}{2}\sum_{j,k=1}^{n} \frac{\partial^2 F}{\partial X_j \partial X_k}dX_j dX_k + \cdots .$$

$$(4.2.56)$$

Substituting (4.2.53) for dX_j, using the Wiener process property (4.2.35), and keeping terms of the order dt and $dB_l(t)$, we have

$$dF(\mathbf{X}, t) = \left(\frac{\partial F}{\partial t} + \sum_{j=1}^{n} m_j \frac{\partial F}{\partial X_j} + \frac{1}{2}\sum_{j,k=1}^{n}\sum_{l=1}^{m} \sigma_{jl}\sigma_{kl}\frac{\partial^2 F}{\partial X_j \partial X_k} \right) dt$$

$$+ \sum_{j=1}^{n}\sum_{l=1}^{m} \sigma_{jl}\frac{\partial F}{\partial X_j}dB_l(t). \qquad (4.2.57)$$

Equation (4.2.57) is known as Ito differential rule or Ito's lemma (Ito, 1951b). For the case of one dimensional system subjected to one Wiener process excitation, (4.2.57) reduces to

$$dF(X, t) = \left(\frac{\partial F}{\partial t} + m\frac{dF}{dX} + \frac{1}{2}\sigma^2\frac{d^2 F}{dX^2} \right) dt + \sigma\frac{dF}{dX}dB(t). \quad (4.2.58)$$

The Ito's lemma shows that the Ito equation for a function of a Markov diffusion process can be derived quite directly and simply. This property is an advantage of using the Ito stochastic differential equation to describe a Markov diffusion process.

4.2.7 *Stratonovich Stochastic Differential Equations*

In contrast to the Ito integral and Ito equation, another is the Stratonovich type which assigns $\tau'_j = \frac{1}{2}(\tau_j + \tau_{j+1})$, denoted by $\tau_{j+1/2}$, in the Stieltjes integral (4.2.48), i.e.,

$$\int_0^t \sigma[X(\tau), \tau] \circ dB(\tau)$$

$$= \lim_{\substack{n \to \infty \\ \Delta_n \to 0}} \sum_{j=1}^n \sigma[X(\tau_{j+1/2}), \tau_{j+1/2}][B(\tau_{j+1}) - B(\tau_j)],$$

$$(4.2.59)$$

where a small circle is inserted between $\sigma[X(\tau), \tau]$ and $dB(\tau)$ to indicate the Stratonovich integral. Correspondingly, the differential equation is called the Stratonovich stochastic differential equation, written as

$$dX(t) = \tilde{m}(X, t)dt + \tilde{\sigma}(X, t) \circ dB(t). \qquad (4.2.60)$$

To find the differences between the Ito and Stratonovich types of differential equations, let's find the first and second derivate moments in the FPK equation corresponding to the Stratonovich type, Eq. (4.2.60). We may write for very small Δt,

$$X(t + \Delta t) - X(t) = \int_t^{t+\Delta t} \tilde{m}[X(\tau), \tau]d\tau + \int_t^{t+\Delta t} \tilde{\sigma}[X(\tau), \tau] \circ dB(\tau)$$

$$\approx \tilde{m}[X(t), t]\Delta t + \tilde{\sigma}\left[X\left(t + \frac{1}{2}\Delta t\right), t + \frac{1}{2}\Delta t\right]$$

$$\times [B(t + \Delta t) - B(t)]. \qquad (4.2.61)$$

Expand the $\tilde{\sigma}$ function at the time instant t,

$$\tilde{\sigma}\left[X\left(t + \frac{1}{2}\Delta t\right), t + \frac{1}{2}\Delta t\right] \quad \tilde{\sigma}\left[\frac{1}{2}X(t + \Delta t) + \frac{1}{2}X(t), t + \frac{1}{2}\Delta t\right]$$

$$= \tilde{\sigma}_t + \frac{1}{2} \left(\frac{\partial \tilde{\sigma}}{\partial t} \right)_t \Delta t + \frac{1}{2} \left(\frac{\partial \tilde{\sigma}}{\partial X} \right)_t$$

$$\times [X(t + \Delta t) - X(t)] + \cdots,$$

$$(4.2.62)$$

where the subscript t denotes that the value is evaluated at the time instant t, i.e.,

$$\tilde{m}_t = \tilde{m}[X(t), t], \quad \tilde{\sigma}_t = \tilde{\sigma}[X(t), t],$$

$$\left(\frac{\partial \tilde{\sigma}}{\partial t} \right)_t = \frac{\partial \tilde{\sigma}[X(t), t]}{\partial t}, \quad \left(\frac{\partial \tilde{\sigma}}{\partial X} \right)_t = \frac{\partial \tilde{\sigma}[X(t), t]}{\partial x}. \qquad (4.2.63)$$

Substitution of (4.2.62) into (4.2.61) results in

$$X(t + \Delta t) - X(t) = \tilde{m}_t \Delta t + \tilde{\sigma}_t [B(t + \Delta t) - B(t)]$$

$$+ \frac{1}{2} \tilde{\sigma}_t \left(\frac{\partial \tilde{\sigma}}{\partial X} \right)_t [B(t + \Delta t) - B(t)]^2 + O(\Delta t^{\frac{3}{2}}),$$

$$(4.2.64)$$

where $O(\cdot)$ denotes the order of magnitude. Using Eq. (4.2.11), the first and second derivate moments can then be obtained as

$$a(x, t) = \lim_{\Delta t \to 0} \frac{1}{\Delta t} E[X(t + \Delta t) - X(t) | X(t) = x]$$

$$= \tilde{m}(x, t) + \frac{1}{2} \tilde{\sigma}(x, t) \frac{\partial \tilde{\sigma}(x, t)}{\partial x}, \qquad (4.2.65)$$

$$b(x, t) = \lim_{\Delta t \to 0} \frac{1}{\Delta t} E\{ [X(t + \Delta t) - X(t)]^2 | X(t) = x \} = \tilde{\sigma}^2(x, t).$$

$$(4.2.66)$$

Thus, the difference between the Ito and Stratonovich equations can be seen by comparing Eqs. (4.2.51) and (4.2.65). With the same form for the two types of equations, i.e., Eqs. (4.2.45) and (4.2.60), the second derivate moments are the same, but the first derivate moments are different if the $\tilde{\sigma}(X, t)$ function in the Stratonovich equation depends on $X(t)$ explicitly. The conversion between the Ito and Stratonovich equations can be made by using the following relationship:

$$m(X,t) = \tilde{m}(X,t) + \frac{1}{2}\tilde{\sigma}(X,t)\frac{\partial\tilde{\sigma}(X,t)}{\partial X}, \quad \sigma(X,t) = \tilde{\sigma}(X,t).$$

$$(4.2.67)$$

The term $\frac{1}{2}\tilde{\sigma}(X,t)\frac{\partial\tilde{\sigma}(X,t)}{\partial X}$ is called the Wong–Zakai correction term (Wong and Zakai, 1965). In other words, the Ito equation equivalent to the Stratonovich equation (4.2.60) is

$$dX(t) = \left[\tilde{m}(X,t) + \frac{1}{2}\tilde{\sigma}(X,t)\frac{\partial\tilde{\sigma}(X,t)}{\partial X}\right]dt + \tilde{\sigma}(X,t)dB(t).$$

$$(4.2.68)$$

In the case of n-dimensional Markov diffusion vector governed by the Stratonovich equations

$$dX_j(t) = \tilde{m}_j(\mathbf{X},t)dt + \sum_{l=1}^{m}\tilde{\sigma}_{jl}(\mathbf{X},t)\circ dB_l(t), \quad j = 1,2,\ldots,n,$$

$$(4.2.69)$$

the first and second derivate moments are

$$a_j(\mathbf{x},t) = \tilde{m}_j(\mathbf{x},t) + \frac{1}{2}\sum_{k=1}^{n}\sum_{l=1}^{m}\tilde{\sigma}_{kl}(\mathbf{x},t)\frac{\partial\tilde{\sigma}_{jl}(\mathbf{x},t)}{\partial X_k},$$

$$b_{jk}(\mathbf{x},t) = \sum_{l=1}^{m}\tilde{\sigma}_{jl}(\mathbf{x},t)\tilde{\sigma}_{kl}(\mathbf{x},t). \quad (4.2.70)$$

Example 4.2.1 Consider the following Ito integral

$$\int_a^b B(t)dB(t) = \lim_{\substack{n\to\infty \\ \Delta_n\to 0}}\sum_{j=1}^{n}B(t_j)[B(t_{j+1}) - B(t_j)]$$

$$= \frac{1}{2}\lim_{\substack{n\to\infty \\ \Delta_n\to 0}}\sum_{j=1}^{n}\{B^2(t_{j+1}) - B^2(t_j) - [B(t_{j+1}) - B(t_j)]^2\}$$

$$= \frac{1}{2}[B^2(b) - B^2(a)] - \frac{1}{2}\sigma^2(b-a), \quad (4.2.71)$$

where $a = t_1 < t_2 < \cdots < t_n < t_{n+1} = b$, $\Delta_n = \max(\tau_{j+1} - \tau_j)$. In deriving (4.2.71), the Levy's oscillation property (4.2.34) has been invoked.

For the same integral in the Stratonovich sense, we have

$$\int_a^b B(t) \circ dB(t) = \lim_{\substack{n \to \infty \\ \Delta_n \to 0}} \sum_{j=1}^n B\left(\frac{t_{j+1} + t_j}{2}\right) [B(t_{j+1}) - B(t_j)]$$

$$= \frac{1}{2} \lim_{\substack{n \to \infty \\ \Delta_n \to 0}} \sum_{j=1}^n [B(t_{j+1}) + B(t_j)][B(t_{j+1}) - B(t_j)]$$

$$= \frac{1}{2}[B^2(b) - B^2(a)]. \qquad (4.2.72)$$

The two integrals have different results. The Stratonovich integral (4.2.72) has the same result as that of the usual integral.

Example 4.2.2 Consider the Ito differential equation

$$dX(t) = KX(t)dB(t), \qquad (4.2.73)$$

where K is a constant, and $B(t)$ is a unit Wiener process. Let $Y(t) = \ln X(t)$. According to the Ito differential rule (4.2.58), we have

$$dY(t) = -\frac{1}{2}K^2 dt + K dB(t). \qquad (4.2.74)$$

Equation (4.2.74) could not be obtained directly from (4.2.73) if it were treated as a regular differential equation.

On the other hand, (4.2.73) can be converted to a Stratonovich stochastic differential equation using (4.2.67),

$$\tilde{\sigma}(X, t) = \sigma(X, t) = KX(t), \qquad (4.2.75)$$

$$\tilde{m}(X, t) = m(X, t) - \frac{1}{2}\tilde{\sigma}(X, t)\frac{\partial \tilde{\sigma}(X, t)}{\partial X} = -\frac{1}{2}K^2 X(t)$$

$$\qquad (4.2.76)$$

and

$$dX(t) = -\frac{1}{2}K^2X(t)dt + KX(t) \circ dB(t). \qquad (4.2.77)$$

Following the usual differential rule, (4.2.77) leads to

$$dY(t) = -\frac{1}{2}K^2dt + K \circ dB(t). \qquad (4.2.78)$$

Equation (4.2.78) is the same as (4.2.74) since the diffusion coefficient K is a constant. Again, it is shown that the usual differential rule is applicable to the Stratonovich type of integrals and differential equations.

Although Eqs. (4.2.67) and (4.2.70) show that the conversion between the Ito and Stratonovich equations is straightforward, each has its own advantage. The Ito equation is related to the FPK equation more directly, while the Stratonovich equation can be treated in the same way as a regular differential equation.

4.3 Systems under Gaussian White-Noise Excitations

In many engineering systems, the excitations are Gaussian distributed and of broad band nature, which may be modeled as Gaussian white noises. The simplest of such systems is a one-dimensional system governed by

$$\frac{d}{dt}X(t) = f(X,t) + g(X,t)W(t), \qquad (4.3.1)$$

where f and g are deterministic functions, and $W(t)$ is a Gaussian white noise with a spectral density K, i.e., $E[W(t)W(t+\tau)] = 2\pi K\delta(\tau)$.

In the last section, it is shown that the Stratonovich stochastic differential equation can be treated in the same way as a regular differential equation. Moreover, a Gaussian white noise can be replaced by the derivative of a Wiener process, as described in Section 4.2.5. Thus, (4.3.1) can be replaced directly by a Stratonovich type equation

$$dX(t) = f(X,t)dt + \sqrt{2\pi K}g(X,t) \circ dB(t), \qquad (4.3.2)$$

where $B(t)$ is a unit Wiener process. According to (4.2.67), (4.3.2) is equivalent to the following Ito type equation

$$dX(t) = \left[f(X,t) + \pi K g(X,t) \frac{\partial}{\partial X} g(X,t) \right] dt + \sqrt{2\pi K} g(X,t) dB(t).$$

$$(4.3.3)$$

The corresponding FPK equation is

$$\frac{\partial}{\partial t} p + \frac{\partial}{\partial x} \left[\left(f + \pi K g \frac{\partial g}{\partial x} \right) p \right] - \pi K \frac{\partial^2}{\partial x^2} (g^2 p) = 0. \qquad (4.3.4)$$

The FPK equation can be derived in an alternative way. Express $X(t)$ from (4.3.1) as

$$X(t + \Delta t) - X(t) = \int_t^{t+\Delta t} f[X(u), u] du + \int_t^{t+\Delta t} g[X(u), u] W(u) du.$$

$$(4.3.5)$$

Expand $f[X(u), u]$ and $g[X(u), u]$ at $u = t$,

$$f[X(u), u] = f[X(t), t] + (u - t) \frac{\partial}{\partial t} f[X(t), t]$$

$$+ [X(u) - X(t)] \frac{\partial}{\partial X} f[X(t), t] + \cdots, \qquad (4.3.6)$$

$$g[X(u), u] = g[X(t), t] + (u - t) \frac{\partial}{\partial t} g[X(t), t]$$

$$+ [X(u) - X(t)] \frac{\partial}{\partial X} g[X(t), t] + \cdots. \qquad (4.3.7)$$

In (4.3.6) and (4.3.7), we substitute $X(u) - X(t)$ using (4.3.5), i.e.,

$$X(u) - X(t) = \int_t^u f[X(v), v] dv + \int_t^u g[X(v), v] W(v) dv. \qquad (4.3.8)$$

Combining (4.3.5) through (4.3.8) and keeping only the leading terms, we have

$$X(t + \Delta t) - X(t) = f[X(t), t]\Delta t + g[X(t), t] \int_t^{t+\Delta t} W(u)du$$

$$+ \left[\frac{\partial}{\partial t}g[X(t), t]\right] \int_t^{t+\Delta t} (u - t)W(u)du$$

$$+ \left[\frac{\partial}{\partial X}g[X(t), t]\right] \int_t^{t+\Delta t} W(u)du$$

$$\times \int_t^u g[X(v), v]W(v)dv + \cdots. \tag{4.3.9}$$

Since $E[W(t)] = 0$, $E[W(t)W(t + \tau)] = 2\pi K\delta(\tau)$, (4.3.9) results in

$$\lim_{\Delta t \to 0} \frac{1}{\Delta t}E[X(t + \Delta t) - X(t)]$$

$$= f[X(t), t] + \frac{2\pi K}{\Delta t} \left[\frac{\partial}{\partial X}g[X(t), t]\right] \tag{4.3.10}$$

$$\times \int_t^{t+\Delta t} du \int_t^u g[X(v), v]\delta(u - v)dv + O(\Delta t).$$

To calculate the integral in (4.3.10), let $\tau = u - v$, and change the order of the double integration to obtain

$$\int_t^{t+\Delta t} du \int_t^u g[X(v), v]\delta(u - v)dv$$

$$= \int_{-\Delta t}^0 \delta(\tau)d\tau \int_{t-\tau}^{t+\Delta t} g[X(u + \tau), u + \tau]du. \tag{4.3.11}$$

Substituting (4.3.11) into (4.3.10), and considering (a) $\delta(\tau) = 0$ for $\tau \neq 0$, and (b) $\int_{-\Delta t}^{0} \delta(\tau)d\tau = \frac{1}{2}$, Eq. (4.3.10) reduces to

$$\lim_{\Delta t \to 0} \frac{1}{\Delta t} E[X(t + \Delta t) - X(t)] = f(X,t) + \pi K g(X,t)\frac{\partial}{\partial X}g(X,t).$$
$$(4.3.12)$$

Using (4.3.5) again, we obtain

$$\lim_{\Delta t \to 0} \frac{1}{\Delta t} E\{[X(t + \Delta t) - X(t)]^2\}$$

$$= \lim_{\Delta t \to 0} \frac{1}{\Delta t} \int_{t}^{t+\Delta t} du \int_{t}^{t+\Delta t} g[X(u), u]g[X(v), v]E[W(u)W(v)]dv$$

$$= 2\pi K \lim_{\Delta t \to 0} \frac{1}{\Delta t} \int_{t}^{t+\Delta t} du \int_{t}^{t+\Delta t} g[X(u), u]g[X(v), v]\delta(u - v)dv$$

$$= 2\pi K g^2(X, t). \qquad (4.3.13)$$

The first and second derivate moments are then obtained from (4.3.12) and (4.3.13) as

$$a = f(x,t) + \pi K g(x,t)\frac{\partial}{\partial x}g(x,t), \quad b = 2\pi K g^2(x,t), \qquad (4.3.14)$$

which lead to the same FPK equation (4.3.4). Therefore, with the criterion of the same FPK equation, the equivalency of the original physical equation (4.3.1), the Stratonovich equation (4.3.2), and the Ito equation (4.3.3) is verified again.

The above analysis can be extended to multi-dimensional cases. Consider a stochastic vector process $\mathbf{X}(t) = [X_1(t), X_1(t), \ldots, X_n(t)]^T$, governed by

$$\frac{d}{dt}X_j(t) = f_j(\mathbf{X}, t) + \sum_{l=1}^{m} g_{jl}(\mathbf{X}, t)W_l(t), \quad j = 1, 2, \ldots, n, \quad (4.3.15)$$

where $W_l(t)$, $l = 1, 2, \ldots, m$, are Gaussian white noises with the correlations

$$E[W_l(t)W_s(t + \tau)] = 2\pi K_{ls}\delta(\tau). \qquad (4.3.16)$$

The FPK equation corresponding to (4.3.15) is

$$\frac{\partial}{\partial t}p + \sum_{j=1}^{n}\frac{\partial}{\partial x_j}(a_j p) - \frac{1}{2}\sum_{j,k=1}^{n}\frac{\partial^2}{\partial x_j \partial x_k}(b_{jk}p) = 0, \qquad (4.3.17)$$

where the first and second derivate moments are obtained as

$$a_j(\mathbf{x}, t) = f_j(\mathbf{x}, t) + \pi \sum_{r=1}^{n}\sum_{l,s=1}^{m}K_{ls}g_{rs}(\mathbf{x}, t)\frac{\partial}{\partial x_r}g_{jl}(\mathbf{x}, t),$$

$$(4.3.18)$$

$$b_{jk}(\mathbf{x}, t) = 2\pi \sum_{l,s=1}^{m}K_{ls}g_{jl}(\mathbf{x}, t)g_{ks}(\mathbf{x}, t). \qquad (4.3.19)$$

The Stratonovich equations equivalent to (4.3.15) can be written directly as

$$dX_j(t) = f_j(\mathbf{X}, t)dt + \sum_{l=1}^{m}g_{jl}(\mathbf{X}, t) \circ d\tilde{B}_l(t), \qquad (4.3.20)$$

where $\tilde{B}_l(t)$ are Wiener processes with

$$E[d\tilde{B}_l(t_1)d\tilde{B}_s(t_2)] = \begin{cases} 2\pi K_{ls}dt, & t_1 = t_2 \\ 0, & t_1 \neq t_2. \end{cases} \qquad (4.3.21)$$

The equivalent Ito equations can be derived from the Stratonovich equation by adding the Wong–Zakai terms, or from the first and second derivate moments given in (4.3.18) and (4.3.19),

$$dX_j(t) = a_j(\mathbf{X}, t)dt + \sum_{l=1}^{m}\sigma_{jl}(\mathbf{X}, t)dB_l(t), \qquad (4.3.22)$$

where $B_l(t)$ are unit Wiener processes with

$$E[dB_l(t_1)dB_s(t_2)] = \begin{cases} dt, & t_1 = t_2 \\ 0, & t_1 \neq t_2 \end{cases} \qquad (4.3.23)$$

and functions $\sigma_{jl}(\mathbf{X},t)$ determined from

$$\sum_{s=1}^{m} \sigma_{js}(\mathbf{X},t)\sigma_{ks}(\mathbf{X},t) = b_{jk}(\mathbf{X},t) = 2\pi \sum_{l,s=1}^{m} K_{ls}g_{jl}(\mathbf{X},t)g_{ks}(\mathbf{X},t).$$

$$(4.3.24)$$

Equation (4.3.24) shows that the equivalent Ito equations are not unique as long as the diffusion coefficient σ_{jl} satisfy Eq. (4.3.24). Thus, it is clear that the equivalency of the physical equations, the Stratonovich equations, and the Ito equations is in the sense that they generate the same FPK equation.

Example 4.3.1 Consider a single-degree-of-freedom oscillator subjected to random excitations, governed by

$$\ddot{X} + h(X,\dot{X}) = XW_1(t) + \dot{X}W_2(t) + W_3(t), \qquad (4.3.25)$$

where $h(X,\dot{X})$ represents the damping force and restoring force, and $W_l(t)$, $l = 1,2,3$, are assumed to be independent Gaussian white noises with the correlations

$$E[W_l(t)W_s(t+\tau)] = 2\pi K_{ls}\delta_{ls}\delta(\tau), \qquad (4.3.26)$$

where δ_{ls} is the Kronecker delta, i.e., $\delta_{ls} = 1$ if $l = s$ and $\delta_{ls} = 0$ if $l \neq s$. By denoting $X_1 = X$ and $X_2 = \dot{X}$, Eq. (4.3.25) can be converted to two equations in the state space

$$\dot{X}_1 = X_2$$
$$\dot{X}_2 = -h(X_1,X_2) + X_1W_1(t) + X_2W_2(t) + W_3(t). \qquad (4.3.27)$$

Using (4.3.18) and (4.3.19), we obtain the first and second derivate moments

$$a_1 = x_2, \quad a_2 = -h(x_1,x_2) + \pi K_{22}x_2,$$
$$b_{11} = 0, \quad b_{12} = 0, \quad b_{21} = 0, \quad b_{22} = 2\pi K_{11}x_1^2 + 2\pi K_{22}x_2^2 + 2\pi K_{33}.$$

$$(4.3.28)$$

Thus, the FPK equation is

$$\frac{\partial}{\partial t}p + \frac{\partial}{\partial x_1}(x_2 p) + \frac{\partial}{\partial x_2}\{[(-h(x_1, x_2) + \pi K_{22}x_2]p\}$$

$$- \pi \frac{\partial^2}{\partial x_2^2}[(K_{11}x_1^2 + K_{22}x_2^2 + K_{33})p] = 0. \tag{4.3.29}$$

The corresponding Stratonovich equations can be written directly from (4.3.27) as

$$dX_1 = X_2 dt,$$
$$dX_2 = -h(X_1, X_2)dt + \sqrt{2\pi K_{11}}X_1 \circ dB_1(t)$$
$$+ \sqrt{2\pi K_{22}}X_2 \circ dB_2(t) + \sqrt{2\pi K_{33}} \circ dB_3(t), \tag{4.3.30}$$

where $B_l(t)$, $l = 1, 2, 3$, are independent unit Wiener processes.

Using (4.3.24), the equivalent Ito equations can be derived from (4.3.28) as follows

$$dX_1 = X_2 dt,$$
$$dX_2 = [-h(X_1, X_2) + \pi K_{22}X_2]dt + \sqrt{2\pi K_{11}}X_1 dB_1(t)$$
$$+ \sqrt{2\pi K_{22}}X_2 dB_2(t) + \sqrt{2\pi K_{33}}dB_3(t) \tag{4.3.31}$$

or

$$dX_1 = X_2 dt,$$
$$dX_2 = [-h(X_1, X_2) + \pi K_{22}X_2]dt + \sqrt{2\pi(K_{11}X_1^2 + K_{22}X_2^2)}dB_1(t)$$
$$+ \sqrt{2\pi K_{33}}dB_2(t) \tag{4.3.32}$$

or

$$dX_1 = X_2 dt,$$
$$dX_2 = [-h(X_1, X_2) + \pi K_{22}X_2]dt$$
$$+ \sqrt{2\pi(K_{11}X_1^2 + K_{22}X_2^2 + K_{33})}dB(t). \tag{4.3.33}$$

Although (4.3.31)–(4.3.33) have different forms, they yield the same FPK equation (4.3.29); therefore, they are equivalent.

4.4 One-Dimensional Diffusion Processes

Many physical phenomena in science and engineering can be modeled as one-dimensional problems, or reduced to those of one-dimensional; thus, the simple case of one-dimensional diffusion process is investigated in detail in this section.

Consider a one-dimensional diffusion process $X(t)$, defined in an interval with a left boundary x_l and a right boundary x_r, and governed by the Ito differential equation

$$dX(t) = m(X)dt + \sigma(X)dB(t), \qquad (4.4.1)$$

where $B(t)$ is a unit Wiener process, and the drift and diffusion coefficients may depend on $X(t)$, but do not depend on tine t explicitly. Such a diffusion process is known as homogeneous in time.

4.4.1 *Probability Density Functions*

The transition probability density $p(x, t|x_0, t_0)$ is governed by the FPK equation

$$\frac{\partial}{\partial t}p + \frac{\partial}{\partial x}[m(x)p] - \frac{1}{2}\frac{\partial^2}{\partial x^2}[\sigma^2(x)p] = 0. \qquad (4.4.2)$$

Equation (4.4.2) can be solved with appropriate initial and boundary conditions. Exact solutions for equation (4.4.2) are known for linear systems and some simple nonlinear systems under additive Gaussian white noise excitations. However, the time-independent stationary probability density $p(x)$ of $X(t)$, if exists, is easy to obtain. This stationary probability density satisfies the reduced FPK equation

$$\frac{d}{dx}G = \frac{d}{dx}\left\{m(x)p - \frac{1}{2}\frac{d}{dx}[\sigma^2(x)p]\right\} = 0. \qquad (4.4.3)$$

Integrating (4.4.3) once, we have

$$G(x) = m(x)p - \frac{1}{2}\frac{d}{dx}[\sigma^2(x)p] = G_c. \qquad (4.4.4)$$

Equation (4.4.4) indicates that the probability flow is a constant everywhere. If a stationary probability density exists, it can be obtained from (4.4.4) as follows

$$p(x) = \frac{\psi(x)}{\sigma^2(x)} \left[C - 2G_c \int_{x_l}^{x} \psi^{-1}(u)du \right], \qquad (4.4.5)$$

where C is a constant, and

$$\psi(x) = \exp \left[\int \frac{2m(x)}{\sigma^2(x)} dx \right]. \qquad (4.4.6)$$

To determine the two integration constant G_c and C, physical phenomena need to be examined. For a stationary probability density to exist, the most common scenario for a finite or infinite boundary is a vanishing probability flow. Then, (4.4.4) implies that the probability flow must also vanish everywhere in the one-dimensional space, i.e., $G_c = 0$, and solution (4.4.5) reduces to

$$p(x) = C\frac{\psi(x)}{\sigma^2(x)} = \frac{C}{\sigma^2(x)} \exp \left[\int \frac{2m(x)}{\sigma^2(x)} dx \right]. \qquad (4.4.7)$$

Another possible scenario is that the state space $[x_l, x_r)$ is a closed loop, the drift and diffusion coefficients $m(x)$ and $\sigma(x)$ are periodic, then

$$p(x_l) = p(x_r), \quad G(x_l) = G(x_r) = G_c. \qquad (4.4.8)$$

Using (4.4.8) for (4.4.5), we obtain

$$G_c = \frac{C}{2} \left[1 - \frac{\psi(x_l)}{\psi(x_r)} \right] \left[\int_{x_l}^{x_r} \psi^{-1}(u)du \right]^{-1}. \qquad (4.4.9)$$

$$p(x) = C\frac{\psi(x)}{\sigma^2(x)} \left\{ 1 - \left[1 - \frac{\psi(x_l)}{\psi(x_r)} \right] \frac{\int_{x_l}^{x} \psi^{-1}(u)du}{\int_{x_l}^{x_r} \psi^{-1}(u)du} \right\}. \qquad (4.4.10)$$

The constant C in (4.4.7) and (4.4.10) can be determined from the normalization condition of the probability density function.

4.4.2 *Classification of Boundaries*

The qualitative behavior of a one-dimensional diffusion process governed by (4.4.1) depends on its characteristics of its drift coefficient $m(X)$ and diffusion coefficient $\sigma(X)$ at two boundaries. Originally proposed by Feller (1952, 1954) and extended by Lin and Cai (1995), classification of the boundaries of one-dimensional diffusion processes are given below.

Regular boundary: the process can either reach the boundary from an interior point or reach an interior point from the boundary.

Exit boundary: the process can reach the boundary from an interior point and then stays at the boundary.

Entrance boundary: the process returns to the interior of the defining interval as soon as it reaches the boundary.

Repulsively natural boundary: the process cannot reach the boundary, and will be repulsed as it approaches the boundary.

Attractively natural boundary: the process cannot reach the boundary, but will approach the boundary.

Strictly natural boundary: the process cannot reach the boundary, and its behavior is uncertain as it approaches the boundary.

According to the above classification, the following conclusions can be drawn of the behaviors of one-dimensional diffusion processes.

1. A non-trivial stationary probability exists within the defining interval if the two boundaries are either entrance or repulsively natural.
2. A trivial stationary probability in the form of a unit delta function exists at an exit boundary if the other boundary is either an entrance or repulsively natural.

To identify the boundary class, several functions were introduced (Ito and McKean, 1965; Karlin and Taylor, 1981) as follows

$$l(x) = \int_{x_0}^{x} \psi^{-1}(u)du, \quad v(x) = \int_{x_0}^{x} \frac{\psi(u)}{\sigma^2(u)}du,$$

$$\Sigma(x) = \int_{x_0}^{x} v(u)dl(u), \quad N(x) = \int_{x_0}^{x} l(u)dv(u), \quad (4.4.11)$$

where $\psi(u)$ is defined in (4.4.6). The classification of a boundary x_b can be identified according to the behaviors of these functions as shown in Table 4.4.1, given originally by Karlin and Taylor (1981) and modified by Lin and Cai (1995).

Table 4.4.1 Classification of boundaries.

Criteria					
$l(x_b)$	$v(x_b)$	$\Sigma(x_b)$	$N(x_b)$	Classifications	
$< \infty^*$	$< \infty^*$	$< \infty$	$< \infty$	Regular	Accessible
$< \infty$	$= \infty^*$	$< \infty^*$	$= \infty$	Exit	
$< \infty^*$	$= \infty^*$	$= \infty^*$	$= \infty$	Attractively natural	Inaccessible
$= \infty^*$	$< \infty^*$	$= \infty$	$= \infty^*$	Repulsively natural	
$= \infty^*$	$= \infty^*$	$= \infty$	$= \infty$	Strictly natural	
$= \infty^*$	$< \infty$	$= \infty$	$< \infty^*$	Entrance	

*The asterisks indicate the minimal sufficient conditions for each type of boundary. For example, the minimal sufficient conditions for a regular boundary are $l(x_b)$ ¡ ∞ and $v(x_b)$ ¡ ∞.

4.4.3 *Singular Boundaries*

Table 4.4.1 and Eq. (4.4.11) show that the boundary class depends on the behaviors of the drift and diffusion coefficients at the boundary. Although the integrations in (4.4.11) may be difficult, only the integrability of these integrals are required for boundary classification, as shown in Table 4.4.1. According to Eq. (4.4.6), the integrability of the integrals in (4.4.11) is easy to identify if $m(x)$ is finite and $\sigma(x)$ does not approach zero as x approaches the boundary. In this case, the boundary is said to be non-singular. On the other hand, a boundary is singular of the first kind if $\sigma(x)$ vanishes at the boundary, and singular of the second kind if $m(x)$ is unbounded. The classification of singular boundaries were investigated by Kozin and Zhang (1990) and Lin and Cai (1995), and will be described in detail below.

Singular boundary of the first kind. Let x_s be a singular boundary of the first kind, $\sigma(x_s) = 0$. The boundary is said to be a shunt if $m(x_s) \neq 0$, and a trap if $m(x_s) = 0$. We introduce three parameters: diffusion exponent α, drift exponent β and character value c. They are determined from the following

$$\sigma^2(x) = O|x - x_s|^\alpha, \quad \text{as } x \to x_s, \tag{4.4.12}$$

$$m(x) = O|x - x_s|^\beta, \quad \text{as } x \to x_s, \tag{4.4.13}$$

$$c = \begin{cases} \lim\limits_{x \to x_s^+} \dfrac{2m(x)(x - x_s)^{\alpha-\beta}}{\sigma^2(x)}, & x_s \text{ is a left boundary} \\[3mm] -\lim\limits_{x \to x_s^-} \dfrac{2m(x)(x_s - x)^{\alpha-\beta}}{\sigma^2(x)}, & x_s \text{ is a right boundary} \end{cases}.$$

$$\tag{4.4.14}$$

In (4.4.12) and (4.4.13), $O|\cdot|$ denotes the order of $|\cdot|$. The classification of singular boundaries of the first kind are then identified from the values of α, β and c, as given in Table 4.4.2.

Singular boundary of the second kind not at infinity. In this case, $|x_s| < \infty$, and $m(x_s) = \infty$. The diffusion exponent α, drift exponent β and character value c are defined as

$$\sigma^2(x) = O|x - x_s|^{-\alpha}, \quad \text{as } x \to x_s, \tag{4.4.15}$$

$$m(x) = O|x - x_s|^{-\beta}, \quad \text{as } x \to x_s, \tag{4.4.16}$$

Table 4.4.2   Classification of singular boundaries of the first kind.

State	Conditions			Class
$\sigma(x_s) = 0$ $(\alpha > 0)$	$\alpha < 1$			Regular
	$\alpha = 1$	$m(x_l) < 0$ or $m(x_r) > 0$		Exit
$m(x_s) \neq 0$ $(\beta = 0)$		$m(x_l) > 0$ or	$0 < c < 1$	Regular
		$m(x_r) < 0$	$c \geq 1$	Entrance
(shunt)	$\alpha > 1$	$m(x_l) < 0$ or $m(x_r) > 0$		Exit
		$m(x_l) > 0$ or $m(x_r) < 0$		Entrance
$\sigma(x_s) = 0$ $(\alpha > 0)$	$\alpha < 1 + \beta$	$\alpha < 1$		Regular
		$1 \leq \alpha < 2$		Exit
$m(x_s) = 0$ $(\beta > 0)$		$\alpha \geq 2$		Attractively natural
	$\alpha > 1 + \beta$	$\beta < 1$	$m(x_l^+) < 0$ or $m(x_r^-) > 0$	Exit
(trap)			$m(x_l^+) > 0$ or $m(x_r^-) < 0$	Entrance
		$\beta \geq 1$	$m(x_l^+) < 0$ or $m(x_r^-) > 0$	Attractively natural
			$m(x_l^+) > 0$ or $m(x_r^-) < 0$	Repulsively natural
	$\alpha = 1 + \beta$	$\beta < 1$	$c \geq 1$	Entrance
			$\beta < c < 1$	Regular
			$c \leq \beta$	Exit
		$\beta = 1$	$c > 1$	Repulsively natural
			$c = 1$	Strictly natural
			$c < 1$	Attractively natural
		$\beta > 1$	$c > \beta$	Repulsively natural
			$1 \leq c \leq \beta$	Strictly natural
			$c < 1$	Attractively natural

$$c = \begin{cases} \lim_{x \to x_s^+} \dfrac{2m(x)(x - x_s)^{\beta - \alpha}}{\sigma^2(x)}, & x_s \text{ is a left bounday} \\[3mm] -\lim_{x \to x_s^-} \dfrac{2m(x)(x_s - x)^{\beta - \alpha}}{\sigma^2(x)}, & x_s \text{ is a right boundary.} \end{cases}$$

$$(4.4.17)$$

The classification of singular boundaries of the second kind not at infinity are identified from the values of α, β and c, as given in Table 4.4.3.

Table 4.4.3   Classification of singular boundaries of the second kind ($|x_s| < \infty$).

State	Conditions			Class
$\|m(x_s)\| = \infty$ $(\beta > 0)$ $\sigma(x_s) < \infty$ $(\alpha = 0)$	$\beta < 1$			Regular
	$\beta = 1$	$c \le -1$		Exit
		$-1 < c < 1$		Regular
		$c \ge 1$		Entrance
	$\beta > 1$	$m(x_l^+) < 0$ or $m(x_r^-) > 0$		Exit
		$m(x_l^+) > 0$ or $m(x_r^-) < 0$		Entrance
$\|m(x_s)\| = \infty$ $(\beta > 0)$ $\sigma(x_s) = \infty$ $(\alpha > 0)$	$\beta < 1 + \alpha$			Regular
	$\beta > 1 + \alpha$	$m(x_l^+) < 0$ or $m(x_r^-) > 0$		Exit
		$m(x_l^+) > 0$ or $m(x_r^-) < 0$		Entrance
	$\beta = 1 + \alpha$	$c \ge -\beta$	$c \ge 1$	Entrance
			$c < 1$	Regular
		$c < -\beta$		Exit

Singular boundary of the second kind at infinity. In this case, $|x_s| = \infty$, and $m(x_s) = \infty$. The diffusion exponent α, drift exponent β and character value c are defined as

$$\sigma^2(x) = O|x|^\alpha, \quad \text{as } |x| \to \infty, \qquad (4.4.18)$$

$$m(x) = O|x|^{\beta}, \quad \text{as } |x| \to \infty, \qquad (4.4.19)$$

$$c = \begin{cases} \lim\limits_{x \to -\infty} \dfrac{2m(x)|x|^{\alpha-\beta}}{\sigma^2(x)} \\[2ex] -\lim\limits_{x \to \infty} \dfrac{2m(x)|x|^{\alpha-\beta}}{\sigma^2(x)}. \end{cases} \qquad (4.4.20)$$

The classification of singular boundaries of the second kind at infinity are identified from the values of α, β and c, as given in Table 4.4.4.

Table 4.4.4 Classification of singular boundaries of the second kind ($|x_s| = \infty$).

State	Condition				Class
$|m(\infty)| = \infty$	$m(-\infty) < 0$	$\beta > 1$			Exit
$(\beta > 0)$	or $m(+\infty) > 0$	$\beta \leq 1$			Attractively natural
$\sigma(\infty) < \infty$	$m(-\infty) > 0$ or	$\beta > 1$			Entrance
$(\alpha = 0)$	$m(+\infty) < 0$	$\beta \leq 1$			Repulsively natural
$|m(\infty)| = \infty$	$\beta > \alpha - 1$	$m(-\infty) < 0$ or	$\beta > 1$		Exit
$(\beta > 0)$		$m(+\infty) > 0$	$\beta \leq 1$		Attractively natural
$\sigma(\infty) = \infty$		$m(-\infty) > 0$ or	$\beta > 1$		Entrance
$(\alpha > 0)$		$m(+\infty) < 0$	$\beta \leq 1$		Repulsively natural
	$\beta < \alpha - 1$				Regular
		$\beta \leq 1$	$c > -\beta$		Repulsively natural
			$c \leq -\beta$	$c \geq -1$	Strictly natural
				$c < -1$	Attractively natural
	$\beta = \alpha - 1$	$\beta > 1$	$c > -\beta$	$c \geq -1$	Entrance
				$c < -1$	Regular
			$c \leq -\beta$		Exit

Example 4.4.1 Consider a one-dimensional diffusion process $X(t)$, defined in $[0,\infty)$, and governed by the Ito equation (4.4.1) with the drift and diffusion coefficients given by

$$m(X) = (-2\zeta\omega_0 + \pi\omega_0^2 K_1)X + \pi K_2, \qquad (4.4.21)$$

$$\sigma(X) = \sqrt{\pi\omega_0^2 K_1 X^2 + 2\pi K_2 X}, \qquad (4.4.22)$$

where $K_1 \geq 0$ and $K_2 \geq 0$, excluding the case of $K_1 = K_2 = 0$. The left boundary $x = 0$ is singular of the first kind since $\sigma(0) = 0$. It can be identified according to Table 4.4.2. Due to the fact that

$m(\infty) = \infty$, the right boundary at ∞ is singular of the second kind, which can be identified using Table 4.4.4. Three cases are considered separately.

1. $K_1 = 0$, $K_2 > 0$

At the left boundary $x = 0$, $\sigma(0) = 0$, $m(0) > 0$, $\alpha = 1$, $\beta = 0$ and $c = 1$; therefore, it is an entrance. At the right boundary $x = \infty$, $m(\infty) = \infty$, $\sigma(\infty) = \infty$, $m(\infty) < 0$, $\alpha = 1$, $\beta = 1$, and it is repulsively natural. The behaviors of sample functions near the two boundaries are illustrated in Fig. 4.4.1. A non-trivial stationary probability exists since both boundaries are unreachable if a sample path begins from an interior point. The stationary probability density is obtained from Eq. (4.4.7) as

$$p(x) = \frac{2\zeta\omega_0}{\pi K_2} \exp\left(-\frac{2\zeta\omega_0}{\pi K_2} x\right). \tag{4.4.23}$$

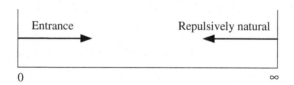

Figure 4.4.1   Boundary behaviors for the case of $K_1 = 0$, $K_2 > 0$.

2. $K_1 > 0$, $K_2 = 0$

At the left boundary $x = 0$, $\sigma(0) = 0$, $m(0) = 0$, $\alpha = 2$, $\beta = 1$, $c = -\frac{4\zeta}{\pi\omega_0 K_1} + 2$, indicating that

$$x = 0 \text{ is } \begin{cases} \text{attractively natural,} & \zeta > \frac{1}{4}\pi\omega_0 K_1 \\[2mm] \text{strictly natural,} & \zeta = \frac{1}{4}\pi\omega_0 K_1 \\[2mm] \text{repulsively natural,} & \zeta < \frac{1}{4}\pi\omega_0 K_1 \end{cases} \tag{4.4.24}$$

At the right boundary $x = \infty$, $m(\infty) = \infty$, $\sigma(\infty) = \infty$, $\alpha = 2$, $\beta = 1$, and $c = \frac{4\zeta}{\pi\omega_0 K_1} - 2$. Thus, we have

$$
x = \infty \text{ is } \begin{cases}
\text{repulsively natural,} & \zeta > \dfrac{1}{4}\pi\omega_0 K_1 \\[2mm]
\text{strictly natural,} & \zeta = \dfrac{1}{4}\pi\omega_0 K_1. \\[2mm]
\text{attractively natural,} & \zeta < \dfrac{1}{4}\pi\omega_0 K_1
\end{cases}
\tag{4.4.25}
$$

The results are represented schematically in Fig. 4.4.2. If $\zeta > \frac{1}{4}\pi\omega_0 K_1$, all sample functions will approach the left boundary $x = 0$. On the other hand, if $\zeta < \frac{1}{4}\pi\omega_0 K_1$, all sample functions will approach infinity. For the case of $\zeta = \frac{1}{4}\pi\omega_0 K_1$ between these two scenarios, the behavior of a sample function is indefinite near the two boundaries.

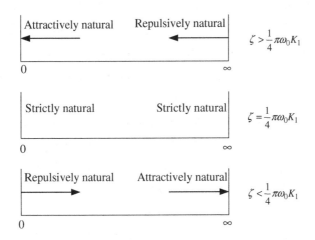

Figure 4.4.2 Boundary behaviors for the case of $K_1 > 0$, $K_2 = 0$.

3. $K_1 > 0$, $K_2 > 0$

The left boundary $x = 0$ is an entrance since $\sigma(0) = 0$, $m(0) > 0$, $\alpha = 2$, $\beta = 0$ and $c = 1$. The behaviors of the right boundary $x = \infty$ are the same as those in the above case 2, given in (4.4.25). The boundary behaviors are shown in Fig. 4.4.3. In the case of

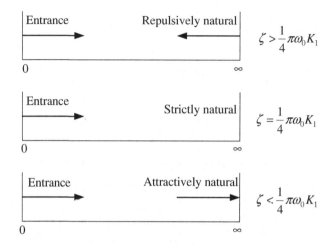

Figure 4.4.3   Boundary behaviors for the case of $K_1 > 0$, $K_2 > 0$.

$\zeta > \frac{1}{4}\pi\omega_0 K_1$, a stationary probability density exists, given by

$$p(x) = \frac{\omega_0^2 K_1}{2K_2} \left(\frac{4\zeta}{\pi\omega_0 K_1} - 1 \right) \left(1 + \frac{\omega_0^2 K_1}{2K_2} x \right)^{-4\zeta/\pi\omega_0 K_1} . \qquad (4.4.26)$$

As expected, the validity of (4.4.26) requires $\zeta > \frac{1}{4}\pi\omega_0 K_1$, which is the same condition for the right boundary at infinity to be repulsively natural. For the case of $\zeta < \frac{1}{4}\pi\omega_0 K_1$, a stationary probability does not exist since all sample functions will approach infinity. If $\zeta = \frac{1}{4}\pi\omega_0 K_1$, the behaviors of a sample function is indefinite at very large value.

The example shows that the qualitative behavior of a one-dimensional diffusion process is determined by the behaviors of its sample functions at two boundaries, which are dictated by the physical phenomenon that the diffusion process describes. It is common for many practical problems that the boundaries are singular, and hence, the boundary conditions cannot be imposed arbitrary. Instead, the class of a singular boundary is determined by the properties of the drift and diffusion coefficients.

4.5 Stochastic Processes Generated from Wiener Processes

4.5.1 *Stochastic Processes Generated from First-Order Filters*

Consider a diffusion process $X(t)$ defined in an interval between x_l and x_r, and governed by the Ito equation

$$dX = -\alpha X dt + D(X)dB(t), \qquad (4.5.1)$$

where α is a positive constant and $B(t)$ is a unit Wiener process. Without loss of generality, we assume that $X(t)$ has zero mean; thus, $x_l < 0$ and $x_r > 0$. Equation (4.5.1) represents a first-order filter, and function $D(X)$ can be linear or nonlinear.

Two most important characteristics of a stochastic process are its probability distribution and power spectral density. Multiplying both sides of (4.5.1) by $X(t - \tau)$ and taking ensemble average, we obtain

$$\frac{d}{dt}R_{XX}(\tau) = -\alpha R_{XX}(\tau), \qquad (4.5.2)$$

where $R_{XX}(\tau)$ is the correlation function, i.e., $R_{XX}(\tau) = E[X(t - \tau)X(t)]$. In deriving (4.5.2), the property of the Ito equation is applied, i.e., $dB(t)$ is independent of $X(t)$ and $X(t - \tau)$. The initial condition for Eq. (4.5.2) is

$$R_{XX}(0) = \sigma_X^2, \qquad (4.5.3)$$

where σ_X^2 is the mean square of $X(t)$. Then the correlation function is solved from (4.5.2) as

$$R_{XX}(\tau) = \sigma_X^2 \exp(-\alpha|\tau|). \qquad (4.5.4)$$

The corresponding spectral density of $X(t)$ is the Fourier transform of (4.5.4), found to be

$$\Phi_{XX}(\omega) = \frac{\alpha\sigma_X^2}{\pi(\omega^2 + \alpha^2)}. \qquad (4.5.5)$$

The correlation functions and spectral densities are shown in Fig. 3.6.5 from different α values. The process is known as low-pass process since the peak frequency is zero, and Eq. (4.5.1) is also called a low-pass filter. The bandwidth of the process is controlled by parameter α. A larger α corresponds to a broader band. It is noted that the diffusion coefficient $D(X)$ in (4.5.1) does not directly affect the spectral density.

The stationary probability density of $X(t)$ is governed by the reduced FPK equation

$$\frac{d}{dx}G = -\frac{d}{dx}\left\{\alpha x p(x) + \frac{1}{2}\frac{d}{dx}[D^2(x)p(x)]\right\} = 0, \qquad (4.5.6)$$

where G is the probability flow. For the present one dimensional case, G must vanish at the two boundaries, and hence, must vanish everywhere. Then (4.5.6) reduces to

$$\alpha x p(x) + \frac{1}{2}\frac{d}{dx}[D^2(x)p(x)] = 0. \qquad (4.5.7)$$

Integration of (4.5.7) results in

$$D^2(x)p(x) = -2\alpha\int_{x_l}^{x} up(u)du + C, \qquad (4.5.8)$$

where C is an integration constant. To determine C, several cases are considered. If $x_l = -\infty$, $p(-\infty) = 0$, leading to $C = 0$. If $x_r = \infty$, $p(\infty) = 0$ and $C = 0$ due to the assumption that the mean of $X(t)$ is zero. If both x_l and x_r are finite, the drift coefficients are not zero at x_l and x_r, $m(x_l) \neq 0$, $m(x_r) \neq 0$. Since a non-trivial stationary probability distribution requires both boundaries to be entrance or repulsively natural (see Section 4.4), the two boundaries must be singular of the first kind, i.e., the diffusion coefficient $D(X)$ must vanish at the two boundaries, leading also to $C = 0$. Therefore, we have from (4.5.8)

$$D^2(x) = -\frac{2\alpha}{p(x)}\int_{x_l}^{x} up(u)du. \qquad (4.5.9)$$

Function $D^2(x)$ in (4.5.9) is non-negative, as it should be, since $p(x) \geq 0$ and the mean of $X(t)$ is zero. Equation (4.5.9) shows that, for any valid probability density $p(x)$, the diffusion coefficient

$D(X)$ can be determined accordingly. Therefore, the Ito equation (4.5.1) can be used to model stochastic processes with a low-pass spectral density (4.5.5) and an arbitrary probability distribution.

Example 4.5.1 Consider a Gaussian process $X(t)$ with a probability density

$$p(x) = \frac{1}{\sqrt{2\pi}\sigma} e^{-\frac{x^2}{2\sigma^2}}, \quad -\infty < x < \infty. \tag{4.5.10}$$

Substituting (4.5.10) into (4.5.9),

$$D^2(x) = 2\alpha\sigma^2, \quad dX = -\alpha X dt + \sqrt{2\alpha}\sigma dB(t). \tag{4.5.11}$$

As expected, the corresponding Ito equation is a linear equation.

Example 4.5.2 Assume that $X(t)$ is uniformly distributed, i.e.,

$$p(x) = \frac{1}{2\Delta}, \quad -\Delta \le x \le \Delta. \tag{4.5.12}$$

In this case, we have

$$D^2(x) = \alpha(\Delta^2 - x^2), \quad dX = -\alpha X dt + \sqrt{\alpha(\Delta^2 - X^2)} dB(t). \tag{4.5.13}$$

Example 4.5.3 Assume that $X(t)$ has the following distribution

$$p(x) = C(\Delta^2 - x^2)^\delta, \quad -\Delta \le x \le \Delta \quad \text{for } \delta \ge 0 \quad \text{or} \quad -\Delta < x < \Delta$$
$$\text{for } -1 < \delta < 0, \tag{4.5.14}$$

where C is a normalization constant. Applying (4.5.9),

$$D^2(x) = \frac{\alpha}{\delta + 1}(\Delta^2 - x^2), \quad dX = -\alpha X dt + \sqrt{\frac{\alpha}{\delta + 1}(\Delta^2 - X^2)} dB(t). \tag{4.5.15}$$

The probability densities (4.5.14) are depicted in Fig. 4.5.1 for several δ values. It is shown that the shapes of the probability densities are diverse for different δ values. The case of $\delta = 0$ corresponds to a uniform distribution. For a fixed α value and different δ values, the processes have different probability distributions, yet they may share the similar spectral density (4.5.5). It is noted that the distribution of (4.5.14) is in fact the λ-distribution as shown in Section 2.5.5.

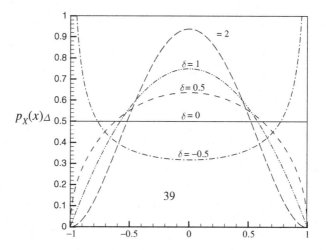

Figure 4.5.1   Stationary probability densities of $X(t)$ given in (4.5.14).

Example 4.5.4 Let $X(t)$ be governed by an exponential distribution as shown in Section 2.5.4

$$p(x) = \lambda \exp(-\lambda x), \quad \lambda > 0, \quad 0 \le x < \infty. \tag{4.5.16}$$

Since its mean is $1/\lambda$, we define the centralized process as $Y(t) = X(t) - 1/\lambda$. The probability density of $Y(t)$ is

$$p(y) = \lambda \exp\left[-\lambda\left(y + \frac{1}{\lambda}\right)\right], \quad -\frac{1}{\lambda} \le y < \infty. \tag{4.5.17}$$

From Eq. (4.5.9),

$$D^2(y) = \frac{2\alpha}{\lambda}\left(y + \frac{1}{\lambda}\right). \tag{4.5.18}$$

The Ito equation for $Y(t)$ is

$$dY = -\alpha Y\,dt + \frac{1}{\lambda}\sqrt{2\alpha(\lambda Y + 1)}\,dB(t) \tag{4.5.19}$$

and the Ito equation for $X(t)$ is

$$dX = -\alpha\left(X - \frac{1}{\lambda}\right)dt + \sqrt{\frac{2\alpha}{\lambda}X}\,dB(t). \tag{4.5.20}$$

Note that the spectral density of $X(t)$ contains a delta function $(1/\lambda^2)\delta(\omega)$ due to the non-zero mean $1/\lambda$.

The above examples show that the first-order filter can generate a low-pass stochastic process defined in a finite or infinite interval and with an arbitrary probability distribution.

4.5.2 *Stochastic Processes Generated from Second-Order Filters*

Now we try to use second-order filters to generate stochastic processes. Extend Eq. (4.5.1) from one-dimensional to two-dimensional, i.e.,

$$
\begin{aligned}
dX_1 &= (-a_{11}X_1 - a_{12}X_2)dt + D_1(X_1, X_2)dB_1(t), \\
dX_2 &= (-a_{21}X_1 - a_{22}X_2)dt + D_2(X_1, X_2)dB_2(t),
\end{aligned}
\tag{4.5.21}
$$

where a_{ij} are parameters, and $B_1(t)$ and $B_2(t)$ are independent unit Wiener processes. Multiplying the two equations in (4.5.21) by $X_1(t-\tau)$, taking the ensemble average, and denoting $R_{ij}(\tau) = E[X_i(t-\tau)X_j(t)]$, we obtain

$$
\begin{aligned}
\frac{d}{d\tau}R_{11}(\tau) &= -a_{11}R_{11}(\tau) - a_{12}R_{12}(\tau), \\
\frac{d}{d\tau}R_{12}(\tau) &= -a_{21}R_{11}(\tau) - a_{22}R_{12}(\tau),
\end{aligned}
\tag{4.5.22}
$$

subjected to initial conditions

$$
R_{11}(0) = E[X_1^2] = \sigma^2, \quad R_{12}(0) = E[X_1 X_2]. \tag{4.5.23}
$$

Equation set (4.5.22) can be solved for the correlation functions. In modeling a stochastic process, its spectral density is usually of interest. It can be calculated from the correlation function using the Fourier transformation. However, it can be derived using an alternative procedure described below. Define the following integral transformation

$$\bar{\Phi}_{ij}(\omega) = \Im[R_{ij}(\tau)] = \frac{1}{\pi} \int_0^\infty R_{ij}(\tau) e^{-i\omega\tau} d\tau. \qquad (4.5.24)$$

The spectral density functions can be derived in terms of $\bar{\Phi}_{ij}(\omega)$ as follows (see Exercise Problem 4.22)

$$\Phi_{ii}(\omega) = \Phi_{X_i X_i}(\omega) = Re[\bar{\Phi}_{ii}(\omega)],$$
$$\Phi_{ij}(\omega) = \Phi_{X_i X_j}(\omega) = \frac{1}{2}[\bar{\Phi}_{ij}(\omega) + \bar{\Phi}_{ji}^*(\omega)]. \qquad (4.5.25)$$

Since $R_{ij}(\tau) \to 0$ as $\tau \to \infty$, it can be shown by using integration by parts that

$$\Im\left[\frac{dR_{ij}(\tau)}{d\tau}\right] = i\omega\Im[R_{ij}(\tau)] - \frac{1}{\pi}R_{ij}(0) = i\omega\bar{\Phi}_{ij}(\omega) - \frac{1}{\pi}m_{ij}. \qquad (4.5.26)$$

Using (4.5.24) and (4.5.26), (4.5.22) can be transformed to

$$i\omega\bar{\Phi}_{11} - \frac{1}{\pi}E[X_1^2] = -a_{11}\bar{\Phi}_{11} - a_{12}\bar{\Phi}_{12}$$
$$i\omega\bar{\Phi}_{12} - \frac{1}{\pi}E[X_1 X_2] = -a_{21}\bar{\Phi}_{11} - a_{22}\bar{\Phi}_{12}. \qquad (4.5.27)$$

Solutions are readily obtained from complex linear algebraic equation set (4.5.27), leading to

$$\Phi_{11}(\omega) = \frac{(a_{11}\omega^2 + A_2 a_{22})E[X_1^2] + a_{12}(\omega^2 - A_2)E[X_1 X_2]}{\pi[(A_2 - \omega^2)^2 + A_1^2\omega^2]}, \qquad (4.5.28)$$

where $A_1 = a_{11} + a_{22}$ and $A_2 = a_{11}a_{22} - a_{12}a_{21}$. By adjusting parameters a_{ij}, (4.5.28) can represent a spectral density with a peak at a specified location and a given bandwidth.

The FPK equation for the joint stationary probability density $p_{X_1 X_2}(x_1, x_2)$ of $X_1(t)$ and $X_2(t)$ corresponding to (4.5.21) is given by

$$\frac{\partial}{\partial x_1}[(-a_{11}x_1 - a_{12}x_2)p] + \frac{\partial}{\partial x_2}[(-a_{21}x_1 - a_{22}x_2)p]$$

$$-\frac{1}{2}\frac{\partial^2}{\partial x_1^2}[D_1^2(x_1, x_2)p] - \frac{1}{2}\frac{\partial^2}{\partial x_2^2}[D_2^2(x_1, x_2)p] = 0.$$

$$(4.5.29)$$

Equation (4.5.29) is satisfied if the following three conditions are met

$$-a_{12}x_2\frac{\partial p}{\partial x_1} - a_{21}x_1\frac{\partial p}{\partial x_2} = 0, \tag{4.5.30}$$

$$-a_{11}x_1 p - \frac{1}{2}\frac{\partial}{\partial x_1}[D_1^2(x_1, x_2)p] = 0, \tag{4.5.31}$$

$$-a_{22}x_2 p - \frac{1}{2}\frac{\partial}{\partial x_2}[D_2^2(x_1, x_2)p] = 0, \tag{4.5.32}$$

indicating that the system belongs to the case of detailed balance (see Chapter 6). The general solution for Eq. (4.5.30) is given by

$$p(x_1, x_2) = \rho(\lambda), \quad \lambda = k_1 x_1^2 + k_2 x_2^2, \tag{4.5.33}$$

where ρ is an arbitrary function of λ, k_1 and k_2 are positive constants satisfying the following condition

$$k_1 a_{12} + k_2 a_{21} = 0. \tag{4.5.34}$$

Substituting (4.5.33) into (4.5.31) and (4.5.32), we obtain

$$D_1^2(x_1, x_2) = -\frac{2a_{11}}{p_{X_1 X_2}(x_1, x_2)}\int_{x_{1_l}}^{x_1} u p_{X_1 X_2}(u, x_2)du$$

$$= \frac{a_{11}}{k_1 \rho(\lambda)}\int_{\lambda}^{\lambda_1} \rho(\lambda)d\lambda, \tag{4.5.35}$$

$$D_2^2(x_1, x_2) = -\frac{2a_{22}}{p_{X_1 X_2}(x_1, x_2)} \int_{x_{2_l}}^{x_2} v p_{X_1 X_2}(x_1, v) dv$$

$$= \frac{a_{22}}{k_2 \rho(\lambda)} \int_{\lambda}^{\lambda_2} \rho(\lambda) d\lambda, \qquad (4.5.36)$$

where λ_1 is the λ value at $x_1 = x_{1_l}$ for a fixed x_2 and λ_2 is the λ value at $x_2 = x_{2_l}$ for a fixed x_1. Equation (4.5.33) shows that, if the probability density of the stochastic process $X_1(t)$ is of a form of $p(x_1) = p(x_1^2)$, function $\rho(\lambda)$ can be constructed, two functions $D_1(x_1, x_2)$ and $D_2(x_1, x_2)$ can be calculated from (4.5.35) and (4.5.36), and (4.5.21) can be used to generate stochastic process $X_1(t)$.

The second-order filter (4.5.21) can be used to generate stochastic process $X_1(t)$ defined in an infinite domain $(-\infty, \infty)$, as well as in a finite range $(-\Delta, \Delta)$.

Example 4.5.5 Assume a Gaussian process $X_1(t)$ with a probability density

$$p(x_1) = \frac{1}{\sqrt{2\pi}\sigma} e^{-\frac{x_1^2}{2\sigma^2}}, \quad -\infty < x_1 < \infty. \qquad (4.5.37)$$

Construct a function $\rho(\lambda)$ according to (4.5.33) and (4.5.34),

$$\lambda = \frac{1}{2\sigma^2} \left(x_1^2 - \frac{a_{12}}{a_{21}} x_2^2 \right). \qquad (4.5.38)$$

The probability density is

$$p(x_1, x_2) = \rho(\lambda) = C \exp(-\lambda) = C \exp \left[-\frac{1}{2\sigma^2} \left(x_1^2 - \frac{a_{12}}{a_{21}} x_2^2 \right) \right]. \qquad (4.5.39)$$

Substituting (4.5.39) into (4.5.35) and (4.5.36), we have

$$D_1^2 = 2a_{11}\sigma^2, \quad D_2^2 = -2\sigma^2 \frac{a_{21}a_{22}}{a_{12}}. \qquad (4.5.40)$$

As expected, the corresponding Ito equation set (4.5.21) is linear. By selecting different parameters of a_{ij} $(i, j = 1, 2)$, different spectral densities can be obtained. The following are two typical cases.

Letting $a_{11} = 0$, $a_{12} = -1$, $a_{21} = \omega_0^2$, $a_{22} = 2\zeta\omega_0$, then $D_1^2 = 0$, $D_2^2 = 4\zeta\omega_0^3\sigma^2$, and

$$\Phi_{11}(\omega) = \frac{2\zeta\omega_0^3 \sigma^2}{\pi[(-\omega^2 + \omega_0^2)^2 + 4\zeta^2\omega_0^2\omega^2]}. \tag{4.5.41}$$

The system is then

$$dX_1 = -X_2 dt,$$

$$dX_2 = (-\omega_0^2 X_1 - 2\zeta\omega_0 X_2)dt + \sqrt{4\zeta\omega_0^3\sigma^2}dB_2(t), \tag{4.5.42}$$

which are the Ito equations corresponding to the second-order linear filter

$$\ddot{X}_1 + 2\zeta\omega_0\dot{X}_1 + \omega_0^2 X_1 = W(t), \tag{4.5.43}$$

with a spectral density $K = 2\zeta\omega_0^3\sigma^2/\pi$ for the white noise $W(t)$.

If we select $a_{11} = 2\zeta\omega_0$, $a_{12} = \omega_0^2$, $a_{21} = -1$, $a_{22} = 0$, then $D_1^2 = 4\zeta\omega_0\sigma^2$, $D_2^2 = 0$, and

$$\Phi_{11}(\omega) = \frac{2\zeta\omega_0\omega^2 \sigma^2}{\pi[(-\omega^2 + \omega_0^2)^2 + 4\zeta^2\omega_0^2\omega^2]}. \tag{4.5.44}$$

The second-order linear filter is then

$$dX_1 = (-2\zeta\omega_0 X_1 - \omega_0^2 X_2)dt + \sqrt{4\zeta\omega_0\sigma^2}dB_1(t),$$

$$dX_2 = -X_1 dt. \tag{4.5.45}$$

In both cases, ζ and ω_0 can be used to adjust the spectral density, and σ^2 is used to match the probability density. Figures 4.5.2 and 4.5.3 show the spectral density functions for the two cases with $\omega_0 = 3$ and several different values of ζ. It is seen that the two processes yield different shapes of spectral densities. The spectral density vanishes at zero-frequency in Fig. 4.5.3, while it does not in Fig. 4.5.2. In both cases, ω_0 determines the peak location and ζ controls the bandwidth.

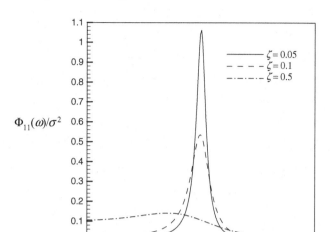

Figure 4.5.2 Spectral densities of $X_1(t)$ generated from system (4.5.2) or (4.5.43) with $\omega_0 = 3$.

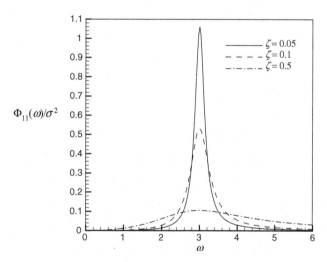

Figure 4.5.3 Spectral densities of $X_1(t)$ generated from system (4.5.2) with $\omega_0 = 3$.

Example 4.5.6 Consider the case in which x_1 and x_2 are bounded by

$$k_1 x_1^2 + k_2 x_2^2 \leq k_1 \Delta^2 \qquad (4.5.46)$$

and

$$\rho(\lambda) = C(k_1 \Delta^2 - \lambda)^{\delta - \frac{1}{2}}, \quad \delta > -\frac{1}{2}. \qquad (4.5.47)$$

The joint stationary probability density is

$$p_{X_1 X_2}(x_1, x_2) = C(k_1 \Delta^2 - k_1 x_1^2 - k_2 x_2^2)^{\delta - \frac{1}{2}} \qquad (4.5.48)$$

and the marginal probability density of X_1 is then obtained as (see Exercise Problem 2.8)

$$p_{X_1}(x_1) = 2 \int_0^{\sqrt{k_1(\Delta^2 - x_1^2)/k_2}} p_{X_1 X_2}(x_1, x_2) dx_2 = C_1(\Delta^2 - x_1^2)^{\delta},$$

$$(4.5.49)$$

where C_1 is normalization constant. Substituting (4.5.48) into (4.5.35) and (4.5.36), we obtain

$$D_1^2(x_1, x_2) = \frac{2a_{11}}{k_1(2\delta + 1)}(k_1 \Delta^2 - k_1 x_1^2 - k_2 x_2^2), \qquad (4.5.50)$$

$$D_2^2(x_1, x_2) = \frac{2a_{22}}{k_2(2\delta + 1)}(k_1 \Delta^2 - k_1 x_1^2 - k_2 x_2^2). \qquad (4.5.51)$$

The probability density (4.5.49) has the same form as (4.5.14), but with a more restrictive range for parameter δ due to the validity of the joint probability density (4.5.48) and the positivity requirement of (4.5.50) and (4.5.51). Thus, equation set (4.5.21), with $D_1(X_1, X_2)$ and $D_1(X_1, X_2)$ given by (4.5.50) and (4.5.51) respectively, can be used to generate a stochastic process $X_1(t)$ with a spectral density (4.5.28) and a probability density (4.5.49). Parameters a_{ij} $(i, j = 1, 2)$ are used to adjust the spectral density, Δ is determined by the allowable range of process $X_1(t)$, and δ is used to match the shape of its probability distribution.

For the spectral density given by (4.5.41), the parameters are:

$$a_{11} = 0, \quad a_{12} = -1, \quad a_{21} = \omega_0^2, \quad a_{22} = 2\zeta\omega_0, \quad D_1^2 = 0,$$

$$D_2^2 = \frac{4\zeta\omega_0^3}{2\delta + 1}\left(\Delta^2 - X_1^2 - \frac{1}{\omega_0^2}X_2^2\right),$$

For the spectral density given by (4.5.44), the parameters are:

$$a_{11} = 2\zeta\omega_0, \quad a_{12} = \omega_0^2, \quad a_{21} = -1, \quad a_{22} = 0,$$

$$D_1^2 = \frac{4\zeta\omega_0}{2\delta + 1}(\Delta^2 - X_1^2 - \omega_0^2 X_2^2), \quad D_2^2 = 0.$$

In both cases, ζ and ω_0 can be used to adjust the spectral density and Δ and δ are used to match the probability density.

4.5.3 *Randomized Harmonic Process*

A type of stochastic processes, known as randomized harmonic processes, is modeled as

$$X(t) = A\sin[\omega_0 t + \sigma B(t) + U], \tag{4.5.52}$$

where A is a positive constant specifying the intensity of the process, ω_0 and σ are also positive constants representing the mean frequency and the randomness level in the phase, respectively, $B(t)$ is a unit Wiener process, and U is a random variable uniformly distributed in $[0, 2\pi)$ and independent of $B(t)$. Physically, the introduction of the random variable U in (4.5.52) means that the initial phase is random. This randomized harmonic process was proposed independently by Dimentberg (1988) and Wedig (1989), and has been used in a variety of engineering problems.

Taking into consideration of the periodicity of $X(t)$ with respect to U, we have

$$E[X(t)] = \int_{-\infty}^{\infty} p_B(b)db \int_0^{2\pi} \frac{A}{2\pi}\sin(\omega_0 t + \sigma b + u)du = 0,$$

$$\tag{4.5.53}$$

$$E[X^2(t)] = \int_{-\infty}^{\infty} p_B(b)db \int_0^{2\pi} \frac{A^2}{2\pi} \sin^2(\omega_0 t + \sigma b + u)du$$

$$= \frac{1}{2}A^2, \tag{4.5.54}$$

$$E[X(t_1)X(t_2)] = A^2 E\{\sin[\omega_0 t_1 + \sigma B(t_1) + U]$$
$$\sin[\omega_0 t_2 + \sigma B(t_2) + U]\}$$

$$= \frac{A^2}{2} E\{\cos[\omega_0(t_2 - t_1) + \sigma B(t_2) - \sigma B(t_1)]\}. \tag{4.5.55}$$

Denote the increment of $B(t)$ as

$$Z = B(t_2) - B(t_1). \tag{4.5.56}$$

According to (4.2.29), its mean and variance are

$$E[Z(t_1, t_2)] = 0, \quad E[Z^2(t_1, t_2)] = E\{[B(t_2) - B(t_1)]^2\} = t_2 - t_1,$$

$$t_2 \geq t_1. \tag{4.5.57}$$

Since the Wiener process $B(t)$ is Gaussian distributed, its increment Z is also Gaussian distributed with a probability density (4.2.33), i.e.,

$$p_Z(z) = \frac{1}{\sqrt{2\pi(t_2 - t_1)}} \exp\left[-\frac{z^2}{2(t_2 - t_1)}\right]. \tag{4.5.58}$$

Continuing the calculation of (4.5.55), we have

$$E[X(t_1)X(t_2)]$$

$$= \frac{A^2}{2} E\{\cos[\omega_0(t_2 - t_1)]\cos(\sigma Z) - \sin[\omega_0(t_2 - t_1)]\sin(\sigma Z)\}$$

$$= \frac{1}{2}A^2 \cos[\omega_0(t_2 - t_1)] \int_{-\infty}^{\infty} \frac{\cos(\sigma z)}{\sqrt{2\pi(t_2 - t_1)}} \exp\left[-\frac{z^2}{2(t_2 - t_1)}\right] dz$$

$$= \frac{1}{2}A^2 \cos(\omega_0 \tau) \exp\left(-\frac{1}{2}\sigma^2 \tau\right), \quad \tau = t_2 - t_1 \geq 0. \tag{4.5.59}$$

Equation (4.5.59) shows that $X(t)$ is a weakly stationary process with an autocorrelation function

$$R_{XX}(\tau) = \frac{1}{2}A^2 \cos(\omega_0\tau) \exp\left(-\frac{1}{2}\sigma^2|\tau|\right). \tag{4.5.60}$$

Carrying the Fourier transform, we have the power spectral density as follows

$$\Phi_{XX}(\omega) = \frac{A^2\sigma^2(\omega^2 + \omega_0^2 + \sigma^4/4)}{4\pi[(\omega^2 - \omega_0^2 - \sigma^4/4)^2 + \sigma^4\omega^2]}. \tag{4.5.61}$$

There is an alternative way to find the correlation function. Let

$$X_1(t) = X(t) = A\sin\Theta(t), \quad X_2(t) = A\cos\Theta(t),$$
$$d\Theta = \omega_0 dt + \sigma dB(t). \tag{4.5.62}$$

Using the Ito differential rule (4.2.58), we obtain

$$dX_1 = \left(-\frac{1}{2}\sigma^2 X_1 + \omega_0 X_2\right)dt + \sigma X_2 dB(t), \tag{4.5.63}$$

$$dX_2 = \left(-\omega_0 X_1 - \frac{1}{2}\sigma^2 X_2\right)dt - \sigma X_1 dB(t). \tag{4.5.64}$$

Multiplying both sides of Eqs. (4.5.63) and (4.5.64) by $X_1(t-\tau)$ and taking ensemble average, we obtain

$$\frac{d}{dt}R_{11}(\tau) = -\frac{1}{2}\sigma^2 R_{11}(\tau) + \omega_0 R_{12}(\tau), \tag{4.5.65}$$

$$\frac{d}{dt}R_{12}(\tau) = -\omega_0 R_{11}(\tau) - \frac{1}{2}\sigma^2 R_{12}(\tau), \tag{4.5.66}$$

where

$$R_{11}(\tau) = R_{X_1X_1}(\tau) = E[X_1(t-\tau)X_1(t)],$$
$$R_{12}(\tau) = R_{X_1X_2}(\tau) = E[X_1(t-\tau)X_2(t)]. \tag{4.5.67}$$

With the initial conditions

$$R_{11}(0) = E[X_1^2(t)] = \frac{1}{2}A^2, \quad R_{12}(0) = E[X_1(t)X_2(t)] = 0, \tag{4.5.68}$$

the ordinary differential equations with constant coefficients, (4.5.65) and (4.5.66), are solved to obtain

$$R_{11}(\tau) = \frac{1}{2}A^2 \cos(\omega_0 \tau) \exp\left(-\frac{1}{2}\sigma^2 \tau\right), \quad \tau \geq 0, \qquad (4.5.69)$$

which is the same as (4.5.60). Figure 4.5.4 depicts the spectral densities in the positive ω range for the case of $\omega_0 = 3$ and several values of σ. It is seen that the spectral densities reach their peaks near $\omega = \omega_0$, and exhibit different band widths for different σ values. In the case of $\sigma = 0$, the stochastic process $X(t)$ reduces to a pure harmonic process with a random initial phase. As σ increases, the band width of the process becomes broader, indicating an increasing randomness.

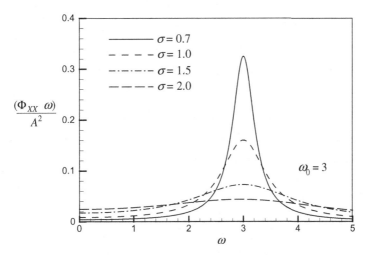

Figure 4.5.4 Spectral density functions of randomized harmonic process $X(t)$ for $\omega_0 = 3$ and different σ values.

As a special case of $\omega_0 = 0$, the correlation function and spectral density are, respectively,

$$R_{XX}(\tau) = \frac{1}{2}A^2 \exp\left(-\frac{1}{2}\sigma^2 |\tau|\right), \qquad (4.5.70)$$

$$\Phi_{XX}(\omega) = \frac{A^2 \sigma^2}{4\pi(\omega^2 + \sigma^4/4)}. \qquad (4.5.71)$$

It is noted that Eqs. (4.5.70) and (4.5.71) have the same forms as those of Eqs. (4.5.4) and (4.5.5), and the process is known as the low-pass process. However, the probability density of the randomized harmonic process is different as those generated from a low-pass filter, as shown below.

To find the probability density of $X(t)$, denote

$$X(t) = A\sin\Theta(t), \quad \Theta = Y + U, \quad Y = \omega_0 t + \sigma B(t). \qquad (4.5.72)$$

Since Y and U are independent, and

$$p_U(u) = \frac{1}{2\pi}, \quad 0 \le u < 2\pi. \qquad (4.5.73)$$

Using (2.7.21), we have the probability density for Θ as follows

$$p_\Theta(\theta) = \int_0^{2\pi} p_U(u)p_Y(\theta - u)du = \frac{1}{2\pi}\int_0^{2\pi} p_Y(\theta - u)du$$

$$= \frac{1}{2\pi}\int_{\theta-2\pi}^{\theta} p_Y(y)dy. \qquad (4.5.74)$$

Note that $\Theta \in (-\infty, \infty)$ according to (4.5.72). Since the harmonic sine function in (4.5.72) is periodic with a period 2π, we can limit the angle within $[0, 2\pi)$ and mark it as Θ_1, and convert $X(t)$ from $A\sin\Theta$ to $A\sin\Theta_1$, i.e.,

$$X(t) = A\sin\Theta_1(t), \quad 0 \le \Theta_1 < 2\pi. \qquad (4.5.75)$$

The value of the probability density $p_{\Theta_1}(\theta_1)$ at $\theta_1 \in [0, 2\pi)$ should be obtained by summing up all values of $p_\Theta(\theta)$ at $\theta_1 + 2K\pi$, where K includes all integers. Therefore,

$$p_{\Theta_1}(\theta_1) = \sum_{\substack{k=-\infty \\ k=\text{integer}}}^{\infty} p_\Theta(\theta_1 + 2k\pi) = \frac{1}{2\pi}\sum_{\substack{k=-\infty \\ k=\text{integer}}}^{\infty} \int_{\theta_1+2(k-1)\pi}^{\theta_1+2k\pi} p_Y(y)dy$$

$$= \frac{1}{2\pi}\int_{-\infty}^{\infty} p_Y(y)dy = \frac{1}{2\pi}. \qquad (4.5.76)$$

In deriving (4.5.76), use has been made of (4.5.74). Equation (4.5.76) shows that Θ_1 is uniformly distributed in $[0, 2\pi)$. According to the transformation rule (2.7.4),

$$p_X(x) = p_{\Theta_1}(\theta_1) \left| \frac{d\theta_1}{dx} \right| = \frac{1}{\pi\sqrt{A^2 - x^2}}, \quad -A < x < A.$$

$$(4.5.77)$$

Figure 4.5.5 depicts the probability density of $X(t)$. It has very large values near the two boundaries. Note that the probability distribution only depends on A, to be determined according to the physical boundary of the underlined phenomenon. The parameters ω_0 and σ have no effect on the probability distribution; however, they can be adjusted to match the spectral density of $X(t)$ to be modeled.

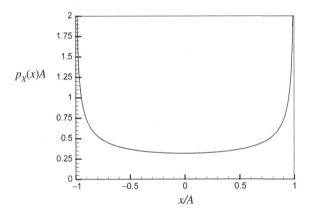

Figure 4.5.5 Probability density of randomized harmonic process $X(t)$.

There are two advantages to use the randomized harmonic process (4.5.52) to model a practical stochastic process: (i) it is more realistic due to its boundedness and (ii) the spectral density can be matched by adjusting ω_0 and σ according to the peak magnitude, peak location, and bandwidth.

For more options in modeling bounded stochastic processes, please refer to the reference by Zhu and Cai (2013).

4.6 Simulation

Simulation of a stochastic process is to generate sample functions according to the characteristics of the stochastic process, among which the power spectral density and the probability density are the most important. For a system subjected to excitations of stationary and ergodic processes, the system response statistics can be obtained from one sample as long as its length is sufficiently long.

4.6.1 *Simulation of Gaussian White Noises*

A mathematically ideal white noise $W(t)$ has a constant spectral density over all frequency band, i.e., $\Phi_{WW}(\omega) = K$, $-\infty < \omega < \infty$. As shown by Eq. (4.2.43) in Section 4.2.5, a Gaussian white noise is the formal derivative of Wiener process with $\sigma^2 = 2\pi K$, where σ^2 is the intensity of the Wiener process. Thus, for a small time step Δt, we may discretize the white noise process as

$$W(t_i) = \frac{\Delta B(t_i)}{\Delta t} = \sqrt{\frac{2\pi K}{\Delta t}} U_i, \qquad (4.6.1)$$

where U_i are N(0,1) identically distributed random variables, and independent for different i. From (4.6.1),

$$E[W(t_i)] = 0, \quad E[W^2(t_i)] = \frac{2\pi K}{\Delta t}, \quad E[W(t_i)W(t_j)] = 0, \quad i \neq j. \qquad (4.6.2)$$

Consider two time instants t_i and $t_i + \tau$, $\tau > 0$. If $\tau \leq \Delta t$, the random variable at $t_i + \tau$ can be obtained by linear interpolation between $W(t_i)$ and $W(t_{i+1})$,

$$W(t_i + \tau) = W(t_i) + [W(t_{i+1}) - W(t_i)]\frac{\tau}{\Delta t} \qquad (4.6.3)$$

and the correlation function is calculated as

$$\begin{aligned} R_{WW}(\tau) &= E[W(t_i)W(t_i + \tau)] \\ &= E\left\{ W(t_i)\left(W(t_i) + [W(t_{i+1}) - W(t_i)]\frac{\tau}{\Delta t} \right) \right\} \\ &= \frac{2\pi K}{\Delta t}\left(1 - \frac{\tau}{\Delta t} \right). \end{aligned} \qquad (4.6.4)$$

For $\tau > \Delta t$, the linear interpolation is carried out on interval other than $[W(t_i), W(t_{i+1})]$, leading to

$$W(t_i + \tau) = W(t_j) + [W(t_{j+1}) - W(t_j)]\frac{\tau}{\Delta t}, \quad j > i. \qquad (4.6.5)$$

The correlation function is then calculated as

$$\begin{aligned} R_{WW}(\tau) &= E[W(t_i)W(t_i + \tau)] \\ &= E\left\{W(t_i)\left[W(t_j) + [W(t_{j+1}) - W(t_j)]\frac{\tau}{\Delta t}\right]\right\} = 0. \end{aligned}$$
$$(4.6.6)$$

Noting that (4.6.4) and (4.6.6) are only for $\tau > 0$ and the correlation function is an even function, we obtain the power spectral density as

$$\begin{aligned} \Phi_{WW}(\omega) &= \frac{1}{2\pi} \int\limits_{-\infty}^{\infty} R_{WW}(\tau)e^{-j\omega\tau}d\tau = \frac{1}{\pi} \int\limits_{0}^{\infty} R_{WW}(\tau)\cos(\omega\tau)d\tau \\ &= \frac{1}{\pi} \int\limits_{0}^{\Delta t} \frac{2\pi K}{\Delta t}\left(1 - \frac{\tau}{\Delta t}\right)\cos(\omega\tau)d\tau = \frac{2K}{(\omega\Delta t)^2}[1 - \cos(\omega\Delta t)]. \end{aligned}$$
$$(4.6.7)$$

Equation (4.6.7) shows that the spectral density of the simulated process given by (4.6.1) is not constant. However, it can be shown that

$$\lim_{\omega\Delta t\to 0} \frac{2K}{(\omega\Delta t)^2}[1 - \cos(\omega\Delta t)] = K, \qquad (4.6.8)$$

indicating that if $\omega\Delta t$ is very small, the spectral density is approximately constant. For a fixed $\omega\Delta t$ value, a smaller Δt allows a larger ω, i.e., a broader frequency band. Figure 4.6.1 depicts the spectral density functions in the positive frequency range for several different Δt values. For the case of $\Delta t = 0.01$, the spectral density remains constant approximately for a quite broad frequency band. With an increasing Δt, the frequency band of approximate constant spectral density becomes narrower.

It is understood that an ideal white noise does not exist since its mean square is unbounded, implying an infinity energy. It is also

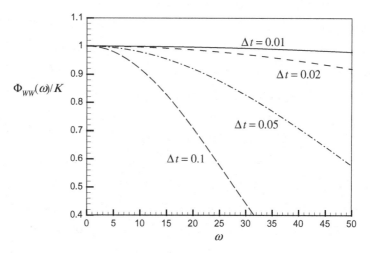

Figure 4.6.1 Power spectral density functions of simulated white noise (4.6.1) for different values of Δt.

true that an ideal white noise cannot be realized in simulation since Δt in (4.6.1) has to be taken as a value other than zero. It is clear that, when simulating a white noise using (4.6.1), the time step Δt is the most important parameter. A smaller Δt leads to a broader frequency band for a constant spectral density. In the case where an oscillatory system is involved, the frequency band of the simulated white noise should be much larger that the important system natural frequencies. Therefore, Δt should be much shorter than all important system natural periods. In general, Δt takes a value no larger than $1/20$ of the smallest system natural period.

Now we try to generate two correlated Gaussian white noises with the following spectral densities

$$E[W_j(t)W_k(t+\tau)] = K_{jk}\delta(\tau), \quad i,j = 1,2. \tag{4.6.9}$$

Consider two N(0,1) identically distributed random variables U and V with a correlation coefficient

$$\rho = \frac{K_{12}}{\sqrt{K_{11}K_{22}}}. \tag{4.6.10}$$

First, we generate two sequences U_i and V_i of the two correlated random variables using the method in Sections 2.8.5 and 2.8.6. In doing that, keep each U_i or V_i sequence independent for different i, and U_i and V_j independent for $i \neq j$. Then the two white noise processes can be simulated as follows

$$W_1(t_i) = \sqrt{\frac{2\pi K_{11}}{\Delta t}} U_i, \quad W_2(t_i) = \sqrt{\frac{2\pi K_{22}}{\Delta t}} V_i. \tag{4.6.11}$$

4.6.2 *Simulation of Ito Equations*

Consider the simplest case of one-dimensional Ito equation

$$dX(t) = m(X, t)dt + \sigma(X, t)dB(t). \tag{4.6.12}$$

The common numerical method to solve an ordinary differential equation is the finite-difference method. Due to the special property of Ito equation as described in Section 4.2.6, the finite difference equation is

$$X(t_i + \Delta t) = X(t_i) + m[X(t_i), t_i]\Delta t + \sigma[X(t_i), t_i]dB(t_i), \tag{4.6.13}$$

where $dB(t_i) = B(t_i + \Delta t) - B(t_i)$ obeys $N(0, \sigma\sqrt{\Delta t})$, i.e., Gaussian distribution with zero mean and variance $\sigma^2 \Delta t$, and $dB(t_i)$ are independent for different i. An important feature of (4.6.13) is that the increment $dB(t_i)$ is independent of $X(t_i)$.

The Ito equation (4.6.12) can be converted to the Stratonovich equation according to (4.2.67)

$$dX(t) = \left[m(X, t) - \frac{1}{2}\sigma(X, t)\frac{\partial\sigma(X, t)}{\partial X} \right] dt + \sigma(X, t) \circ dB(t), \tag{4.6.14}$$

which is equivalent to the regular differential equation with a Gaussian white noise excitation

$$\frac{d}{dt}X(t) = m(X, t) - \frac{1}{2}\sigma(X, t)\frac{\partial\sigma(X, t)}{\partial X} + \sigma(X, t)W(t), \tag{4.6.15}$$

where the white noise $W(t)$ has a spectral density of $K = 1/(2\pi)$. Instead of the Ito equation (4.6.12) and its difference equation (4.6.13), (4.6.15) can now be used for simulation. Since (4.6.15) can be treated as a regular ordinary differential equation, various numerical algorithms, such as Runge–Kutta method, can be applied to speed the computation with higher accuracy.

4.6.3 *Simulation of Stationary Gaussian Processes*

Consider a stationary Gaussian process $X(t)$ with zero mean and spectral density $\Phi_{XX}(\omega)$. For a practical stochastic process, there always exists a frequency band within which $\Phi_{XX}(\omega)$ has significant values. Assume that ω_l and ω_r are the left and right boundaries of the frequency band in the positive ω range, and $\Phi_{XX}(\omega)$ is negligible if $0 \leq \omega < \omega_l$ and $\omega > \omega_r$. Divide $[\omega_l, \omega_r]$ into N intervals, and denote

$$\Delta\omega = \frac{1}{N}(\omega_r - \omega_l), \quad \omega_j = \omega_l + \left(j - \frac{1}{2}\right)\Delta\omega,$$

$$\sigma_j^2 = 2\Phi_{XX}(\omega_j)\Delta\omega, \quad j = 1, 2, \ldots, N, \tag{4.6.16}$$

where $\Delta\omega$ is the length of interval, ω_j is the central frequency of the jth interval, and σ_j^2 is the contribution of the spectral density in the jth interval toward total mean-square value of $X(t)$, as shown in Fig. 4.6.2. Note that the total area under the $\Phi_{XX}(\omega)$ curve is the mean-square value. The factor 2 in the last equation of (4.6.16) is due to the positive and negative ranges of frequency ω. The following

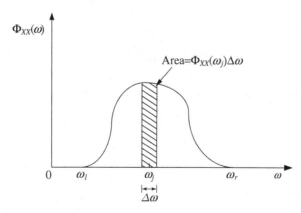

Figure 4.6.2 Contribution of the spectral density in the jth interval toward total mean-square value of $X(t)$.

three algorithms can be used to simulate a stationary Gaussian process $X(t)$ with a given spectral density $\Phi_{XX}(\omega)$.

Random amplitude representation. The equation to generate a sample function of $X(t)$ is

$$X_1(t) = \sum_{j=1}^{N} \sigma_j(U_j \cos \omega_j t + V_j \sin \omega_j t), \qquad (4.6.17)$$

where U_j and V_j are mutually independent $N(0,1)$ random variables, and they are also independent for different j. The correlation function of $X_1(t)$ is obtained as

$$
\begin{aligned}
R_{X_1 X_1}(\tau) &= E[X_1(t) X_1(t+\tau)] \\
&= \sum_{j,k=1}^{N} E\{\sigma_j \sigma_k (U_j \cos \omega_j t + V_j \sin \omega_j t) \\
&\qquad \times [U_k \cos \omega_k(t+\tau) + V_k \sin \omega_k(t+\tau)]\} \\
&= \sum_{j=1}^{N} \sigma_j^2 \cos(\omega_j \tau). \qquad (4.6.18)
\end{aligned}
$$

The corresponding spectral density is

$$
\begin{aligned}
\Phi_{X_1 X_1}(\omega) &= \frac{1}{2\pi} \int_{-\infty}^{\infty} R_{X_1 X_1}(\tau) \exp(-i\omega\tau) d\tau \\
&= \frac{1}{2\pi} \int_{-\infty}^{\infty} \sum_{j=1}^{N} \sigma_j^2 \cos(\omega_j \tau) \exp(-i\omega\tau) d\tau \\
&= \frac{1}{2} \sum_{j=1}^{N} \sigma_j^2 [\delta(\omega - \omega_j) + \delta(\omega + \omega_j)] \\
&= \sum_{j=1}^{N} [\Phi_{XX}(\omega_j)\delta(\omega - \omega_j) + \Phi_{XX}(-\omega_j)\delta(\omega + \omega_j)] \Delta\omega.
\end{aligned}
$$

$$(4.6.19)$$

In deriving (4.6.19), the following identity is used

$$\int_{-\infty}^{\infty} \exp(ix\tau)d\tau = 2\pi\delta(x).$$ (4.6.20)

When $N \to \infty$ and $\Delta\omega \to 0$, the summation in (4.6.19) becomes integration, leading to

$$\lim_{N\to\infty} \Phi_{X_1 X_1}(\omega) = \Phi_{XX}(\omega).$$ (4.6.21)

Therefore, the spectral density of the simulation equation (4.6.17) is approaching that of simulated process $X(t)$ when the length of interval $\Delta\omega$ is approaching zero.

The probability distribution of $X_1(t)$ is Gaussian since both U_j and V_j are Gaussian and $X_1(t)$ is a linear transformation of them. The mean and mean square are

$$E[X_1(t)] = 0, \quad E[X_1^2(t)] = \sum_{j=1}^{N} \sigma_j^2 = \sum_{j=1}^{N} 2\Phi_{XX}(\omega_j)\Delta\omega.$$ (4.6.22)

When $\Delta\omega \to 0$,

$$E[X_1^2(t)] = \sum_{j=1}^{N} 2\Phi_{XX}(\omega_j)\Delta\omega \to \int_{-\infty}^{\infty} \Phi_{XX}(\omega)d\omega.$$ (4.6.23)

For the spectral density of the simulation representation (4.6.17) to be the given $\Phi_{XX}(\omega)$ approximately, a large N and a small $\Delta\omega$ are required depending on the system under investigation. In general, $\Delta\omega$ should be less than $1/20$ of the lowest natural circular frequency of the system if it is an oscillatory system.

Random amplitude and phase representation. A sample function of $X(t)$ can be also generated from

$$X_2(t) = \sum_{j=1}^{N} A_j \cos(\omega_j t + \theta_j),$$ (4.6.24)

where A_j are random variables with identical Rayleigh distribution, i.e.,

$$p_{A_j}(a_j) = \frac{a_j}{\sigma_j^2} \exp\left(-\frac{a_j^2}{2\sigma_j^2}\right), \tag{4.6.25}$$

θ_j are random variables obeying an identically uniform distribution in $[0, 2\pi)$, A_j and θ_j are mutually independent, and they are also independent for different j. Since the mean and mean square of the Rayleigh distribution are $\sigma_j\sqrt{\pi/2}$ and $2\sigma_j^2$, respectively, the correlation function of $X_2(t)$ is obtained as

$$R_{X_2X_2}(\tau) = \sum_{j=1}^{N} \sigma_j^2 \cos(\omega_j\tau), \tag{4.6.26}$$

which is the same as that of $X_1(t)$ given in (4.6.18). Thus, the spectral density of $X_2(t)$ has the same behavior as that of $X_1(t)$, i.e., it approaches the given spectral density $\Phi_{XX}(\omega)$ as $N \to \infty$ and $\Delta\omega \to 0$.

To prove that the probability distribution of $X_2(t)$ is also Gaussian, consider one term in the summation of (4.6.24) and denote

$$Y = A\cos\Phi, \quad Z = A\sin\Phi, \tag{4.6.27}$$

where Y, Z, A, Φ are random variables, A obeys the Rayleigh distribution (4.6.25), Φ is uniformly distributed in a 2π interval, and A and Φ are independent. According to (2.7.12),

$$p_{YZ}(y, z) = p_{A\Phi}(a, \varphi)|J| = \frac{a}{2\pi\sigma^2} \exp\left(-\frac{a^2}{2\sigma^2}\right)|J|, \tag{4.6.28}$$

where

$$J = \begin{vmatrix} \dfrac{\partial a}{\partial y} & \dfrac{\partial a}{\partial z} \\[2mm] \dfrac{\partial \varphi}{\partial y} & \dfrac{\partial \varphi}{\partial z} \end{vmatrix} = \frac{1}{a}. \tag{4.6.29}$$

Substituting (4.6.29) into (4.6.28), and letting $a^2 = y^2 + z^2$, we have

$$p_{YZ}(y, z) = \frac{1}{2\pi\sigma^2} \exp\left(-\frac{y^2 + z^2}{2\sigma^2}\right). \tag{4.6.30}$$

Equation (4.6.30) shows that Y and Z are independent, and each follows $N(0, \sigma)$ distribution. Since each term in the summation of (4.6.24) is Gaussian distributed, $X_2(t)$ is also Gaussian distributed with a zero mean and the same mean square as that of $X_1(t)$, given in (4.6.22) and (4.6.23).

Random phase representation. Alternatively, a sample function of $X(t)$ can be generated from

$$X_3(t) = \sum_{j=1}^{N} \sqrt{2}\sigma_j \cos(\omega_j t + \theta_j), \qquad (4.6.31)$$

where θ_j are independent random variables uniformly distributed in $[0, 2\pi)$. Taking into consideration that only θ_j are random, the correlation function of $X_3(t)$, and hence the spectral density can be calculated. It turns out that they are the same as those of $X_1(t)$ and $X_2(t)$ (see Exercise Problem 4.24).

Each term in the summation of (4.6.31), i.e., $\cos(\omega_j t + \theta_j)$, represents a random variable. For different j, these random variables are independent and have identical probability distribution. According to the central limit theorem, the probability distribution of $X_3(t)$ converges to a normal distribution as the number of terms, N, approaches infinity.

Simulation of a stochastic process with a non-Gaussian distribution and a given spectral density is much more complicated (Winterstein, 1988; Grigoriu, 1998; Deodatis and Micaletti, 2001).

4.6.4 *Simulation of Randomized Harmonic Process*

As shown above, the randomized harmonic process modeled in (4.5.52) is a stationary process due to the existence of the random variable U. But it is this random variable U that renders the process not ergodic (see Exercise Problem 4.21), and a large number of samples must be generated in Monte Carlo simulation. If the system under investigation is complex with many degrees of freedom, the computational time for the simulation may be prohibitively long. To reduce the computational time, it is advantageous to use the equivalent representation (4.5.63) and (4.5.64) of Ito stochastic differential equations. An alternative way is to transform the Ito equations (4.5.63) and (4.5.64) into the Stratonovich stochastic differential equations taking into account the Wong–Zakai correction. i.e., to use (4.2.70),

$$\dot{X}_1 = \omega_0 X_2 + X_2 W(t), \quad \dot{X}_2 = -\omega_0 X_1 - X_1 W(t), \qquad (4.6.32)$$

where $W(t)$ is a Gaussian white noise with a spectral density $K = \sigma^2/2\pi$. It can be shown that the stochastic process $X_1(t)$ modeled in (4.5.63) and (4.5.64) or in (4.6.32) is equivalent to the randomized harmonic process $X(t)$ in (4.5.52) with the same probability density and the spectral density. The advantage of using $X_1(t)$ is that it is ergodic, and only one sample is needed for simulation, which reduces the computational time significantly.

4.6.5 *Simulation of Bounded Processes Generated from First-Order Nonlinear Filters*

Consider the bounded processes generated from the first-order nonlinear filters (4.5.15) of Example 4.5.3,

$$dX = -\alpha X dt + \sqrt{\frac{\alpha}{\delta + 1}(\Delta^2 - X^2)}dB(t). \qquad (4.6.33)$$

It is not suitable for carrying our Monte Carlo simulation directly since $X(t)$ may exceed its boundaries $\pm\Delta$ during the numerical calculations. To overcome the difficulty, make the transformation

$$X(t) = \Delta \sin \varphi(t) \qquad (4.6.34)$$

and obtain

$$\frac{d\varphi}{dX} = \frac{1}{\Delta \cos \varphi}, \quad \frac{d^2\varphi}{dX^2} = \frac{\sin \varphi}{\Delta^2 \cos^3 \varphi}. \qquad (4.6.35)$$

Applying the Ito differential rule (4.2.58) and using (4.6.33), we obtain an Ito equation for the new variable φ

$$d\varphi = -\frac{2\delta + 1}{2(\delta + 1)}\alpha \tan \varphi dt + \sqrt{\frac{\alpha}{\delta + 1}}sgn(\cos \varphi)dB(t), \qquad (4.6.36)$$

where sgn$(\cdot)$ denotes the sign function.

The Ito equation (4.6.33) is equivalent to a stochastic differential equation in Stratonovich sense

$$\dot{X} = -\frac{2\delta + 1}{2(\delta + 1)}\alpha X dt + \sqrt{(\Delta^2 - X^2)}W(t), \qquad (4.6.37)$$

where $W(t)$ is a Gaussian white noise with a spectral density $\alpha/2\pi(\delta + 1)$. Then we have from (4.6.37)

$$\dot{\varphi} = -\frac{2\delta + 1}{2(\delta + 1)}\alpha \tan \varphi dt + sgn(\cos \varphi)W(t). \qquad (4.6.38)$$

Either (4.6.36) or (4.6.38) can be used for simulation conveniently and effectively.

4.6.6 *Simulation of Bounded Processes Generated from Second-Order Nonlinear Filters*

For the bounded processes generated from the second-order nonlinear filter in Example 4.5.6 of Section 4.5.2, the system is

$$dX_1 = X_2 dt$$

$$dX_2 = (-\omega_0^2 X_1 - 2\zeta\omega_0 X_2)dt + \sqrt{\frac{4\zeta\omega_0^3}{2\delta + 1}\left(\Delta^2 - X_1^2 - \frac{1}{\omega_0^2}X_2^2\right)}dB(t),$$
$$(4.6.39)$$

consider the transformations

$$X_1 = \Delta \sin \varphi \cos \theta, \quad X_2 = -\Delta\omega_0 \sin \varphi \sin \theta \qquad (4.6.40)$$

The following partial derivatives can be obtained from (4.6.39) and (4.6.40),

$$\frac{\partial \varphi}{\partial X_1} = \frac{\cos \theta}{\Delta \cos \varphi}, \quad \frac{\partial \varphi}{\partial X_2} = -\frac{\sin \theta}{\Delta\omega_0 \cos \varphi},$$

$$\frac{\partial \theta}{\partial X_1} = -\frac{\sin \theta}{\Delta \sin \varphi}, \quad \frac{\partial \theta}{\partial X_2} = -\frac{\cos \theta}{\Delta\omega_0 \sin \varphi},$$

$$\frac{\partial^2 \varphi}{\partial X_2^2} = \frac{1}{\Delta^2 \omega_0^2} \left(\frac{\cos^2 \theta}{\sin \varphi \cos \varphi} + \frac{\sin \varphi \sin^2 \theta}{\cos^3 \varphi} \right),$$

$$\frac{\partial^2 \theta}{\partial X_2^2} = -\frac{2 \sin \theta \cos \theta}{\Delta^2 \omega_0^2 \sin^2 \varphi}. \tag{4.6.41}$$

The Ito differential equations for the new processes $\varphi(t)$ and $\theta(t)$ can be derived using the Ito differential rule

$$d\varphi = [-(2\zeta\omega_0 - h) \tan \varphi \sin^2 \theta + h \cot \varphi \cos^2 \theta]dt$$
$$- \sqrt{2h} \, sgn(\cos \varphi) \sin \theta \, dB(t), \tag{4.6.42}$$

$$d\theta = [\omega_0 - 2\zeta\omega_0 \sin \theta \cos \theta - 2h \cot^2 \varphi \sin \theta \cos \theta]dt$$
$$- \sqrt{2h} \frac{|\cos \varphi|}{\sin \varphi} \cos \theta \, dB(t), \tag{4.6.43}$$

where $h = 2\zeta\omega_0/(2\delta + 1)$. On the other hand, taking into account the Wong–Zakai correction terms, the two Ito equations in (4.6.39) are equivalent to the following two Stratonovich stochastic differential equations

$$\dot{X}_1 = X_2$$

$$\dot{X}_2 = -\omega_0^2 X_1 - (2\zeta\omega_0 - h)X_2 + \sqrt{\Delta^2 - X_1^2 - \frac{1}{\omega_0^2} X_2^2} W(t),$$

$$\tag{4.6.44}$$

where $W(t)$ is a Gaussian white noise with a spectral density $K = \omega_0^2 h/\pi$. The corresponding equations for the new variables are

$$\dot{\varphi} = -(2\zeta\omega_0 - h) \tan \varphi \sin^2 \theta - \frac{1}{\omega_0} sgn(\cos \varphi) \sin \theta W(t), \quad (4.6.45)$$

$$\dot{\theta} = \omega_0 - (2\zeta\omega_0 - h) \sin \theta \cos \theta - \frac{|\cos \varphi|}{\omega_0 \sin \varphi} \cos \theta W(t). \tag{4.6.46}$$

Either the set of Ito equations (4.6.42) and (4.6.43) or the set of Stratonovich equations (4.6.45) and (4.6.46) can be used for simulation.

Exercise Problems

4.1 Show that the process $X(t)$ in Problem 3.30 with the following correlation function

$$R_{XX}(\tau) = \begin{cases} \lambda E[Y^2]\frac{\Delta-\tau}{\Delta^2}, & \tau \le \Delta \\ 0, & \tau > \Delta \end{cases},$$

approaches a white noise as Δ tends to zero.

4.2 Let $B(t)$ be a Wiener process with a correlation function

$$E[B(t_1)B(t_2)] = \sigma^2 \min(t_1, t_2).$$

Consider its mean-square derivative process, $\dot{B}(t) = dB(t)/dt$. Show that the correlation function of $\dot{B}(t)$ is given by

$$E[\dot{B}(t_1)\dot{B}(t_2)] = \sigma^2 \delta(t_1 - t_2).$$

4.3 Let $B(t)$ be a unit Wiener process defined on $[a, b]$. Show that the integral $I = \int_a^b B(t)dB(t)$ may achieve the following three different results:

(a) $I_1 = \displaystyle\int_a^b B(t)dB(t) = \lim_{\substack{n \to \infty \\ \Delta_n \to 0}} \sum_{j=1}^n B(t_j)[B(t_{j+1}) - B(t_j)]$

$$= \frac{1}{2}[B^2(b) - B^2(a)] - \frac{1}{2}(b - a),$$

(b) $I_2 = \displaystyle\int_a^b B(t)dB(t) = \lim_{\substack{n \to \infty \\ \Delta_n \to 0}} \sum_{j=1}^n B(t_{j+1})[B(t_{j+1}) - B(t_j)]$

$$= \frac{1}{2}[B^2(b) - B^2(a)] + \frac{1}{2}(b - a)$$

(c) $I_3 = \displaystyle\int_a^b B(t)dB(t) = \lim_{\substack{n \to \infty \\ \Delta_n \to 0}} \sum_{j=1}^n B[rt_{j+1} + (1 - r)t_j]$

$$\times [B(t_{j+1}) - B(t_j)]$$

$$= \frac{1}{2}[B^2(b) - B^2(a)] + \left(r - \frac{1}{2}\right)(b - a),$$

$$0 \le r \le 1,$$

where $a = t_1 < t_2 < \cdots < t_n < t_{n+1} = b$, and $\Delta_n = \max(\tau_{j+1} - \tau_j)$.

4.4 Random process $X(t)$ is defined as

$$X(t) = A\sin[\omega_0 t + \sigma B(t)],$$

where A, ω_0 and σ are positive constants, and $B(t)$ is a unit Wiener process. Show that $X(t)$ is not a stationary process.

4.5 The well-known Ornstein–Uhlenbeck process $X(t)$ is a stationary Markov diffusion process with the probability density function

$$p(x, t) = \frac{1}{\sqrt{2\pi}\sigma} e^{-x^2/(2\sigma^2)}$$

and transition probability density function

$$p(x_2, t_2 | x_1, t_1) = \frac{1}{\sqrt{2\pi(1 - e^{-2\alpha\tau})}\sigma} \exp\left[-\frac{(x_2 - x_1 e^{-\alpha\tau})^2}{2\sigma^2(1 - e^{-2\alpha\tau})}\right],$$

$$\tau = t_2 - t_1.$$

Determine the drift and diffusion coefficients of $X(t)$, and write the Ito equation and FPK equation.

4.6 The Ornstein–Uhlenbeck process $X(t)$ is governed by the Ito differential equation

$$dX = -\alpha X dt + \sqrt{2\alpha}\sigma dB(t).$$

Assuming that $X(0) = x_0$, determine the mean and variance by using the Ito differential rule.

4.7 The equation of motion for an angular process $\Theta(t)$ is given by

$$\dot{\Theta} + P[1 + W_1(t)]\Theta = PW_2(t).$$

Find the Ito differential equation for $\Theta(t)$.

4.8 The Ito differential equation of the phase process $\Theta(t)$ is

$$d\Theta = \omega_0 dt + \sigma dB(t).$$

Use the Ito differential rule to derive the Ito equations for $X_1(t)$ and $X_2(t)$ defined as

$$X_1(t) = A \sin \Theta(t), \qquad X_2(t) = A \cos \Theta(t),$$

where A is a constant.

4.9 The Ito equation for a stochastic process $X(t)$ is given by

$$dX = -\alpha X dt + \sqrt{\frac{\alpha}{\delta + 1}(\Delta^2 - X^2)} dB(t).$$

Use the Ito differential rule to derive the Ito equations for the phase process $\varphi(t)$, defined by

$$X(t) = \Delta \sin \varphi(t).$$

4.10 Derive the Stratonovich type equations from the Ito equations.

(a) $dX = -\alpha \left(X - \dfrac{1}{\lambda} \right) dt + \sqrt{\dfrac{2\alpha}{\lambda} X} dB(t), \quad \lambda > 0, \quad X \geq 0.$

(b) $dX = -\alpha \left(X - \dfrac{2}{\gamma} \right) dt + \sqrt{\dfrac{2\alpha}{\gamma} X} dB(t), \quad \gamma > 0, \quad X \geq 0.$

(c) $dX = -\alpha X dt$

$$+ \sqrt{\frac{\alpha}{\gamma(\sigma - 1)}(1 + \gamma X^2)} dB(t), \quad \gamma > 0, \quad \sigma > 1.$$

4.11 The Markov diffusion vector process $[X_1(t), X_2(t)]$ is governed by the Ito differential equations

$$dX_1 = X_2 dt$$

$$dX_2 = -\omega_0^2 - h(X_1, X_2)dt + g(X_1, X_2)dB(t).$$

Derive the Ito equations for amplitude process $A(t)$ and phase process $\Theta(t)$ defined as follows

$$X_1 = A \cos \Theta, X_2 = -A\omega_0 \sin \Theta.$$

4.12 Derive the Ito equations for energy process $\Lambda(t)$ and phase process $\Theta(t)$ defined as follows

$$X_1 = \frac{\sqrt{2\Lambda}}{\omega_0} \cos \Theta, X_2 = -\sqrt{2\Lambda} \sin \Theta,$$

where $[X_1(t), X_2(t)]$ is the same Markov vector process as defined in Problem 4.11.

4.13 The amplitude process $A(t)$ in Problem 4.11 and the energy process $\Lambda(t)$ in Problem 4.12 have a relationship

$$\Lambda = \frac{1}{2}\omega_0^2 A^2.$$

Use the Ito differential rule to obtain the Ito equation for $\Lambda(t)$ directly from that for $A(t)$ found in Problem 4.11.

4.14 A equation of motion is given by

$$\dot{\varphi} = -\frac{2\delta + 1}{2(\delta + 1)}\alpha \tan \varphi + sgn(\cos \varphi)W(t),$$

where $sgn(\cdot)$ denotes the sign function, $\delta > -1$ and $W(t)$ is a Gaussian white noise with a spectral density $K = \alpha/2\pi(\delta+1)$. Find the corresponding Ito differential equation.

4.15 Consider the following system

$$\dot{X}_1 = a_{11}X_1 + a_{12}X_2 + W_1(t),$$
$$\dot{X}_2 = a_{21}X_1 + a_{22}X_2 + W_2(t),$$

where $W_1(t)$ and $W_2(t)$ are Gaussian white noises with the following correlations

$$E[W_l(t)W_s(t + \tau)] = 2\pi K_{ls}\delta(\tau), \quad l, s = 1, 2.$$

(a) Derive the FPK equation for the system.
(b) Write the corresponding Ito equations.

4.16 For the system in Problem 4.15, find the corresponding Ito equations in the following form

$$dX_1 = (a_{11}X_1 + a_{12}X_2)dt + \sigma_{11}dB_1(t) + \sigma_{12}dB_2(t),$$
$$dX_2 = (a_{21}X_1 + a_{22}X_2)dt + \sigma_{21}dB_1(t) + \sigma_{22}dB_2(t),$$

where $B_1(t)$ and $B_2(t)$ are two unit independent Wiener processes.

4.17 A stochastic model of the predator–prey type ecosystem is given by

$$\dot{X}_1 = X_1 \left[a - bX_2 - \frac{s}{f}(-c + fX_1) + W_1(t) \right],$$

$$\dot{X}_2 = X_2[-c + fX_1 + W_2(t)],$$

where X_1 and X_2 are the population densities of preys and predators, respectively, a, b, c and f are positive constants, and $W_1(t)$ and $W_2(t)$ are two independent Gaussian white noises. Derive the Ito equations for X_1 and X_2, and function $R(X_1, X_2)$ defined as

$$R(X_1, X_2) = fX_1 - c - c\ln\frac{fX_1}{c} + bX_2 - a - a\ln\frac{bX_2}{a}.$$

4.18 For Problem 4.17, let $Y_1 = \ln X_1$ and $Y_1 = \ln X_1$. Obtain the stochastic differential equations for Y_1 and Y_2 of both Stratonovich type and Ito type.

4.19 Find the stationary probability density function of stochastic process $X(t)$ for each case of Problem 4.10, i.e.,

(a) $dX = -\alpha\left(X - \frac{1}{\lambda}\right) dt + \sqrt{\frac{2\alpha}{\lambda}} X dB(t), \quad \lambda > 0, \quad X \geq 0.$

(b) $dX = -\alpha\left(X - \frac{2}{\gamma}\right) dt + \sqrt{\frac{2\alpha}{\gamma}} X dB(t), \quad \gamma > 0, \quad X \geq 0.$

(c) $dX = -\alpha X dt$

$$+ \sqrt{\frac{\alpha}{\gamma(\sigma - 1)}}(1 + \gamma X^2) dB(t), \quad \gamma > 0, \quad \sigma > 1.$$

4.20 A one-dimensional diffusion process $X(t)$, defined in $[0, \infty)$, is governed by the Ito equation

$$dX = \left[\left(-\zeta\omega_0 + \frac{3\pi}{8}\omega_0^2 K_1\right) X + \frac{\pi K_2}{2\omega_0^2 X} - \delta X^3\right] dt$$

$$+ \sqrt{\frac{\pi}{4}\omega_0^2 K_1 X^2 + \frac{\pi K_2}{\omega_0^2}} \, dB(t),$$

where $\zeta, \delta > 0$. Identify the left boundary at $x = 0$ and the right boundary at $x = \infty$ for the cases of (a) $K_1 = 0$ and $K_2 > 0$, (b) $K_1 > 0$ and $K_2 = 0$ and (c) $K_1 > 0$ and $K_2 > 0$. Find the stationary probability density, if it exists.

4.21 Determine the ergodicity of the randomized harmonic process $X(t)$ given by

$$X(t) = A\sin[\omega_0 t + \sigma B(t) + U],$$

where A, ω_0 and σ are positive constants, $B(t)$ is a unit Wiener process, and U is a random variable uniformly distributed in $[0, 2\pi)$ and independent of $B(t)$.

4.22 In Section 4.5.2, the following integral transformation is defined

$$\bar{\Phi}_{ij}(\omega) = \Im\left[R_{ij}(\tau)\right] = \frac{1}{\pi}\int_{-\infty}^{\infty} R_{X_i X_j}(\tau)\, e^{-i\omega\tau}\, d\tau.$$

Show that the spectral density functions can be determined from

$$\Phi_{X_i X_i}(\omega) = \text{Re}[\bar{\Phi}_{ii}(\omega)]$$

$$\Phi_{X_i X_j}(\omega) = \frac{1}{2}[\bar{\Phi}_{ij}(\omega) + \bar{\Phi}_{ij}^*(\omega)].$$

4.23 Use the method in Section 4.5, find the correlation function and the spectral density function for the Ornstein–Uhlenbeck process $X(t)$, governed by the Ito differential equation

$$dX = -\alpha X dt + \sqrt{2\alpha}\sigma dB(t).$$

4.24 Calculate the correlation function and the spectral density of the process $X_3(t)$ defined by (4.6.31), i.e.,

$$X_3(t) = \sum_{j=1}^{N} \sqrt{2}\sigma_j \cos(\omega_j t + \theta_j),$$

where θ_j are independent random variables uniformly distributed in $[0, 2\pi)$, and σ_j and ω_j are given in Eq. (4.6.16).

4.25 A stochastic process has a low-pass spectral density given by Eq. (4.5.5), and one of the following probability distributions

(a) $p(x) = Cx^2 \exp(-ax^2)$, $a > 0$.
(b) $p(x) = Cx \exp(-\gamma x)$, $\gamma > 0$, $x \geq 0$.
(c) $p(x) = C \exp(-ax^2 - bx^4)$, $b > 0$.
(d) $p(x) = C(1 + \gamma x^2)^{-\sigma}$, $-\infty < x < \infty$, $\gamma > 0$.

For each case, derive the nonlinear filter in terms of an Ito equation to generate $X(t)$.

4.26 Simulate the Gaussian white noise process for the cases of $\Delta t = 0.005$, 0.01 and 0.02. Calculate and draw the respective probability density functions and power spectral density functions.

4.27 Use three different algorithms given in Eqs. (4.6.17), (4.6.18) and (4.6.30) to simulate a Gaussian process with a power spectral density given by

$$\Phi_{XX}(\omega) = \frac{\alpha \sigma_X^2}{\pi(\omega^2 + \alpha^2)}.$$

Calculate the probability density functions and power spectral density functions for the three algorithms.

4.28 Simulate the Wiener process $B(t)$ using the relationship (4.2.43), i.e.,

$$\frac{dB(t)}{dt} = W(t).$$

Calculate the transient probability density, and compare with that given in Eq. (4.2.33).

4.29 Simulate the stochastic process governed by

$$dX = -\alpha X dt + \sqrt{\frac{\alpha}{\delta+1}(\Delta^2 - X^2)}dB(t).$$

For $\delta = -0.5$, 0, 0.5, 1 and 2, calculate the respective probability density functions and power spectral density functions.

4.30 Simulate stochastic process $X(t)$ of an exponential distribution (2.5.10) using both the Stratonovich and Ito equations, derived in Problem 4.10(a). Calculate the probability density functions.

4.31 Generate a sample of the randomized harmonic process from

$$X(t) = A\sin[\omega_0 t + \sigma B(t) + U].$$

Show that the process is not ergodic in correlation.

4.32 Generate a sample function and calculate the stationary correlation function of the randomized harmonic process $X(t)$ from (a) the equations in (4.5.62), i.e.,

$$X_1(t) = X(t) = A\sin\Theta(t),$$
$$X_2(t) = A\cos\Theta(t), \quad d\Theta = \omega_0 dt + \sigma dB(t)$$

and (b) the equations in Eq. (4.6.32), i.e.,

$$\dot{X}_1 = \omega_0 X_2 + X_2 W(t), \quad \dot{X}_2 = -\omega_0 X_1 - X_1 W(t).$$

Compare the results in (a) and (b).

CHAPTER 5

RESPONSES OF LINEAR SYSTEMS TO STOCHASTIC EXCITATIONS

Based on the fundamental knowledge of random variables and stochastic processes, we are able to explore its applications to stochastic dynamical systems encountered in various areas, such as engineering, biology, economy, ecology, physiology, etc. For many practical problems, the dynamical system can be modeled as a linear system, or approximated as a linear system when the system motion is small and the system properties are still in the linear range. For a deterministic linear system, the response can be obtained exactly, either analytically or numerically, if the system properties and the excitations are known. Since responses of a stochastically excited system are stochastic processes, only their probabilistic and/or statistical properties can be acquired. It is noted that a stochastic process is a collective of sample functions, and the system becomes deterministic for each sample function. Therefore, a brief review of certain theories of deterministic linear systems is given first, and then the methods to obtain probabilistic and statistical properties of responses of stochastically excited linear systems are introduced and described.

5.1 Review of Deterministic Theory of Linear Systems

Consider the following linear ordinary differential equation

$$m\ddot{x} + c\dot{x} + kx = f(t), \qquad (5.1.1)$$

which is a typical mass-spring-damper mechanical system. Equation (5.1.1) can describe the dynamics of various single-degree-of-freedom (SDOF) linear systems, such as a resistance-inductor-capacitor electric circuit. Equation (5.1.1) can be written as a standard form

$$\ddot{x} + 2\zeta\omega_0\dot{x} + \omega_0^2 x = \frac{1}{m}f(t), \tag{5.1.2}$$

where

$$\omega_0 = \sqrt{\frac{k}{m}}, \quad \zeta = \frac{c}{2\omega_0 m} = \frac{c}{2\sqrt{km}}. \tag{5.1.3}$$

The system parameter ω_0 and ζ are known as the undamped natural frequency and damping ratio, respectively. In the present, we only consider the case of $\zeta < 1$, i.e., the underdamped system, which is true for most engineering problems. To investigate the response $x(t)$ of the system to a general force $f(t)$, it is useful to find out the characteristics of the system response to two typical forces: a sinusoidal force and an impulsive force.

For a deterministic multi-degree-of-freedom (MDOF) linear system, governed by

$$\mathbf{M}\ddot{\mathbf{x}} + \mathbf{C}\dot{\mathbf{x}} + \mathbf{K}\mathbf{x} = \mathbf{f}(t), \tag{5.1.4}$$

where $\mathbf{x} = [x_1(t), x_2(t), \ldots, x_n(t)]^T$ is the response vector, representing a set of independent physical quantities, $\mathbf{f}(t) = [f_1(t), f_2(t), \ldots, f_n(t)]^T$ is a force vector, and $\mathbf{M}$, $\mathbf{C}$ and $\mathbf{K}$ are the mass matrix, damping matrix and stiffness matrix respectively. It is known by the nature that the mass matrix $\mathbf{M}$ is symmetric and positive definite, the stiffness matrix $\mathbf{K}$ is symmetric and positive definite if the rigid-body motion is removed, and the damping matrix $\mathbf{C}$ is also positive definite. For a MDOF system, it is also desired to acquire the characteristics of the system response if all components of $\mathbf{f}(t)$ are sinusoidal forces and if all component are impulsive forces, respectively.

5.1.1 *Frequency Response Function*

Consider a sinusoidal force of a unit amplitude

$$f(t) = e^{i\omega t}. \tag{5.1.5}$$

The solution of system (5.1.1) consists of two parts, the complementary solution of the corresponding homogeneous equation and the particular solution, as given below

$$x(t) = Ce^{-\zeta\omega_0 t + i\omega_d t} + H(\omega)e^{i\omega t}, \tag{5.1.6}$$

where C is a complex constant determined from the initial conditions, ω_d is known as the damped natural frequency, given by

$$\omega_d = \sqrt{1 - \zeta^2}\omega_0 \tag{5.1.7}$$

and

$$H(\omega) = \frac{1}{m(\omega_0^2 - \omega^2 + 2i\zeta\omega_0\omega)}. \tag{5.1.8}$$

The particular solution, i.e., the second term on the right-hand side of (5.1.6), is known as the steady-state response of the system, i.e., the solution as $t \to \infty$. Function $H(\omega)$ is called the frequency response function of system (5.1.1). In a practical problem, only the real part of (5.1.5) represents the actual force, and accordingly, only the real part of (5.1.6) is the system response.

5.1.2 *Impulse Response Function*

Another type of excitation is the unit impulsive force represented by a Dirac delta function,

$$f(t) = \delta(t). \tag{5.1.9}$$

Equation (5.1.9) indicates that the force is very strong, but the acting time is very short, and the total impulse is unit. Assuming that the system is at rest prior to the impulse excitation, then it is described by the equation and initial conditions as follows:

$$\ddot{x} + 2\zeta\omega_0\dot{x} + \omega_0^2 x = \frac{1}{m}\delta(t); \quad x(0) = 0, \dot{x}(0) = 0. \tag{5.1.10}$$

According to the Newton's second law, the impulse of a force is equal to the change in the system momentum. Thus, the net effect of the unit impulsive force $\delta(t)$ is a gain of the system momentum, hence, an initial velocity. Thus, (5.1.10) can be replaced by

$$\ddot{x} + 2\zeta\omega_0\dot{x} + \omega_0^2 x = 0; \quad x(0) = 0, \dot{x}(0) = \frac{1}{m}. \tag{5.1.11}$$

The solution of (5.1.11) is called the impulse response function of system (5.1.1), denoted by $h(t)$, obtained as

$$h(t) = \begin{cases} \dfrac{1}{m\omega_d}e^{-\zeta\omega_0 t}\sin\omega_d t, & t \geq 0 \\ 0, & t < 0. \end{cases} \tag{5.1.12}$$

An arbitrary forcing function $f(t)$ applied from $t = 0$ can be constructed in term of a sequence of impulses, i.e.,

$$f(t) = \int_0^\infty f(\tau)\delta(t - \tau)d\tau. \tag{5.1.13}$$

Since the principle of superposition is applicable to linear systems, the solution of system (5.1.1) initially at rest can then be written as

$$x(t) = \int_0^\infty f(\tau)h(t - \tau)d\tau = \int_0^t f(\tau)h(t - \tau)d\tau. \tag{5.1.14}$$

The change of the integration up-limit from ∞ to t is due to the fact that $h(t) = 0$ for $t < 0$. The integral in (5.1.14) is known as the convolution integral, or the Duhamel integral.

It is noted that (5.1.14) is not a steady-state solution, but the complete solution of system (5.1.1) under zero initial conditions. The complete solution for initial conditions $x(0) = x_0$ and $\dot{x}(0) = \dot{x}_0$ is found to be

$$x(t) = x_i(t) + \int_0^t f(\tau)h(t - \tau)d\tau, \qquad (5.1.15)$$

where the first term $x_i(t)$ in the solution is due to the initial conditions, given by (Sun, 2006)

$$x_i(t) = [2m\zeta\omega_0 h(t) + m\dot{h}(t)]x_0 + mh(t)\dot{x}_0. \qquad (5.1.16)$$

The steady-state solution can be obtained from (5.1.14) or (5.1.15) by letting $t \to \infty$. However, if we assume that the excitation force $f(t)$ begins from the infinite past, i.e., from $t = -\infty$, the steady-state solution can then be obtained from (5.1.15) as

$$x_{\text{steady}}(t) = \int_{-\infty}^{\infty} f(\tau)h(t - \tau)d\tau, \quad t \geq 0. \qquad (5.1.17)$$

5.1.3 *Relationship between the Frequency Response Function and the Impulse Response Function*

Assume an arbitrary excitation force $f(t)$ begins at $t = -\infty$. It can be expressed as a sum of infinitesimal sinusoidal forces with different frequencies, i.e.,

$$f(t) = \int_{-\infty}^{\infty} F(\omega)e^{i\omega t}d\omega, \qquad (5.1.18)$$

where

$$F(\omega) = \frac{1}{2\pi} \int_{-\infty}^{\infty} f(t)e^{-i\omega t}dt, \qquad (5.1.19)$$

is the Fourier transform of $f(t)$, assuming that it exists. Using the principle of superposition and Eqs. (5.1.6) and (5.1.18), we obtain the steady-state solution

$$x_{steady}(t) = \int\limits_{-\infty}^{\infty} H(\omega)F(\omega)e^{i\omega t}d\omega. \qquad (5.1.20)$$

Substitution of (5.1.19) into (5.1.20) leads to

$$x_{steady}(t) = \int\limits_{-\infty}^{\infty} H(\omega)\left[\frac{1}{2\pi}\int\limits_{-\infty}^{\infty} f(\tau)e^{-i\omega\tau}d\tau\right]e^{i\omega t}d\omega. \qquad (5.1.21)$$

Changing the order of the integrations in (5.1.21), we have

$$x_{steady}(t) = \int\limits_{-\infty}^{\infty} f(\tau)\left[\frac{1}{2\pi}\int\limits_{-\infty}^{\infty} H(\omega)e^{i\omega(t-\tau)}d\omega\right]d\tau. \qquad (5.1.22)$$

Comparing (5.1.22) and (5.1.17) and taking into account that $f(t)$ is an arbitrary function, we have

$$h(t) = \frac{1}{2\pi}\int\limits_{-\infty}^{\infty} H(\omega)e^{i\omega t}d\omega. \qquad (5.1.23)$$

Equation (5.1.23) is a Fourier transform, and its inverse is

$$H(\omega) = \int\limits_{-\infty}^{\infty} h(t)e^{-i\omega t}dt. \qquad (5.1.24)$$

Therefore, the impulse response function $h(t)$ and the frequency response function $H(\omega)$ constitute a Fourier transform pair.

5.1.4 *MDOF Systems*

Extension of the deterministic analysis of SDOF linear systems to MDOF linear systems is straightforward. Consider first the frequency response of the MDOF linear system governed by (5.1.4). Let $f_k(t) = e^{i\omega t}$, $f_l(t) = 0$ for $l \neq k$. The steady-state solutions of (5.1.4) can be expressed as

$$\mathbf{X} = [H_{1k}(\omega), H_{2k}(\omega), \ldots, H_{nk}(\omega)]^T e^{i\omega t}, \qquad (5.1.25)$$

where $H_{jk}(\omega)$ can be derived from a set of complex linear algebraic equations obtained by substituting (5.1.25) into (5.1.4). Equation (5.1.25) indicates that each $H_{jk}(\omega)$ is the ratio of the steady-state response of the jth coordinate to the harmonic excitation acting at the kth coordinate. Thus, instead of a single frequency response function for a SDOF linear system, a frequency response matrix is formed for a MDOF linear system as $\mathbf{H}(\omega) = [H_{jk}(\omega)]$. It can be shown that

$$\mathbf{H}(\omega) = (-\omega^2 \mathbf{M} + i\omega \mathbf{C} + \mathbf{K})^{-1}. \qquad (5.1.26)$$

For a single-frequency harmonic force system $\mathbf{f}(t) = \mathbf{F}e^{i\omega t}$, where $\mathbf{F} = [F_1, F_2, \ldots, F_n]^T$, representing the force amplitudes, the system steady-state solutions are

$$x_j(t) = \sum_{k=1}^{n} H_{jk}(\omega) F_k e^{i\omega t}, \quad j = 1, 2, \ldots, n. \qquad (5.1.27)$$

In matrix form,

$$\mathbf{x} = \mathbf{H}(\omega)\mathbf{F}e^{i\omega t}. \qquad (5.1.28)$$

Similarly, the impulse response function is replaced by the impulse response matrix $\mathbf{h}(t) = [h_{jk}(t)]$ for MDOF linear systems. Each element $h_{jk}(t)$ is the impulse response (transient response under zero initial conditions) of the jth coordinate to a unit impulse excitation acting at the kth coordinate, i.e., $f_k(t) = \delta(t)$, $f_l(t) = 0$ for $l \neq k$. The system response to the excitations of general force vector $\mathbf{f}(t)$ under zero initial conditions is now obtained by the Duhamel integral of a matrix form

$$\mathbf{x}(t) = \int_0^t \mathbf{h}(t - \tau)\mathbf{f}(\tau)d\tau. \qquad (5.1.29)$$

It is emphasized again that the system is at rest prior to the action of the excitation, i.e., prior to $t = 0$. For each response component,

we have

$$x_j(t) = \sum_{k=1}^{n} \int_0^t h_{jk}(t-\tau)f_k(\tau)d\tau. \qquad (5.1.30)$$

Analogous to (5.1.23) and (5.1.24), the frequency response matrix and the impulse response matrix are related as follows

$$\mathbf{h}(t) = \frac{1}{2\pi} \int_{-\infty}^{\infty} \mathbf{H}(\omega)e^{i\omega t}d\omega. \qquad (5.1.31)$$

$$\mathbf{H}(\omega) = \int_{-\infty}^{\infty} \mathbf{h}(t)e^{-i\omega t}dt. \qquad (5.1.32)$$

Equations (5.1.31) and (5.1.32) state that the $(j,\,k)$ elements of $\mathbf{h}(t)$ and $\mathbf{H}(\omega)$ are a Fourier transform pair.

Theoretically, above analysis procedure is applicable to any MDOF linear system. But it is tedious to obtain the frequency response matrix or impulse response matrix even for a two- or three-degree-of-freedom system. An alternative method known as the normal mode analysis has been widely used to treat MDOF linear systems, which will be introduced in the next section.

Example 5.1.1 Shown in Fig. 5.1.1 is a two-degree-of-freedom system. The equations of motion are

$$m_1\ddot{x}_1 + (c_1 + c_2)\dot{x}_1 - c_2\dot{x}_2 + (k_1 + k_2)x_1 - k_2x_2 = f_1(t),$$
$$m_2\ddot{x}_2 - c_2\dot{x}_1 + (c_2 + c_3)\dot{x}_2 - k_2x_1 + (k_2 + k_3)x_2 = f_2(t).$$
$$(5.1.33)$$

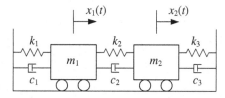

Figure 5.1.1   A two-degree-of-freedom system.

Letting $f_1(t) = e^{i\omega t}$, $f_2(t) = 0$, $x_1 = H_{11}(\omega)e^{i\omega t}$ and $x_2 = H_{21}(\omega)e^{i\omega t}$ in (5.1.33), we obtain

$$H_{11}(\omega) = \frac{1}{\Delta}[-m_2\omega^2 + i(c_2 + c_3)\omega + (k_2 + k_3)], \qquad (5.1.34)$$

$$H_{21}(\omega) = \frac{1}{\Delta}(ic_2\omega + k_2), \qquad (5.1.35)$$

where

$$\Delta = [-m_1\omega^2 + i(c_1 + c_2)\omega + (k_1 + k_2)]$$
$$\times [-m_2\omega^2 + i(c_2 + c_3)\omega + (k_2 + k_3)] - (ic_2\omega + k_2)^2. \qquad (5.1.36)$$

Next, letting $f_1(t) = 0$ and $f_2(t) = e^{i\omega t}$, $x_1 = H_{12}(\omega)e^{i\omega t}$ and $x_2 = H_{22}(\omega)e^{i\omega t}$ in (5.1.33), we obtain

$$H_{12}(\omega) = \frac{1}{\Delta}(ic_2\omega + k_2), \qquad (5.1.37)$$

$$H_{22}(\omega) = \frac{1}{\Delta}[-m_1\omega^2 + i(c_1 + c_2)\omega + (k_1 + k_2)]. \qquad (5.1.38)$$

For the case in which $m_1 = m_2 = m$, $c_1 = c_2 = c_3 = c$ and $k_1 = k_2 = k_3 = k$,

$$H_{11}(\omega) = H_{22}(\omega) = \frac{1}{m\Delta_1}(-\omega^2 + 4i\zeta_0\omega_0\omega + 2\omega_0^2), \qquad (5.1.39)$$

$$H_{12}(\omega) = H_{21}(\omega) = \frac{1}{m\Delta_1}(2i\zeta_0\omega_0\omega + \omega_0^2), \qquad (5.1.40)$$

where

$$\Delta_1 = (-\omega^2 + 2i\zeta_1\omega_1\omega + \omega_1^2)(-\omega^2 + 2i\zeta_2\omega_2\omega + \omega_2^2) \qquad (5.1.41)$$

and

$$\omega_0 = \sqrt{\frac{k}{m}}, \quad \zeta_0 = \frac{c}{2m\omega_0}, \quad \omega_1 = \omega_0, \quad \omega_2 = \sqrt{3}\omega_0,$$

$$\zeta_1 = \zeta_0, \quad \zeta_2 = \sqrt{3}\zeta_0. \qquad (5.1.42)$$

The impulse response functions can be obtained from the Fourier transform,

$$h_{11}(t) = \frac{1}{2\pi} \int_{-\infty}^{\infty} H_{11}(\omega)e^{i\omega t}d\omega = \frac{1}{2\pi m} \int_{-\infty}^{\infty} \frac{2\omega_0^2 - \omega^2 + 4i\zeta\omega_0\omega}{\Delta_1} e^{i\omega t}d\omega.$$

$$(5.1.43)$$

Consider the complex function

$$F(z) = \frac{(2\omega_0^2 - z^2 + 4i\zeta_0\omega_0 z)e^{izt}}{(z^2 - \omega_1^2 - 2i\zeta_1\omega_1 z)(z^2 - \omega_2^2 - 2i\zeta_2\omega_2 z)}, \qquad (5.1.44)$$

which has four poles

$$z_{1,2} = \pm\omega_{d1} + i\zeta_1\omega_1, \quad z_{3,4} = \pm\omega_{d2} + i\zeta_2\omega_2, \qquad (5.1.45)$$

where

$$\omega_{di} = \omega_i\sqrt{1 - \zeta_i^2}, \quad i = 1, 2. \qquad (5.1.46)$$

All four poles are on the upper half-plane. Their residues can be found from

$$\operatorname*{Res}_{z=z_j} F(z) = [(z - z_j)F(z)]_{z\to z_j}. \qquad (5.1.47)$$

According to the residue theory,

$$h_{11}(t) = \frac{1}{m}i\sum_{j=1}^{4}\operatorname{Res}[F(z)]$$

$$= \frac{1}{2m}\left[\frac{1}{\omega_{d1}}e^{-\zeta_1\omega_1 t}\sin(\omega_{d1}t) + \frac{1}{\omega_{d2}}e^{-\zeta_2\omega_2 t}\sin(\omega_{d2}t)\right], \quad t > 0.$$

$$(5.1.48)$$

Since no poles are on the lower half-plane, $h_{11}(t) = 0$ when $t < 0$. Similarly,

$$h_{12}(t) = h_{21}(t)$$

$$= \frac{1}{2m}\left[\frac{1}{\omega_{d1}}e^{-\zeta_1\omega_1 t}\sin(\omega_{d1}t) - \frac{1}{\omega_{d2}}e^{-\zeta_2\omega_2 t}\sin(\omega_{d2}t)\right], \quad t > 0.$$

$$(5.1.49)$$

Another way to obtain the impulse response functions is to use the Laplace transform. Letting $f_1(t) = \delta(t)$ and $f_2(t) = 0$ in (5.1.33), taking Laplace transform, and denoting the Laplace transforms of $x_1(t)$ and $x_2(t)$ as $X_1(s)$ and $X_2(s)$, respectively, we have

$$[ms^2 + (c_1 + c_2)s + (k_1 + k_2)]X_1(s) - (c_2s + k_2)X_2(s) = 1$$
$$-(c_2s + k_2)X_1(s) + [ms^2 + (c_1 + c_2)s + (k_1 + k_2)]X_2(s) = 0.$$
$$(5.1.50)$$

$X_1(s)$ and $X_2(s)$ can then be solved from (5.1.50), and $x_1(t) = h_{11}(t)$ and $x_2(t) = h_{21}(t)$ are obtained through the inverse Laplace transform.

For the special case of $m_1 = m_2 = m$, $c_1 = c_2 = c_3 = c$ and $k_1 = k_2 = k_3 = k$, we obtain

$$X_1(s) = \frac{1}{m\Delta}(s^2 + 4\zeta_0\omega_0 s + 2\omega_0^2), \qquad (5.1.51)$$

$$X_2(s) = \frac{1}{m\Delta}(2\zeta_0\omega_0 s + \omega_0^2), \qquad (5.1.52)$$

where

$$\Delta = (s^2 + 2\zeta_1\omega_1 s + \omega_1^2)(s^2 + 2\zeta_2\omega_2 s + \omega_2^2). \qquad (5.1.53)$$

Using the method of partial-fraction expansion, we assume

$$X_1(s) = \frac{1}{m}\left(\frac{A_1 s + B_1}{s^2 + 2\zeta_1\omega_1 s + \omega_1^2} + \frac{A_2 s + B_2}{s^2 + 6\zeta_2\omega_2 s + 3\omega_2^2}\right). \qquad (5.1.54)$$

Equating (5.1.54) and (5.1.51), we obtain

$$A_1 = A_2 = 0, \quad B_1 = B_2 = \frac{1}{2}. \qquad (5.1.55)$$

Using the following formulas

$$L^{-1}\left[\frac{1}{s^2 + 2\zeta_j\omega_j s + \omega_j^2}\right] = \frac{1}{\omega_{dj}}e^{-\zeta_j\omega_j t}\sin\omega_{dj}t, \qquad (5.1.56)$$

we obtain the impulse response function

$$h_{11}(t) = x_1(t) = L^{-1}[X_1(s)]$$

$$= \frac{1}{2m} \left[\frac{1}{\omega_{d1}} e^{-\zeta_1 \omega_1 t} \sin(\omega_{d1} t) + \frac{1}{\omega_{d2}} e^{-\zeta_2 \omega_2 t} \sin(\omega_{d2} t) \right].$$

$$(5.1.57)$$

By a similar procedure, we obtain

$$h_{21}(t) = x_2(t) = L^{-1}[X_2(s)]$$

$$= \frac{1}{2m} \left[\frac{1}{\omega_{d1}} e^{-\zeta_1 \omega_1 t} \sin(\omega_{d1} t) - \frac{1}{2\omega_{d2}} e^{-\zeta_2 \omega_2 t} \sin(\omega_{d2} t) \right].$$

$$(5.1.58)$$

From the symmetry, we know that $h_{12}(t) = h_{21}(t)$, $h_{11}(t) = h_{22}(t)$.

5.1.5 *Normal Mode Analysis*

A popular method to treat MDOF linear system is the normal mode analysis. Consider the free motion of a MDOF linear system without damping and excitation, governed by the following linear homogeneous equations and initial conditions

$$\mathbf{M\ddot{x} + Kx = 0}; \quad \mathbf{x}(0) = \mathbf{x_0}, \dot{\mathbf{x}}(0) = \dot{\mathbf{x}}_0. \qquad (5.1.59)$$

Assume that the solution has the form

$$\mathbf{x}(t) = \mathbf{A} \cos(\omega t + \phi). \qquad (5.1.60)$$

Substitution of (5.1.60) into the equation of motion in (5.1.59) leads to

$$(\mathbf{K} - \omega^2 \mathbf{M})\mathbf{A} = \mathbf{0}. \qquad (5.1.61)$$

For a non-trivial solution $\mathbf{A} \neq \mathbf{0}$, the determinant of the coefficient matrix must be zero, i.e.,

$$|\mathbf{K} - \omega^2 \mathbf{M}| = 0. \qquad (5.1.62)$$

Equation (5.1.62) is known as the characteristic equation. It can be shown that all solutions (roots) ω_j^2 of (5.1.62) are positive due to the symmetric and positive-definite nature of $\mathbf{M}$ and $\mathbf{K}$. All ω_j are called the natural frequencies of the system. Arrange ω_j in ascending order, $\omega_1 \leq \omega_2 \leq \cdots \leq \omega_n$, and the lowest one ω_1 is called the fundamental frequency.

For each ω_j, a normalized vector $\mathbf{A}_j$ can be found from (5.1.61), which is called the natural mode or normal mode. The natural modes are orthogonal with the mass matrix and stiffness matrix in the sense of

$$\mathbf{A}_j^T \mathbf{M} \mathbf{A}_k = \delta_{jk} = \begin{cases} 1, & j = k \\ 0, & j \neq k \end{cases}, \quad \mathbf{A}_j^T \mathbf{K} \mathbf{A}_k = \omega_j^2 \delta_{jk} = \begin{cases} \omega_j^2, & j = k \\ 0, & j \neq k. \end{cases}$$
(5.1.63)

Although the solution $\mathbf{A}_j$ of (5.1.61) corresponding to each ω_j is not unique, the orthogonal property (5.1.63) establishes the normalization condition for $\mathbf{A}_j$ as

$$\mathbf{A}_j^T \mathbf{M} \mathbf{A}_j = 1.$$
(5.1.64)

The model matrix $\mathbf{\Psi}$ is formed from the natural modes as follows

$$\mathbf{\Psi} = [\mathbf{A}_1, \mathbf{A}_1, \ldots, \mathbf{A}_n].$$
(5.1.65)

Then (5.1.63) can be rewritten in matrix form

$$\mathbf{\Psi}^T \mathbf{M} \mathbf{\Psi} = \mathbf{I}, \quad \mathbf{\Psi}^T \mathbf{K} \mathbf{\Psi} = \mathbf{\Omega},$$
(5.1.66)

where $\mathbf{I}$ is an identity matrix, and $\mathbf{\Omega} = \text{diag}(\omega_j^2)$. In Eq. (5.1.59), letting

$$\mathbf{x} = \mathbf{\Psi} \mathbf{q}$$
(5.1.67)

and pre-multiplying $\mathbf{\Psi}^T$, we obtain

$$\ddot{\mathbf{q}} + \mathbf{\Omega} \mathbf{q} = \mathbf{0}; \quad \mathbf{q}(0) = \mathbf{\Psi}^T \mathbf{M} \mathbf{x}_0, \dot{\mathbf{q}}(0) = \mathbf{\Psi}^T \mathbf{M} \dot{\mathbf{x}}_0,$$
(5.1.68)

where $\mathbf{q} = [q_1(t), q_2(t), \ldots, q_n(t)]^T$ is called the vector of model coordinates. In deriving (5.1.68), use has been made of the first equation in (5.1.66), which implies

$$\mathbf{\Psi}^{-1} = \mathbf{\Psi}^T \mathbf{M} \tag{5.1.69}$$

and then the transformation (5.1.67) leads to

$$\mathbf{q} = \mathbf{\Psi}^{-1} \mathbf{x} = \mathbf{\Psi}^T \mathbf{M} \mathbf{x}. \tag{5.1.70}$$

The equations in (5.1.68) are uncoupled, and each is a SDOF system. With the initial conditions, each $q_j(t)$ can be solved individually, and the original physical quantities $x_j(t)$ can be obtained from the transformation (5.1.67).

Now consider a damped MDOF linear system under external excitations, governed by (5.1.4). Carry out the model transformation to obtain

$$\ddot{\mathbf{q}} + \mathbf{\Psi}^T \mathbf{C} \mathbf{\Psi} + \mathbf{\Omega} \mathbf{q} = \mathbf{g}(t) = \mathbf{\Psi}^T \mathbf{f}(t). \tag{5.1.71}$$

If the damping term $\mathbf{\Psi}^T \mathbf{C} \mathbf{\Psi}$ in (5.1.71) is diagonal, the system damping is called the classical damping. Although it is a special case, $\mathbf{\Psi}^T \mathbf{C} \mathbf{\Psi}$ can be approximated as diagonal in some practical problems. In these cases, the equations in (5.1.71) become uncoupled, and can be written in the scalar form

$$\ddot{q}_j + 2\zeta_j \omega_j \dot{q}_j + \omega_j^2 q_j = g_j(t) \tag{5.1.72}$$

With the initial conditions given in (5.1.68), either the transient solutions or the steady-state solutions can be obtained for the model coordinates $q_j(t)$ first, and then transformed back to the original coordinates $x_j(t)$.

Consider the case of zero initial conditions, the solution for each model coordinate q_j is

$$q_j(t) = \int_0^\infty g_j(\tau) h_{q,j}(t - \tau) d\tau = \int_0^t g_j(\tau) h_{q,j}(t - \tau) d\tau, \tag{5.1.73}$$

where $h_{q,j}(t)$ is the jth model impulse response function

$$h_{q,j}(t) = \begin{cases} \dfrac{1}{\omega_{d,j}} e^{-\zeta_j \omega_j t} \sin \omega_{d,j} t, & t \geq 0 \\ 0, & t < 0, \end{cases} \tag{5.1.74}$$

$$\omega_{d,j} = \sqrt{1 - \zeta_j^2} \, \omega_j. \tag{5.1.75}$$

Equation (5.1.73) can be written as the matrix form

$$\mathbf{q}(t) = \int_0^t \mathbf{h_q}(t - \tau) \mathbf{g}(\tau) d\tau, \tag{5.1.76}$$

where $\mathbf{h_q}(t)$ is a diagonal matrix, called the model impulse response matrix, given by

$$\mathbf{h_q}(t) = \text{diag}[h_{q,j}(t)]. \tag{5.1.77}$$

The solution of the original physical coordinates is then

$$\mathbf{x}(t) = \mathbf{\Psi q}(t) = \int_0^t \mathbf{\Psi h_q}(t - \tau) \mathbf{\Psi}^T \mathbf{f}(\tau) d\tau. \tag{5.1.78}$$

Comparison of (5.1.29) and (5.1.78) reveals the relationship between the impulse response matrices for the model coordinates and for the original coordinates as

$$\mathbf{h}(t) = \mathbf{\Psi h_q}(t) \mathbf{\Psi}^T. \tag{5.1.79}$$

The model frequency response matrix $\mathbf{H_q}(\omega)$ is the Fourier transform of the model impulse response matrix $\mathbf{h_q}(t)$

$$\mathbf{H_q}(\omega) = \int_{-\infty}^{\infty} \mathbf{h_q}(\tau) e^{i\omega\tau} d\tau, \tag{5.1.80}$$

where $\mathbf{H_q}(\omega)$ is also a diagonal matrix with the diagonal elements being the model frequency response functions

$$H_{q,j}(\omega) = \frac{1}{\omega_j^2 - \omega^2 + 2i\zeta_j \omega_j \omega}. \tag{5.1.81}$$

The frequency response functions for the model coordinates and for the original coordinates are related according to

$$\mathbf{H}(\omega) = \int_{-\infty}^{\infty} \mathbf{h}(\tau)e^{i\omega\tau}d\tau = \int_{-\infty}^{\infty} \mathbf{\Psi}\mathbf{h_q}(\tau)\mathbf{\Psi}^T e^{i\omega\tau}d\tau = \mathbf{\Psi}\mathbf{H_q}(\omega)\mathbf{\Psi}^T.$$

$$(5.1.82)$$

It is emphasized again that the above analysis is applicable only for the case in which the damping matrix can be transformed to a diagonal one exactly or approximately.

Example 5.1.2 For the system in Example 5.1.1, the impulse and frequency response functions can also be found using the method of normal modes. The procedure is illustrated below for the special case, $m_1 = m_2 = m$, $c_1 = c_2 = c_3 = c$ and $k_1 = k_2 = k_3 = k$.

The mass, damping and stiffness matrices are obtained from (5.1.33), respectively,

$$\mathbf{M} = \begin{bmatrix} 1 & 0 \\ 0 & 1 \end{bmatrix} m, \quad \mathbf{C} = \begin{bmatrix} 2 & -1 \\ -1 & 2 \end{bmatrix} c, \quad \mathbf{K} = \begin{bmatrix} 2 & -1 \\ -1 & 2 \end{bmatrix} k.$$

$$(5.1.83)$$

Since the damping matrix $\mathbf{C}$ and the stiffness matrix $\mathbf{K}$ have the same form, the system belongs to the category of classical damping. The characteristic equation is

$$|\mathbf{K} - \omega^2\mathbf{M}| = \begin{vmatrix} 2k - \omega^2 m & -k \\ -k & 2k - \omega^2 m \end{vmatrix} = 0. \qquad (5.1.84)$$

From (5.1.84), the two natural frequencies are solved as

$$\omega_1 = \omega_0, \quad \omega_2 = \sqrt{3}\omega_0. \qquad (5.1.85)$$

The normalized model matrix $\mathbf{\Psi}$ can be found from the normal modes as follows

$$\mathbf{\Psi} = \frac{1}{\sqrt{2m}} \begin{bmatrix} 1 & 1 \\ 1 & -1 \end{bmatrix}. \qquad (5.1.86)$$

Then, we have

$$\Psi^T M \Psi = \begin{bmatrix} 1 & 0 \\ 0 & 1 \end{bmatrix}, \quad \Psi^T C \Psi = \frac{c}{m} \begin{bmatrix} 1 & 0 \\ 0 & 3 \end{bmatrix}, \quad \Psi^T K \Psi = \frac{k}{m} \begin{bmatrix} 1 & 0 \\ 0 & 3 \end{bmatrix}.$$

(5.1.87)

In the equations of motion in (5.1.33), letting

$$\begin{Bmatrix} x_1 \\ x_2 \end{Bmatrix} = \Psi \begin{Bmatrix} q_1 \\ q_2 \end{Bmatrix}$$

(5.1.88)

and pre-multiplying Ψ^T, we obtain

$$\ddot{q}_1 + 2\zeta_1 \omega_1 \dot{q}_1 + \omega_1^2 q_1 = g_1(t), \qquad (5.1.89)$$

$$\ddot{q}_2 + 2\zeta_2 \omega_2 \dot{q}_2 + \omega_2^2 q_2 = g_2(t), \qquad (5.1.90)$$

where

$$g_1(t) = \frac{1}{\sqrt{2m}} [f_1(t) + f_2(t)], \quad g_2(t) = \frac{1}{\sqrt{2m}} [f_1(t) - f_2(t)].$$

(5.1.91)

The model impulse response functions are, respectively,

$$h_{q1}(t) = \frac{1}{\omega_{d1}} e^{-\zeta_1 \omega_1 t} \sin \omega_{d1} t, \qquad (5.1.92)$$

$$h_{q2}(t) = \frac{1}{\omega_{d2}} e^{-\zeta_2 \omega_2 t} \sin \omega_{d2} t. \qquad (5.1.93)$$

According to Eq. (5.1.79), we have

$$h_{11}(t) = h_{22}(t) = \frac{1}{2m} [h_{q1}(t) + h_{q2}(t)]$$

$$= \frac{1}{2m} \left[\frac{1}{\omega_{d1}} e^{-\zeta_1 \omega_1 t} \sin(\omega_{d1} t) + \frac{1}{\omega_{d2}} e^{-\zeta_2 \omega_2 t} \sin(\omega_{d2} t) \right].$$

(5.1.94)

$$h_{12}(t) = h_{21}(t) = \frac{1}{2m} [h_{q1}(t) - h_{q2}(t)]$$

$$= \frac{1}{2m} \left[\frac{1}{\omega_{d1}} e^{-\zeta_1 \omega_1 t} \sin(\omega_{d1} t) - \frac{1}{\omega_{d2}} e^{-\zeta_2 \omega_2 t} \sin(\omega_{d2} t) \right].$$

(5.1.95)

As expected, the results are the same as those obtained in Example 5.1.1.

The model frequency response functions corresponding to Eqs. (5.1.89) and (5.1.90) are, respectively,

$$H_{q1}(\omega) = \frac{1}{\omega_1^2 - \omega^2 + 2i\zeta_1\omega_1\omega}, \tag{5.1.96}$$

$$H_{q2}(\omega) = \frac{1}{\omega_2^2 - \omega^2 + 2i\zeta_2\omega_2\omega}. \tag{5.1.97}$$

Then from Eq. (5.1.82),

$$\begin{aligned} H_{11}(\omega) = H_{11}(\omega) &= \frac{1}{2m}[H_{q1}(\omega) + H_{q2}(\omega)] \\ &= \frac{1}{m\Delta_1}(-\omega^2 + 4i\zeta_0\omega_0\omega + 2\omega_0^2), \end{aligned} \tag{5.1.98}$$

$$\begin{aligned} H_{12}(\omega) = H_{21}(\omega) &= \frac{1}{2m}[H_{q1}(\omega) - H_{q2}(\omega)] \\ &= \frac{1}{m\Delta_1}(2i\zeta_0\omega_0\omega + \omega_0^2], \end{aligned} \tag{5.1.99}$$

where Δ_1 is given in (5.1.41). Also, the results in Eqs. (5.1.98) and (5.1.99) are the same as those in Example 5.1.1.

5.2 Response of Stochastically Excited Linear Systems

In a deterministic SDOF linear system, the system parameters, such as the mass m, damping coefficient c and stiffness k in Eq. (5.1.1), are precisely known, the excitation $f(t)$ is a known function of time, and the initial conditions are also given precisely. Then the response of the system to the excitation can be found as a deterministic function, either analytically or numerically. If any of the system parameters, excitation, and initial state is random, the system response will also be random. In many practical engineering problems, the system parameters and the initial state can be reasonably considered to be deterministic, while the excitations may be random in nature. In what follows, we will limit our topics to this case.

We now rewrite Eq. (5.1.1) as a stochastic differential equation by replacing the force $f(t)$ by a stochastic process $F(t)$ and the motion $x(t)$ by its stochastic counterpart $X(t)$,

$$m\ddot{X} + c\dot{X} + kX = F(t), \tag{5.2.1}$$

with initial conditions

$$X(0) = x_0, \quad \dot{X}(0) = \dot{x}_0. \tag{5.2.2}$$

In writing Eq. (5.2.1), it is implied that the derivatives of $X(t)$ exist in a certain sense defined in Section 3.5.3. As mentioned before, all differentiations in this book are in L_2 sense, i.e., in the mean-square sense.

The solution of Eq. (5.2.1) under initial conditions (5.2.2) is obtained from (5.1.14) as

$$X(t) = x_i(t) + \int_0^t F(\tau)h(t - \tau)d\tau, \tag{5.2.3}$$

where

$$x_i(t) = [ch(t) + m\dot{h}(t)]x_0 + mh(t)\dot{x}_0. \tag{5.2.4}$$

Similar to the differentiation in L_2 sense, the stochastic integral in (5.2.3) and also those appearing hereafter in the book are also defined in L_2 sense. Although $x_i(t)$ is important in transient solution, it will be damped out eventually; therefore, we may neglect it by letting $x_0 = 0$ and $\dot{x}_0 = 0$ throughout in the rest of this chapter.

The mean value of the response $X(t)$ is obtained by taking ensemble averaging of (5.2.3) as

$$\mu_X(t) = \int_0^t \mu_F(\tau)h(t - \tau)d\tau = \int_0^t \mu_F(t - \tau)h(\tau)d\tau. \tag{5.2.5}$$

The covariance of $X(t)$ is given by

$$\kappa_{XX}(t_1, t_2) = E\{[X(t_1) - \mu_X(t_1)][X(t_2) - \mu_X(t_2)]\}$$

$$= E\left\{ \int_0^{t_1} [F(\tau_1) - \mu_F(\tau_1)]h(t_1 - \tau_1)d\tau_1 \right.$$

$$\left. \int_0^{t_2} [F(\tau_2) - \mu_F(\tau_2)]h(t_2 - \tau_2)d\tau_2 \right\}$$

$$= \int_0^{t_1}\int_0^{t_2} \kappa_{FF}(\tau_1, \tau_2)h(t_1 - \tau_1)h(t_2 - \tau_2)d\tau_1 d\tau_2$$

$$= \int_0^{t_1}\int_0^{t_2} \kappa_{FF}(t_1 - \tau_1, t_2 - \tau_2)h(\tau_1)h(\tau_2)d\tau_1 d\tau_2.$$

$$(5.2.6)$$

The higher-order moments and cumulants can be obtained in a similar way, as given below

$$E[X(t_1)X(t_2)\cdots X(t_n)]$$

$$= \int_0^{t_1}\int_0^{t_2}\cdots\int_0^{t_n} E[F(\tau_1)F(\tau_2)\cdots F(\tau_n)]h(t_1 - \tau_1)$$

$$\times h(t_2 - \tau_2)\cdots h(t_n - \tau_n)d\tau_1 d\tau_2 \cdots d\tau_n, \qquad (5.2.7)$$

$$\kappa_n[X(t_1)X(t_2)\cdots X(t_n)]$$

$$= \int_0^{t_1}\int_0^{t_2}\cdots\int_0^{t_n} \kappa_n[F(\tau_1)F(\tau_2)\cdots F(\tau_n)]h(t_1 - \tau_1)$$

$$\times h(t_2 - \tau_2)\cdots h(t_n - \tau_n)d\tau_1 d\tau_2 \cdots d\tau_n. \qquad (5.2.8)$$

Equations (5.2.7) and (5.2.8) show that the level of a solution for the response process $X(t)$ depends on the knowledge known from the excitation process $F(t)$. In many cases, only the second-order properties are known for the excitation process, such as the correlation function, then we can only obtain the second-order properties of the response process.

If the excitation $F(t)$ is Gaussian, all its cumulants of orders higher that two are zeros. Thus, (5.2.8) shows the same nature for the response $X(t)$, i.e., $X(t)$ is also Gaussian. In this case, the first two moments, the mean and covariance function specify the first-order properties of the process completely. However, if $F(t)$ is non-Gaussian, the response process is in general also non-Gaussian and cannot be obtained.

The deterministic analysis for MDOF linear deterministic systems will now be applied to stochastically excited systems, governed by

$$\mathbf{M\ddot{X}} + \mathbf{C\dot{X}} + \mathbf{KX} = \mathbf{F}(t). \tag{5.2.9}$$

Assume that the system is at rest initially, i.e., $\mathbf{X}(0) = \mathbf{0}$. Using (5.1.30) and (5.1.29), we have the response mean in scalar form and in vector form, respectively,

$$\mu_{X_j}(t) = \sum_{k=1}^{n} \int_0^t h_{jk}(t - \tau) \mu_{F_k}(\tau) d\tau. \tag{5.2.10}$$

$$\boldsymbol{\mu_x}(t) = \int_0^t \mathbf{h}(t - \tau) \boldsymbol{\mu_F}(\tau) d\tau. \tag{5.2.11}$$

The covariance function is then calculated as

$$\kappa_{X_i X_j}(t_1, t_2) = E\{[X_i(t_1) - \mu_{X_i}(t_1)][X_j(t_2) - \mu_{X_j}(t_2)]\}$$

$$= E \left\{ \sum_{k=1}^{n} \int_0^{t_1} [F_k(\tau_1) - \mu_{F_k}(\tau_1)] h_{ik}(t_1 - \tau_1) d\tau_1 \right.$$

$$\left. \sum_{l=1}^{n} \int_0^{t_2} [F_l(\tau_2) - \mu_{F_l}(\tau_2)] h_{jl}(t_2 - \tau_2) d\tau_2 \right\}$$

$$= \sum_{l,k=1}^{n} \int_0^{t_1} \int_0^{t_2} \kappa_{F_k F_l}(\tau_1, \tau_2) h_{ik}(t_1 - \tau_1) h_{jl}(t_2 - \tau_2) d\tau_1 d\tau_2.$$

$$\tag{5.2.12}$$

In matrix form, (5.2.12) is

$$\boldsymbol{\kappa_{XX}}(t_1, t_2) = \int_0^{t_1} \int_0^{t_2} \mathbf{h}(t_1 - \tau_1)\boldsymbol{\kappa_{FF}}(\tau_1, \tau_2)\mathbf{h}^T(t_2 - \tau_2)d\tau_1 d\tau_2$$

$$= \int_0^{t_1} \int_0^{t_2} \mathbf{h}(\tau_1)\boldsymbol{\kappa_{FF}}(t_1 - \tau_1, t_2 - \tau_2)\mathbf{h}^T(\tau_2)d\tau_1 d\tau_2.$$

$$(5.2.13)$$

Although the mean and covariance functions of the response can be expressed in (5.2.10) through (5.2.13), the impulse response matrix $\mathbf{h}(t)$ is difficult to calculate. However, if the model analysis is applicable, their calculations are more feasible. In this case, the mean and variance functions of the response can be calculated from Eq. (5.1.77) as

$$\boldsymbol{\mu_x}(t) = \int_0^t \boldsymbol{\Psi} \mathbf{h_q}(t - \tau)\boldsymbol{\Psi}^{\mathbf{T}}\boldsymbol{\mu_F}(\tau)d\tau, \qquad (5.2.14)$$

$$\boldsymbol{\kappa_{XX}}(t_1, t_2) = \int_0^{t_1} \int_0^{t_2} \boldsymbol{\Psi} \mathbf{h_q}(t_1 - \tau_1)\boldsymbol{\Psi}^T \boldsymbol{\kappa_{FF}}(\tau_1, \tau_2)\boldsymbol{\Psi} \mathbf{h_q}(t_2 - \tau_2)$$

$$\times \boldsymbol{\Psi}^T d\tau_1 d\tau_2. \qquad (5.2.15)$$

Note that the model impulsive response matrix $\mathbf{h_q}(t)$ is diagonal with elements given by (5.1.74).

5.3 Stationary Stochastic Excitations

Consider the case in which the excitation $\mathbf{F}(t)$ is a vector stationary stochastic process. Its mean is a constant vector $\boldsymbol{\mu_F}$, and covariance matrix depends only on the time difference, i.e.,

$$\boldsymbol{\kappa_{FF}}(t_1, t_2) = \boldsymbol{\kappa_{FF}}(\tau), \quad \tau = t_2 - t_1 \qquad (5.3.1)$$

In this case, the system responses will also be stationary after the system is exposed to the excitations for a long period of time.

In the stationary stage, all the first-order statistics, such as means, variances and higher-order moments, are constants, and the second-order statistics, such as correlation functions, depend only on the time difference. The results in Eqs. (5.2.5) through (5.2.15) give the stationary response statistics as the time instants approach infinity.

5.3.1 *Time-Domain Analysis*

Equations (5.2.5) through (5.2.15) give the transient response statistics in time domain under zero initial conditions.

Example 5.3.1 Consider a white-noise excitation with zero mean and a correlation function $R_{FF}(\tau) = 2\pi K \delta(\tau)$. The transient solution of the covariance function is obtained from (5.2.6),

$$\kappa_{XX}(t_1, t_2) = \int_0^{t_1}\int_0^{t_2} 2\pi K \delta(\tau_2 - \tau_1)h(t_1 - \tau_1)h(t_2 - \tau_2)d\tau_1 d\tau_2.$$

(5.3.2)

Without loss of generality, assume that $t_1 \leq t_2$, and we have

$$\int_0^{t_2} \delta(\tau_2 - \tau_1)h(t_2 - \tau_2)d\tau_2 = \int_0^{t_1} \delta(\tau_2 - \tau_1)h(t_2 - \tau_2)d\tau_2$$

$$+ \int_{t_1}^{t_2} \delta(\tau_2 - \tau_1)h(t_2 - \tau_2)d\tau_2$$

$$= h(t_2 - \tau_1).$$ (5.3.3)

Substituting (5.3.3) into (5.3.2) and applying (5.1.11), we obtain the transient autocovariance function

$$\kappa_{XX}(t_1, t_2)$$

$$= 2\pi K \int_0^{t_1} h(t_1 - \tau_1)h(t_2 - \tau_1)d\tau_1$$

$$= \frac{2\pi K}{m^2 \omega_d^2} \int_0^{t_1} e^{-\zeta \omega_0 (t_1 + t_2 - 2\tau_1)} \sin \omega_d (t_1 - \tau_1) \sin \omega_d (t_2 - \tau_1) d\tau_1,$$

$$= \frac{\pi K}{2m^2 \zeta \omega_0^3} \left\{ e^{-\zeta \omega_0 (t_2 - t_1)} \left[\cos \omega_d (t_2 - t_1) + \frac{\zeta \omega_0}{\omega_d} \sin \omega_d (t_2 - t_1) \right] \right.$$

$$- e^{-\zeta \omega_0 (t_2 + t_1)} \left[\frac{\omega_0^2}{\omega_d^2} \cos \omega_d (t_2 - t_1) - \frac{\zeta^2 \omega_0^2}{\omega_d^2} \cos \omega_d (t_2 + t_1) \right.$$

$$\left. \left. + \frac{\zeta \omega_0}{\omega_d} \sin \omega_d (t_2 + t_1) \right] \right\}. \tag{5.3.4}$$

By letting $t = t_1 = t_2$, we obtain the transient mean square response

$$\sigma_X^2(t) = \frac{\pi K}{2m^2 \zeta \omega_0^3} \left[1 - e^{-2\zeta \omega_0 t} \left(\frac{\omega_0^2}{\omega_d^2} - \frac{\zeta^2 \omega_0^2}{\omega_d^2} \cos 2\omega_d t + \frac{\zeta \omega_0}{\omega_d} \sin 2\omega_d t \right) \right]. \tag{5.3.5}$$

As $t_1, t_2 \to \infty$, we have the stationary mean-square value and autocovariance function

$$\sigma_X^2 = \frac{\pi K}{2m^2 \zeta \omega_0^3}, \tag{5.3.6}$$

$$\kappa_{XX}(\tau) = \frac{\pi K}{2m^2 \zeta \omega_0^3} \left[e^{-\zeta \omega_0 \tau} \left(\cos \omega_d \tau + \frac{\zeta \omega_0}{\omega_d} \sin \omega_d \tau \right) \right], \tag{5.3.7}$$

where $\tau = t_2 - t_1 \geq 0$. The cross-covariance function of $X(t)$ and $\dot{X}(t)$, and the auto-covariance function of $\dot{X}(t)$ can be obtained from (3.5.23) and (3.5.24) as follows

$$\kappa_{\dot{X}X}(\tau) = -\frac{d}{d\tau} \kappa_{XX}(\tau), \quad \kappa_{\dot{X}\dot{X}}(\tau) = -\frac{d^2}{d\tau^2} \kappa_{XX}(\tau). \tag{5.3.8}$$

Also from (3.5.25), we have $R_{\dot{X}X}(0) = 0$, indicating that the displacement $X(t)$ and velocity $\dot{X}(t)$ are uncorrelated.

Mathematically, infinitely long time is required for the effect of the initial state to diminish and for the response to become stationary. But practically, after a sufficiently long period of time, the response can be considered as stationary. It is observed from

Eqs. (5.3.4) and (5.3.5) that the time required should satisfy $t \gg \tau_{rel} = 1/(\zeta \omega_0)$, where τ_{rel} is known as the system relaxation time defined as the time required for the response amplitude to decrease by a factor of e^{-1}.

For a MDOF linear system, we only consider the case that the damping matrix is diagonal after the model transformation. From Eqs. (5.2.14) and (5.2.15), we have

$$\lim_{t \to \infty} \boldsymbol{\mu_X}(t) = \boldsymbol{\mu_F} \int_0^\infty \mathbf{h}(\tau) d\tau = \boldsymbol{\mu_F} \int_0^\infty \boldsymbol{\Psi} \mathbf{h_q}(\tau) \boldsymbol{\Psi}^T d\tau,$$

$$(5.3.9)$$

$$\lim_{t_1, t_2 \to \infty} \boldsymbol{\kappa_{XX}}(t_1, t_2) = \boldsymbol{\kappa_{XX}}(\tau)$$

$$= \int_0^\infty \int_0^\infty \mathbf{h}(\tau_1) \boldsymbol{\kappa_{FF}}(\tau + \tau_1 - \tau_2) \mathbf{h}^T(\tau_2) d\tau_1 d\tau_2$$

$$= \int_0^\infty \int_0^\infty \boldsymbol{\Psi} \mathbf{h_q}(\tau_1) \boldsymbol{\Psi}^T \boldsymbol{\kappa_{FF}}(\tau + \tau_1 - \tau_2)$$

$$\times \boldsymbol{\Psi} \mathbf{h_q}(\tau_2) \boldsymbol{\Psi}^T d\tau_1 d\tau_2, \quad \tau = t_2 - t_1. \quad (5.3.10)$$

Example 5.3.2 Consider the special case, $m_1 = m_2 = m$, $c_1 = c_2 = c_3 = c$ and $k_1 = k_2 = k_3 = k$, for the system in Examples 5.1.1 and 5.1.2, and assume $f_1(t)$ is a Gaussian white noise with a spectral density K, and $f_2(t) = 0$. From Example 5.1.2, we found that

$$h_{11}(t) = \frac{1}{2m}[h_{q1}(t) + h_{q2}(t)], \quad (5.3.11)$$

$$h_{21}(t) = \frac{1}{2m}[h_{q1}(t) - h_{q2}(t)], \quad (5.3.12)$$

where $h_{q1}(t)$ and $h_{q2}(t)$ are given in Eqs. (5.1.92) and (5.1.93). Following the same procedure as in Example 5.3.1,

$$\kappa_{X_1 X_1}(t_1, t_2) = 2\pi K \int_0^{t_1} h_{11}(t_1 - \tau_1) h_{11}(t_2 - \tau_1) d\tau_1$$

$$= \frac{\pi K}{2m^2} \int_0^{t_1} [h_{q1}(t_1 - \tau_1) + h_{q2}(t_1 - \tau_1)][h_{q1}(t_2 - \tau_1)$$

$$+ h_{q2}(t_2 - \tau_1)]d\tau_1, \tag{5.3.13}$$

$$\kappa_{X_2 X_2}(t_1, t_2) = 2\pi K \int_0^{t_1} h_{21}(t_1 - \tau_1)h_{21}(t_2 - \tau_1)d\tau_1$$

$$= \frac{\pi K}{2m^2} \int_0^{t_1} [h_{q1}(t_1 - \tau_1) - h_{q2}(t_1 - \tau_1)]$$

$$\times [h_{q1}(t_2 - \tau_1) - h_{q2}(t_2 - \tau_1)]d\tau_1. \tag{5.3.14}$$

Substituting $h_{q1}(t)$ and $h_{q2}(t)$ into (5.3.13) and (5.3.14), we can obtain the transient autocovariance functions of responses $X_1(t)$ and $X_2(t)$, respectively; and hence, the transient mean square responses by letting $t = t_1 = t_2$.

For the stationary state, i.e., letting $t_1, t_2 \rightarrow \infty$, we have the autocovariance functions

$$\kappa_{X_1 X_1}(\tau) = \frac{\pi K}{2m^2}[h_1(\tau) + h_2(\tau) + h_3(\tau) + h_4(\tau)], \quad (5.3.15)$$

$$\kappa_{X_2 X_2}(\tau) = \frac{\pi K}{2m^2}[h_1(\tau) - h_2(\tau) - h_3(\tau) + h_4(\tau)], \quad (5.3.16)$$

where $\tau = t_2 - t_1$, and

$$h_1(\tau) = \lim_{t_1, t_2 \rightarrow \infty} \int_0^{t_1} h_{q1}(t_1 - \tau_1)h_{q1}(t_2 - \tau_1)d\tau_1$$

$$= \frac{1}{4\zeta_1 \omega_1^3} \left\{ e^{-\zeta_1 \omega_1 \tau} \left[\cos(\omega_{d1}\tau) + \frac{\zeta_1 \omega_1}{\omega_{d1}} \sin(\omega_{d1}\tau) \right] \right\},$$

$$\tag{5.3.17}$$

$$h_2(\tau) = \lim_{t_1,t_2\to\infty} \int_0^{t_1} h_{q1}(t_1 - \tau_1)h_{q2}(t_2 - \tau_1)d\tau_1$$

$$= \frac{2}{\Delta_1\Delta_2}\left\{e^{-\zeta_2\omega_2\tau}\left[(\zeta_1\omega_1 + \zeta_2\omega_2)\cos(\omega_{d2}\tau)\right.\right.$$

$$\left.\left. + \frac{1}{\omega_{d1}}\left[\omega_{d1}^2 - \frac{1}{4}(\Delta_1 + \Delta_2)\right]\sin(\omega_{d2}\tau)\right]\right\}, \qquad (5.3.18)$$

$$h_3(\tau) = \lim_{t_1,t_2\to\infty} \int_0^{t_1} h_{q2}(t_1 - \tau_1)h_{q1}(t_2 - \tau_1)d\tau_1$$

$$= \frac{2}{\Delta_1\Delta_2}\left\{e^{-\zeta_1\omega_1\tau}\left[(\zeta_1\omega_1 + \zeta_2\omega_2)\cos(\omega_{d1}\tau)\right.\right.$$

$$\left.\left. + \frac{1}{\omega_{d2}}[\omega_{d2}^2 - \frac{1}{4}(\Delta_1 + \Delta_2)]\sin(\omega_{d1}\tau)\right]\right\}, \qquad (5.3.19)$$

$$h_4(\tau) = \lim_{t_1,t_2\to\infty} \int_0^{t_1} h_{q2}(t_1 - \tau_1)h_{q2}(t_2 - \tau_1)d\tau_1$$

$$= \frac{1}{4\zeta_2\omega_2^3}\left\{e^{-\zeta_2\omega_2\tau}\left[\cos(\omega_{d2}\tau) + \frac{\zeta_2\omega_2}{\omega_{d2}}\sin(\omega_{d2}\tau)\right]\right\},$$

$$(5.3.20)$$

in which ζ_1, ζ_2, ω_1, ω_2, ω_{d1} and ω_{d2} are given in (5.1.4) and (5.1.46), and

$$\begin{aligned}\Delta_1 &= (\zeta_1\omega_1 + \zeta_2\omega_2)^2 + (\omega_{d2} - \omega_{d1})^2, \\ \Delta_2 &= (\zeta_1\omega_1 + \zeta_2\omega_2)^2 + (\omega_{d2} + \omega_{d1})^2.\end{aligned} \qquad (5.3.21)$$

The stationary mean-square values are obtained from (5.3.15) and (5.3.16) by letting $\tau = 0$,

$$\sigma_{X_1}^2 = \frac{\pi K}{2m^2}\left[\frac{1}{4\zeta_1\omega_1^3} + \frac{4(\zeta_1\omega_1 + \zeta_2\omega_2)}{\Delta_1\Delta_2} + \frac{1}{4\zeta_2\omega_2^3}\right], \qquad (5.3.22)$$

$$\sigma_{X_2}^2 = \frac{\pi K}{2m^2}\left[\frac{1}{4\zeta_1\omega_1^3} - \frac{4(\zeta_1\omega_1 + \zeta_2\omega_2)}{\Delta_1\Delta_2} + \frac{1}{4\zeta_2\omega_2^3}\right]. \qquad (5.3.23)$$

5.3.2 *Frequency-Domain Analysis*

Assume that the forcing function $F(t)$ is stationary with zero mean. Letting $\tau = t_2 - t_1$, and taking into consideration that $h(t) = 0$ for $t < 0$, then Eq. (5.2.6) can be written for the stationary stage as

$$\kappa_{XX}(\tau) = \int_{-\infty}^{\infty}\int_{-\infty}^{\infty} \kappa_{FF}(\tau + \tau_1 - \tau_2)h(\tau_1)h(\tau_2)d\tau_1 d\tau_2. \qquad (5.3.24)$$

The power spectral density $\Phi_{XX}(\omega)$, as the Fourier transform of the covariance function, is obtained from (5.3.24) as

$$\Phi_{XX}(\omega) = \frac{1}{2\pi}\int_{-\infty}^{\infty}\kappa_{XX}(\tau)e^{-i\omega\tau}d\tau$$

$$= \frac{1}{2\pi}\int_{-\infty}^{\infty}\left[\int_{-\infty}^{\infty}\int_{-\infty}^{\infty}\kappa_{FF}(\tau + \tau_1 - \tau_2)h(\tau_1)h(\tau_2)d\tau_1 d\tau_2\right]e^{-i\omega\tau}d\tau$$

$$= \frac{1}{2\pi}\int_{-\infty}^{\infty}\kappa_{FF}(\tau + \tau_1 - \tau_2)e^{-i\omega(\tau+\tau_1-\tau_2)}d\tau$$

$$\times\left[\int_{-\infty}^{\infty}h(\tau_1)e^{i\omega\tau_1}d\tau_1\right]\left[\int_{-\infty}^{\infty}h(\tau_2)e^{-i\omega\tau_2}d\tau_2\right]. \qquad (5.3.25)$$

Note that the Fourier transform of the impulse response function $h(t)$ is the frequency response function $H(\omega)$, given in (5.1.23), and that the Fourier transform of the covariance function of the force $F(t)$ is its power spectral density, i.e.,

$$\Phi_{FF}(\omega) = \frac{1}{2\pi}\int_{-\infty}^{\infty}\kappa_{FF}(\tau)e^{-i\omega\tau}d\tau. \qquad (5.3.26)$$

Therefore, (5.3.25) can be written as

$$\Phi_{XX}(\omega) = |H(\omega)|^2\Phi_{FF}(\omega), \qquad (5.3.27)$$

where $|H(\omega)|^2$ can be found from (5.1.7) as follows

$$|H(\omega)|^2 = \frac{1}{m^2[(\omega_0^2 - \omega^2)^2 + 4\zeta^2\omega^2\omega_0^2]}. \tag{5.3.28}$$

According to Eq. (3.6.3), the stationary mean-square value of $X(t)$ can be calculated from their power spectral density function (5.3.27) as

$$\sigma_X^2 = \int\limits_{-\infty}^{\infty} \Phi_{XX}(\omega)d\omega = \int\limits_{-\infty}^{\infty} |H(\omega)|^2\Phi_{FF}(\omega)d\omega. \tag{5.3.29}$$

The power spectral densities of the velocity $\dot{X}(t)$ and acceleration $\ddot{X}(t)$ can be obtained from (3.6.11) as

$$\Phi_{\dot{X}\dot{X}}(\omega) = \omega^2 |H(\omega)|^2 \Phi_{FF}(\omega), \quad \Phi_{\ddot{X}\ddot{X}}(\omega) = \omega^4 |H(\omega)|^2 \Phi_{FF}(\omega). \tag{5.3.30}$$

Similarly, the stationary mean-square values of $\dot{X}(t)$ and $\ddot{X}(t)$ can be calculated from their respective power spectral density functions.

Example 5.3.3 As the same as in Example 5.3.1, the excitation is a white-noise excitation with zero mean and a spectral density K. The response spectral density is then

$$\Phi_{XX}(\omega) == \frac{K}{m^2[(\omega_0^2 - \omega^2)^2 + 4\zeta^2\omega^2\omega_0^2]}. \tag{5.3.31}$$

The stationary mean square is calculated from

$$\sigma_X^2 = \int\limits_{-\infty}^{\infty} \frac{K}{m^2[(\omega_0^2 - \omega^2)^2 + 4\zeta^2\omega^2\omega_0^2]}d\omega. \tag{5.3.32}$$

The integral in (5.3.32) can be evaluated by using the method of residues. Treat the integrand as a complex function $g(z)$ by replacing

ω by the complex variable z. i.e.,

$$g(z) = \frac{K}{m^2[(\omega_0^2 - z^2)^2 + 4\zeta^2\omega_0^2 z^2]}. \tag{5.3.33}$$

It has two poles in the upper half complex plane

$$z_{1,2} = \pm\sqrt{1 - \zeta^2}\,\omega_0 + i\zeta\omega_0. \tag{5.3.34}$$

Applying the method of residues, we have

$$\sigma_X^2 = 2\pi i \left\{ \sum \text{ residues in the upper complex plane} \right\}$$

$$= 2\pi i\{[(z - z_1)g(z)]_{z=z_1} + [(z - z_2)g(z)]_{z=z_2}\}$$

$$= \frac{\pi K}{2m^2\zeta\omega_0^3}, \tag{5.3.35}$$

which is the same as in (5.3.6) obtained in the time-domain analysis.

It is known that the white noise is a mathematical ideation of a real noise with a flat power spectral density over a broad frequency band. Its mean square is infinite since its energy is unbounded. However, the response of a SDOF linear system has a finite mean square provided certain damping mechanism exists. This is because the system response diminishes faster with increasing frequency. While this is still true for the velocity process $\dot{X}(t)$, the acceleration process $\ddot{X}(t)$ has an infinite mean square, shown clearly from Eqs. (5.3.30) and (5.3.28).

For a MDOF linear system, the power spectral density matrix $\Phi_{\mathbf{XX}}(\omega)$, as the Fourier transform of the covariance matrix, is obtained from (5.3.10) as

$$\Phi_{\mathbf{XX}}(\omega) = \frac{1}{2\pi} \int_{-\infty}^{\infty} \boldsymbol{\kappa}_{\mathbf{XX}}(\tau)e^{-i\omega\tau}\,d\tau$$

$$= \frac{1}{2\pi} \int_{-\infty}^{\infty} \left[\int_0^{\infty}\int_0^{\infty} \mathbf{h}(\tau_1)\boldsymbol{\kappa}_{\mathbf{FF}}(\tau + \tau_1 - \tau_2)\mathbf{h}^T(\tau_2)d\tau_1 d\tau_2 \right]e^{-i\omega\tau}\,d\tau$$

$$= \left[\int_{-\infty}^{\infty} \mathbf{h}(\tau_1) e^{i\omega\tau_1} d\tau_1 \right]$$

$$\times \left[\frac{1}{2\pi} \int_{-\infty}^{\infty} \boldsymbol{\kappa}_{\mathbf{FF}}(\tau + \tau_1 - \tau_2) e^{-i\omega(\tau+\tau_1-\tau_2)} d\tau \right]$$

$$\times \left[\int_{-\infty}^{\infty} \mathbf{h}^T(\tau_2) e^{-i\omega\tau_2} d\tau_2 \right]$$

$$= \mathbf{H}^*(\omega) \boldsymbol{\Phi}_{\mathbf{FF}}(\omega) \mathbf{H}^T(\omega), \tag{5.3.36}$$

where an asterisk denotes the complex conjugate, and $\mathbf{H}(\omega)$ can be calculated from (5.1.81) from the modal frequency response matrix.

Example 5.3.4 For the same problem as in Example 5.3.2, the stationary mean-square values of the responses can also be calculated using the frequency response functions, which are shown in Eqs. (5.1.39) and (5.1.40), i.e.,

$$H_{11}(\omega) = \frac{1}{m\Delta_1}(-\omega^2 + 4i\zeta_0\omega_0\omega + 2\omega_0^2), \tag{5.3.37}$$

$$H_{12}(\omega) = \frac{1}{m\Delta_1}(2i\zeta_0\omega_0\omega + \omega_0^2], \tag{5.3.38}$$

where

$$\Delta_1 = (-\omega^2 + 2i\zeta_1\omega_1\omega + \omega_1^2)(-\omega^2 + 2i\zeta_2\omega_2\omega + \omega_2^2). \tag{5.3.39}$$

The stationary mean square is calculated from

$$\sigma_{X_1}^2 = \int_{-\infty}^{\infty} |H_{11}(\omega)|^2 d\omega$$

$$= \int_{-\infty}^{\infty} \frac{(\omega^2 - 2\omega_0^2) + (4\zeta_0\omega_0\omega)^2}{m^2[(\omega^2 - \omega_1^2)^2 + (2\zeta_1\omega_1\omega)^2][(\omega^2 - \omega_2^2)^2 + (2\zeta_2\omega_2\omega)^2]} d\omega.$$

$$\tag{5.3.40}$$

$$\sigma_{X_2}^2 = \int\limits_{-\infty}^{\infty} |H_{21}(\omega)|^2 \, d\omega$$

$$= \int\limits_{-\infty}^{\infty} \frac{\omega_0^4 + (2\zeta_0\omega_0\omega)^2}{m^2[(\omega^2 - \omega_1^2)^2 + (2\zeta_1\omega_1\omega)^2][(\omega^2 - \omega_2^2)^2 + (2\zeta_2\omega_2\omega)^2]} \, d\omega.$$

$$(5.3.41)$$

Applying the method of residue theory, the stationary mean square responses can be obtained, although the calculation is tedious. The results should be the same as those obtained in Example 5.3.2.

5.4 Non-Stationary Stochastic Excitations

For a non-stationary excitation $F(t)$, the system response $X(t)$ is also non-stationary, and its covariance function can be calculated from Eq. (5.2.6). The spectral density of the response is obtained as

$$\Phi_{XX}(\omega_1, \omega_2) = \frac{1}{(2\pi)^2} \int\limits_{-\infty}^{\infty} \kappa_{XX}(t_1, t_2) e^{-i(\omega_1 t_1 - \omega_2 t_2)} dt_1 dt_2$$

$$= \frac{1}{2\pi} \int\limits_{-\infty}^{\infty} \int\limits_{-\infty}^{\infty}$$

$$\times \left[\int\limits_{-\infty}^{\infty} \int\limits_{-\infty}^{\infty} \kappa_{FF}(\tau_1, \tau_2) h(t_1 - \tau_1) h(t_2 - \tau_2) d\tau_1 d\tau_2 \right]$$

$$\times e^{-i(\omega_1 t_1 - \omega_2 t_2)} dt_1 dt_2$$

$$= \Phi_{FF}(\omega_1, \omega_2) H(\omega_1) H^*(\omega_2). \qquad (5.4.1)$$

Equation (5.4.1) can be used to calculate the general spectral density of the response if that of the excitation is given.

A useful type of non-stationary stochastic process is the evolutionary stochastic processes introduced in Section 3.9.3. Assume the excitation is an evolutionary process with zero mean, defined in

(3.9.10), i.e.,

$$F(t) = \int_{-\infty}^{\infty} a(t,\omega)e^{i\omega t}dZ(\omega), \tag{5.4.2}$$

where $a(t,\omega)$ is a deterministic function, and $Z(\omega)$ is an orthogonal-increment process, namely,

$$E[dZ(\omega_1)dZ^*(\omega_2)] = \begin{cases} d\Psi(\omega) = \Phi(\omega)d\omega, & \omega_1 = \omega_2 \\ 0, & \omega_1 \neq \omega_2 \end{cases}. \tag{5.4.3}$$

It is known from Section 3.9.3 that the evolutionary spectral density and the covariance function of $F(t)$ are

$$\Phi_{FF}(t,\omega) = |a(t,\omega)|^2\Phi(\omega), \tag{5.4.4}$$

$$\kappa_{FF}(t_1,t_2) = \int_{-\infty}^{\infty} a(t_1,\omega)a^*(t_2,\omega)e^{i\omega(t_1-t_2)}\Phi(\omega)d\omega. \tag{5.4.5}$$

Substitution of (5.4.5) into (5.2.6) leads to

$$\kappa_{XX}(t_1,t_2) = \int_0^{t_1}\int_0^{t_2} \kappa_{FF}(t_1-\tau_1, t_2-\tau_2)h(\tau_1)h(\tau_2)d\tau_1 d\tau_2$$

$$= \int_0^{t_1}\int_0^{t_2} \left[\int_{-\infty}^{\infty} a(t_1-\tau_1,\omega)a^*(t_2-\tau_2,\omega)\right.$$

$$\left. e^{i\omega(t_1-\tau_1-t_2+\tau_2)}\Phi(\omega)d\omega\right] h(\tau_1)h(\tau_2)d\tau_1 d\tau_2.$$

$$\tag{5.4.6}$$

Exchanging the order of integration and denoting

$$\hat{H}(t,\omega) = \int_0^t a(t-u,\omega)h(u)e^{-i\omega u}du, \tag{5.4.7}$$

we have

$$\kappa_{XX}(t_1, t_2) = \int\limits_{-\infty}^{\infty} \hat{H}(t_1, \omega)\hat{H}^*(t_2, \omega)e^{i\omega(t_1-t_2)}\Phi(\omega)d\omega. \qquad (5.4.8)$$

The mean square of $X(t)$ is obtained by letting $t_1 = t_1 = t$.

$$E[X^2(t)] = \int\limits_{-\infty}^{\infty} \left|\hat{H}(t, \omega)\right|^2 \Phi(\omega)d\omega. \qquad (5.4.9)$$

Therefore, the response is also an evolutionary process with a spectral density as

$$\Phi_{XX}(t, \omega) = \left|\hat{H}(t, \omega)\right|^2 \Phi(\omega). \qquad (5.4.10)$$

It is of interest to examine the special case in which $a(t, \omega)$ only depends on ω. In this case, the excitation process $F(t)$ degenerates to a stationary process, and the response process $X(t)$ approached the stationary state as $t \to \infty$.

Example 5.4.1 A type of the evolution stochastic process is the random pulse train defined in (3.8.8), i.e.,

$$F(t) = \sum_{j=1}^{N(T)} Y_j w(t - \tau_j), \quad 0 < t \le T, \qquad (5.4.11)$$

where $N(T)$ is a Poisson process with a random arrival rate $\Lambda(t)$, τ_j is the random arrival time, $w(t - \tau)$ represents a deterministic pulse shape, and Y_j are the magnitudes which are identical and independent random variables with zero mean. The covariance function of $F(t)$ is given in (3.9.17) as

$$\kappa_{FF}(t_1, t_2) = \frac{1}{2\pi}E[Y^2]\int\limits_{-\infty}^{\infty} a(t_1, \omega)a^*(t_2, \omega)e^{i\omega(t_1-t_2)}d\omega, \qquad (5.4.12)$$

where

$$a(t, \omega) = \int\limits_{-\infty}^{\infty} \sqrt{\mu_\Lambda(t - u)}w(u)e^{-i\omega u}du \qquad (5.4.13)$$

and $\mu_\Lambda(t)$ is the mean pulse arrival rate. Equation (5.4.12) show that $\Phi(\omega) = E[Y^2]/(2\pi)$. The evolutionary spectral density of $F(t)$ is

$$\Phi_{FF}(t,\omega) = \frac{1}{2\pi}E[Y^2]|a(t,\omega)|^2. \qquad (5.4.14)$$

With given forms of the shape function $w(t)$ and the mean pulse arrival rate $\mu_\Lambda(t)$, the covariance function and the mean square function of the response process $X(t)$ can be calculated from (5.4.8) and (5.4.9).

If the Poisson process $N(t)$ in (5.4.11) is homogeneous, then the pulse arrival rate $\Lambda(t)$ is a stationary process, $\mu_\Lambda(t) = \mu_\Lambda$, $a(t,\omega) = a(\omega)$, $\Phi_{FF}(t,\omega) = \Phi_{FF}(\omega)$, and the random pulse train $F(t)$ becomes a stationary process.

5.5 Method of Diffusion Process

Consider a SDOF linear system governed by (5.2.1) with a Gaussian white noise excitation. Denoting $X_1(t) = X(t)$ and $X_2(t) = \dot{X}(t)$, system (5.2.1) can be rewritten in the state space

$$\begin{aligned}\dot{X}_1 &= X_2 \\ \dot{X}_2 &= -\omega_0^2 X_1 - 2\zeta\omega_0 X_2 + \tfrac{1}{m}W(t),\end{aligned} \qquad (5.5.1)$$

where $W(t)$ is a Gaussian white noise with spectral density K. As stated in Section 4.2, the vector $\boldsymbol{X}(t) = [X_1(t), X_2(t)]^T$ is a vector Markov diffusion process, and the theory of Markov process presented in Chapter 4 can then be applied. The Ito equations corresponding to (5.5.1) is

$$\begin{aligned}dX_1 &= X_2 dt, \\ dX_2 &= (-\omega_0^2 X_1 - 2\zeta\omega_0 X_2)dt + \sigma dB(t),\end{aligned} \qquad (5.5.2)$$

where $B(t)$ is a unit Wiener process, and

$$\sigma = \frac{1}{m}\sqrt{2\pi K}. \qquad (5.5.3)$$

Consider a MDOF linear system excited by Gaussian white noises, i.e., $\mathbf{F}(t) = \mathbf{W}(t)$ in (5.2.9), and $\mathbf{W}(t)$ satisfies

$$E[\mathbf{W}(t)] = \mathbf{0}, \quad E[\mathbf{W}(t)\mathbf{W}^T(t+\tau)] = 2\pi\mathbf{K}\delta(\tau), \qquad (5.5.4)$$

where $\mathbf{K}$ is the spectral density matrix of $\mathbf{W}(t)$. In this case, the response $\mathbf{X}(t)$ and its derivative $\dot{\mathbf{X}}(t)$ constitute a Markov diffusion vector process, and the Markov process approach can be applied. Write system (5.2.9) in the state space

$$\frac{d}{dt}\mathbf{Y} = A\mathbf{Y} + \mathbf{G}\mathbf{W}(t), \qquad (5.5.5)$$

where

$$\mathbf{Y} = \begin{bmatrix} \mathbf{X} \\ \dot{\mathbf{X}} \end{bmatrix}, \quad A = \begin{bmatrix} \mathbf{0} & \mathbf{I} \\ -\mathbf{M}^{-1}\mathbf{K} & -\mathbf{M}^{-1}\mathbf{C} \end{bmatrix}, \quad \mathbf{G} = \begin{bmatrix} \mathbf{0} \\ \mathbf{M}^{-1} \end{bmatrix}. \quad (5.5.6)$$

Equation (5.5.5) can be written in vector form of Ito differential equation

$$d\mathbf{Y}(t) = A\mathbf{Y}(t)dt + \sigma d\mathbf{B}(t), \qquad (5.5.7)$$

where σ is a matrix calculated from $\mathbf{G}$ and $\mathbf{K}$, and the components of the unit Wiener vector process are independent, i.e.,

$$E[d\mathbf{B}(t)] = \mathbf{0}, \quad E[d\mathbf{B}(t_1)d\mathbf{B}^T(t_2)] = \begin{cases} \mathbf{I}dt, & t_1 = t_2 \\ \mathbf{0}, & t_1 \neq t_2. \end{cases} \qquad (5.5.8)$$

5.5.1 *Moment Equations*

Consider the SDOF linear system (5.5.1) or (5.5.2) first. Denote

$$M(X_1, X_2) = X_1^i X_2^j. \qquad (5.5.9)$$

Carrying out the Ito differential rule to $M(X_1, X_2)$, we obtain

$$\frac{dM}{dt} = X_2\frac{\partial M}{\partial X_1} + (-\omega_0^2 X_1 - 2\zeta\omega_0 X_2)\frac{\partial M}{\partial X_2} + \sigma\frac{\partial M}{\partial X_2}dB(t) + \frac{1}{2}\sigma^2\frac{\partial^2 M}{\partial X_2^2}.$$
$$(5.5.10)$$

Taking the ensemble averaging on (5.5.10), and denoting

$$m_{ij} = E[X_1^i X_2^j], \tag{5.5.11}$$

we have

$$\frac{d}{dt} m_{ij} = im_{i-1,j+1} - j\omega_0^2 m_{i+1,j-1} - 2j\zeta\omega_0 m_{i,j} + \frac{1}{2}j(j-1)\sigma^2 m_{i,j-2}. \tag{5.5.12}$$

In deriving (5.5.12), the characteristic of Ito differential equation is taken into consideration that the increment $dB(t)$ of the excitation is independent of the response $X(t)$, as described in Section 4.2.6. The left-hand side of (5.5.12) is a time derivative of a nth $(n = i+j)$ moment, and the right-hand side contains only the nth and lower-order moments. Thus, the equations of the nth order moments are closed, and can be solved from lower order to higher order sequentially.

Consider the cases of the first and second orders by letting $M = X_1, X_2, X_1^2, X_1X_2$ and X_2^2. We obtain from (5.5.12)

$$\frac{dm_{10}}{dt} = m_{01},$$

$$\tag{5.5.13}$$

$$\frac{dm_{01}}{dt} = -\omega_0^2 m_{10} - 2\zeta\omega_0 m_{01},$$

$$\frac{dm_{20}}{dt} = 2m_{11},$$

$$\frac{dm_{11}}{dt} = m_{02} - \omega_0^2 m_{20} - 2\zeta\omega_0 m_{11}, \tag{5.5.14}$$

$$\frac{dm_{02}}{dt} = -2\omega_0^2 m_{11} - 4\zeta\omega_0 m_{02} + \sigma^2.$$

Equations (5.5.13) and (5.5.14) are linear ordinary differential equations with constant coefficients, and can be solved for given initial conditions. The stationary solutions and transient solutions under zero initial conditions for (5.5.13) are zeros, i.e., $m_{10}(t) = 0$ and

$m_{01}(t) = 0$. The stationary solutions for (5.5.14) are obtained by letting all derivative terms be zero,

$$m_{20} = D, \quad m_{11} = 0,$$

$$m_{02} = \omega_0^2 D, \quad D = \frac{\sigma^2}{4\zeta\omega_0^3} = \frac{\pi K}{2m^2\zeta\omega_0^3}. \tag{5.5.15}$$

For the transient solution, consider the case that the system is at rest initially, i.e., $m_{20}(0) = 0$, $m_{11}(0) = 0$, and $m_{02}(0) = 0$. Denoting

$$\mathbf{m} = \left\{ \begin{matrix} m_{20}(t) \\ m_{11}(t) \\ m_{02}(t) \end{matrix} \right\}, \quad \mathbf{A} = \begin{bmatrix} 0 & 2 & 0 \\ -\omega_0^2 & -2\zeta\omega_0 & 1 \\ 0 & -2\omega_0^2 & -4\zeta\omega_0 \end{bmatrix}. \tag{5.5.16}$$

The homogeneous equation set corresponding to (5.5.14) can be written in matrix form

$$\frac{d}{dt}\mathbf{m} = \mathbf{A}\mathbf{m}. \tag{5.5.17}$$

The eigenvalues and eigenvectors of matrix $\mathbf{A}$ can be found as follows

$$\lambda_1 = -2\zeta\omega_0, \quad \lambda_{2,3} = -2\zeta\omega_0 \pm 2\omega_d i, \tag{5.5.18}$$

$$\mathbf{V}_1 = \left\{ \begin{matrix} 1 \\ -\zeta\omega_0 \\ \omega_0^2 \end{matrix} \right\}, \quad \mathbf{V}_2 = \left\{ \begin{matrix} 1 \\ -\zeta\omega_0 + \omega_d i \\ (2\zeta^2 - 1)\omega_0^2 - 2i\zeta\omega_0\omega_d \end{matrix} \right\},$$

$$\mathbf{V}_3 = \left\{ \begin{matrix} 1 \\ -\zeta\omega_0 - \omega_d i \\ (2\zeta^2 - 1)\omega_0^2 + 2i\zeta\omega_0\omega_d \end{matrix} \right\}. \tag{5.5.19}$$

Constructing a transformation matrix $\mathbf{D}$ as $\mathbf{D} = [\mathbf{V}_1 \, \mathbf{V}_2 \, \mathbf{V}_3]$, we have

$$\mathbf{D}^{-1}\mathbf{A}\mathbf{D} = \mathbf{\Lambda} = \begin{bmatrix} \lambda_1 & 0 & 0 \\ 0 & \lambda_2 & 0 \\ 0 & 0 & \lambda_3 \end{bmatrix}. \tag{5.5.20}$$

In (5.5.17), carrying the transformation

$$\mathbf{m} = \mathbf{D}\mathbf{x}$$

$$= \begin{bmatrix} 1 & 1 & 1 \\ -\zeta\omega_0 & -\zeta\omega_0 + \omega_d i & -\zeta\omega_0 - \omega_d i \\ \omega_0^2 & (2\zeta^2 - 1)\omega_0^2 - 2i\zeta\omega_0\omega_d & (2\zeta^2 - 1)\omega_0^2 + 2i\zeta\omega_0\omega_d \end{bmatrix}$$

$$\times \begin{Bmatrix} x_1 \\ x_2 \\ x_3 \end{Bmatrix} \tag{5.5.21}$$

and pre-multiplying $\mathbf{D}^{-1}$, we obtain a set of decoupled equations

$$\frac{d}{dt}\mathbf{x} = \mathbf{\Lambda}\mathbf{x}. \tag{5.5.22}$$

Solutions of (5.5.22) are

$$x_1(t) = C_1 e^{-2\zeta\omega_0 t}, \quad x_2(t) = C_2 e^{(-2\zeta\omega_0 + 2\omega_d i)t},$$

$$x_3(t) = C_3 e^{(-2\zeta\omega_0 - 2\omega_d i)t}. \tag{5.5.23}$$

Substituting (5.5.23) into (5.5.21), we have the general solutions of the homogeneous equations (5.5.17)

$$m_{20g}(t) = x_1(t) + x_2(t) + x_3(t),$$

$$m_{11g}(t) = -\zeta\omega_0 x_1(t) + (-\zeta\omega_0 + \omega_d i)x_2(t) + (-\zeta\omega_0 - \omega_d i)x_3(t),$$

$$m_{02g}(t) = \omega_0^2 x_1(t) + [(2\zeta^2 - 1)\omega_0^2 - 2i\zeta\omega_0\omega_d]x_2(t) + [(2\zeta^2 - 1)\omega_0^2$$

$$+ 2i\zeta\omega_0\omega_d]x_3(t). \tag{5.5.24}$$

Since all m_{20g}, m_{11g} and m_{02g} approach zero as $t \to \infty$, the particular solution must be the stationary solutions given in (5.5.15); thus,

$$m_{20}(t) = x_1(t) + x_2(t) + x_3(t) + D,$$

$$m_{11}(t) = -\zeta\omega_0 x_1(t) + (-\zeta\omega_0 + \omega_d i)x_2(t) + (-\zeta\omega_0 - \omega_d i)x_3(t),$$

$$m_{02}(t) = \omega_0^2 x_1(t) + [(2\zeta^2 - 1)\omega_0^2 - 2i\zeta\omega_0\omega_d]x_2(t)$$

$$+ [(2\zeta^2 - 1)\omega_0^2 + 2i\zeta\omega_0\omega_d]x_3(t) + \omega_0^2 D. \tag{5.5.25}$$

Use the zero initial conditions to obtain the following equations for the three constants

$$C_1 + C_2 + C_3 + D = 0,$$
$$-\zeta\omega_0 C_1 - \zeta\omega_0(C_2 + C_3) + \omega_d i(C_2 - C_3) = 0,$$
$$-\omega_0^2 C_1 + (2\zeta^2 - 1)\omega_0^2(C_2 + C_3) - 2i\zeta\omega_0\omega_d(C_2 - C_3) + \omega_0^2 D = 0.$$
$$(5.5.26)$$

The solutions of (5.5.26) are

$$C_1 = -\frac{\omega_0^2}{\omega_d^2}D, \quad C_2 + C_3 = \frac{\zeta^2\omega_0^2}{\omega_d^2}D, \quad C_2 - C_3 = i\frac{\zeta\omega_0}{\omega_d}D.$$
$$(5.5.27)$$

Substitution of (5.5.27) into (5.5.23) and (5.5.25) results in the final solutions

$$m_{20}(t) = \sigma_X^2(t)$$
$$= \frac{\pi K}{2m^2\zeta\omega_0^3}$$
$$\times \left[1 - e^{-2\zeta\omega_0 t}\left(\frac{\omega_0^2}{\omega_d^2} - \frac{\zeta^2\omega_0^2}{\omega_d^2}\cos 2\omega_d t + \frac{\zeta\omega_0}{\omega_d}\sin 2\omega_d t\right)\right],$$
$$(5.5.28)$$

$$m_{11}(t) = E[X(t)\dot{X}(t)] = \frac{\pi K}{2m^2\omega_d^2}e^{-2\zeta\omega_0 t}(1 - \cos 2\omega_d t), \qquad (5.5.29)$$

$$m_{02}(t) = \sigma_{\dot{X}}^2(t)$$
$$= \frac{\pi K}{2m^2\zeta\omega_0}$$
$$\times \left[1 - e^{-2\zeta\omega_0 t}\left(\frac{\omega_0^2}{\omega_d^2} - \frac{\zeta^2\omega_0^2}{\omega_d^2}\cos 2\omega_d t - \frac{\zeta\omega_0}{\omega_d}\sin 2\omega_d t\right)\right].$$
$$(5.5.30)$$

As expected, result (5.5.28) is the same as the one given in (5.3.5).

The higher-order moments can be calculated following the same procedure. However, since the response is Gaussian, the higher-order moments can be calculated from the first- and second-order moments; thus, application of (5.5.12) for determining the higher-order moments is not necessary.

It is noted that the moment functions, either transient or stationary, are the first-order properties of the response since they only involve one time instant.

For a MDOF linear system, taking the ensemble average of Eq. (5.5.7), we have

$$\frac{d}{dt}\boldsymbol{\mu_Y} = \mathbf{A}\boldsymbol{\mu_Y}, \tag{5.5.31}$$

where $\boldsymbol{\mu_Y} = [\mu_{Y_1}, \mu_{Y_1}, \ldots, \mu_{Y_{2n}}]^T$, is the mean vector. Applying the Ito differential rule (4.2.57), we obtain

$$d(\mathbf{Y}\mathbf{Y}^T) = (d\mathbf{Y})\mathbf{Y}^T + \mathbf{Y}(d\mathbf{Y}^T) + (d\mathbf{Y})(d\mathbf{Y}^T). \tag{5.5.32}$$

Substituting Eq. (5.5.7) into (5.5.32) and taking the ensemble average, we obtain

$$\frac{d}{dt}\mathbf{m_{YY}} = \mathbf{A}\mathbf{m_{YY}} + \mathbf{m_{YY}}\mathbf{A}^T + \boldsymbol{\sigma}\boldsymbol{\sigma}^T, \tag{5.5.33}$$

where $\mathbf{m_{YY}}$ is the second-moment matrix with the elements defined as

$$(\mathbf{m_{YY}})_{i,j} = m_{Y_iY_j} = E[Y_i(t)Y_j(t)]. \tag{5.5.34}$$

Equations (5.5.31) and (5.5.33) are linear ordinary differential equations with constant coefficients, and can be solved either analytically or numerically for given initial conditions. Note that $\mathbf{m_{YY}}$ is symmetric, and contains only $n(2n+1)$ independent elements.

By letting all derivative terms be zero in Eqs. (5.5.31) and (5.5.33), the stationary first- and second-order moments can be solved from the linear algebraic equations

$$\boldsymbol{\mu_Y} = \mathbf{0}, \quad \mathbf{A}\mathbf{m_{YY}} + \mathbf{m_{YY}}\mathbf{A}^T + \boldsymbol{\sigma}\boldsymbol{\sigma}^T = \mathbf{0}. \tag{5.5.35}$$

For MDOF systems, analytical solutions for response moments, especially in the transient state, are tedious to obtain, if not impossible. However, numerical solutions are straightforward if all system parameters are given.

5.5.2 *Correlation and Spectral Density Functions*

The correlation functions and spectral density functions are the representative statistics of the second-order properties. Consider only the stationary response. Applying the same method as that in Section 4.5.2, i.e., multiplying the two equations in (5.5.2) by $X_1(t - \tau)$, and taking the ensemble average, we obtain

$$\frac{d}{d\tau} R_{11}(\tau) = R_{12}(\tau),$$

$$\frac{d}{d\tau} R_{12}(\tau) = -\omega_0^2 R_{11}(\tau) - 2\zeta\omega_0 R_{12}(\tau), \tag{5.5.36}$$

where $R_{ij}(\tau) = R_{X_i X_j}(\tau) = E[X_i(t - \tau)X_j(t)]$ are the correlation functions. Subjected to the initial conditions

$$R_{11}(0) = E[X_1^2] = \frac{\pi K}{2m^2 \zeta\omega_0^3}, \quad R_{12}(0) = E[X_1 X_2] = 0.$$

$R_{11}(\tau)$ can be solved from (5.5.36) as

$$R_{11}(\tau) = \frac{\pi K}{2m^2 \zeta\omega_0^3} \left[e^{-\zeta\omega_0\tau} \left(\cos \omega_d\tau + \frac{\zeta\omega_0}{\omega_d} \sin \omega_d\tau \right) \right]. \tag{5.5.37}$$

The result of (5.5.37) is the same as that of (5.3.7).

The power spectral density of $X_1(t)$ can be calculated by taking the Fourier transform of the correlation function (5.5.37). However, it can be obtained directly by using the same procedure as in Section 4.5.2. For convenience, we rewrite the integral transformation (4.5.24).

$$\bar{\Phi}_{ij}(\omega) = \Im[R_{ij}(\tau)] = \frac{1}{\pi} \int_0^\infty R_{ij}(\tau)e^{-i\omega\tau} d\tau \tag{5.5.38}$$

and the property (4.5.26)

$$\Im\left[\frac{dR_{ij}(\tau)}{d\tau}\right] = i\omega\bar{\Phi}_{ij}(\omega) - \frac{1}{\pi}m_{ij}. \tag{5.5.39}$$

Using Eqs. (5.5.38) and (5.5.39), (5.5.36) can be transformed to

$$i\omega\bar{\Phi}_{11} - \frac{1}{\pi}m_{20} = \bar{\Phi}_{12},$$

$$i\omega\bar{\Phi}_{12} - \frac{1}{\pi}m_{11} = -\omega_0^2\bar{\Phi}_{11} - 2\zeta\omega_0\bar{\Phi}_{12}.$$

(5.5.40)

Equation (5.5.40) is a set of complex algebraic equations, and can be solved to obtain

$$\bar{\Phi}_{11}(\omega) = \frac{(i\omega + 2\zeta\omega_0)m_{20}}{\pi(-\omega^2 + \omega_0^2 - 2i\zeta\omega_0\omega)}.$$

(5.5.41)

Taking the real part, we obtain the power spectral density $\Phi_{11}(\omega) = \Phi_{XX}(\omega)$, as shown in (5.3.31).

Following the same idea and procedure for SDOF linear systems, the correlation functions and the spectral densities for MDOF linear systems in the stationary state can be calculated. Since the equations for the general MDOF case are tedious to formulate, only a two-degree-of-freedom system is considered for illustration. Consider the following system with coupled damping terms,

$$\ddot{X}_1 + c_{11}\dot{X}_1 + c_{12}\dot{X}_2 + \omega_1^2 X_1 = W_1(t),$$
$$\ddot{X}_2 + c_{21}\dot{X}_1 + c_{22}\dot{X}_2 + \omega_2^2 X_2 = W_2(t),$$

(5.5.42)

where $W_1(t)$ and $W_2(t)$ are two independent Gaussian white noises. The corresponding Ito equations are

$$dZ_1 = Z_2 dt,$$
$$dZ_2 = (-\omega_1^2 Z_1 - c_{11}Z_2 - c_{12}Z_4)dt + \sigma_1 dB_1(t),$$
$$dZ_3 = Z_4 dt,$$
$$dZ_4 = (-\omega_2^2 Z_3 - c_{21}Z_2 - c_{22}Z_4)dt + \sigma_2 dB_2(t),$$

(5.5.43)

where $Z_1 = X_1, Z_2 = \dot{X}_1, Z_3 = X_2$, and $Z_4 = \dot{X}_2$ and $B_1(t)$ and $B_2(t)$ are independent unit Wiener processes. Multiplying the equations in

(5.5.43) by $Z_1(t - \tau)$ and taking the ensemble average, we obtain

$$\frac{d}{d\tau}R_{11}(\tau) = R_{12}(\tau),$$

$$\frac{d}{d\tau}R_{12}(\tau) = -\omega_1^2 R_{11}(\tau) - c_{11}R_{12}(\tau) - c_{12}R_{14}(\tau),$$

$$\frac{d}{d\tau}R_{13}(\tau) = R_{14}(\tau),$$

$$\frac{d}{d\tau}R_{14}(\tau) = -\omega_2^2 R_{13}(\tau) - c_{21}R_{12}(\tau) - c_{22}R_{14}(\tau),$$

$$(5.5.44)$$

where $R_{1j}(\tau) = R_{Z_1 Z_j}(\tau) = E[Z_1(t - \tau)Z_j(t)]$ are the correlation functions. The initial conditions for (5.5.44) are

$$R_{1j}(0) = E[X_1 X_j], \quad j = 1, 2, 3, 4, \quad (5.5.45)$$

which are assumed to be found using the method of moment equations. Equation (5.5.44) is a set of ordinary differential equations with constant coefficients, and can be solved both analytically and numerically. It is noted that $R_{11}(\tau)$ is the autocorrelation function of $X_1(t)$.

The power spectral density of $X_1(t)$ can be calculated by taking the Fourier transform of the correlation function $R_{11}(\tau)$. However, it can be obtained directly by using the transformation (5.5.38) and the property (5.5.39),

$$i\omega\bar{\Phi}_{11} - \frac{1}{\pi}E[X_1^2] = \bar{\Phi}_{12},$$

$$i\omega\bar{\Phi}_{12} - \frac{1}{\pi}E[X_1 X_2] = -\omega_1^2\bar{\Phi}_{11} - c_{11}\bar{\Phi}_{12} - c_{12}\bar{\Phi}_{14},$$

$$i\omega\bar{\Phi}_{13} - \frac{1}{\pi}E[X_1 X_3] = \bar{\Phi}_{14},$$

$$i\omega\bar{\Phi}_{14} - \frac{1}{\pi}E[X_1 X_4] = -\omega_2^2\bar{\Phi}_{13} - c_{21}\bar{\Phi}_{12} - c_{22}\bar{\Phi}_{14}. \quad (5.5.46)$$

Equation (5.5.46) is a set of complex algebraic equations, and can be solved either analytically or numerically. The power spectral density of $X_1(t)$, $\Phi_{XX}(\omega) = \Phi_{11}(\omega)$, is the real part of $\bar{\Phi}_{11}(\omega)$ according to (4.5.25).

Following the same procedure, the correlation functions and spectral densities of other state processes, if of interest, can be calculated.

5.5.3 *Fokker–Planck–Kolmogorov (FPK) Equation*

As shown in Section 4.3, the probability density function of the response for system (5.5.1) is governed by the FPK equation as follows

$$\frac{\partial}{\partial t}p + \frac{\partial}{\partial x_1}(x_2 p) + \frac{\partial}{\partial x_2}[(-\omega_0^2 x_1 - 2\zeta\omega_0 x_2)p] - \frac{\pi K}{m^2}\frac{\partial^2 p}{\partial x_2^2} = 0,$$

(5.5.47)

where $p = p(\mathbf{x}, t|\mathbf{x}_0, t_0)$. For fixed initial state, we have the following initial condition

$$p(\mathbf{x}, t_0|\mathbf{x}_0, t_0) = \delta(\mathbf{x} - \mathbf{x}_0) = \delta(x_1 - x_{10})\delta(x_2 - x_{20}).$$ (5.5.48)

Equation (5.5.47) can be solved mathematically. However, since the response $X_1(t)$ and $X_2(t)$ are jointly Gaussian, the probability density function is expressed in Eq. (2.6.28) as

$$p(x_1, x_2) = \frac{1}{2\pi\sigma_{X_1}\sigma_{X_2}\sqrt{1 - \rho_{X_1 X_2}^2}}$$

$$\times \exp\left[-\frac{\sigma_{X_2}^2(x_1 - \mu_{X_1})^2 - 2\sigma_{X_1}\sigma_{X_2}\rho_{X_1 X_2}(x_1 - \mu_{X_1})(x_2 - \mu_{X_2}) + \sigma_{X_1}^2(x_2 - \mu_{X_2})^2}{2\sigma_{X_1}^2\sigma_{X_2}^2(1 - \rho_{X_1 X_2}^2)}\right].$$

(5.5.49)

The five parameters in (5.5.49) can be determined more easily from the moment equations (5.5.13) and (5.5.14) for given initial conditions.

Although the transient solution of (5.5.47) is tedious, the stationary solution $p(x_1, x_2)$ can be obtained relatively easily. It is governed by the reduced FPK equation, i.e., Eq. (5.5.47) without the time-derivative $\partial p / \partial t$ term,

$$x_2 \frac{\partial p}{\partial x_1} + \frac{\partial}{\partial x_2}[(-\omega_0^2 x_1 - 2\zeta\omega_0 x_2)p] - \frac{\pi K}{m^2} \frac{\partial^2 p}{\partial x_2^2} = 0. \qquad (5.5.50)$$

Since the probability density $p(x_1, x_2)$ is non-negative and normalizable, it may be written in the form of

$$p(x_1, x_2) = C \exp[-\phi(x_1, x_2)]. \qquad (5.5.51)$$

Equation (5.5.50) is satisfied if the following two equations are met

$$x_2 \frac{\partial \phi}{\partial x_1} - \omega_0^2 x_1 \frac{\partial \phi}{\partial x_2} = 0, \qquad (5.5.52)$$

$$-2\zeta\omega_0 x_2 - \frac{\pi K}{m^2} \frac{\partial \phi}{\partial x_2} = 0. \qquad (5.5.53)$$

Solving $\phi(x_1, x_2)$ from (5.5.52) and (5.5.53) and substituting into (5.5.51), we have

$$p(x_1, x_2) = C \exp\left[-\frac{m^2 \xi \omega_0}{\pi K}(\omega_0^2 x_1^2 + x_2^2)\right]. \qquad (5.5.54)$$

Mathematically, the set of Eqs. (5.5.52) and (5.5.53) is a sufficient condition for (5.5.50), and it may not be equivalent to Eq. (5.5.50). However, they are indeed equivalent for the present case. We say that system (5.5.1) belong to the case of detailed balance (see, Section 6.2).

Exercise Problems

5.1 Use Laplace transform method to derive the impulse response function of a linear SDOF system

$$m\ddot{x} + c\dot{x} + kx = \delta(t).$$

5.2 Let a linear SDOF system be excited by an external force $f(t)$. Use Laplace transform method to show that, under zero initial conditions, the response is given by the convolution integral

$$x(t) = \int_0^t f(\tau)h(t-\tau)d\tau$$

$$= \frac{1}{m\omega_d} \int_0^t f(\tau)e^{-\zeta\omega_0(t-\tau)} \sin \omega_d(t-\tau)d\tau.$$

5.3 Let a linear SDOF system be excited by an external force $f(t)$. Use Laplace transform method to show that the complete solution is given by Eqs. (5.1.15) and (5.1.16), i.e.,

$$x(t) = [2m\zeta\omega_0 h(t) + m\dot{h}(t)]x_0 + mh(t)\dot{x}_0$$

$$+ \int_0^t f(\tau)h(t-\tau)d\tau.$$

5.4 It is known that the impulse response function and the frequency response function of a SDOF system are, respectively,

$$h(t) = \begin{cases} \dfrac{1}{m\omega_d}e^{-\zeta\omega_0 t} \sin \omega_d t, & t \geq 0 \\[2mm] 0, & t < 0 \end{cases},$$

$$H(\omega) = \frac{1}{m(\omega_0^2 - \omega^2 + 2i\zeta\omega_0\omega)}.$$

Use the Fourier transform to calculate (a) $H(\omega)$ from $h(t)$, and (b) $h(t)$ from $H(\omega)$.

5.5 A model to describe the ground motion of earthquakes is known as the Kanai–Tajimi model, given by

$$\ddot{G} + 2\zeta\omega_0\dot{G} + \omega_0^2 G = 2\zeta\omega_0\dot{R} + \omega_0^2 R,$$

where G and R are the displacements of the ground and the base, respectively. Find the frequency response function and impulse response function of response G to excitation R.

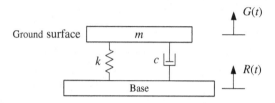

5.6 A modified version of the evolutionary Kanai–Tajimi earthquake model is represented by a random pulse train, given by

$$G(t) = \sum_{j=1}^{N(T)} Y_j w(t - \tau_j), \quad 0 < t \le T,$$

where the shape function is the impulse response function of the system shown in the figure. The input and output are the base displacement x_0 and the ground displacement x_2, respectively. Find the impulse response function using the Laplace transform method.

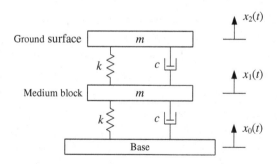

5.7 In the above problem, find the frequency response function.

5.8 For the same system as that in Problems 5.6 and 5.7, use the method of normal modes to find (a) the impulse response function, and (b) the frequency response function.

5.9 A linear SDOF system is excited by a stationary process $F(t)$ with zero mean and a autocovariance function

$$\kappa_{FF}(\tau) = \sigma_F^2 e^{-\alpha|\tau|}, \quad D, \alpha > 0$$

Determine the stationary mean-square value of the system response.

5.10 A linear SDOF system is excited by a stationary process $F(t)$ with zero mean and a spectral density

$$\Phi_{FF}(\omega) = \frac{\sigma_F^2 \alpha}{\pi(\omega^2 + \alpha^2)}.$$

Use the method of residues to calculate the mean-square value of the system response in the stationary state.

5.11 The transient mean square response of a linear SDOF system to a white noise excitation is given by Eq. (5.3.5), i.e.,

$$\sigma_X^2(t) = \frac{\pi K}{2m^2 \zeta \omega_0^3}$$

$$\times \left[1 - e^{-2\zeta \omega_0 t} \left(\frac{\omega_0^2}{\omega_d^2} - \frac{\zeta^2 \omega_0^2}{\omega_d^2} \cos 2\omega_d t + \frac{\zeta \omega_0}{\omega_d} \sin 2\omega_d t \right) \right].$$

Show that, as $\zeta \to 0$, it reduces to

$$\sigma_X^2(t) = \frac{\pi K}{2m^2 \omega_0^3} (2\omega_0 t - \sin 2\omega_0 t.$$

5.12 The transient covariance function, $\kappa_{XX}(t_1, t_2)$, of a linear SDOF system response to a white noise is given by Eq. (5.3.4). Find the transient covariance function $\kappa_{\dot{X}X}(t_1, t_2)$, and show that $X(t)$ and $\dot{X}(t)$ are uncorrelated in stationary state.

5.13 A linear SDOF system is subjected to a white noise excitation. Find the transient covariance function $\kappa_{\dot{X}X}(t_1, t_2)$, stationary covariance function $\kappa_{\dot{X}\dot{X}}(\tau)$ and the transient mean square of $\dot{X}(t)$.

5.14 Consider a linear SDOF system subjected to a band-limited Gaussian white noise $F(t)$ with a power spectral density given by

$$\Phi_{FF}(\omega) = \begin{cases} K, & |\omega| \le \omega_b \\ 0, & |\omega| > \omega_b \end{cases}.$$

Derive the expressions of the power spectral density and the correlation function of the system response in the stationary state. Investigate the effect of the band width ω_b on the correlation function numerically, i.e., calculate the correlation function for a specific system with several ω_b, and show that it approaches Eq. (5.3.7) when ω_b is increasing.

5.15 Consider the following linear system

$$\dot{X}_1 + \alpha_1 X_1 = W_1(t),$$
$$\dot{X}_2 + \alpha_2 X_2 = W_2(t),$$

where $W_1(t)$ and $W_2(t)$ are Gaussian white noises with correlation functions

$$E[W_l(t)W_s(t+\tau)] = 2\pi K_{ls}\delta(\tau), \quad l, s = 1, 2.$$

Find the transient and stationary covariance functions of $X_1(t)$ and $X_2(t)$.

5.16 A two-story building can be modeled as a 2DOF system shown in the figure. For the response $x_2(t)$ of the second story to excitation $f(t)$ applied also to the second story, use the normal mode analysis to find (a) the impulse response function, and (b) the frequency response function.

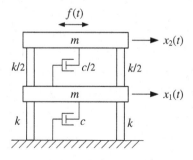

5.17 For the system in Problem 5.16, letting $f(t)$ be a Gaussian white noise, find the autocorrelation function and mean-square value of $X_2(t)$ in the stationary state.

5.18 The equation of motion is given by

$$\ddot{X} + 2\zeta\omega_0\dot{X} + \omega_0^2 X = a(t)W(t),$$

where $a(t)$ is a deterministic function and $W(t)$ is a Gaussian white noise with a spectral density K. Determine the autocorrelation function of response $X(t)$ for the following two cases: (a) $a(t) = e^{-\alpha t}$, $\alpha > 0$, (b) $a(t) = e^{-\alpha t} - e^{-\beta t}$, $\beta > \alpha > 0$.

5.19 Give an equation of motion

$$\ddot{X} + 2\zeta\omega_0\dot{X} + \omega_0^2 X = e^{-\alpha t}S(t),$$

where $S(t)$ is a stationary process with zero mean and a spectral density $\Phi_{SS}(\omega)$ given by

$$\Phi_{SS}(\omega) = \frac{c}{\pi(c^2 + \omega^2)},$$

calculate the mean square of $X(t)$.

5.20 Let $F(t)$ be an evolutionary process given by

$$F(t) = \int\limits_{-\infty}^{\infty} a(t, \omega)e^{i\omega t} dZ(\omega),$$

where

$$a(t, \omega) = \exp(-|\omega^2 - \omega_1^2|t),$$

$$E[dZ(\omega_1)dZ^*(\omega_2)] = \begin{cases} K d\omega, & \omega_1 = \omega_2 \\ 0, & \omega_1 \neq \omega_2 \end{cases}.$$

Find the evolutionary spectral density of $X(t)$ governed by

$$\ddot{X} + 2\zeta\omega_0\dot{X} + \omega_0^2 X = F(t).$$

5.21 An evolutionary Kanai–Tajimi earthquake model is a random pulse train defined in Section 3.9.4, given by

$$G(t) = \sum_{j=1}^{N(T)} Y_j h(t - \tau_j), \quad 0 < t \leq T,$$

where Y_j are independent, identically distributed random variables with a zero mean and a finite mean-square value $E[Y^2]$, the shape function is given by a linear Kainai–Tajimi model in Problem 5.5, i.e.,

$$h(t) = \omega_0 e^{-\zeta \omega_0 t} \left(\frac{1 - 2\zeta^2}{\sqrt{1 - \zeta^2}} \sin \omega_d t + 2\zeta \cos \omega_d t \right).$$

Determine the covariance function of $G(t)$, assuming that the average pulse arrival rate is

$$\mu_\Lambda(t) = e^{-\alpha t} - e^{-\beta t}, \quad \beta > \alpha > 0.$$

5.22 Use the method of moment equations described in Section 5.5.1 to find the transient mean square functions of response displacement and velocity of a linear SDOF system under a Gaussian white noise excitation, assuming zero initial conditions.

5.23 Use the method in Section 5.5.2 to find the stationary correlation function and spectral density function of response velocity of a linear SDOF system under a Gaussian white noise excitation.

5.24 A linear SDOF system is excited by a stochastic process $F(t)$ generated from the following first-order filter

$$\dot{F} + \alpha F = W(t),$$

where $W(t)$ is a Gaussian white noise. Determine the stationary mean-square values of the system response using the method of moment equations described in Section 5.5.1.

5.25 For the same system and excitation as those in the previous problem, use the method described in Section 5.5.2 to find the correlation function of $X(t)$ in the stationary state.

5.26 For the same system and excitation as those in the previous problem, use the method described in Section 5.5.2 to find the spectral density of $X(t)$ in the stationary state.

5.27 Consider the following linear system

$$\dot{X}_1 = -\omega_1 X_1 + \gamma X_2 + W_1(t),$$
$$\dot{X}_2 = \gamma X_1 - \omega_2 X_2 + W_2(t),$$

where ω_1, ω_2 and γ are positive constants, and $W_1(t)$ and $W_2(t)$ are independent Gaussian white noises with an identical spectral density.

(a) Derive the reduced FPK equation for the system.
(b) Use the method in Section 5.5.3 to solve the FPK equation for the stationary probability density.
(c) Find the condition for the probability density to exist.
(d) From the probability density function, find the mean-square values of $X_1(t)$ and $X_2(t)$, and their correlation coefficient.

5.28 For the same system as in the previous problem, use the methods in Sections 5.5.1 and 5.5.2 to find for the stationary state, (a) the mean-square values of $X_1(t)$ and $X_2(t)$, and their correlation coefficient, (b) the autocovariance function of $X_1(t)$ and (c) the power spectral density of $X_1(t)$.

5.29 Consider the following linear system

$$\dot{X}_1 = a_{11} X_1 + a_{12} X_2 + W_1(t),$$
$$\dot{X}_2 = a_{21} X_1 + a_{22} X_2 + W_2(t),$$

where $W_1(t)$ and $W_2(t)$ are Gaussian white noises with correlation functions

$$E[W_l(t) W_s(t+\tau)] = 2\pi K_{ls}\delta(\tau), \quad l, s = 1, 2.$$

(a) Derive the reduced FPK equation for the system.
(b) Use the method in Section 5.5.3 to solve the FPK equation for the stationary probability density.

5.30 Consider the following linear system

$$\dot{X}_1 + \alpha_1 X_1 = W_1(t),$$
$$\dot{X}_2 + \alpha_2 X_2 = W_2(t),$$

where $W_1(t)$ and $W_2(t)$ are Gaussian white noises with correlation functions

$$E[W_l(t)W_s(t+\tau)] = 2\pi K_{ls}\delta(\tau), \quad l, s = 1, 2.$$

(a) Derive the reduced FPK equation for the system, and solve it.

(b) Use the method in Section 5.5.2 to find the second-order moments.

(c) Use the method in Section 5.5.3 to find the cross-correlation function and cross-spectral density of $X_1(t)$ and $X_2(t)$.

5.31 Give a linear SDOF system

$$\ddot{X} + 2\zeta\omega_0\dot{X} + \omega_0^2 X = F(t),$$

where $F(t)$ is a stationary process governed by a one-dimensional filter

$$dF = -\alpha F dt + \sqrt{\frac{\alpha}{\delta+1}(\Delta^2 - F^2)}dB(t),$$

simulate the system to obtain the probability density and the mean square of $X(t)$ in the stationary state for cases of (a) $\delta = 1$, (b) $\delta = 0$ and (c) $\delta = -0.5$. For all three cases, use the same mean-square value for $F(t)$.

CHAPTER 6

EXACT STATIONARY SOLUTIONS OF NONLINEAR STOCHASTIC SYSTEMS

For linear systems, exact analytical solutions are obtainable based on the known knowledge of the excitation processes, as shown in Chapter 5. However, many practical dynamical systems are nonlinear and also cannot be approximated as linear systems, then the methods described in Chapter 5 are not applicable.

Exact solutions for stochastically excited nonlinear dynamical systems are difficult to obtain. The possibility for exact solutions does exist, however, when the stochastic excitations can be idealized as Gaussian white noises, in which case the response of a system is a Markov diffusion process. The probability density of a diffusion process is governed by the Fokker–Planck–Kolmogorov (FPK) equation which can be derived from the system equations of motion. Still, the full solution to the FPK equation, which shows how the probability structure evolves with time, can be obtained only for very special first-order systems for which the system response is a scalar diffusion process (Caughey and Dienes, 1961).

In many practical cases, only stationary states are of concern. For a nonlinear dynamical system subjected to Gaussian white-noise excitations, the stationary probability density, if exists, is governed by the reduced FPK equation, i.e., FPK equation without the time-derivative term. This chapter will discuss several classes of such nonlinear stochastic dynamical systems for which exact stationary solution are obtainable.

When the excitations appear only as homogeneous terms on the right-hand sides of the equations of motions, they are known as the external or additive excitations. On the other hand, if the excitations appear in the coefficients of the unknowns in the equations of motion, they are called the parametric or multiplicative excitations. The existence of parametric excitations destroys the superposition principle. Therefore, a linear system only allows external excitations, and a system with parametric excitations is essentially nonlinear, even the system properties, such as damping and stiffness, are linear. Exact stationary first- and second-order statistics will be explored for linear systems under external and/or parametric stochastic excitations.

6.1 Stationary Potential

A dynamical system under Gaussian white-noise excitations can be described by the equations of motion in the state space,

$$\frac{d}{dt}X_j = f_j(\mathbf{X}) + \sum_{l=1}^{m} g_{jl}(\mathbf{X})W_l(t), \quad j = 1, 2, \ldots, n, \qquad (6.1.1)$$

where $\mathbf{X} = [X_1, X_2, \ldots, X_n]^T$, and $W_j(t)$ are Gaussian white noises with

$$E[W_l(t)W_s(t+\tau)] = 2\pi K_{ls}\delta(\tau), \quad l, s = 1, 2, \ldots, m. \qquad (6.1.2)$$

Equation (6.1.1) implies that the system properties are time invariant. The system response $\mathbf{X}$ is a Markov diffusion vector, and its stationary probability density $p(\mathbf{x})$, if exists, is governed by the reduced FPK equation

$$\sum_{j=1}^{n} \frac{\partial}{\partial x_j} G_j = 0, \qquad (6.1.3)$$

where G_j is known as the probability flow in the jth direction given by

$$G_j = a_j(\mathbf{x})p - \frac{1}{2}\sum_{k=1}^{n}\frac{\partial}{\partial x_k}[b_{jk}(\mathbf{x})p]. \qquad (6.1.4)$$

In (6.1.4), the first and second derivate moments a_j and b_{jk} are obtained from the equations of motion (6.1.1) as follows

$$a_j(\mathbf{x}) = f_j(\mathbf{x}) + \pi\sum_{k=1}^{n}\sum_{l,s=1}^{m}K_{ls}g_{ks}(\mathbf{x})\frac{\partial}{\partial x_k}g_{jl}(\mathbf{x}). \qquad (6.1.5)$$

$$b_{jk}(\mathbf{x}) = 2\pi\sum_{l,s=1}^{m}K_{ls}g_{jl}(\mathbf{x})g_{ks}(\mathbf{x}). \qquad (6.1.6)$$

For practical cases, the boundaries are either reflective or natural, indicating that the probability flows vanish at boundaries,

$$G_j = 0, \quad \text{at boundaries} \qquad (6.1.7)$$

Consider a set of sufficient conditions for Eq. (6.1.3), namely, all probability flows vanish not only at boundaries but everywhere for every j, i.e.,

$$G_j = a_j(\mathbf{x})p - \frac{1}{2}\sum_{k=1}^{n}\frac{\partial}{\partial x_k}[b_{jk}(\mathbf{x})p] = 0. \qquad (6.1.8)$$

In this case, system (6.1.1) is said to belong to the class of stationary potential. Express the probability density in the form

$$p(\mathbf{x}) = C\exp[-\phi(\mathbf{x})], \qquad (6.1.9)$$

where C is a normalization constant, and function $\phi(\mathbf{x})$ is called the probability potential function. Expressing the probability density function $p(\mathbf{x})$ in the form of negative exponential functional in Eq. (6.1.9) is based on two characteristics of the probability density function $p(\mathbf{x})$, non-negative and normalizable. Substituting (6.1.9) into (6.1.8), we obtain

$$\sum_{k=1}^{n} b_{jk}(\mathbf{x}) \frac{\partial \phi(\mathbf{x})}{\partial x_k} = \sum_{k=1}^{n} \frac{\partial}{\partial x_k} b_{jk}(\mathbf{x}) - 2a_j(\mathbf{x}). \tag{6.1.10}$$

There are n equations in (6.1.10). If system (6.1.1) belongs to the class of stationary potential, a function $\phi(\mathbf{x})$ can be found to satisfy all the n equations.

Consider the case of a non-singular matrix of the second derivate moments, $\mathbf{B} = [b_{jk}]$, and denote its inverse as $\mathbf{B}^{-1} = \mathbf{D} = [d_{jk}]$. Then Eq. (6.1.10) can be rewritten as

$$\frac{\partial \phi(\mathbf{x})}{\partial x_l} = \sum_{j=1}^{n} d_{lj}(\mathbf{x}) \sum_{k=1}^{n} \left[\frac{\partial}{\partial x_k} b_{jk}(\mathbf{x}) - 2a_j(\mathbf{x}) \right]. \tag{6.1.11}$$

A special case known as isotropic diffusion (Stratonovich, 1963) is defined as

$$b_{ij}(\mathbf{x}) = \begin{cases} K(\mathbf{x}), & i = j \\ 0, & i \neq j \end{cases}. \tag{6.1.12}$$

Substitution of (6.1.12) into (6.1.11) leads to

$$\frac{\partial \phi(\mathbf{x})}{\partial x_l} = \frac{1}{K(\mathbf{x})} \left[\frac{\partial K(\mathbf{x})}{\partial x_l} - 2a_l(\mathbf{x}) \right]. \tag{6.1.13}$$

If the following compatibility conditions are satisfied

$$\frac{\partial}{\partial x_m} \left[\frac{a_l(\mathbf{x})}{K(\mathbf{x})} \right] = \frac{\partial}{\partial x_l} \left[\frac{a_m(\mathbf{x})}{K(\mathbf{x})} \right], \tag{6.1.14}$$

a consistent $\phi(\mathbf{x})$ function can be derived from (6.1.13), and the stationary probability density is then obtained as

$$p(\mathbf{x}) = \frac{C}{K(\mathbf{x})} \exp \left[\sum_{i=1}^{n} \int \frac{2a_i(\mathbf{x})}{K(\mathbf{x})} dx_i \right]. \tag{6.1.15}$$

One-dimensional diffusion process is a special case of the isotropic diffusion with $K(\mathbf{x}) = b(x)$. The stationary probability density is then obtained from (6.1.15) as

$$p(x) = \frac{C}{b(x)} \exp \int \frac{2a(x)}{b(x)} dx. \tag{6.1.16}$$

In this case, we have from (6.1.3)

$$G = a(x)p - \frac{1}{2}\frac{\partial}{\partial x}[b(x)p] = G_c = \text{constant.} \tag{6.1.17}$$

According to (6.1.7), $G_c = 0$, indicating that the probability flow G vanishes everywhere.

Consider a SDOF oscillatory system governed by

$$\ddot{X} + h(X, \dot{X}) + u(X) = W(t), \tag{6.1.18}$$

where $h(X, \dot{X})$ and $u(X)$ represent the damping force and restoring force, respectively, and $W(t)$ is a Gaussian white noise. Denoting $X_1 = X$ and $X_2 = \dot{X}$, (6.1.18) can be written in state space as

$$\begin{aligned} \dot{X}_1 &= X_2, \\ \dot{X}_2 &= -h(X_1, X_2) - u(X_1) + W(t). \end{aligned} \tag{6.1.19}$$

Thus, the probability flows are

$$G_1 = x_2, \quad G_2 = -[h(x_1, x_2) + u(x_1)]p - \pi K \frac{\partial p}{\partial x}. \tag{6.1.20}$$

Equation (6.1.20) shows that the probability flows do not vanish everywhere in the defining domain of X_1 and X_2; hence, a SDOF oscillatory system (6.1.19) does not belong to the class of stationary potential. This statement can also be extended to MDOF oscillatory systems.

Example 6.1.1 Consider the system

$$\begin{aligned} \dot{X}_1 &= -\alpha_1 X_1 - \beta_1 X_1^3 + \gamma X_2 + W_1(t), \\ \dot{X}_2 &= \gamma X_1 - \alpha_2 X_2 - \beta_2 X_2^3 + W_2(t), \end{aligned} \tag{6.1.21}$$

where α_1, β_1, α_2, β_2, and γ are positive constants and $W_1(t)$ and $W_2(t)$ are independent Gaussian white noises with an identical spectral density. The first and second derivate moments are

$$a_1 = -\alpha_1 x_1 - \beta_1 x_1^3 + \gamma x_2, \quad a_2 = \gamma x_1 - \alpha_2 x_2 - \beta_2 x_2^3,$$
$$b_{11} = b_{22} = 2\pi K, b_{12} = b_{21} = 0. \tag{6.1.22}$$

From (6.1.10), we obtain

$$\pi K \frac{\partial \phi}{\partial x_1} = \alpha_1 x_1 + \beta_1 x_1^3 - \gamma x_2, \tag{6.1.23}$$

$$\pi K \frac{\partial \phi}{\partial x_2} = -\gamma x_1 + \alpha_2 x_2 + \beta_2 x_2^3. \tag{6.1.24}$$

Function ϕ can be obtained to satisfy both Eqs. (6.1.23) and (6.1.24) as follows

$$\phi(x_1, x_2) = \frac{1}{\pi K} \left(\frac{1}{2} \alpha_1 x_1^2 + \frac{1}{4} \beta_1 x_1^4 - \gamma x_1 x_2 + \frac{1}{2} \alpha_2 x_2^2 + \frac{1}{4} \beta_2 x_2^4 \right). \tag{6.1.25}$$

Therefore, system (6.1.21) belongs to the class of stationary potential.

6.2 Detailed Balance

It is clear that system (6.1.1) does not belong to the class of stationary potential in general. However, to satisfy the FPK equation, we may split each first derivate moment into two parts as

$$a_j(\mathbf{x}) = a_j^R(\mathbf{x}) + a_j^I(\mathbf{x}), \tag{6.2.1}$$

where $a_j^R(\mathbf{x})$ and $a_j^I(\mathbf{x})$ are known as the reversible and irreversible components, respectively. The terms of "reversible" and "irreversible" were originated in a reference (Graham and Haken, 1971) from mathematical point of view. Substituting (6.2.1) into the reduced FPK equation of (6.1.4) and (6.1.3), we have

$$\sum_{j=1}^{n} \frac{\partial}{\partial x_j} G_j = \sum_{j=1}^{n} \frac{\partial}{\partial x_j} a_j^R(\mathbf{x}) p$$

$$+ \sum_{j=1}^{n} \frac{\partial}{\partial x_j} \left\{ a_j^I(\mathbf{x}) p - \frac{1}{2} \sum_{k=1}^{n} \frac{\partial}{\partial x_k} [b_{jk}(\mathbf{x}) p] \right\} = 0. \tag{6.2.2}$$

A set of sufficient conditions for the reduced FPK equation (6.2.2) to be satisfied are

$$a_j^I(\mathbf{x})p - \frac{1}{2}\sum_{k=1}^{n}\frac{\partial}{\partial x_k}b_{jk}(\mathbf{x})p = 0, \quad j = 1, 2, \ldots, n, \qquad (6.2.3)$$

$$\sum_{j=1}^{n}\frac{\partial}{\partial x_j}a_j^R(\mathbf{x})p = 0. \qquad (6.2.4)$$

There are total $n + 1$ equations in (6.2.3) and (6.2.4). If a valid probability density function p can be found to satisfy all the $n + 1$ equations, the system is said to belong to the class of detailed balance. The concept of detailed balance was first applied by van Kampen (1957), and its mathematical implication and derivation, and its application to solve the FPK equation were due to Graham and Haken (1971) and Yong and Lin (1987).

Equation (6.2.3) is similar to Eq. (6.1.8), and the left-hand side of (6.2.3) is known as the potential component of the probability flow in the jth direction. Eq. (6.2.3) indicates that every potential component vanishes in the detailed balance case. On the other hand, each term in the summation in Eq. (6.2.4) is called the circulatory component of the probability flow, and Eq. (6.2.4) states that all circulatory components are balanced. Therefore, the reversible portion a_j^R of the drift coefficient a_j contributes to the circulatory probability flow, and the remaining irreversible portion a_j^I is used to balance the differential diffusion and maintain a zero probability potential flow in the jth direction.

It is noticed that the stationary potential is a special case of the detailed balance with $a_j^I(\mathbf{x}) = a_j(\mathbf{x})$ and $a_j^R(\mathbf{x}) = 0$, and then the balancing of the circulatory components of the probability flow is guaranteed.

In terms of the probability potential function $\phi(\mathbf{x})$ defined in (6.1.9), (6.2.3) and (6.2.4) are written as

$$a_j^I(\mathbf{x}) = \frac{1}{2}\sum_{k=1}^{n}\left[\frac{\partial}{\partial x_k}b_{jk}(\mathbf{x}) - b_{jk}(\mathbf{x})\frac{\partial\phi}{\partial x_k}\right], \qquad (6.2.5)$$

$$\sum_{j=1}^{n}\frac{\partial}{\partial x_j}a_j^R(\mathbf{x}) = \sum_{j=1}^{n}a_j^R(\mathbf{x})\frac{\partial\phi}{\partial x_j}. \qquad (6.2.6)$$

If the system belongs to the class of detailed balance, a consistent $\phi(\mathbf{x})$ function exists to satisfy all the equations.

In the following sections, it is illustrated how to split the first derivate moments and how to use Eqs. (6.2.5) and (6.2.6) to obtain exact stationary solutions for several different types of nonlinear stochastic systems.

6.2.1 A SDOF System under an External Excitation

Consider the system

$$\ddot{X} + h(\Lambda)\dot{X} + u(X) = W(t), \tag{6.2.7}$$

where $u(X)$ is the restoring force, $W(t)$ is a Gaussian white noise with a spectral density K, and Λ is the system total energy

$$\Lambda = \frac{1}{2}\dot{X}^2 + \int_0^X u(z)dz. \tag{6.2.8}$$

Equation (6.2.7) indicates that the damping force is energy dependent.

Letting $X_1 = X$ and $X_2 = \dot{X}$, the Ito equations corresponding to (6.2.7) are

$$\begin{aligned} dX_1 &= X_2 dt, \\ dX_2 &= -[h(\Lambda)X_2 + u(X_1)]dt + \sqrt{2\pi K}dB(t). \end{aligned} \tag{6.2.9}$$

The first and second derivate moments are

$$a_1 = x_2, \quad a_2 = -h(\lambda)x_2 - u(x_1), \quad b_{11} = b_{12} = b_{21} = 0, \quad b_{22} = 2\pi K. \tag{6.2.10}$$

Split each first derivate moments into the reversible and irreversible parts as

$$a_1^R = x_2, \quad a_1^I = 0, \quad a_2^R = -u(x_1), \quad a_2^I = -h(\lambda)x_2. \tag{6.2.11}$$

The splitting in (6.2.11) indicates that the irreversible parts are associated with the damping forces, and the reversible parts with the inertia and restoring forces. This is generally true for systems in

Lagrange format (1.0.2) and Hamiltonian format (1.0.4). Substituting (6.2.10) and (6.2.11) into (6.2.5) and (6.2.6), we have

$$\pi K \frac{\partial \phi}{\partial x_2} = h(\lambda)x_2, \tag{6.2.12}$$

$$x_2 \frac{\partial \phi}{\partial x_1} = u(x_1)\frac{\partial \phi}{\partial x_2}. \tag{6.2.13}$$

The general solution of (6.2.12) is

$$\phi = \frac{1}{\pi K} \int_0^\lambda h(z)dz + g(x_1). \tag{6.2.14}$$

By substituting (6.2.14) into (6.2.13), it is found that $g(x_1)$ must be a constant. Thus,

$$p(x_1, x_2) = C \exp\left[-\frac{1}{\pi K}\int_0^\lambda h(z)dz\right], \quad \lambda = \frac{1}{2}x_2^2 + \int_0^{x_1} u(z)dz. \tag{6.2.15}$$

Therefore, system (6.2.7) belongs to the class of detailed balance. For the special case of linear damping, $h(\Lambda)\dot{X} = \alpha\dot{X}$,

$$p(x_1, x_2) = C \exp\left\{-\frac{\alpha}{\pi K}\left[\int_0^{x_1} u(z)dz + \frac{1}{2}x_2^2\right]\right\}. \tag{6.2.16}$$

6.2.2 *A SDOF System under both External and Parametric Excitations*

Consider a nonlinear system under both external and parametric excitations

$$\ddot{X} + (\alpha + \beta X^2)\dot{X} + \omega_0^2 X = XW_1(t) + W_2(t), \tag{6.2.17}$$

where $W_1(t)$ and $W_2(t)$ are assumed to be independent Gaussian white noises with spectral densities K_{11} and K_{22}, respectively. Denoting X by X_1 and $\dot{X}$ by X_2, the Ito equations for system (6.2.17) are

$$dX_1 = X_2 dt,$$
$$dX_2 = -[(\alpha + \beta X_1^2)X_2 + \omega_0^2 X_1]dt + \sqrt{2\pi(K_{11}X_1^2 + K_{22})}dB(t).$$
$$(6.2.18)$$

The first and second derivate moments are

$$a_1 = x_2, \quad a_2 = -(\alpha + \beta x_1^2)x_2 - \omega_0^2 x_1,$$
$$b_{11} = b_{12} = b_{21} = 0, \quad b_{22} = 2\pi(K_{11}x_1^2 + K_{22}). \tag{6.2.19}$$

By splitting the first derivate moments into

$$a_1^R = x_2, \quad a_1^I = 0, \quad a_2^R = -\omega_0^2 x_1, \quad a_2^I = -(\alpha + \beta x_1^2)x_2. \tag{6.2.20}$$

Equations (6.2.5) and (6.2.6) become

$$\pi(K_{11}x_1^2 + K_{22})\frac{\partial \phi}{\partial x_2} = (\alpha + \beta x_1^2)x_2, \tag{6.2.21}$$

$$x_2 \frac{\partial \phi}{\partial x_1} = \omega_0^2 x_1 \frac{\partial \phi}{\partial x_2}. \tag{6.2.22}$$

The general solution of (6.2.22) is

$$\phi(x_1, x_2) = \phi(\lambda), \quad \lambda = \frac{1}{2}\omega_0^2 x_1^2 + \frac{1}{2}x_2^2. \tag{6.2.23}$$

Substituting (6.2.23) into (6.2.21), we obtain

$$\frac{d\phi}{d\lambda} = \frac{\alpha + \beta x_1^2}{\pi(K_{22} + K_{11}x_1^2)}. \tag{6.2.24}$$

Since ϕ is a function of λ, the right-hand side of (6.2.24) must also be a function of λ, leading to

$$\frac{\alpha}{\beta} = \frac{K_{22}}{K_{11}}. \tag{6.2.25}$$

When condition (6.2.25) is met, we have

$$\phi = \frac{\alpha}{\pi K_{22}}\lambda \tag{6.2.26}$$

and

$$p(x_1, x_2) = C\exp\left[-\frac{\alpha}{2\pi K_{22}}(\omega_0^2 x_1^2 + x_2^2)\right]. \tag{6.2.27}$$

Equation (6.2.25) is a condition for the system to belong to the class of detailed balance and have an exact solution for the stationary probability distribution. It relates the system properties α and β with the excitations K_{11} and K_{22}. For practical problems, such condition will not be satisfied.

It is noted that (6.2.27) is a Gaussian distribution, which is also obtained for the linear system without the nonlinear damping and the parametric excitation in (6.2.17).

6.2.3 *A SDOF System with Parametric Excitations in both Damping and Stiffness*

A more complicated nonlinear system is given by

$$\ddot{X} + f(X, \dot{X})\dot{X} + \omega_0^2 X = X W_1(t) + \dot{X} W_2(t) + W_3(t), \qquad (6.2.28)$$

where $W_1(t)$, $W_2(t)$ and $W_3(t)$ are independent Gaussian white noises with spectral densities K_{11}, K_{22} and K_{33}, respectively. Denoting X by X_1 and $\dot{X}$ by X_2, the Ito equations for system (6.2.28) are

$$\begin{aligned}
dX_1 &= X_2 dt, \\
dX_2 &= \{[-f(X_1, X_2) + \pi K_{22}]X_2 - \omega_0^2 X_1\}dt \\
&\quad + \sqrt{2\pi(K_{11}X_1^2 + K_{22}X_2^2 + K_{33})}dB(t),
\end{aligned} \qquad (6.2.29)$$

where the term $\pi K_{22}X_2$ arises from the Wong–Zakai correction. Separating the first derivate moments into the reversible and irreversible parts,

$$a_1^R = x_2, \quad a_1^I = 0, \quad a_2^R = -\omega_0^2 x_1, \quad a_2^I = [-f(x_1, x_2) + \pi K_{22}]x_2. \qquad (6.2.30)$$

The second derivate moments are

$$b_{11} = b_{12} = b_{21} = 0, \quad b_{22} = 2\pi(K_{11}x_1^2 + K_{22}x_2^2 + K_{33}). \qquad (6.2.31)$$

Equations (6.2.5) and (6.2.6) now become

$$\pi(K_{11}x_1^2 + K_{22}x_2^2 + K_{33})\frac{\partial\phi}{\partial x_2} = [f(x_1, x_2) + \pi K_{22}]x_2, \qquad (6.2.32)$$

$$x_2\frac{\partial\phi}{\partial x_1} = \omega_0^2 x_1\frac{\partial\phi}{\partial x_2}. \qquad (6.2.33)$$

The general solution of (6.2.33) is the same as that of the system in Section 6.2.2, given in (6.2.23). Substituting (6.2.23) into (6.2.32), we obtain

$$\frac{d\phi}{d\lambda} = \frac{f(x_1, x_2) + \pi K_{22}}{\pi(K_{11}x_1^2 + K_{22}x_2^2 + K_{33})}. \tag{6.2.34}$$

The right-hand side of (6.2.34) must be a function of λ. We exclude the special case of absence of the parametric excitation in the damping, i.e., the case of $K_{22} = 0$, as it is discussed in Section 6.2.2. The validity of (6.2.34) leads to

$$f(x_1, x_2) = f(\lambda) \quad \text{and} \quad K_{11} = \omega_0^2 K_{22} \tag{6.2.35}$$

and (6.2.34) is then written as

$$\frac{d\phi}{d\lambda} = \frac{f(\lambda) + \pi K_{22}}{\pi(2K_{22}\lambda + K_{33})}. \tag{6.2.36}$$

Integrate (6.2.36) to obtain $\phi(\lambda)$ and hence the stationary probability density

$$p(x_1, x_2) = C(2K_{22}\lambda + K_{33})^{-1/2} \exp\left[-\int_0^\lambda \frac{f(z)}{\pi(2K_{22}z + K_{33})} dz\right]. \tag{6.2.37}$$

If $f(\lambda)$ is a constant α, i.e., the system damping is linear, (6.2.37) is simplified as

$$p(x_1, x_2) = C(2K_{22}\lambda + K_{33})^{-\gamma_1}, \tag{6.2.38}$$

where

$$\gamma_1 = \frac{1}{2}\left(1 + \frac{\alpha}{\pi K_{22}}\right). \tag{6.2.39}$$

The stationary probability density (6.2.38) is integrable only if $\gamma_1 > 1$, i.e., $\alpha > \pi K_{22}$, indicating that the system damping must be large enough so that the system is not divergent when x_1 and/or x_2 approach infinity.

If $f(\lambda)$ is a linear function, i.e., $f(\lambda) = \alpha + \beta\lambda$ $(\beta > 0)$, (6.2.37) results in

$$p(x_1, x_2) = C(2K_{22}\lambda + K_{33})^{-\gamma_2} \exp\left(-\frac{\beta\lambda}{2\pi K_{22}}\right), \qquad (6.2.40)$$

where

$$\gamma_2 = \frac{1}{2}\left(1 + \frac{\alpha}{\pi K_{22}} - \frac{\beta K_{33}}{2\pi K_{22}^2}\right). \qquad (6.2.41)$$

It can be shown that (6.2.40) is always integrable as long as the external excitation is present, i.e., $K_{33} \neq 0$. In this case, the existence of the external excitation moves the system away from the trivial solution $(0, 0)$, and the strong damping prevents the system from divergence at infinite boundaries. If the external excitation is absent, (6.2.40) is integrable only if $\gamma_2 < 1$, i.e., $\alpha < \pi K_{22}$. Otherwise, the system will approach the origin, a trivial solution, due to strong damping.

More cases of $f(\lambda)$ as different functions of energy λ can be discussed similarly.

6.2.4 *A Two-Degree-of-Freedom System with Coupled Restoring Forces*

Now consider a two-degree-of-freedom (two-DOF) system with coupled restoring forces, governed by

$$\begin{aligned}
\ddot{Z}_1 + \alpha_1\dot{Z}_1 + \frac{\partial U(Z_1, Z_2)}{\partial Z_1} &= W_1(t), \\
\ddot{Z}_2 + \alpha_2\dot{Z}_2 + \frac{\partial U(Z_1, Z_2)}{\partial Z_2} &= W_2(t),
\end{aligned} \qquad (6.2.42)$$

where $U(Z_1, Z_2)$ is the potential energy of the entire system and $W_1(t)$ and $W_2(t)$ are independent Gaussian white noises with spectral densities K_{11} and K_{22}, respectively. Let $X_1 = Z_1$, $X_2 = \dot{Z}_1$, $X_3 = Z_2$ and $X_4 = \dot{Z}_2$. The reversible and irreversible parts of the first derivate moments, and the second derivate moments are

$$a_1^R = x_2, \quad a_1^I = 0, \quad a_2^R = -\frac{\partial U(x_1, x_3)}{\partial x_1}, \quad a_2^I = -\alpha_1 x_2,$$

$$a_3^R = x_4, \quad a_3^I = 0, \quad a_4^R = -\frac{\partial U(x_1, x_3)}{\partial x_3}, \quad a_4^I = -\alpha_2 x_4, \tag{6.2.43}$$

$$b_{11} = 2\pi K_{11}, \quad b_{12} = b_{21} = 0, \quad b_{22} = 2\pi K_{22}. \tag{6.2.44}$$

Substituting (6.2.43) and (6.2.44) into (6.2.5) and (6.2.6), we obtain

$$\pi K_{11} \frac{\partial \phi}{\partial x_2} = \alpha_1 x_2, \quad \pi K_{22} \frac{\partial \phi}{\partial x_4} = \alpha_2 x_4, \tag{6.2.45}$$

$$x_2 \frac{\partial \phi}{\partial x_1} - \frac{\partial U(x_1, x_3)}{\partial x_1} \frac{\partial \phi}{\partial x_2} + x_4 \frac{\partial \phi}{\partial x_3} - \frac{\partial U(x_1, x_3)}{\partial x_3} \frac{\partial \phi}{\partial x_4} = 0. \tag{6.2.46}$$

The general solution of (6.2.4) is

$$\phi(x_1, x_2, x_3, x_4) = \phi(\lambda), \quad \lambda = U(x_1, x_3) + \frac{1}{2}x_2^2 + \frac{1}{2}x_4^2. \tag{6.2.47}$$

Substituting (6.2.47) into (6.2.45), we have

$$\frac{d\phi}{d\lambda} = \frac{\alpha_1}{\pi K_{11}}, \quad \frac{d\phi}{d\lambda} = \frac{\alpha_2}{\pi K_{22}}. \tag{6.2.48}$$

In order for the two equations in (6.2.48) to be consistent, the following condition must be imposed

$$\frac{\alpha_1}{K_{11}} = \frac{\alpha_2}{K_{22}}. \tag{6.2.49}$$

With this restriction, the stationary probability density is

$$p(x_1, x_2, x_3, x_4) = C \exp\left\{ -\frac{\alpha_1}{\pi K_{11}} \left[\frac{1}{2}x_2^2 + \frac{1}{2}x_4^2 + U(x_1, x_3) \right] \right\}. \tag{6.2.50}$$

Equation (6.2.50) shows that the kinetic energy in each coordinate is identically distributed, known as the case of equipartitioned energy (Lin, 1967). The above solution procedure can be extended to cases of systems with more degrees-of-freedom.

6.2.5 *A Two-DOF System with Coupled Damping Forces*

Consider another two-DOF system with coupled damping forces, governed by

$$\ddot{Z}_1 + \alpha_{11}\dot{Z}_1 + \alpha_{12}\dot{Z}_2 + u_1(Z_1) = W_1(t),$$
$$\ddot{Z}_2 + \alpha_{21}\dot{Z}_1 + \alpha_{22}\dot{Z}_2 + u_2(Z_2) = W_2(t), \tag{6.2.51}$$

where $W_1(t)$ and $W_2(t)$ are Gaussian white noises with correlation functions

$$E[W_i(t)W_j(t+\tau)] = 2\pi K_{ij}\delta(\tau), \quad i, j = 1, 2. \tag{6.2.52}$$

Let $X_1 = Z_1$, $X_2 = \dot{Z}_1$, $X_3 = Z_2$ and $X_4 = \dot{Z}_2$. The reversible and irreversible parts of the first derivate moments and the non-zero second derivate moments are

$$a_1^R = x_2, \quad a_1^I = 0, \quad a_2^R = -u_1(x_1), \quad a_2^I = -\alpha_{11}x_2 - \alpha_{12}x_4,$$
$$a_3^R = x_4, \quad a_3^I = 0, \quad a_4^R = -u_2(x_3), \quad a_4^I = -\alpha_{21}x_2 - \alpha_{22}x_4, \tag{6.2.53}$$

$$b_{22} = 2\pi K_{11}, \quad b_{24} = b_{42} = 2\pi K_{12}, \quad b_{44} = 2\pi K_{22}. \tag{6.2.54}$$

Now (6.2.5) and (6.2.6) are specifically,

$$\pi K_{11}\frac{\partial \phi}{\partial x_2} + \pi K_{12}\frac{\partial \phi}{\partial x_4} = \alpha_{11}x_2 + \alpha_{12}x_4,$$
$$\pi K_{12}\frac{\partial \phi}{\partial x_2} + \pi K_{22}\frac{\partial \phi}{\partial x_4} = \alpha_{21}x_2 + \alpha_{22}x_4, \tag{6.2.55}$$

$$x_2\frac{\partial \phi}{\partial x_1} - u_1(x_1)\frac{\partial \phi}{\partial x_2} + x_4\frac{\partial \phi}{\partial x_3} - u_2(x_3)\frac{\partial \phi}{\partial x_4} = 0. \tag{6.2.56}$$

The general solution of (6.2.56) is

$$\phi(x_1, x_2, x_3, x_4) = \phi(\lambda_1, \lambda_2)$$
$$\lambda_1 = \frac{1}{2}x_2^2 + \int u_1(x_1)dx_1, \quad \lambda_2 = \frac{1}{2}x_4^2 + \int u_2(x_3)dx_3. \tag{6.2.57}$$

Substituting (6.2.57) into (6.2.55) and solving for $\partial\phi/\partial\lambda_1$ and $\partial\phi/\partial\lambda_2$, we have

$$\frac{\partial\phi}{\partial\lambda_1} = \frac{(\alpha_{11}K_{22} - \alpha_{21}K_{12})x_2 + (\alpha_{12}K_{22} - \alpha_{22}K_{12})x_4}{\pi(K_{11}K_{22} - K_{12}^2)x_2}, \qquad (6.2.58)$$

$$\frac{\partial\phi}{\partial\lambda_2} = \frac{(\alpha_{21}K_{11} - \alpha_{11}K_{12})x_2 + (\alpha_{22}K_{11} - \alpha_{12}K_{12})x_4}{\pi(K_{11}K_{22} - K_{12}^2)x_4}. \qquad (6.2.59)$$

The conditions for (6.2.58) and (6.2.59) to be valid are

$$\frac{\alpha_{21}}{\alpha_{11}}K_{11} = \frac{\alpha_{12}}{\alpha_{22}}K_{22} = K_{12}. \qquad (6.2.60)$$

The condition in (6.2.60) is a sufficient condition for system (6.2.51) to belong to the class of detailed balance. Under this condition, the exact stationary probability density of the response is

$$p(x_1, x_2, x_3, x_4) = C \exp\left\{-\frac{\alpha_{11}}{\pi K_{11}}\left[\frac{1}{2}x_2^2 + \int u_1(x_1)dx_1\right]\right.$$
$$\left. - \frac{\alpha_{22}}{\pi K_{22}}\left[\frac{1}{2}x_4^2 + \int u_2(x_3)dx_3\right]\right\}. \qquad (6.2.61)$$

Compared with (6.2.50), it is found that in (6.2.61) the kinetic energies in different coordinates are not identically distributed, in contrast to the case of equipartitioned energy in (6.2.50). This is because the restoring forces are decoupled in system (6.2.51) while coupled in (6.2.42). According to Hamiltonian formulation in Section 6.4, system (6.2.42) belongs to non-integrable case while system (6.2.51) to integrable case. So, the former has energy equipartition exact stationary solution while the later energy non-equipartition exact stationary solution.

6.3 Generalized Stationary Potential

It is shown in Section 6.2 that, by splitting the first derivate moments, the exactly solvable classes are expanded from stationary potential to detailed balance. Following the similar idea, a more general scheme was developed by splitting both the first and second derivate moments (Lin and Cai, 1988). Let

$$a_j(\mathbf{x}) = a_j^{(1)}(\mathbf{x}) + a_j^{(2)}(\mathbf{x}), \tag{6.3.1}$$

$$b_{jk}(\mathbf{x}) = b_{jk}^{(j)}(\mathbf{x}) + b_{kj}^{(k)}(\mathbf{x}). \tag{6.3.2}$$

In (6.3.1), the two terms denoted by the superscripts (1) and (2) are not restricted to the reversible and irreversible parts. The splitting in (6.3.2) does retains the symmetric property of the second derivate moments $b_{jk} = b_{kj}$. Substituting (6.3.1) and (6.3.2) into the reduced FPK equation (6.1.3), we have

$$\sum_{j=1}^{n} \frac{\partial}{\partial x_j} \left[a_j^{(1)} p - \sum_{k=1}^{n} \frac{\partial}{\partial x_k} b_{jk}^{(j)} p \right] + \sum_{j=1}^{n} \frac{\partial}{\partial x_j} a_j^{(2)} p = 0. \tag{6.3.3}$$

Equation (6.3.3) is satisfied if following more restrictive conditions are met

$$\sum_{j=1}^{n} \frac{\partial}{\partial x_j} a_j^{(2)} p = 0, \tag{6.3.4}$$

$$a_j^{(1)} p - \sum_{k=1}^{n} \frac{\partial}{\partial x_k} b_{jk}^{(j)} p = 0, \quad j = 1, 2, \ldots, n. \tag{6.3.5}$$

Similar to Eq. (6.2.4) in the detailed balance case, Eq. (6.3.4) describes the balancing of the circulatory components of the probability flow caused by one portion $a_j^{(2)}$ of each drift coefficient. The remaining portion $a_j^{(1)}$ is used to balance the differential diffusion and maintain a zero probability flow in the jth direction, i.e., each potential component of the probability flow vanishes. It is noticed that the splitting of the second derivate moments provides an extra flexibility in constructing the potential components of the probability flow, as shown in Eq. (6.3.5).

In terms of the probability potential function $\phi(\mathbf{x})$, (6.3.4) and (6.3.5) are replaced by

$$\sum_{j=1}^{n} \frac{\partial}{\partial x_j} a_j^{(2)}(\mathbf{x}) = \sum_{j=1}^{n} a_j^{(2)}(\mathbf{x}) \frac{\partial \phi(\mathbf{x})}{\partial x_j}, \tag{6.3.6}$$

$$a_j^{(1)}(\mathbf{x}) = \sum_{k=1}^{n} \left[\frac{\partial}{\partial x_k} b_{jk}^{(j)}(\mathbf{x}) - b_{jk}^j(\mathbf{x}) \frac{\partial \phi(\mathbf{x})}{\partial x_k} \right], \quad j = 1, 2, \ldots, n.$$

$$(6.3.7)$$

If a consistent ϕ function can be found to satisfy all $n + 1$ equations in (6.3.6) and (6.3.7), it is the exact solution of the problem, and the system is said to belong to the class of generalized stationary potential.

The case of detailed balance is a special case of generalized stationary potential with

$$a_j^{(1)} = a_j^I, \quad a_j^{(2)} = a_j^R, \quad b_{jk}^{(j)} = b_{kj}^{(k)} = \frac{1}{2} b_{jk}. \qquad (6.3.8)$$

It can be shown that any two-dimensional linear systems belong to the class of generalized stationary potential (see Exercise Problem 6.4). Applications of generalized stationary potential to find possible exact stationary solutions for nonlinear stochastic oscillatory systems will be illustrated in the following two sections.

6.3.1 *SDOF Nonlinear Systems*

A general form of SDOF nonlinear stochastic systems can be written as

$$\ddot{X} + h(X, \dot{X}) + v(X) = \sum_{l=1}^{m} g_l(X, \dot{X}) W_l(t), \qquad (6.3.9)$$

where $W_l(t)$ are Gaussian white noises with correlation functions given by (6.1.2). Function $h(X, \dot{X})$ and $v(X)$ are the damping and restoring forces, respectively. Denoting X by X_1 and $\dot{X}$ by X_2, the Ito equations for system (6.3.9) are

$$dX_1 = X_2 dt$$

$$dX_2 = \left[-h(X_1, X_2) - v(X_1) + \sum_{l,s=1}^{m} \pi K_{ls} g_l(X_1, X_2) \right. $$

$$\left. \frac{\partial}{\partial X_2} g_s(X_1, X_2) \right] dt$$

$$+ \sqrt{2\pi \sum_{l,s=1}^{m} K_{ls} g_l(X_1, X_2) g_s(X_1, X_2) dB(t)}. \qquad (6.3.10)$$

Note that additional terms in the drift coefficient are caused by the Wong–Zakai correction. The first and second derivate moments are

$$a_1 = x_2,$$

$$a_2 = -h(x_1, x_2) - v(x_1) + \sum_{l,s=1}^{m} \pi K_{ls} g_l(x_1, x_2) \frac{\partial}{\partial x_2} g_s(x_1, x_2),$$

$$(6.3.11)$$

$$b_{11} = b_{12} = b_{21} = 0, \quad b_{22} = 2\pi \sum_{l,s=1}^{m} K_{ls} g_l(x_1, x_2) g_s(x_1, x_2).$$

$$(6.3.12)$$

The corresponding reduced FPK equation is

$$x_2 \frac{\partial p}{\partial x_1} + \frac{\partial}{\partial x_2} \left\{ \left[-h(x_1, x_2) - v(x_1) + \sum_{l,s=1}^{m} \pi K_{ls} g_l(x_1, x_2) \right. \right.$$

$$\left. \left. \times \frac{\partial}{\partial x_2} g_s(x_1, x_2) \right] p \right\}$$

$$- \frac{\partial^2}{\partial x_2^2} \left\{ \left[\sum_{l,s=1}^{m} \pi K_{ls} g_l(x_1, x_2) g_s(x_1, x_2) \right] p \right\} = 0. \quad (6.3.13)$$

The last term on the left-hand side can be written as

$$
\frac{\partial^2}{\partial x_2^2} \left\{ \left[\sum_{l,s=1}^{m} \pi K_{ls} g_l(x_1, x_2) g_s(x_1, x_2) \right] p \right\}
$$

$$
= \frac{\partial}{\partial x_2} \left\{ \left[\sum_{l,s=1}^{m} 2\pi K_{ls} g_l(x_1, x_2) \frac{\partial}{\partial x_2} g_s(x_1, x_2) \right] p \right\}
$$

$$
+ \frac{\partial}{\partial x_2} \left\{ \left[\sum_{l,s=1}^{m} \pi K_{ls} g_l(x_1, x_2) g_s(x_1, x_2) \right] \frac{\partial p}{\partial x_2} \right\}. \qquad (6.3.14)
$$

Substitution of (6.3.14) into (6.3.13) results in

$$
x_2 \frac{\partial p}{\partial x_1} + \frac{\partial}{\partial x_2} \left\{ \left[-h(x_1, x_2) - v(x_1) - \pi \sum_{l,s=1}^{m} K_{ls} g_l(x_1, x_2) \right. \right.
$$

$$
\left. \left. \times \frac{\partial}{\partial x_2} g_s(x_1, x_2) \right] p \right\}
$$

$$
- \frac{\partial}{\partial x_2} \left\{ \left[\sum_{l,s=1}^{m} \pi K_{ls} g_l(x_1, x_2) g_s(x_1, x_2) \right] \frac{\partial p}{\partial x_2} \right\} = 0.
$$

$$
(6.3.15)
$$

The Wong–Zakai term in (6.3.15) may be split into two parts

$$
\pi \sum_{l,s=1}^{m} K_{ls} g_l(x_1, x_2) \frac{\partial}{\partial x_2} g_s(x_1, x_2) = u^*(x_1) + h^*(x_1, x_2).
$$

$$
(6.3.16)
$$

The terms $u^*(x_1)$ and $h^*(x_1, x_2)$ are called the additional restoring force and the additional damping force, respectively. These two terms are contributions from the Wong–Zakai correction. Thus, when parametric excitations are present, the possible Wong–Zakai correction terms may give rise to two different effects: to change the stiffness and the damping. Define the effective restoring force as

$$
u(x_1) = v(x_1) + u^*(x_1). \qquad (6.3.17)
$$

Now the FPK equation is cast in the form

$$
x_2 \frac{\partial p}{\partial x_1} - [u(x_1)] \frac{\partial}{\partial x_2} - \frac{\partial}{\partial x_2} \{[h(x_1, x_2) + h^*(x_1, x_2)] p\}
$$

$$
- \frac{\partial}{\partial x_2} \left\{ \left[\sum_{l,s=1}^{m} \pi K_{ls} g_l(x_1, x_2) g_s(x_1, x_2) \right] \frac{\partial p}{\partial x_2} \right\} = 0. \qquad (6.3.18)
$$

Equation (6.3.18) is satisfied under the following sufficient conditions

$$
x_2 \frac{\partial p}{\partial x_1} - u(x_1) \frac{\partial p}{\partial x_2} = 0, \qquad (6.3.19)
$$

$$
[h(x_1, x_2) + h^*(x_1, x_2)] p + \left[\sum_{l,s=1}^{m} \pi K_{ls} g_l(x_1, x_2) g_s(x_1, x_2) \right] \frac{\partial p}{\partial x_2} = 0. \qquad (6.3.20)
$$

In terms of the probability potential function ϕ, (6.3.19) and (6.3.20) are equivalent to

$$
x_2 \frac{\partial \phi}{\partial x_1} - u(x_1) \frac{\partial \phi}{\partial x_2} = 0, \qquad (6.3.21)
$$

$$
h(x_1, x_2) + h^*(x_1, x_2) - \left[\sum_{l,s=1}^{m} \pi K_{ls} g_l(x_1, x_2) g_s(x_1, x_2) \right] \frac{\partial \phi}{\partial x_2} = 0. \qquad (6.3.22)
$$

The general solution of (6.3.21) is

$$
\phi(x_1, x_2) = \phi(\lambda), \qquad (6.3.23)
$$

where function ϕ is to be determined, and

$$
\lambda = \frac{1}{2} x_2^2 + \int u(x_1) dx_1. \qquad (6.3.24)
$$

It is seen from (6.3.24) that λ is the total effective energy of the system. Substituting (6.3.23) into (6.3.22) and using (6.3.16), we obtain

$$
h(x_1, x_2) = \left[\pi x_2 \sum_{l,s=1}^{m} K_{ls} g_l g_s \right] \frac{d\phi}{d\lambda} - h^*(x_1, x_2). \qquad (6.3.25)
$$

System (6.3.9) is said to belong to the class of generalized stationary potential if function h can be cast in the form of (6.3.25). In this case, the exact stationary probability density of the system response can be obtained. Also Eq. (6.3.25) can be interpreted as a restriction between the system properties, represented by $h(x_1, x_2)$ and $g_l(x_1, x_2)$, and the spectral densities K_{ls} of the random excitations in order for the SDOF nonlinear stochastic system to possess an exact stationary probability density.

Example 6.3.1 Consider the following system

$$\ddot{X} + [\delta + (\alpha X + \beta \dot{X})^2]\dot{X} + v(X) = (aX + b\dot{X})W_1(t) + W_2(t),$$
$$(6.3.26)$$

where $W_1(t)$ and $W_2(t)$ are independent Gaussian white noises with spectral densities K_{11} and K_{22}, respectively. Denoting X by X_1 and $\dot{X}$ by X_2, and comparing (6.3.26) with the standard form (6.3.9), we see that

$$\begin{aligned} h(X_1, X_2) &= [\delta + (\alpha X_1 + \beta X_2)^2]X_2 + v(X_1), \\ g_1(X_1, X_2) &= aX_1 + bX_2, \quad g_2(X_1, X_2) = 1. \end{aligned} \qquad (6.3.27)$$

We need to identify the additional restoring force and additional damping force. Application of (6.3.16) leads to

$$\pi \sum_{l,s=1}^{m} K_{ls}g_l(x_1, x_2)\frac{\partial}{\partial x_2}g_s(x_1, x_2) = \pi K_{11}abx_1 + \pi K_{11}b^2 x_2. \quad (6.3.28)$$

Therefore,

$$u^*(x_1) = \pi K_{11}abx_1, \quad h^*(x_1, x_2) = \pi K_{11}b^2 x_2. \qquad (6.3.29)$$

The effective restoring force and the effective total energy are

$$u(x_1) = v(x_1) + \pi K_{11}abx_1^2, \quad \lambda = \frac{1}{2}x_2^2 + \int v(x_1)dx_1 + \frac{1}{2}\pi K_{11}abx_1^2.$$
$$(6.3.30)$$

Substituting (6.3.29) into (6.3.25), we have

$$\pi[K_{11}(ax_1 + bx_2)^2 + K_{22}]\frac{d\phi}{d\lambda} = \delta + \pi K_{11}b^2 + (\alpha x_1 + \beta x_2)^2. \quad (6.3.31)$$

In order to satisfy (6.3.31), the following conditions must be imposed

$$\beta = \alpha \frac{b}{a}, \quad \alpha^2 = \frac{a^2 K_{11}}{K_{22}} (\delta + \pi K_{11} b^2). \qquad (6.3.32)$$

Under conditions in (6.3.32), the system belongs to the class of generalized stationary potential, and we obtain

$$p(x_1, x_2) = C \exp \left\{ -\frac{\delta + \pi K_{11} b^2}{2\pi K_{22}} \left[x_2^2 + \pi K_{11} abx_1^2 + 2 \int_0^{x_1} v(z) dz \right] \right\}. \qquad (6.3.33)$$

Equation (6.3.33) shows that, due to the Wong–Zakai correction, the presence of the parametric excitation $W_1(t)$ adds an extra linear damping force $\pi K_{11} b^2 X_2$ and an extra linear restoring force $\pi K_{11} abX_1$ to the system. The damping is always positive, but the stiffness depends on the signs of parameter a and b. It can be shown that the system does not belong to the class of detailed balance (Cai and Lin, 1988).

6.3.2 *MDOF Nonlinear Systems*

We extend the above procedure to MDOF systems. Consider the following random oscillatory system of N-degrees-of-freedom

$$\ddot{Z}_j + h_j(\mathbf{Z}, \dot{\mathbf{Z}}) + v_j(\mathbf{Z}) = \sum_{l=1}^{m} g_{jl}(\mathbf{Z}, \dot{\mathbf{Z}}) W_l(t), \quad j = 1, 2, \ldots, n, \qquad (6.3.34)$$

where $\mathbf{Z} = [Z_1, Z_2, \ldots, Z_n]^T$ and $\dot{\mathbf{Z}} = [\dot{Z}_1, \dot{Z}_2, \ldots, \dot{Z}_n]^T$ are vectors of displacements and velocities, respectively, and $W_l(t)$ are Gaussian white noises with correlation functions given by (6.1.2). Letting $X_{2j-1} = Z_j, X_{2j} = \dot{Z}_j, \mathbf{X} = [X_1, X_2, \ldots, X_{2n}]^T$, the reduced FPK equation is given by (6.1.3) and (6.1.4) with the first and second derivate moments obtained from (6.1.5) and (6.1.6) as follows

$$a_{2j-1} = x_{2j},$$

$$a_{2j} = -h_j(\mathbf{x}) - v_j(\mathbf{x}_d) + \sum_{k=1}^{n} \sum_{l,s=1}^{m} \pi K_{ls} g_{ks}(\mathbf{x}) \frac{\partial}{\partial x_{2k}} g_{jl}(\mathbf{x}),$$

$$(6.3.35)$$

$$b_{2j-1,k} = b_{k,2j-1} = 0, \quad b_{2j,2k} = 2\pi \sum_{l,s=1}^{m} K_{ls} g_{ks}(\mathbf{x}) g_{jl}(\mathbf{x}), \quad (6.3.36)$$

where $\mathbf{x}_d = [x_1, x_3, \ldots, x_{2n-1}]^T$ is the displacement vector. Split the second derivate moments as follows

$$b_{2j-1,k}^{(2j-1)} = b_{k,2j-1}^{(k)} = 0, \quad b_{2j,2k} = b_{2j,2k}^{(2j)} + b_{2k,2j}^{(2k)}. \tag{6.3.37}$$

Substituting (6.3.37) into the FPK equation (6.1.3), we obtain

$$\sum_{j=1}^{n} \left(x_{2j} \frac{\partial p}{\partial x_{2j-1}} \right)$$

$$+ \sum_{j=1}^{n} \frac{\partial}{\partial x_{2j}} \left[\left(-h_j - v_j + \pi \sum_{k=1}^{n} \sum_{l,s=1}^{m} K_{ls} g_{ks} \frac{\partial}{\partial x_{2k}} g_{jl} \right) p \right]$$

$$- \sum_{j,k=1}^{n} \frac{\partial^2}{\partial x_{2j} \partial x_{2k}} \left[b_{2j,2k}^{(2j)} p \right] = 0. \tag{6.3.38}$$

The last term on the left-hand side can be written as

$$\sum_{j,k=1}^{n} \frac{\partial^2}{\partial x_{2j} \partial x_{2k}} \left[b_{2j,2k}^{(2j)} p \right] = \sum_{j,k=1}^{n} \frac{\partial}{\partial x_{2j}} \left[p \frac{\partial}{\partial x_{2k}} b_{2j,2k}^{(2j)} \right]$$

$$+ \sum_{j,k=1}^{n} \frac{\partial}{\partial x_{2j}} \left[b_{2j,2k}^{(2j)} \frac{\partial p}{\partial x_{2k}} \right]. \tag{6.3.39}$$

It can be shown that

$$\frac{\partial}{\partial x_{2k}} b_{2j,2k}^{(2j)} = 2\pi \sum_{l,s=1}^{m} K_{ls} g_{ks} \frac{\partial}{\partial x_{2k}} g_{jl}. \tag{6.3.40}$$

Substituting (6.3.39) into (6.3.38) and using (6.3.40), we have

$$\sum_{j=1}^{n}\left(x_{2j}\frac{\partial p}{\partial x_{2j-1}}\right)$$

$$+\sum_{j=1}^{n}\frac{\partial}{\partial x_{2j}}\left[\left(-h_j - v_j - \pi\sum_{k=1}^{n}\sum_{l,s=1}^{m}K_{ls}g_{ks}\frac{\partial}{\partial x_{2k}}g_{jl}\right)p\right]$$

$$-\sum_{k=1}^{n}\left[b_{2j,2k}^{(2j)}\frac{\partial p}{\partial x_{2k}}\right] = 0. \qquad (6.3.41)$$

We now consider the case in which we can identify the following

$$\pi\sum_{k=1}^{n}\sum_{l,s=1}^{m}K_{ls}g_{ks}\frac{\partial}{\partial x_{2k}}g_{jl} = u_j^*(\mathbf{x}_d) + h_j^*(\mathbf{x}). \qquad (6.3.42)$$

The first term on the right-hand side depends only on vector $\mathbf{x}_d$, and it is the additional restoring force in the jth coordinate, while the second term is the additional damping force in the jth coordinate. The effective restoring force in the jth coordinate is then

$$u_j(\mathbf{x}_d) = v_j(\mathbf{x}_d) + u_j^*(\mathbf{x}_d). \qquad (6.3.43)$$

In this case, Eq. (6.3.41) may be cast in the form

$$\sum_{j=1}^{n}\left(x_{2j}\frac{\partial p}{\partial x_{2j-1}}\right) - \sum_{j=1}^{n}u_j\frac{\partial p}{\partial x_{2j}}$$

$$-\sum_{j=1}^{n}\frac{\partial}{\partial x_{2j}}\left\{(h_j + h_j^*)p + \sum_{k=1}^{n}\left[b_{2j,2k}^{(2j)}\frac{\partial p}{\partial x_{2k}}\right]\right\} = 0. \quad (6.3.44)$$

Equation (6.3.44) is satisfied if the following sufficient conditions are met

$$\sum_{j=1}^{n}\left(x_{2j}\frac{\partial p}{\partial x_{2j-1}}\right) - \sum_{j=1}^{n}u_j\frac{\partial p}{\partial x_{2j}} = 0, \qquad (6.3.45)$$

$$(h_j + h_j^*)p + \sum_{k=1}^{n}\left[b_{2j,2k}^{(2j)}\frac{\partial p}{\partial x_{2k}}\right] = 0, \quad j = 1, 2, \ldots, n. \qquad (6.3.46)$$

In terms of the probability potential function ϕ, (6.3.45) and (6.3.46) can be written as

$$\sum_{j=1}^{n}\left(x_{2j}\frac{\partial\phi}{\partial x_{2j-1}}\right) - \sum_{j=1}^{n}u_{j}\frac{\partial\phi}{\partial x_{2j}} = 0, \qquad (6.3.47)$$

$$h_{j} + h_{j}^{*} - \sum_{k=1}^{n}\left[b_{2j,2k}^{(2j)}\frac{\partial\phi}{\partial x_{2k}}\right] = 0, \quad j = 1, 2, \ldots, n. \qquad (6.3.48)$$

Consider two situations. In the first one, each function u_{j} is dependent only on x_{2j-1}, indicating that the restoring forces are decoupled, i.e., the restoring force in each coordinate is independent of the other coordinates. Then the general solution of (6.3.47) is

$$\phi(\mathbf{x}) = \phi(\lambda_{1}, \lambda_{2}, \ldots, \lambda_{n}), \quad \lambda_{j} = \frac{1}{2}x_{2j}^{2} + \int_{0}^{x_{2j-1}} u_{j}(z)dz, \qquad (6.3.49)$$

where λ_{j} is the total effective energy of the system in the jth coordinate. Substituting (6.3.49) into (6.3.48), we have

$$h_{j} + h_{j}^{*} - \sum_{k=1}^{n}\left[x_{2k}b_{2j,2k}^{(2j)}\frac{\partial\phi}{\partial\lambda_{k}}\right] = 0, \quad j = 1, 2, \ldots, n. \qquad (6.3.50)$$

Equation (6.3.50) provides n restrictions between the system dampings and the excitation spectral densities. If a consistent ϕ function can be found to satisfy all n equations in (6.3.50), the system belongs to the class of generalized stationary potential, and the stationary solution is obtained as

$$p(\mathbf{x}) = C\exp[-\phi(\lambda_{1}, \lambda_{2}, \ldots, \lambda_{n})]. \qquad (6.3.51)$$

Another situation is that u_{j} functions can be expressed as

$$u_{j}(\mathbf{x}_{d}) = \frac{\partial U(\mathbf{x}_{d})}{\partial x_{2j-1}}. \qquad (6.3.52)$$

Equation (6.3.52) implies that $U(\mathbf{x}_d)$ is the total effective potential energy of the system. In this case, the general solution of (6.3.47) is

$$\phi(\mathbf{x}) = \phi(\lambda), \quad \lambda = \sum_{j=1}^{n} \frac{1}{2}x_{2j}^2 + U(\mathbf{x}_d), \tag{6.3.53}$$

where λ is the total effective energy of the system. Substituting (6.3.53) into (6.3.48), we have

$$h_j + h_j^* - \sum_{k=1}^{n} \left[x_{2k} b_{2j,2k}^{(2j)} \right] \frac{d\phi}{d\lambda} = 0, \quad j = 1, 2, \dots, n. \tag{6.3.54}$$

If a consistent ϕ function can be found to satisfy all n equations in (6.3.54), an exact stationary solution is obtained as

$$p(\mathbf{x}) = C \exp[-\phi(\lambda)]. \tag{6.3.55}$$

Equation (6.3.53) indicates that the kinetic energies in different coordinates are equally distributed, while (6.3.49) shows they may not be identically distributed in another situation. According to Hamiltonian formulation in Section 6.4, the first situation is completely integrable while the second situation completely non-integrable.

The above procedure will be illustrated by the following example.

Example 6.3.2 Consider the two-DOF nonlinear stochastic system

$$\begin{aligned} \ddot{Z}_1 + \alpha_{11}\dot{Z}_1 + \alpha_{12}\dot{Z}_2 + u_1(Z_1, Z_2) &= W_1(t), \\ \ddot{Z}_2 + \alpha_{21}\dot{Z}_1 + \alpha_{22}\dot{Z}_2 + u_2(Z_1, Z_2) &= W_2(t). \end{aligned} \tag{6.3.56}$$

Let $X_1 = Z_1, X_2 = \dot{Z}_1, X_3 = Z_2$ and $X_4 = \dot{Z}_2$. The non-zero second derivate moments are

$$b_{22} = 2\pi K_{11}, \quad b_{24} = b_{42} = 2\pi K_{12}, \quad b_{44} = 2\pi K_{22}. \tag{6.3.57}$$

Split them into

$$b_{22}^{(2)} = \pi K_{11}, \quad b_{24}^{(2)} = \pi K_{12} - D, \quad b_{42}^{(4)} = \pi K_{12} + D, \quad b_{44}^{(4)} = \pi K_{22}, \tag{6.3.58}$$

where D is an arbitrary constant. Since no parametric excitations are present, there is no Wong–Zakai correction. Corresponding to the two situations described above, there are two possibilities. If

$u_1(X_1, X_3) = u_1(X_1)$ and $u_2(X_1, X_3) = u_2(X_3)$, the probability potential function is found from (6.3.49) as

$$\phi(x_1, x_2, x_3, x_4) = \phi_1(\lambda_1) + \phi_2(\lambda_2),$$

$$\lambda_1 = \frac{1}{2}x_2^2 + \int u_1(x_1)dx_1, \quad \lambda_2 = \frac{1}{2}x_4^2 + \int u_2(x_3)dx_3.$$

(6.3.59)

The two equations in (6.3.50) become

$$\pi K_{11}x_2 \frac{\partial\phi}{\partial\lambda_1} + (\pi K_{12} - D)x_4 \frac{\partial\phi}{\partial\lambda_2} = \alpha_{11}x_2 + \alpha_{12}x_4,$$

$$(\pi K_{12} + D)x_2 \frac{\partial\phi}{\partial\lambda_1} + \pi K_{22}x_4 \frac{\partial\phi}{\partial\lambda_2} = \alpha_{21}x_2 + \alpha_{22}x_4,$$

(6.3.60)

from which $\partial\phi_1/\partial\lambda_1$ and $\partial\phi_2/\partial\lambda_2$ can be solved as

$$\frac{\partial\phi}{\partial\lambda_1} = \frac{[\alpha_{11}K_{22} - \alpha_{21}(K_{12} - D)]x_2 + [\alpha_{12}K_{22} - \alpha_{22}(K_{12} - D)]x_4}{\pi(K_{11}K_{22} - K_{12}^2 + D^2)x_2}.$$

(6.3.61)

$$\frac{\partial\phi}{\partial\lambda_2} = \frac{[\alpha_{21}K_{11} - \alpha_{11}(K_{12} + D)]x_2 + [\alpha_{22}K_{11} - \alpha_{12}(K_{12} + D)]x_4}{\pi(K_{11}K_{22} - K_{12}^2 + D^2)x_4}.$$

(6.3.62)

In order for the right-hand sides of (6.3.61) and (6.3.62) to be functions of just λ_1 and λ_2, respectively, we must have

$$\alpha_{12}K_{22} - \alpha_{22}(K_{12} - D) = 0, \quad \alpha_{21}K_{11} - \alpha_{11}(K_{12} + D) = 0. \quad (6.3.63)$$

Since D is arbitrary, the two equations in (6.3.63) can be combined by eliminating D as

$$\frac{\alpha_{21}}{\alpha_{11}}K_{11} + \frac{\alpha_{12}}{\alpha_{22}}K_{22} = 2K_{12}. \quad (6.3.64)$$

Under condition (6.3.64), the exact stationary probability density is

$$p(x_1, x_2, x_3, x_4) = C \exp\left\{-\frac{\alpha_{11}}{\pi K_{11}}\left[\frac{1}{2}x_2^2 + \int u_1(x_1)dx_1\right]\right.$$

$$\left. -\frac{\alpha_{22}}{\pi K_{22}}\left[\frac{1}{2}x_4^2 + \int u_2(x_3)dx_3\right]\right\}. \quad (6.3.65)$$

It is noted that the solution (6.3.65) is the same as that in (6.2.61). However, the condition (6.3.64) is less stringent than that given by (6.2.60) for the case of detailed balance, indicating that the class of generalized stationary potential is wider than the class of detailed balance.

If the two restoring forces in (6.3.56) can be expressed as

$$u_1(x_1, x_3) = \frac{\partial U(x_1, x_3)}{\partial x_1}, \quad u_3(x_1, x_3) = \frac{\partial U(x_1, x_3)}{\partial x_3}, \qquad (6.3.66)$$

then the two equations in (6.3.54) become

$$[\pi K_{11} x_2 + (\pi K_{12} - D) x_4] \frac{d\phi}{d\lambda} = \alpha_{11} x_2 + \alpha_{12} x_4,$$

$$[(\pi K_{12} + D) x_2 + \pi K_{22} x_4] \frac{d\phi}{d\lambda} = \alpha_{21} x_2 + \alpha_{22} x_4. \qquad (6.3.67)$$

Satisfaction of the two equations in (6.3.67) requires

$$\frac{\alpha_{12} + \alpha_{21}}{2\alpha_{11}} K_{11} = \frac{\alpha_{12} + \alpha_{21}}{2\alpha_{22}} K_{22} = K_{12}. \qquad (6.3.68)$$

Under condition (6.3.68), the exact stationary probability density is

$$p(x_1, x_2, x_3, x_4) = C \exp \left\{ -\frac{\alpha_{11}}{\pi K_{11}} \left[\frac{1}{2} x_2^2 + \frac{1}{2} x_4^2 + U(x_1, x_3) \right] \right\}. \qquad (6.3.69)$$

It is noted that the stationary probability density in (6.3.69) is different from that in (6.3.65), and condition (6.3.68) is different from either (6.2.60) or (6.3.64).

6.4 Stochastically Excited and Dissipated Hamiltonian Systems

As mentioned in Chapter 1, nonlinear stochastic systems can be formulated as stochastically excited and dissipated Hamiltonian systems, and this formulation is especially suitable to deal with MDOF strongly nonlinear stochastic systems. Systematic procedures have been developed for obtaining the exact stationary solutions of stochastically excited and dissipated Hamiltonian systems (Zhu and

Yang, 1996; Ying and Zhu, 2000; Huang and Zhu, 2000; Zhu and Huang, 2001).

6.4.1 *Hamiltonian Systems and Their Classification*

Without energy-dissipation mechanisms and excitations, a n-DOF Hamiltonian system has the equations of motion of the form

$$\dot{q}_j = \frac{\partial H}{\partial p_j}, \quad \dot{p}_j = -\frac{\partial H}{\partial q_j}, \quad i = 1, 2, \ldots, n, \tag{6.4.1}$$

where q_j and p_j are generalized displacements and momenta, respectively, and $H = H(\mathbf{q}, \mathbf{p})$ is a Hamiltonian with continuous first-order derivatives. The Hamiltonian systems can be classified into three main categories: completely integrable, partially integrable and completely non-integrable. The criteria for the classification are rather complicated (Arnold, 1989; Tabor, 1989; Zhu and Huang, 2001). In this book, only simple and practical subclasses of the three categories are discussed.

Consider a Hamiltonian system of n degrees of freedom. If there exist n independent integrals of motion, $H_1, H_2, \ldots, H_n$, which are in involution, then the Hamiltonian system is said to be completely integrable. Here only the following special case of completely integrable Hamiltonian system is considered, i.e., the system Hamiltonian $H(\mathbf{q}, \mathbf{p})$ can be expressed as

$$H(\mathbf{q}, \mathbf{p}) = \sum_{j=1}^{n} H_j(q_j, p_j). \tag{6.4.2}$$

However, if only $r+1$ ($r + 1 < n$) independent integrals of motion exist, and the Hamiltonian $H(\mathbf{q}, \mathbf{p})$ can be expressed as

$$H(\mathbf{q}, \mathbf{p}) = \sum_{j=1}^{r} H_j(q_j, p_j) + \tilde{H}(q_{r+1}, q_{r+2}, \ldots, q_n; p_{r+1}, p_{r+2}, \ldots, p_n),$$

$$\tag{6.4.3}$$

then the system belongs to the partially integrable class. The extreme case, in which only one Hamiltonian exists, i.e., $r = 0$, is said to be completely non-integrable.

For the completely integrable case and partially integrable case, an internal resonance may occur between two degrees of freedom if the frequencies of the two degrees of freedom have a rational relationship, such as that the two frequencies are equal or one is twice as another, etc. Therefore, each of the two classes has two subclasses: resonant and non-resonant. With or without excitations, the resonant cases are more complex than non-resonant cases. Overall, the Hamiltonian systems can be divided into five classes: completely non-integrable, completely integrable and non-resonant, completely integrable and resonant, partially integrable and non-resonant and partially integrable and resonant. It is noted that, for the completely non-integrable case, no internal resonance occurs since there is only one motion involving all dimensions.

6.4.2 *Exact Stationary Solutions of Stochastically Excited and Dissipated Hamiltonian Systems*

Now consider dissipated Hamiltonian systems excited by Gaussian white noises, governed by

$$\dot{Q}_j = \frac{\partial H'}{\partial P_j},$$

$$\dot{P}_j = -\frac{\partial H'}{\partial Q_j} - \sum_{k=1}^{n} c_{jk}(\mathbf{Q}, \mathbf{P})\frac{\partial H'}{\partial P_k} + \sum_{l=1}^{m} g_{jl}(\mathbf{Q}, \mathbf{P})W_l(t),$$

$$(6.4.4)$$

where the term $\sum_{k=1}^{n} c_{jk}(\mathbf{Q}, \mathbf{P})\frac{\partial H'}{\partial P_k}$ in the second equation represents the energy-dissipation mechanism, $W_l(t)$ are Gaussian white noises and $c_{jk}(\mathbf{Q}, \mathbf{P})$ and $g_{jl}(\mathbf{Q}, \mathbf{P})$ are differentiable functions.

Equation set (6.4.4) has the same form as that in (6.1.1) with Q_j and P_j identified as X_{2j-1} and X_{2j}, respectively. The first and second derivate moments for the Hamiltonian system (6.4.4) are obtained from (6.1.5) and (6.1.6) as follows:

$$a_{2j-1} = \frac{\partial H'}{\partial p_j},$$

$$a_{2j} = -\frac{\partial H'}{\partial q_j} - \sum_{k=1}^{n} c_{jk} \frac{\partial H'}{\partial p_k} + \pi \sum_{k=1}^{n} \sum_{l,s=1}^{m} K_{ls} g_{ks} \frac{\partial}{\partial p_k} g_{jl}, \quad (6.4.5)$$

$$b_{2j-1,k} = b_{k,2j-1} = 0, b_{2j,2k} = B_{jk} = 2\pi \sum_{l,s=1}^{m} K_{ls} g_{jl} g_{ks}. \quad (6.4.6)$$

The last term in (6.4.5) is due to the Wong–Zakai correction. Split the second derivate moments as

$$b_{2j-1,k}^{(2j-1)} = b_{k,2j-1}^{(k)} = 0, \quad b_{2j,2k} = b_{2j,2k}^{(2j)} + b_{2k,2j}^{(2k)} = B_{jk}^{(j)} + B_{kj}^{(k)}. \quad (6.4.7)$$

Substituting (6.4.7) into the reduced FPK equation (6.1.3), we obtain

$$\sum_{j=1}^{n} \frac{\partial}{\partial q_j} \left(\frac{\partial H'}{\partial p_j} p \right) + \sum_{j=1}^{n} \frac{\partial}{\partial p_j}$$

$$\times \left[\left(-\frac{\partial H'}{\partial q_j} - \sum_{k=1}^{n} c_{jk} \frac{\partial H'}{\partial p_k} + \pi \sum_{k=1}^{n} \sum_{l,s=1}^{m} K_{ls} g_{ks} \frac{\partial}{\partial p_k} g_{jl} \right) p \right]$$

$$- \sum_{j,k=1}^{n} \frac{\partial^2}{\partial p_j \partial p_k} \left[B_{jk}^{(j)} p \right] = 0. \quad (6.4.8)$$

The last term on the left-hand side can be written as

$$\sum_{j,k=1}^{n} \frac{\partial^2}{\partial p_j \partial p_k} \left[B_{jk}^{(j)} p \right] = \sum_{j,k=1}^{n} \frac{\partial}{\partial p_j} \left[p \frac{\partial}{\partial p_k} B_{jk}^{(j)} \right] + \sum_{j,k=1}^{n} \frac{\partial}{\partial p_j} \left[B_{jk}^{(j)} \frac{\partial p}{\partial p_k} \right]. \quad (6.4.9)$$

Substituting (6.4.9) into (6.4.8) and noticing

$$\frac{\partial}{\partial p_k} B_{jk}^{(j)} = 2\pi \sum_{l,s=1}^{m} K_{ls} g_{ks} \frac{\partial}{\partial p_k} g_{jl}, \quad (6.4.10)$$

we have

$$
\sum_{j=1}^{n} \frac{\partial}{\partial q_j} \left(\frac{\partial H'}{\partial p_j} p \right) + \sum_{j=1}^{n} \frac{\partial}{\partial p_j}
$$

$$
\times \left[\left(-\frac{\partial H'}{\partial q_j} - \sum_{k=1}^{n} c_{jk} \frac{\partial H'}{\partial p_k} - \pi \sum_{k=1}^{n} \sum_{l,s=1}^{m} K_{ls} g_{ks} \frac{\partial}{\partial p_k} g_{jl} \right) p \right]
$$

$$
- \sum_{j,k=1}^{n} \frac{\partial}{\partial p_j} \left[B_{jk}^{(j)} \frac{\partial p}{\partial p_k} \right] = 0. \tag{6.4.11}
$$

The term in Eq. (6.4.11), $\pi \sum_{l,s=1}^{m} K_{ls} g_{ks} \frac{\partial}{\partial p_k} g_{jl}$, may be separated into two parts: one plays the role of additional conservative force and another the role of additional damping force. These two parts may be combined with the original conservative and damping forces, and Eq. (6.4.11) may be cast in the form

$$
\sum_{j=1}^{n} \frac{\partial}{\partial q_j} \left(\frac{\partial H}{\partial p_j} p \right) + \sum_{j=1}^{n} \frac{\partial}{\partial p_j} \left[\left(-\frac{\partial H}{\partial q_j} - \sum_{k=1}^{n} m_{jk} \frac{\partial H}{\partial p_k} \right) p \right]
$$

$$
- \sum_{j,k=1}^{n} \frac{\partial}{\partial p_j} \left[B_{jk}^{(j)} \frac{\partial p}{\partial p_k} \right] = 0, \tag{6.4.12}
$$

where H is the new Hamiltonian function, taking into account the additional conservative force, and m_{jk} is the new damping coefficients including those from the additional dampings.

Now split the first derivate moments

$$
a_{2j-1}^{(1)} = 0, \qquad a_{2j-1}^{(2)} = \frac{\partial H}{\partial p_j},
$$

$$
a_{2j}^{(1)} = -\sum_{k=1}^{n} m_{jk} \frac{\partial H}{\partial p_k}, \qquad a_{2j}^{(2)} = -\frac{\partial H}{\partial q_j}. \tag{6.4.13}
$$

Substitution of (6.4.7) and (6.4.13) into (6.3.6) and (6.3.7) results in

$$
\sum_{j=1}^{n} \left(\frac{\partial H}{\partial p_j} \frac{\partial \phi}{\partial q_j} - \frac{\partial H}{\partial q_j} \frac{\partial \phi}{\partial p_j} \right) = 0, \tag{6.4.14}
$$

$$\sum_{k=1}^{n} B_{jk}^{(j)} \frac{\partial \phi}{\partial p_k} = \sum_{k=1}^{n} \frac{\partial}{\partial p_k} B_{jk}^{(j)} + \sum_{k=1}^{n} m_{jk} \frac{\partial H}{\partial p_k}, \quad j = 1, 2, \ldots, n,$$

$$(6.4.15)$$

where ϕ is the probability potential function, defined in (6.1.9). If a consistent function $\phi(\mathbf{q}, \mathbf{p})$ can be found to satisfy all $n + 1$ equations of (6.4.14) and (6.4.15), then the system belongs to the class of the generalized stationary potential, and an exact stationary probability density can be found. However, the formats of exact stationary probability densities are different for the five classes of the Hamiltonian systems. For illustrating the solution procedures more clearly, only completely non-integrable case, completely integrable and non-resonant case, and partially integrable and non-resonant case are considered here.

6.4.3 *Completely Non-Integrable Case*

As stated previously, no internal resonance occurs in this case. It has been proved that in this case the general solution of (6.4.14) is of the form (Zhu, Cai, and Lin, 1990)

$$\phi = \phi\left[H(\mathbf{q}, \mathbf{p})\right]. \qquad (6.4.16)$$

Substituting (6.4.16) into (6.4.15), we obtain

$$\frac{d\phi}{dH} \sum_{k=1}^{n} B_{jk}^{(j)} \frac{\partial H}{\partial p_k} = \sum_{k=1}^{n} \frac{\partial}{\partial p_k} B_{jk}^{(j)} + \sum_{k=1}^{n} m_{jk} \frac{\partial H}{\partial p_k}, \quad j = 1, 2, \ldots, n.$$

$$(6.4.17)$$

There are n equations in (6.4.17) for function ϕ. If a consistent function ϕ can be found to satisfy all these equations, the exact stationary probability density is given by

$$p(\mathbf{q}, \mathbf{p}) = C \exp\{-\phi[H(\mathbf{q}, \mathbf{p})]\}. \qquad (6.4.18)$$

Some special cases are discussed below.

If c_{jk} and g_{jl} are functions only of $\mathbf{q}$ in system equations (6.4.4), i.e., $c_{jk} = c_{jk}(\mathbf{q})$ and $g_{jl} = g_{jl}(\mathbf{q})$, then both the additional restoring forces and damping forces do not exist; therefore, $H = H'$, $m_{jk} = c_{jk}$, and (6.4.17) is simplified as

$$\frac{d\phi}{dH} \sum_{k=1}^{n} B_{jk}^{(j)} \frac{\partial H}{\partial p_k} = \sum_{k=1}^{n} c_{jk} \frac{\partial H}{\partial p_k}, \quad j = 1, 2, \ldots, n. \tag{6.4.19}$$

If the following conditions are satisfied

$$c_{jk} = \eta B_{jk}^{(j)}, \text{ or equivalently, } c_{jk} + c_{kj} = \eta B_{jk}, \quad j = 1, 2, \ldots, n, \tag{6.4.20}$$

where η is a constant, then

$$\phi = \eta H, \quad p(\mathbf{q}, \mathbf{p}) = C \exp[-\eta H(\mathbf{q}, \mathbf{p})]. \tag{6.4.21}$$

A more general case is that $c_{jk} = c(H)\bar{c}_{jk}(\mathbf{q})$ and $g_{jl} = g(H)\bar{g}_{jl}(\mathbf{q})$, in which case the additional conservative force does not exist, $H = H'$, then (6.4.17) can be written as

$$\frac{d\phi}{dH} g^2(H) \sum_{k=1}^{n} \bar{B}_{jk}^{(j)} \frac{\partial H}{\partial p_k} = g(H) \frac{dg(H)}{dH} \sum_{k=1}^{n} \bar{B}_{jk}^{(j)} \frac{\partial H}{\partial p_k} + c(H) \sum_{k=1}^{n} \bar{c}_{jk} \frac{\partial H}{\partial p_k},$$
$$j = 1, 2, \ldots, n, \tag{6.4.22}$$

where

$$\bar{B}_{jk}^{(j)} + \bar{B}_{kj}^{(k)} = 2\pi \sum_{l,s=1}^{m} K_{ls}\bar{g}_{jl}\bar{g}_{ks}. \tag{6.4.23}$$

If the following conditions are met

$$\bar{c}_{jk} = \bar{c}_{kj} = \frac{1}{2}\eta\bar{B}_{jk}, \quad j = 1, 2, \ldots, n, \tag{6.4.24}$$

Eq. (6.4.22) is simplified to

$$\frac{d\phi}{dH} = \frac{1}{g(H)} \frac{dg(H)}{dH} + \frac{\eta c(H)}{g^2(H)}, \tag{6.4.25}$$

and the stationary probability potential and density are, respectively,

$$\phi = \ln g(H) + \eta \int_0^H \frac{c(u)}{g^2(u)} du, \qquad (6.4.26)$$

$$p(\mathbf{q}, \mathbf{p}) = \frac{C}{g(H)} \exp\left[-\eta \int_0^H \frac{c(u)}{g^2(u)} du \right]. \qquad (6.4.27)$$

6.4.4 *Completely Integrable and Non-Resonant Case*

For this case, the Hamiltonian function has the form of (6.4.2), and the general solution of (6.4.14) is proved to be of the form (Zhu and Yang, 1996)

$$\phi(\mathbf{q}, \mathbf{p}) = \phi(H_1, H_2, \dots, H_n), \quad H_j = H_j(q_j, p_j). \qquad (6.4.28)$$

Substituting (6.4.28) into (6.4.15), we obtain

$$\sum_{k=1}^n B_{jk}^{(j)} \frac{\partial H_k}{\partial p_k} \frac{\partial \phi}{\partial H_k} = \sum_{k=1}^n \frac{\partial B_{jk}^{(j)}}{\partial p_k} + \sum_{k=1}^n m_{jk} \frac{\partial H_k}{\partial p_k}. \qquad (6.4.29)$$

Denoting

$$\alpha_{jk} = B_{jk}^{(j)} \frac{\partial H_k}{\partial p_k}, \quad \beta_j = \sum_{k=1}^n \frac{\partial B_{jk}^{(j)}}{\partial p_k} + \sum_{k=1}^n m_{jk} \frac{\partial H_k}{\partial p_k}, \qquad (6.4.30)$$

equation (6.4.29) can be rewritten as a set of linear partial differential equations as follows

$$\sum_{k=1}^n \alpha_{jk} \frac{\partial \phi}{\partial H_k} = \beta_j, \quad j = 1, 2, \dots, n. \qquad (6.4.31)$$

Consider the case in which matrix $\mathbf{A} = [\alpha_{jk}]$ is non-singular so that its inverse $\mathbf{A}^{-1} = \mathbf{D} = [d_{jk}]$ exists. Then (6.4.31) can be simplified to

$$\frac{\partial \phi}{\partial H_k} = \sum_{i=1}^n d_{ki} \beta_i, \quad k = 1, 2, \dots, n. \qquad (6.4.32)$$

If for each k, the right-hand side of (6.4.32) is a function of H_k only, then a function ϕ can be found to satisfy all n equations in (6.4.31), and an exact probability density exists. In another situation, a consistent ϕ function can be solved from (6.4.32) if the following compatibility conditions are satisfied

$$\frac{\partial}{\partial H_j}\left(\sum_{i=1}^{n} d_{ki}\beta_i\right) = \frac{\partial}{\partial H_k}\left(\sum_{i=1}^{n} d_{ji}\beta_i\right), \quad j,k = 1,2,\ldots,n. \quad (6.4.33)$$

Example 6.4.1 As an example, consider the following Hamiltonian

$$H = H_1 + H_2, \quad H_1 = \frac{1}{2}P_1^2 + \int u_1(Q_1)dQ_1,$$

$$H_2 = \frac{1}{2}P_2^2 + \int u_2(Q_2)dQ_2. \quad (6.4.34)$$

It follows from (6.4.34) that

$$\frac{\partial H}{\partial P_j} = P_j, \quad \frac{\partial H}{\partial Q_j} = u_j(Q_j), \quad i = 1,2. \quad (6.4.35)$$

The corresponding stochastically excited and dissipated Hamiltonian system is governed by

$$\dot{Q}_j = P_j,$$

$$\dot{P}_j = -u_j(Q_j) - \sum_{k=1}^{2} c_{jk}P_k + W_j(t), \quad j = 1,2, \quad (6.4.36)$$

where c_{jk} are constants, and $W_1(t)$ and $W_2(t)$ are Gaussian white noises with correlation functions

$$E[W_l(t)W_s(t+\tau)] = 2\pi K_{ls}\delta(\tau), \quad l,s = 1,2. \quad (6.4.37)$$

In this case, (6.4.29) is expressed as

$$\pi K_{11}p_1\frac{\partial\phi}{\partial H_1} + (\pi K_{12} + D)p_2\frac{\partial\phi}{\partial H_2} = c_{11}p_1 + c_{12}p_2,$$

$$(\pi K_{12} - D)p_1\frac{\partial\phi}{\partial H_1} + \pi K_{22}p_2\frac{\partial\phi}{\partial H_2} = c_{21}p_1 + c_{22}p_2. \quad (6.4.38)$$

In deriving (6.4.38), use has been made of (6.4.6), i.e.,

$$B_{11}^{(1)} = \frac{1}{2}b_{22} = \pi K_{11}, \quad B_{22}^{(2)} = \frac{1}{2}b_{44} = \pi K_{22},$$

$$B_{12}^{(1)} = b_{24}^{(2)} = \pi K_{12} + D, \quad B_{21}^{(2)} = b_{42}^{(4)} = \pi K_{12} - D,$$

(6.4.39)

where D is an arbitrary constant. Equation (6.4.38) leads to

$$\frac{\partial \phi}{\partial H_1} = \frac{[c_{11}K_{22} - c_{21}(K_{12} + D)]p_1 + [c_{12}K_{22} - c_{22}(K_{12} + D)]p_2}{\pi(K_{11}K_{22} - K_{12}^2 + D^2)p_1},$$

$$\frac{\partial \phi}{\partial H_2} = \frac{[c_{21}K_{11} - c_{11}(K_{12} - D)]p_1 + [c_{22}K_{11} - c_{12}(K_{12} - D)]p_2}{\pi(K_{11}K_{22} - K_{12}^2 + D^2)p_2}.$$

(6.4.40)

For the right and left sides of (6.4.40) to be consistent, the following conditions must be met

$$c_{12}K_{22} - c_{22}(K_{12} + D) = 0, \quad c_{21}K_{11} - c_{11}(K_{12} - D) = 0. \quad (6.4.41)$$

Elimination of D results in

$$\frac{c_{21}}{c_{11}}K_{11} + \frac{c_{12}}{c_{22}}K_{22} = 2K_{12}. \tag{6.4.42}$$

If condition (6.4.42) is satisfied, the exact stationary probability density is

$$p(\mathbf{q}, \mathbf{p}) = C \exp\left(-\frac{c_{11}H_1}{\pi K_{11}} - \frac{c_{22}H_2}{\pi K_{22}}\right)$$

$$= C \exp\left\{-\frac{c_{11}}{\pi K_{11}}\left[\frac{1}{2}p_1^2 + \int u_1(q_1)dq_1\right]\right.$$

$$\left. -\frac{c_{22}}{\pi K_{22}}\left[\frac{1}{2}p_2^2 + \int u_2(q_2)dq_2\right]\right\}, \tag{6.4.43}$$

which is of the same form as that of Eq. (6.2.61).

6.4.5 *Partially Integrable and Non-Resonant Case*

For this case, the Hamiltonian has the form of (6.4.3), and it is proved that the general solution of (6.4.14) is of the form (Zhu and Huang,

2001)

$$\phi(\mathbf{q}, \mathbf{p}) = \phi(H_1, H_2, \ldots, H_{r-1}, H_r, \tilde{H}), \qquad (6.4.44)$$

where

$$
\begin{aligned}
H_j &= H_j(q_j, p_j), \quad j = 1, 2, \ldots, r \\
\tilde{H} &= H(q_{r+1}, q_{r+2}, \ldots, q_n; p_{r+1}, p_{r+2}, \ldots, p_n).
\end{aligned} \qquad (6.4.45)
$$

Substituting (6.4.44) into (6.4.15), we obtain

$$
\sum_{k=1}^{r} B_{jk}^{(j)} \frac{\partial H_k}{\partial p_k} \frac{\partial \phi}{\partial H_k} + \frac{\partial \phi}{\partial \tilde{H}} \sum_{k=r+1}^{n} B_{jk}^{(j)} \frac{\partial \tilde{H}}{\partial p_k}
$$

$$
= \sum_{k=1}^{n} \frac{\partial B_{jk}^{(j)}}{\partial p_k} + \sum_{k=1}^{r} m_{jk} \frac{\partial H_k}{\partial p_k} + \sum_{k=r+1}^{n} m_{jk} \frac{\partial \tilde{H}}{\partial p_k} \quad j = 1, 2, \ldots, n.
$$

$$(6.4.46)$$

If a consistent ϕ function can be found to satisfy all n equations in (6.4.46), the system possesses a stationary probability density.

Example 6.4.2 Consider the following system (Zhu and Huang, 2001):

$$
m_1 \ddot{X}_1 + c_{11} \dot{X}_1 + c_{12} \dot{X}_2 + c_{13} \dot{\varphi} + \frac{\partial U(X_1, X_2)}{\partial X_1} = W_1(t),
$$

$$
(m_2 + m_3) \ddot{X}_2 + c_{21} \dot{X}_1 + c_{22} \dot{X}_2 + c_{23} \dot{\varphi} + \frac{\partial U(X_1, X_2)}{\partial X_2} = W_2(t),
$$

$$
I \ddot{\varphi} + c_{31} \dot{X}_1 + c_{32} \dot{X}_2 + c_{33} \dot{\varphi} = W_3(t), \qquad (6.4.47)
$$

where $W_l(t)$, $l = 1, 23$, are Gaussian white noises with correlation functions

$$
E[W_l(t) W_s(t + \tau)] = 2\pi K_{ls} \delta(\tau), \quad l, s = 1, 2, 3. \qquad (6.4.48)
$$

The Hamiltonian of system (6.4.47) is the system total energy, given by

$$
H = H_1 + \tilde{H}, \quad H_1 = \frac{1}{2} I \dot{\varphi}^2,
$$

$$(6.4.49)$$

$$
\tilde{H} = \frac{1}{2} m_1 \dot{X}_1^2 + \frac{1}{2}(m_2 + m_3) \dot{X}_2^2 + U(X_1, X_2).
$$

Equation (6.4.49) shows that the system is partially integrable. For this system, we have from (6.4.6) and (6.4.7)

$$
\begin{aligned}
&b_{2j,2k} = 2\pi K_{jk}, && B_{jj}^{(j)} = \pi K_{jj}, \quad j,k = 1,2,3 \\
&B_{12}^{(1)} = \pi K_{12} - D_{12}, && B_{21}^{(2)} = \pi K_{12} + D_{12}, \\
&B_{13}^{(1)} = \pi K_{13} - D_{13}, && B_{31}^{(3)} = \pi K_{13} + D_{13}, \\
&B_{23}^{(2)} = \pi K_{23} - D_{23}, && B_{32}^{(3)} = \pi K_{23} + D_{23},
\end{aligned}
\tag{6.4.50}
$$

where D_{jk} are constants. Substituting (6.4.49) and (6.4.50) into (6.4.46), we obtain for $j = 1$

$$
\begin{aligned}
&c_{11} m_1 \dot{x}_1 + c_{12}(m_2 + m_3)\dot{x}_2 + c_{13} I \dot{\phi} \\
&= \frac{\partial \phi}{\partial \tilde{H}} \left[\pi K_{11} m_1 \dot{x}_1 + \pi (K_{12} - D_{12})(m_2 + m_3)\dot{x}_2 \right] \\
&\quad + \frac{\partial \phi}{\partial H_1} \left[\pi (K_{13} - D_{13}) I \dot{\phi} \right].
\end{aligned}
\tag{6.4.51}
$$

In order for (6.4.51) to be satisfied, the following conditions must be met

$$
\frac{c_{11}}{\pi K_{11}} = \frac{c_{12}}{\pi K_{12} - D_{12}} = \alpha_1, \quad \frac{c_{13}}{\pi K_{13} - D_{13}} = \alpha_2.
\tag{6.4.52}
$$

Applying the same procedure to (6.4.46) for $j = 2$ and 3, we obtain more conditions

$$
\frac{c_{21}}{\pi K_{12} + D_{12}} = \frac{c_{22}}{\pi K_{22}} = \alpha_1, \quad \frac{c_{23}}{\pi K_{23} - D_{23}} = \alpha_2, \tag{6.4.53}
$$

$$
\frac{c_{31}}{\pi K_{13} + D_{13}} = \frac{c_{32}}{\pi K_{23} + D_{23}} = \alpha_1, \quad \frac{c_{33}}{\pi K_{33}} = \alpha_2. \tag{6.4.54}
$$

Combining (6.4.52), (6.4.53) and (6.4.54), and eliminating constants D_{jk}, we obtain

$$
\begin{aligned}
&\frac{c_{11}}{\pi K_{11}} = \frac{c_{22}}{\pi K_{22}} = \frac{c_{12} + c_{21}}{2\pi K_{12}} = \alpha_1, \quad \frac{c_{33}}{\pi K_{33}} = \alpha_2, \\
&\frac{c_{13}}{\alpha_2} + \frac{c_{31}}{\alpha_1} = 2\pi K_{13}, \quad \frac{c_{23}}{\alpha_2} + \frac{c_{32}}{\alpha_1} = 2\pi K_{23}.
\end{aligned}
\tag{6.4.55}
$$

If conditions in (6.4.55) are met, the stationary probability density of the system is

$$p(x_1, \dot{x}_1, x_2, \dot{x}_2, \dot{\phi}) = C \exp(-\alpha_1 \tilde{H} - \alpha_2 H_1)$$

$$= C \exp \left\{ -\frac{c_{11}}{\pi K_{11}} \left[\frac{1}{2} m_1 \dot{x}_1^2 + \frac{1}{2}(m_2 + m_3)\dot{x}_2^2 + U(x_1, x_2) \right] \right.$$

$$\left. -\frac{c_{33}}{2\pi K_{33}} I \dot{\varphi}^2 \right\}. \tag{6.4.56}$$

6.5 Parametrically Excited Linear Systems

As mentioned before, a system under parametric excitations is non-linear in nature even if the system itself is linear, namely, the system properties such as the restoring and damping forces are linear. This is because the superposition principle, as the signature of linearity, is not valid anymore. Nevertheless, exact solutions of some statistical properties, such as statistical moments, correlation functions, and spectral densities, may be obtainable for such type of systems.

Consider the following system

$$\ddot{Z}_j + \sum_{k=1}^{n} (\alpha_{jk} Z_k + \beta_{jk} \dot{Z}_k) = \sum_{k=1}^{n} [a_{jk} Z_k W_{1k}(t) + b_{jk} \dot{Z}_k W_{2k}(t)]$$

$$+ W_{3j}(t), \quad j = 1, 2, \ldots, n, \tag{6.5.1}$$

where $W_{1k}(t)$, $W_{2k}(t)$ and $W_{3j}(t)$ are Gaussian white noises. The system (6.5.1) has the following characteristics: (i) the system damping forces and restoring forces are linear, (ii) the parametric excitations are multiplicative on the linear terms of the system variables and (iii) the superposition principle is not applicable. Letting $X_{2j-1} = Z_j$ and $X_{2j} = \dot{Z}_j$, a set of Ito differential equations is derived from (6.5.1) as follows

$$dX_{2j-1} = X_{2j} dt$$

$$dX_{2j} = \sum_{k=1}^{2n} C_{jk} X_k dt + \sigma_j(\mathbf{X}) dB_j(t), \tag{6.5.2}$$

where $\mathbf{X} = [X_1, X_2, \ldots, X_{2n}]^{\mathrm{T}}$, $B_j(t)$ are independent unit Wiener processes, and C_{jk} and $\sigma_j(\mathbf{X})$ can be derived from Eq. (6.5.1) by incorporating the possible Wong–Zakai correction terms. Now the methods to calculate moments, correlation functions and spectral densities of system response $\mathbf{X} = [X_1, X_2, \ldots, X_{2n}]^{\mathrm{T}}$, described in Sections 5.5.1 and 5.5.2 for linear systems, can be applied for system (6.5.2). Although the parametric excitations are present, each set of these equations are closed due to the nature of the system, namely, only linear terms of X_j appear in the parametric excitations, and analytical solutions can be obtained from the equations.

Example 6.5.1 As an example, consider the following system

$$\ddot{X} + 2\zeta\omega_0\dot{X} + \omega_0^2 X = W_1(t) + \dot{X}W_2(t), \qquad (6.5.3)$$

where $W_1(t)$ and $W_2(t)$ are independent Gaussian white noises with spectral densities K_1 and K_2, respectively. Denoting X by X_1 and $\dot{X}$ by X_2, the Ito equations for system (6.5.3) are

$$\begin{aligned}
dX_1 &= X_2 dt, \\
dX_2 &= (-2\zeta\omega_0 X_2 + \pi K_2 X_2 - \omega_0^2 X_1)dt \\
&\quad + \sqrt{2\pi K_1}dB_1(t) + \sqrt{2\pi K_2}X_2 dB_2(t),
\end{aligned} \qquad (6.5.4)$$

where $B_1(t)$ and $B_2(t)$ are independent unit Wiener processes. Using the Ito differential rule to find dX_1^2, $dX_1 X_2$ and dX_2^2, and then taking ensemble averaging, we obtain the equations for the second-order moments

$$\frac{dm_{20}}{dt} = 2m_{11},$$

$$\frac{dm_{11}}{dt} = m_{02} - \omega_0^2 m_{20} - (2\zeta\omega_0 - \pi K_2)m_{11},$$

$$\frac{dm_{11}}{dt} = -2\omega_0^2 m_{11} - 2(2\zeta\omega_0 - \pi K_2)m_{02} + 2\pi(K_2 m_{02} + K_1),$$

$$(6.5.5)$$

where $m_{ij} = E[X_1^i X_2^j]$. Equation (6.5.5) is a set of ordinary differential equations with constant coefficients, and can be solved exactly subjected to given initial conditions. In particular, the stationary solutions of (6.5.5) are

$$m_{20} = \frac{\pi K_1}{2\omega_0^2(\zeta\omega_0 - \pi K_2)}, \quad m_{02} = \omega_0^2 m_{20}, \quad m_{11} = 0. \qquad (6.5.6)$$

The condition for (6.5.6) to be valid is

$$\zeta\omega_0 > \pi K_2, \qquad (6.5.7)$$

which is the condition for the response to be bounded due to large enough damping. For the special case of $K_1 = 0$, i.e., absence of the external excitation, only trivial solutions exist for the second-order moments under the condition (6.5.7), indicating the system is eventually approaching the trivial solution.

Using the same procedure, the higher-order moments can be obtained. Two fourth-order moments are given below

$$m_{40} = \frac{3\pi K_1 m_{20}}{\Delta_m}[2\omega_0^2 + 3(\zeta\omega_0 - 2\pi K_2)(2\zeta\omega_0 - 3\pi K_2)], \qquad (6.5.8)$$

$$m_{04} = \frac{3\pi K_1 \omega_0^4 m_{20}}{\Delta_m}[2\omega_0^2 + 3(2\zeta\omega_0 - 3\pi K_2)(\zeta\omega_0 - \pi K_2)], \qquad (6.5.9)$$

where

$$\Delta_m = \omega_0^4(4\zeta\omega_0 - 7\pi K_2)$$
$$+ 6\omega_0^2(\zeta\omega_0 - \pi K_2)(\zeta\omega_0 - 2\pi K_2)(2\zeta\omega_0 - 3\pi K_2). \qquad (6.5.10)$$

Similar to the case of the second-order moments, certain condition must be satisfied for (6.5.8) and (6.5.9) to be valid.

Multiplying the two equations in (6.5.4) by $X_1(t-\tau)$ and taking the ensemble average, we obtain

$$\frac{d}{d\tau}R_{11}(\tau) = R_{12}(\tau),$$

$$\frac{d}{d\tau}R_{12}(\tau) = -\omega_0^2 R_{11}(\tau) - (2\zeta\omega_0 - \pi K_2)R_{12}(\tau), \qquad (6.5.11)$$

where $R_{ij}(\tau) = R_{X_i X_j}(\tau) = E[X_i(t-\tau)X_j(t)]$ are the correlation functions. The initial conditions are $R_{11}(0) = m_{20}$ and $R_{12}(0) = m_{11} = 0$. Denote

$$\zeta_R = \frac{2\zeta\omega_0 - \pi K_2}{2\omega_0}. \tag{6.5.12}$$

According to (6.5.7), ζ_R is positive. For three different situations of ζ_R values, $R_{11}(\tau)$ can be solved from (6.5.11) for the range of $\tau \geq 0$ as

$$R_{11}(\tau) = m_{20}e^{-\zeta_R\omega_0\tau}$$

$$\times \left[\cos\omega_0\sqrt{1-\zeta_R^2}\,\tau + \frac{\zeta_R}{\sqrt{1-\zeta_R^2}}\sin\omega_0\sqrt{1-\zeta_R^2}\tau\right],$$

$$0 < \zeta_R < 1. \tag{6.5.13}$$

$$R_{11}(\tau) = m_{20}e^{-\zeta_R\omega_0\tau}(1+\omega_0\tau), \quad \zeta_R = 1. \tag{6.5.14}$$

$$R_{11}(\tau) = m_{20}e^{-\zeta_R\omega_0\tau}$$

$$\times \left[\cosh\omega_0\sqrt{\zeta_R^2-1}\tau + \frac{\zeta_R}{\sqrt{\zeta_R^2-1}}\sinh\omega_0\sqrt{\zeta_R^2-1}\tau\right],$$

$$\zeta_R > 1, \tag{6.5.15}$$

where m_{20} is given in (6.5.6). $R_{12}(\tau)$ can then be obtained from the first equation of (6.5.11). The other correlation functions, $R_{21}(\tau)$ and $R_{22}(\tau)$, can be derived similarly. For the negative range of τ, the symmetry properties in Eq. (3.3.5), i.e., $R_{ii}(\tau) = R_{ii}(-\tau)$, $R_{ij}(\tau) = R_{ji}(-\tau)$, can be used.

The spectral densities of $X_1(t)$ and $X_2(t)$ can be calculated by taking the Fourier transform of the correlation functions. However, they can be obtained directly without performing tedious Fourier transformation. Following the same procedure described in Sections 4.5.2 and 5.5.2, the equations for $\bar{\Phi}_{ij}(\omega)$ defined in (4.5.24) or (5.5.38) are obtained as follows

$$i\omega\bar{\Phi}_{11} - \bar{\Phi}_{12} = \frac{1}{\pi}m_{20},$$

$$\omega_0^2\bar{\Phi}_{11} + (2\zeta\omega_0 - \pi K_2 + i\omega)\bar{\Phi}_{12} = 0, \qquad (6.5.16)$$

$$i\omega\bar{\Phi}_{12} - \bar{\Phi}_{22} = 0,$$

$$\omega_0^2\bar{\Phi}_{12} + (2\zeta\omega_0 - \pi K_2 + i\omega)\bar{\Phi}_{22} = \frac{1}{\pi}m_{02}. \qquad (6.5.17)$$

The complex functions $\bar{\Phi}_{ij}(\omega)$ are then solved from (6.5.16) and (6.5.17). According to (4.5.25), we obtain the spectral densities from $\bar{\Phi}_{ij}(\omega)$

$$\Phi_{11}(\omega) = \frac{\omega_0^2 m_{20}}{\pi\Delta_s}(2\zeta\omega_0 - \pi K_2), \quad \Phi_{22}(\omega) = \omega^2\Phi_{11}(\omega),$$

$$\Phi_{12}(\omega) = \frac{\omega_0^2 m_{20}}{\pi\Delta_s}(\omega_0^2 - \omega^2), \qquad\qquad (6.5.18)$$

where

$$\Delta_s = (\omega^2 - \omega_0^2)^2 + \omega^2(2\zeta\omega_0 - \pi K_2)^2. \qquad (6.5.19)$$

Equations. (6.5.13), (6.5.14), (6.5.15) and (6.5.18) show that the existence of the correlation functions and spectral densities requires the existence of the second-order moments.

Exercise Problems

6.1 Consider the following system

$$\dot{X}_1 = f_1(X_1) - \gamma X_2 + W_1(t),$$
$$\dot{X}_2 = \gamma X_1 + f_2(X_2) + W_2(t),$$

where $W_1(t)$ and $W_2(t)$ are two independent Gaussian white noises with spectral densities K_1 and K_2, respectively. Find the condition for the system to belong to the class of stationary potential.

6.2 Find the condition for the following system to belong to the class of detailed balance

$$\dot{X}_1 = f_1(X_1, X_2) - \gamma X_2 + W_1(t),$$
$$\dot{X}_2 = \gamma X_1 + f_2(X_1, X_2) + W_2(t),$$

where $\gamma \neq 0$ and $W_1(t)$ and $W_2(t)$ are two independent Gaussian white noises with spectral densities K_1 and K_2, respectively.

6.3 A two-dimensional system is governed by

$$\dot{X}_1 = \alpha X_1 - \gamma X_2 + W_1(t),$$
$$\dot{X}_2 = \gamma X_1 + \beta X_2 + W_2(t),$$

where $\gamma \neq 0$ and $W_1(t)$ and $W_2(t)$ are two independent Gaussian white noises. Show that

(a) the system does not belong to the class of stationary potential,

(b) the system does not belong to the class of detailed balance in general,

(c) the system belongs to the class of generalized stationary potential.

6.4 Show that a general two-dimensional linear system

$$\dot{X}_1 = a_{11}X_1 + a_{12}X_2 + W_1(t),$$
$$\dot{X}_2 = a_{21}X_1 + a_{22}X_2 + W_2(t),$$

where $W_1(t)$ and $W_2(t)$ are two Gaussian white noises with

$$E[W_i(t)W_j(t + \tau)] = 2\pi K_{ij}\delta(\tau), \quad i, j = 1, 2,$$

belongs to the class of generalized stationary potential.

6.5 Consider the following nonlinear system

$$\ddot{X} + \alpha\dot{X} + \beta\dot{X}^3 + \gamma X^2\dot{X} + X = W(t),$$

where $W(t)$ is a Gaussian white noise with a spectral density K.

(a) Find the first and second derivate moments, and separate the first derivate moments into reversible and irreversible parts.

(b) Determine the condition for the system to belong to be in detailed balance, and find the stationary probability density $p(x, \dot{x})$ if the condition is satisfied.

6.6 For the following system

$$\ddot{X} + 2\zeta\omega_0\dot{X} + \omega_0^2[1 + W_1(t)]X = W_2(t),$$

where $W_1(t)$ and $W_2(t)$ are independent Gaussian white noises with spectral densities K_{11} and K_{22}, respectively, determine the condition for the system to belong to the class of generalized stationary potential.

6.7 Consider the system

$$\ddot{X} + \alpha\dot{X} + \beta\dot{X}^3 + X = W_1(t) + W_2(t)\dot{X},$$

where $W_1(t)$ and $W_2(t)$ are Gaussian white noises with correlation functions

$$E[W_i(t)W_j(t + \tau)] = 2\pi K_{ij}\delta(\tau), \quad i, j = 1, 2.$$

(a) Letting $X = X_1$ and $\dot{X} = X_2$, derive the FPK equation for the Markov diffusion vector $\{X_1, X_2\}$.

(b) Determine the condition for the system to belong to the class of generalized stationary potential.

(c) Obtain the exact stationary probability density when the condition in (b) is satisfied.

6.8 Determine the conditions under which the following systems belong to the class of generalized stationary potential:

(a) $\ddot{X} + 2\zeta\omega_0\dot{X}(1 + \alpha X^2) + \omega_0^2[1 + W_1(t)]X + f(X) = W_2(t),$

(b) $\ddot{X} + 2\zeta\omega_0\dot{X}(1 + \alpha X^2 + \beta\dot{X}^2 + \gamma|X^3|) + \omega_0^2(1 + \delta|X|)X = W(t),$

where $W_1(t)$, $W_2(t)$ and $W_2(t)$ are independent Gaussian white noises, and find the exact stationary probability density functions.

6.9 Find at least three stochastic systems which have the same stationary probability densities as that of the system

$$\ddot{X} + \alpha\dot{X} + \beta(X^2 + \dot{X}^2)\dot{X} + X = W(t).$$

6.10 Consider the following nonlinear system under a parametric random excitation

$$\ddot{X} + X^2\left(\beta + \frac{\alpha}{X^2 + \dot{X}^2}\right)\dot{X} + [1 + W(t)]X = 0, \quad \beta > 0.$$

(a) Show that the system possesses an exact stationary solution.

(b) Find the condition for the stationary probability density to be non-trivial.

6.11 Consider a SDOF linear system under a random excitation

$$\ddot{X} + 2\zeta\omega_0\dot{X} + \omega_0^2 X = F(t),$$

where $F(t)$ is a randomized harmonic process generated from

$$F(t) = A\sin[\omega_F t + \sigma B(t) + U],$$

where $B(t)$ is a unit Wiener process, and U is a random variable uniformly distributed in $[0, 2\pi]$ and independent of $B(t)$. Use the method in Section 6.5 to find the stationary second-order moments of the system.

6.12 Consider the following system

$$\ddot{X} + \alpha\dot{X} + \omega_0^2 X = XW_1(t) + \dot{X}W_2(t) + W_3(t),$$

where $W_1(t)$, $W_2(t)$ and $W_2(t)$ are Gaussian white noises with spectral densities K_{ij} $(i, j = 1, 2, 3)$. Use the method in Section 6.5 to find the first- and second-order moments.

6.13 Consider the following system

$$\ddot{X} + 2\xi\omega_0\dot{X} + \omega_0^2 X = XW_1(t) + W_2(t),$$

where $W_1(t)$ and $W_2(t)$ are Gaussian white noises with spectral densities K_{ij} $(i, j = 1, 2)$. For the stationary state, (a) calculate the second-order moments, (b) calculate the fourth-order moments, (c) determine the correlation functions $R_{ij}(\tau)$ and (d) find the spectral density functions $\Phi_{ij}(\omega)$.

6.14 Consider the following system

$$\ddot{X} + 2\alpha\dot{X} + \omega_0^2 X = XW_1(t) + \dot{X}W_2(t) + W_3(t),$$

where $W_1(t)$, $W_2(t)$ and $W_3(t)$ are independent Gaussian white noises with spectral densities K_{11}, K_{22} and K_{33}, respectively. For the stationary state, (a) calculate the second-order moments, (b) calculate the fourth-order moments, (c) determine the correlation functions $R_{ij}(\tau)$ and (d) find the spectral density functions $\Phi_{ij}(\omega)$.

6.15 Consider a two-DOF system with coupled parametric excitations, governed by

$$\ddot{Z}_1 + 2\zeta_1\omega_1\dot{Z}_1 + \omega_1^2 Z_1 = \omega_1\omega_2 Z_2 W_1(t) + W_2(t),$$

$$\ddot{Z}_2 + 2\zeta_2\omega_2\dot{Z}_2 + \omega_2^2 Z_2 = \omega_1\omega_2 Z_1 W_1(t),$$

where $0 < \zeta_1,\ \zeta_2 < 1$ and $W_1(t)$ and $W_2(t)$ are independent Gaussian white noises with spectral densities K_{11} and K_{22}, respectively. Find the stationary second-order moments of system response, and the power spectral densities of Z_1 and Z_2.

CHAPTER 7

APPROXIMATE SOLUTIONS OF NONLINEAR STOCHASTIC SYSTEMS

In Chapter 5, it is shown that certain probabilistic and/or statistical solutions of responses of linear systems subjected to external random excitations can be obtained based on given knowledge of the excitations. For nonlinear stochastic systems, exact solutions are obtainable only for some special types, such as a SDOF system with a linear damping force and a nonlinear restoring force subjected to a Gaussian white noise excitation. In general, exact stationary solutions of nonlinear stochastic systems are possible only when the excitations are Gaussian white noises and certain highly restrictive relations between the system parameters and the spectral densities of the excitations are satisfied, as shown in Chapter 6. In practical cases, such restrictive conditions are rarely met; therefore, approximate solution methods are needed to solve the problems.

In this chapter, several approximation techniques for obtaining probabilistic and/or statistical solutions of nonlinear stochastic systems are presented. For each method, its procedure for SDOF systems will be introduced first. Extension to MDOF systems then follows if the method is also applicable. It is pointed out that these approximation methods are by no means comprehensive.

7.1 Equivalent Linearization

7.1.1 *Equivalent Linearization*

One of the most frequently used approximation method is the equivalent linearization, in which the original nonlinear stochastic system is replaced by an equivalent linear stochastic system. The parameters in the replacing linear system are determined by minimizing the discrepancy between the nonlinear and linear stochastic systems in certain statistical sense. The most common criterion is the least mean square difference. For this reason, the method is also called statistical linearization. Consider the general form of SDOF nonlinear oscillatory systems subjected to an external random excitation

$$\ddot{X} + g(X, \dot{X}) = F(t), \tag{7.1.1}$$

where $g(X, \dot{X})$ is a nonlinear function, representing the damping force and restoring force, and $F(t)$ is a random excitation. Now we replace system (7.1.1) by a linear stochastic system

$$\ddot{X} + \alpha_e \dot{X} + k_e X = F(t). \tag{7.1.2}$$

The difference between (7.1.1) and (7.1.2) is

$$\delta = g(X, \dot{X}) - \alpha_e \dot{X} - k_e X. \tag{7.1.3}$$

The criterion of the least mean square error requires

$$\frac{\partial}{\partial \alpha_e} E[\delta^2] = 0, \quad \frac{\partial}{\partial k_e} E[\delta^2] = 0. \tag{7.1.4}$$

Substituting (7.1.3) into (7.1.4), exchanging the order of differentiation and expectation, and solving for α_e and k_e, we obtain

$$\alpha_e = \frac{E[X^2]E[\dot{X}g(X, \dot{X})] - E[X\dot{X}]E[Xg(X, \dot{X})]}{E[X^2]E[\dot{X}^2] - (E[X\dot{X}])^2}, \tag{7.1.5}$$

$$k_e = \frac{E[\dot{X}^2]E[Xg(X, \dot{X})] - E[X\dot{X}]E[\dot{X}g(X, \dot{X})]}{E[X^2]E[\dot{X}^2] - (E[X\dot{X}])^2}. \tag{7.1.6}$$

The ensemble averaging of the right-hand sides may be calculated according to the replacement linear system (7.1.2) with linearization parameters α_e and k_e. Thus, (7.1.5) and (7.1.6) are not explicit expressions for α_e and k_e. Even if there is enough information of the excitation $F(t)$ to calculate all expectations in (7.1.5) and (7.1.6) for given α_e and k_e, as described in Chapter 5, iterations are required to solve the problem.

To illustrate more clearly the essence of the linearization procedure, we assume that the excitation $F(t)$ is stationary process with zero mean, and that only stationary responses are of concern. It is known from Section 3.5.4 that the stationary responses X and $\dot{X}$ are uncorrelated, i.e., $E[X\dot{X}] = 0$, and (7.1.5) and (7.1.6) are simplified to

$$\alpha_e = \frac{E[\dot{X}g(X,\dot{X})]}{E[\dot{X}^2]}, \quad k_e = \frac{E[Xg(X,\dot{X})]}{E[X^2]}. \tag{7.1.7}$$

Furthermore, If $F(t)$ is a Gaussian white noise with a spectral density K, the stationary probability density of the linear stochastic system (7.1.2) is

$$p(x,\dot{x}) = C\exp\left[-\frac{\alpha_e}{2\pi K}(\dot{x}^2 + k_e x^2)\right], \tag{7.1.8}$$

which is treated as an approximate stationary solution for the original system (7.1.1). Equation (7.1.8) shows that

$$E[X^2] = \frac{\pi K}{\alpha_e k_e}, \quad E[\dot{X}^2] = \frac{\pi K}{\alpha_e}. \tag{7.1.9}$$

Then (7.1.7) can be written as

$$E[Xg(X,\dot{X})] = \frac{\pi K}{\alpha_e}, \quad E[\dot{X}g(X,\dot{X})] = \pi K. \tag{7.1.10}$$

The ensemble averaging on the left-hand sides of (7.1.10) is calculated using the approximate stationary probability density given by (7.1.8), leading to

$$\int\limits_{-\infty}^{\infty}\int\limits_{-\infty}^{\infty} xg(x,\dot{x})\exp\left[-\frac{\alpha_e}{2\pi K}(\dot{x}^2 + k_e x^2)\right]dxd\dot{x} = \frac{\pi K}{\alpha_e}, \tag{7.1.11}$$

$$\int\limits_{-\infty}^{\infty}\int\limits_{-\infty}^{\infty} \dot{x}g(x,\dot{x})\exp\left[-\frac{\alpha_e}{2\pi K}(\dot{x}^2 + k_e x^2)\right]dxd\dot{x} = \pi K. \tag{7.1.12}$$

Equations (7.1.11) and (7.1.12) are nonlinear algebraic equations for the unknown α_e and k_e, and can be solved either analytically or numerically. If function $g(x, \dot{x})$ is a polynomial function of x and $\dot{x}$, α_e and k_e can be solved from (7.1.11) and (7.1.12) analytically in closed form.

Consider the case in which the damping force and restoring force can be separated as

$$g(X, \dot{X}) = h(\dot{X}) + u(X). \tag{7.1.13}$$

Assume that the restoring force $u(X)$ is an odd function of X, and the damping force is an odd function of $\dot{X}$. In this case, (7.1.11) and (7.1.12) are simplified to

$$\int_{-\infty}^{\infty} xu(x) \exp\left(-\frac{\alpha_e k_e}{2\pi K} x^2\right) dx = \frac{\pi K}{\alpha_e}, \tag{7.1.14}$$

$$\int_{-\infty}^{\infty} \dot{x} h(\dot{x}) \exp\left(-\frac{\alpha_e}{2\pi K} \dot{x}^2\right) d\dot{x} = \pi K. \tag{7.1.15}$$

The equivalent damping coefficient α_e can be determined from (7.1.15) first, and the equivalent stiffness coefficient k_e is then found from (7.1.14).

Example 7.1.1 For the nonlinear system (7.1.1) with a Gaussian white noise excitation and

$$g(X, \dot{X}) = \alpha\dot{X} + \beta\dot{X}^3 + kX + \delta X^3. \tag{7.1.16}$$

The two equations in (7.1.10) become

$$kE[X^2] + \delta E[X^4] = \frac{\pi K}{\alpha_e}, \quad \alpha E[\dot{X}^2] + \beta E[\dot{X}^4] = \pi K. \tag{7.1.17}$$

Since X and $\dot{X}$ are Gaussian distributed, $E[X^4] = 3(E[X^2])^2$ and $E[\dot{X}^4] = 3(E[\dot{X}^2])^2$, then α_e and k_e are solved form (7.1.17) and (7.1.9) as

$$\alpha_e = \frac{1}{2}\left(\alpha + \sqrt{\alpha^2 + 12\pi\beta K}\right), \quad k_e = \frac{1}{2}\left(k + \sqrt{k^2 + \frac{12\pi\delta K}{\alpha_e}}\right).$$
$$\tag{7.1.18}$$

Equation (7.1.18) shows that the equivalent linearization parameters α_e and k_e depend not only on the system properties α, β, k and δ, but also on the excitation level K. The stronger the excitation is, the higher the system nonlinearity occurs.

The equivalent linearization method can also be applied to MDOF nonlinear systems subjected to external stochastic excitations. Consider the following MDOF nonlinear stochastic system

$$\ddot{Z}_j + g_j(\mathbf{Z}, \dot{\mathbf{Z}}) = \sum_{l=1}^{m} b_{jl} W_l(t), \quad j = 1, 2, \ldots, n, \qquad (7.1.19)$$

where b_{jl} are constant, $\mathbf{Z} = [Z_1, Z_2, \ldots, Z_n]^T$, $\dot{\mathbf{Z}} = [\dot{Z}_1, \dot{Z}_2, \ldots, \dot{Z}_n]^T$, and $W_l(t)$ are Gaussian white noises. Now replace system (7.1.19) by an equivalent linear stochastic system

$$\ddot{Z}_j + \sum_{k=1}^{n} \alpha_{jk} \dot{Z}_k + \sum_{k=1}^{n} k_{jk} Z_k = \sum_{l=1}^{m} b_{jl} W_l(t), \quad j = 1, 2, \ldots, n. \quad (7.1.20)$$

The difference between (7.1.19) and (7.1.20) is denoted as

$$\delta = \sum_{j=1}^{n} \left[g_j(\mathbf{Z}, \dot{\mathbf{Z}}) - \sum_{k=1}^{n} \alpha_{jk} \dot{Z}_k - \sum_{k=1}^{n} k_{jk} Z_k \right]. \qquad (7.1.21)$$

Applying the criterion of the least mean square error, i.e.,

$$\frac{\partial}{\partial k_{rs}} E[\delta^2] = 0, \quad \frac{\partial}{\partial \alpha_{rs}} E[\delta^2] = 0, \quad r, s = 1, 2, \ldots, n, \qquad (7.1.22)$$

we obtain

$$E\left\{ Z_s \sum_{j=1}^{n} \left[g_j(\mathbf{Z}, \dot{\mathbf{Z}}) - \sum_{k=1}^{n} \alpha_{jk} \dot{Z}_k - \sum_{k=1}^{n} k_{jk} Z_k \right] \right\} = 0, \quad (7.1.23)$$

$$E\left\{ \dot{Z}_s \sum_{j=1}^{n} \left[g_j(\mathbf{Z}, \dot{\mathbf{Z}}) - \sum_{k=1}^{n} \alpha_{jk} \dot{Z}_k - \sum_{k=1}^{n} k_{jk} Z_k \right] \right\} = 0. \quad (7.1.24)$$

Equations (7.1.23) and (7.1.24) can be written in the compact matrix form

$$E[\mathbf{X}\mathbf{X}^T]\begin{bmatrix}\mathbf{k}^T\\ \boldsymbol{\alpha}^T\end{bmatrix} = E[\mathbf{X}\mathbf{g}^T], \qquad (7.1.25)$$

where $\mathbf{X} = [\mathbf{Z}^T, \dot{\mathbf{Z}}^T]^T = [Z_1, Z_2, \ldots, Z_n, \dot{Z}_1, \dot{Z}_2, \ldots, \dot{Z}_n]^T$, $\mathbf{k} = [k_{jk}]$, $\boldsymbol{\alpha} = [\alpha_{jk}]$ and $\mathbf{g} = [g_1, g_2, \ldots, g_n]^T$. There are $2n^2$ equations in (7.1.25), which can be used to calculate the $2n^2$ unknowns k_{jk} and α_{jk}, as well as the second-order moments of system response $\mathbf{X}$. The iteration procedure could be listed as follows: (i) assume initial trail values of k_{jk} and α_{jk}, (ii) construct a Gaussian probability density for $\mathbf{X}$ from the linear equivalent system (7.1.20), (iii) calculate $E[\mathbf{X}\mathbf{X}^T]$ and $E[\mathbf{X}\mathbf{g}^T]$ in (7.1.25), (iv) solve a new set of k_{jk} and α_{jk} from (7.1.25) and (v) iterate until meeting accuracy requirement.

There are some shortcomings of the equivalent linearization method. It is known from the procedure that only external excitations are allowed, and it cannot be applied to systems under parametric excitations. Moreover, if the system nonlinearity is strong, the discrepancy between the original nonlinear system and the replacing linear system may be large. In this case, the method may result in large errors.

7.1.2 *Partial Linearization*

Consider a SDOF stochastic system in which the nonlinear damping force and nonlinear restoring force can be separated, i.e., the system

$$\ddot{X} + h(X, \dot{X}) + u(X) = W(t), \qquad (7.1.26)$$

where $W(t)$ is a Gaussian white noise with a spectral density K. As shown in Section 6.2.1, the system

$$\ddot{X} + \alpha_e \dot{X} + u(X) = W(t), \qquad (7.1.27)$$

possesses an exact stationary probability density

$$p(x, \dot{x}) = C \exp\left\{-\frac{\alpha_e}{\pi K}\left[\int_0^x u(z)dz + \frac{1}{2}\dot{x}^2\right]\right\}. \qquad (7.1.28)$$

Thus, linearization of the restoring force is not necessary. The procedure in which only the nonlinear damping force is linearized is called the partial linearization (Elishakoff and Cai, 1992). Using the least mean square criterion, parameter α_e for the linearized damping force is still calculated from (7.1.7), i.e.,

$$\alpha_e = \frac{E[\dot{X}h(X, \dot{X})]}{E[\dot{X}^2]}. \tag{7.1.29}$$

Equation (7.1.28) shows that the marginal stationary probability densities of X and $\dot{X}$ are separable, and $\dot{X}$ is Gaussian distributed; therefore, $E[\dot{X}^2] = \pi K/\alpha_e$ in (7.1.9) is still valid, and (7.1.29) can be written as

$$\int\limits_{-\infty}^{\infty} \int\limits_{-\infty}^{\infty} \dot{x}h(x, \dot{x}) \exp\left\{-\frac{\alpha_e}{\pi K}[\frac{1}{2}\dot{x}^2 + \int\limits_0^x u(x)dx]\right\} dx d\dot{x} = \pi K.$$

$$\tag{7.1.30}$$

The linearization coefficient α_e can be solved from the nonlinear algebraic equation (7.1.30). It is expected that the partial linearization is more accurate than the equivalent linearization since it retains the nonlinear restoring force.

It is noted, however, that the partial linearization method is applicable only when the external excitation is a Gaussian white noise, in which case the exact solution is available for the partially linearized system. The equivalent linearization, however, can be applied even if the excitations are not Gaussian white noises.

Example 7.1.2 For the same nonlinear stochastic system given by (7.1.16) in Example 7.1.1, we apply the partial linearization to obtain the approximate stationary probability density from (7.1.28)

$$p(x, \dot{x}) = C \exp\left[-\frac{\alpha_e}{2\pi K}\left(\dot{x}^2 + kx^2 + \frac{1}{2}\delta x^4\right)\right], \tag{7.1.31}$$

where α_e is given in (7.1.18). This result is more accurate than the one obtained in Example 7.1.1 since the nonlinearity in the restoring force is reserved.

7.1.3 *Linearization for Parametrically Excited Nonlinear Systems*

The equivalent linearization and partial linearization are applicable only for systems under external excitations, in which cases the response probability distributions can be obtained for the replacing systems. If parametric excitations are also present, these methods may result in large error, even in qualitative change of system topology. However, it is known in Section 6.5 that the exact statistical moments, correlation functions, and spectral densities can be found for parametrically excited linear systems under both external and parametric Gaussian white noise excitation. Based on this knowledge, the linearization method has been extended for approximate solution of nonlinear stochastic systems subjected to both external and parametric excitations (Cai, 2003, 2004; Cai and Suzuki, 2005).

The nonlinear stochastic system of concern is governed by

$$\ddot{X} + g(X, \dot{X}) = XW_1(t) + \dot{X}W_2(t) + W_3(t), \qquad (7.1.32)$$

where $g(X, \dot{X})$ is assumed to be a polynomial of X and $\dot{X}$, and $W_1(t)$, $W_2(t)$ and $W_3(t)$ are Gaussian white noises with correlations functions

$$E[W_l(t)W_s(t+\tau)] = 2\pi K_{ls}\delta(\tau), \quad l, s = 1, 2, 3. \qquad (7.1.33)$$

The equivalent linear system is of the form

$$\ddot{X} + \alpha_e \dot{X} + k_e X = XW_1(t) + \dot{X}W_2(t) + W_3(t). \qquad (7.1.34)$$

It is noted that all excitation terms remain unchanged in the replacement system (7.1.34). If the criterion of the least mean square error is used, and only stationary response is of concern, we obtain the equations for α_e and k_e from (7.1.7), i.e.,

$$\alpha_e = \frac{E[\dot{X}g(X, \dot{X})]}{E[\dot{X}^2]}, \quad k_e = \frac{E[Xg(X, \dot{X})]}{E[X^2]}. \qquad (7.1.35)$$

The ensemble averages on the right sides of (7.1.35) can be calculated exactly from the equivalent system (7.1.34), as shown in Section 6.5. Therefore, the Gaussian assumption for the response in the classical linearization procedure is not needed. Still, iteration is needed to calculate α_e and k_e.

It is known in Section 6.5 that, besides the response moments, the correlation functions and spectral densities can be obtained exactly for linear systems under both external and parametric excitations. Consequently, they can serve as approximate correlation functions and spectral densities for the original nonlinear stochastic systems, which is an advantage of this procedure.

Example 7.1.3 Consider the following system with nonlinear damping and stiffness, and under both external and parametric excitations

$$\ddot{X} + \alpha\dot{X} + \beta\dot{X}^3 + kX + \delta X^3 = W_1(t) + \dot{X}W_2(t), \qquad (7.1.36)$$

where Gaussian white noises $W_1(t)$ and $W_2(t)$ are assumed to be independent with spectral densities K_1 and K_2, respectively. System (7.1.36) is now approximated by the following linearized system

$$\ddot{X} + \alpha_e\dot{X} + k_eX = W_1(t) + \dot{X}W_2(t). \qquad (7.1.37)$$

The two linearization coefficients α_e and k_e are obtained from (7.1.35) as follows

$$\alpha_e = \alpha + \frac{\beta E[\dot{X}^4]}{E[\dot{X}^2]}, \quad k_e = k + \frac{\delta E[X^4]}{E[X^2]}. \qquad (7.1.38)$$

Note that system (7.1.37) is identical as system (6.5.3) in Example 6.5.1 with different notations. We have the second-order and some of the fourth-order moments for the linearized system (7.1.37) obtained in Example 6.5.1 as

$$m_{20} = \frac{\pi K_1}{k_e(\alpha_e - 2\pi K_2)}, \quad m_{02} = k_e m_{20}, \quad m_{11} = 0, \qquad (7.1.39)$$

$$m_{40} = \frac{3\pi K_1 m_{20}}{\Delta}\left[2k_e + \frac{3}{2}(\alpha_e - 4\pi K_2)(\alpha_e - 3\pi K_2)\right], \qquad (7.1.40)$$

$$m_{04} = \frac{3\pi K_1 k_e^2 m_{20}}{\Delta}\left[2k_e + \frac{3}{2}(\alpha_e - 2\pi K_2)(\alpha_e - 3\pi K_2)\right], \qquad (7.1.41)$$

where $m_{ij} = E[X^i \dot{X}^j]$ and

$$\Delta = k_e^2(2\alpha_e - 7\pi K_2) + \frac{3}{2}k_e(\alpha_e - 2\pi K_2)(\alpha_e - 3\pi K_2)(\alpha_e - 4\pi K_2).$$
$$(7.1.42)$$

The two linearization coefficients α_e and k_e, as well as the second- and fourth-order moments can be solved iteratively from (7.1.38).

Since m_{20}, m_{02}, m_{40} and m_{04} must be non-negative, the validity of (7.1.39) through (7.1.41) requires certain conditions, which are in fact the stability conditions for the second- and fourth-order moments of system (7.1.37). However, for the original system, the trivial solution is unstable due to the presence of the external excitation, also the system will not be divergent due to the nonlinear stiffness and damping. Thus, a non-trivial stationary probability density exists without any conditions required. It is concluded, therefore, that the conditions required from Eqs. (7.1.39) through (7.1.41) arises merely from the linearization procedure.

The correlation function $R(\tau)$ of $X(t)$ can be found according to Example 6.5.1 for the range of $\tau \geq 0$ as follows

$$R(\tau) = m_{20}e^{-\zeta_R\omega_0\tau}\left[\cos\omega_0\sqrt{1-\zeta_R^2}\tau + \frac{\zeta_R}{\sqrt{1-\zeta_R^2}}\sin\omega_0\sqrt{1-\zeta_R^2}\tau\right],$$

$$0 < \zeta_R < 1, \tag{7.1.43}$$

$$R(\tau) = m_{20}e^{-\zeta_R\omega_0\tau}(1+\omega_0\tau), \quad \zeta_R = 1, \tag{7.1.44}$$

$$R(\tau) = m_{20}e^{-\zeta_R\omega_0\tau}$$

$$\times\left[\cosh\omega_0\sqrt{\zeta_R^2-1}\tau + \frac{\zeta_R}{\sqrt{\zeta_R^2-1}}\sinh\omega_0\sqrt{\zeta_R^2-1}\tau\right],$$

$$\zeta_R > 1, \tag{7.1.45}$$

where

$$\omega_0 = \sqrt{k_e}, \quad \zeta = \frac{\alpha_e}{2\omega_0}, \quad \zeta_R = \frac{2\zeta\omega_0 - \pi K_2}{2\omega_0}. \tag{7.1.46}$$

The power spectral density $\Phi(\omega)$ is

$$\Phi(\omega) = \frac{\omega_0^2(2\zeta\omega_0 - \pi K_2)m_{20}}{\pi[(\omega^2 - \omega_0^2)^2 + \omega^2(2\zeta\omega_0 - \pi K_2)^2]}. \qquad (7.1.47)$$

Numerical investigation was conducted in (Cai, 2003) to substantiate the accuracy of the procedure.

Extension of the procedure from SDOF to MDOF nonlinear stochastic systems is straightforward. Consider the following MDOF system

$$\ddot{Z}_j + g_j(\mathbf{Z}, \dot{\mathbf{Z}}) = \sum_{k=1}^{n} [a_{jk}Z_k W_{1k}(t) + b_{jk}\dot{Z}_k W_{2k}(t)] + W_{3j}(t),$$

$$j = 1, 2, \ldots, n, \qquad (7.1.48)$$

where $\mathbf{Z} = [Z_1, Z_2, \ldots, Z_n]^T$, $\dot{\mathbf{Z}} = [\dot{Z}_1, \dot{Z}_2, \ldots, \dot{Z}_n]^T$, and $W_{1k}(t)$, $W_{2k}(t)$ and $W_{3j}(t)$ are Gaussian white noises. In system (7.1.48), the damping forces and restoring forces may be nonlinear, but the parametric excitations are multiplicative on the linear terms of Z_k or $\dot{Z}_k$. It is assumed that the nonlinear functions g_j are polynomial of Z_k and $\dot{Z}_k$. The linearization procedure replaces nonlinear system (7.1.48) by the following system

$$\ddot{Z}_j + \sum_{k=1}^{n} \alpha_{jk}\dot{Z}_k + \sum_{k=1}^{n} k_{jk}Z_k$$

$$= \sum_{k=1}^{n} [a_{jk}Z_k W_{1k}(t) + b_{jk}\dot{Z}_k W_{2k}(t)] + W_{3j}(t), \quad j = 1, 2, \ldots, n.$$

$$(7.1.49)$$

The equivalent damping and stiffness coefficients α_{jk} and k_{jk} can be calculated using the same method as given in Section 7.1, i.e., using the equation (7.1.25). The advantage of the present linearization procedure is that the terms in $E[\mathbf{XX}^T]$ and $E[\mathbf{Xg}^T]$ in (7.1.25) can be now calculated from (7.1.49) exactly for given α_{jk} and k_{jk} using the procedure described in Section 6.5. Thus, the assumption of Gaussian distribution for the response is not needed.

After the linearization coefficients α_{jk} and k_{jk} are obtained, the correlation functions and spectral densities can be calculated following the same procedure in Section 6.5.

Example 7.1.4 Consider a nonlinear system subjected to a sinusoidal function with a random amplitude and a random phase disturbed by two independent noises, respectively, as suggested by Hou *et al.* (1996). The system is governed by

$$\ddot{X} + \alpha \dot{X} + \beta \dot{X}^3 + kX + \delta X^3 = \xi(t), \qquad (7.1.50)$$

where

$$\xi(t) = [A + W_1(t)] \cos \theta(t), \quad \frac{d\theta}{dt} = \nu + W_2(t). \qquad (7.1.51)$$

In the noise model (7.1.51), A and ν are constants representing the mean amplitude and mean frequency, respectively, and $W_1(t)$ and $W_2(t)$ are two independent Gaussian white noises with spectral density K_{11} and K_{22}, respectively. The case of $K_{11} = 0$ and $K_{22} \neq 0$ corresponds to the randomized harmonic process described in Section 4.5.3.

Letting $X_1 = X$, $X_2 = \dot{X}$, $X_3 = \cos \theta$ and $X_4 = \sin \theta$, the system equation (7.1.50) can be written as

$$\begin{aligned}
\dot{X}_1 &= X_2, \\
\dot{X}_2 &= -\alpha X_2 - \beta X_2^3 - kX_1 - \delta X_1^3 + AX_3 + X_3 W_1(t), \\
\dot{X}_3 &= -\nu X_4 - X_4 W_2(t), \\
\dot{X}_4 &= \nu X_3 + X_3 W_2(t).
\end{aligned} \qquad (7.1.52)$$

It is noted that the SDOF system (7.1.50) is subjected to an external non-Gaussian random excitation, and it is transformed to system (7.1.52) under parametric excitations of Gaussian white noises. As a price for the simplification from a non-Gaussian excitation to Gaussian white noises, the dimension of the system is increased, and an external excitation is replaced by multiple parametric excitations. Nevertheless, the advantage of the transformation lies in the applicability of the linearization approach, and the accuracy of the approximate results may be improved.

Carry out the linearization procedure to obtain the equivalent system

$$
\begin{aligned}
\dot{X}_1 &= X_2, \\
\dot{X}_2 &= -\alpha_e X_2 - k_e X_1 + A X_3 + X_3 W_1(t), \\
\dot{X}_3 &= -\nu X_4 - X_4 W_2(t), \\
\dot{X}_4 &= \nu X_3 + X_3 W_2(t),
\end{aligned}
\tag{7.1.53}
$$

where the linearization coefficients are expressed in (7.1.35). To solve (7.1.53) following the procedure in Section 6.5, system (7.1.53) is transformed to the corresponding Ito equations as follows

$$
\begin{aligned}
dX_1 &= X_2 dt, \\
dX_2 &= (-\alpha_e X_2 - k_e X_1 + A X_3)dt + \sqrt{2\pi K_{11}}X_3 dB_1(t), \\
dX_3 &= (-\pi K_{22} - \nu X_4)dt - \sqrt{2\pi K_{22}}X_4 dB_2(t), \\
dX_4 &= (\nu X_3 - \pi K_{22}X_4) + \sqrt{2\pi K_{22}}X_3 dB_2(t).
\end{aligned}
\tag{7.1.54}
$$

Use the Ito stochastic differential rule, linear algebraic equations can be obtained and solved for stationary moments of second- and fourth-order, which can be used in (7.1.35) to calculate the linearization coefficients α_e and k_e, as well as the approximate stationary moments for the original nonlinear system. Numerical calculations were carried out in (Cai and Suzuki, 2005). Some numerical results are given below. The system parameters are: $\alpha = 0.6$, $k = 36$, $A = 1$ and $\nu = 6$. Figure 7.1.1 depicts the stationary mean values of the displacement $X(t)$ for two sets of β and δ values. The spectral density of $W_1(t)$ is fixed as $K_{11} = 0.1$, while that of $W_2(t)$, K_{22}, controlling the bandwidth of the excitation, is varying. A larger K_{22} indicates a broader band, and $K_{22} = 0$ corresponds to a harmonic excitation with a random amplitude. Figure 7.1.2 shows the calculated spectral densities of $X(t)$. In these figures, solid lines are the results from the present linearization procedure, dashed lines from conventional linearization applied directly to system (7.1.50), and symbols from Monte Carlo simulations. It is seen that the present linearization procedure yields more accurate results than the conventional linearization.

It is noted that only an external excitation is present in the original system (7.1.50), and the equivalent linearization can be

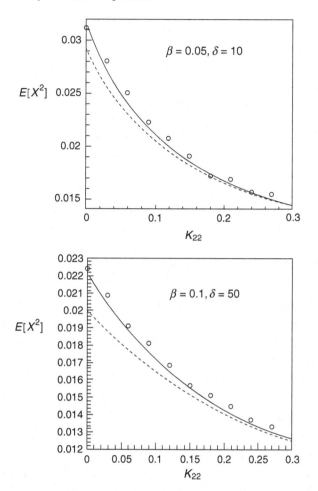

Figure 7.1.1 Stationary mean-square values of displacement $X(t)$ of system (7.1.50) (from Cai and Suzuki, 2005).

applied directly in which Gaussian distribution has to be assumed for the system response. However, the system response may be highly non-Gaussian due to the highly non-Gaussian excitation $\xi(t)$ and the system nonlinearity. The assumption of response Gaussian distribution may cause large errors as shown in Figs. 7.1.1 and 7.1.2. The present linearization provides a solution method for higher accuracy.

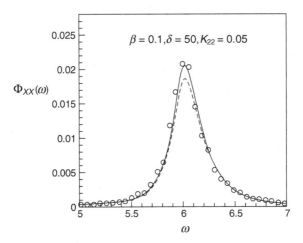

Figure 7.1.2   Spectral densities of displacement $X(t)$ of system (7.1.50) (from Cai and Suzuki, 2005).

Example 7.1.5 Consider a two-DOF nonlinear system with coupled parametric excitations, governed by

$$\ddot{Z}_1 + \alpha\dot{Z}_1 + \beta\dot{Z}_1^3 + \omega_1^2 Z_1 = \omega_1\omega_2 Z_2 W_1(t) + W_2(t),$$
$$\ddot{Z}_2 + 2\zeta_2\omega_2\dot{Z}_2 + \omega_2^2 Z_2 = \omega_1\omega_2 Z_1 W_1(t), \tag{7.1.55}$$

where $W_1(t)$ and $W_2(t)$ are independent Gaussian white noises with spectral densities K_{11} and K_{22}, respectively. The equation set (7.1.55) describes the fundamental modes of the transverse deflection and the angle of twist for a simply supported beam of narrow cross-section subjected to random transverse force and end moments, and undergoing bending and torsion (Ariaratnam and Srikantaiah, 1978). The damping force for the bending motion Z_1 is assumed to be nonlinear.

Applying the linearization procedure, the first equation in (7.1.55) is now replaced by

$$\ddot{Z}_1 + 2\zeta_1\omega_1\dot{Z}_1 + \omega_1^2 Z_1 = \omega_1\omega_2 Z_2 W_1(t) + W_2(t). \tag{7.1.56}$$

Letting $X_1 = Z_1$, $X_2 = \dot{Z}_1$, $X_3 = Z_2$ and $X_4 = \dot{Z}_2$, and $m_{ijkl} = E[X_1^i X_2^j X_3^k X_4^l]$. The linearization damping ratio ζ_1 is given by, according to (7.1.35),

$$\zeta_1 = \frac{1}{2\omega_1} \left(\alpha + \frac{\beta m_{0400}}{m_{0200}} \right). \tag{7.1.57}$$

The second-order moments are obtained for the linearized system as (see Exercise Problem 6.15)

$$E[Z_1^2] = E[X_1^2] = m_{2000} = \frac{2\pi\zeta_2 K_{22}}{\omega_1^3 \Delta},$$

$$E[\dot{Z}_1^2] = E[X_2^2] = m_{0200} = \frac{2\pi\zeta_2 K_{22}}{\omega_1 \Delta},$$

$$E[Z_2^2] = E[X_3^2] = m_{0020} = \frac{\pi^2 K_{11} K_{22}}{\omega_1 \omega_2 \Delta},$$

$$E[\dot{Z}_2^2] = E[X_4^2] = m_{0002} = \frac{\pi^2 \omega_2 K_{11} K_{22}}{\omega_1 \Delta}, \tag{7.1.58}$$

where

$$\Delta = 4\zeta_1\zeta_2 - \pi^2 \omega_1 \omega_2 K_{11}^2. \tag{7.1.59}$$

The other second-order moments are zero. The validity of (7.1.58) requires

$$4\zeta_1\zeta_2 > \pi^2 \omega_1 \omega_2 K_1^2. \tag{7.1.60}$$

The analytical expression for the fourth-order moments can also be derived exactly, but rather tediously. However, their numerical solutions are quite simple. It is noted that the linearization damping ratio ζ_1, given in (7.1.57), should be determined by iteration.

The spectral densities of Z_1 and Z_2 are obtained as

$$\Phi_{11}(\omega) = \frac{2\zeta_1\omega_1^3 m_{2000}}{\pi[(\omega_1^2 - \omega^2)^2 + 4\zeta_1^2\omega_1^2\omega^2]},$$

$$\Phi_{22}(\omega) = \frac{2\zeta_2\omega_2^3 m_{0020}}{\pi[(\omega_2^2 - \omega^2)^2 + 4\zeta_2^2\omega_2^2\omega^2]}. \tag{7.1.61}$$

The expressions in (7.1.61) show that each spectral density has the same form as that of a SDOF linear system. The coupling effects between the two modes are accounted for in the second-order moments, as shown in (7.1.58).

Numerical calculations were performed to show the accuracy of the proposed linearization procedure for the example (Cai, 2004). In calculation, the system parameters were chosen as: $\alpha = 1.2$, $\omega_1 = 6$, $\omega_2 = 20$, $\xi_2 = 0.1$ and $K_{11} = 0.004$. The mean-square values and spectral densities of $Z_1(t)$ are shown in Figs. 7.1.3 and 7.1.4, respectively. In the figures, lines are results calculated from the present linearization procedure, while symbols are from Monte Carlo simulations. The figures show that the analytical results are quite accurate compared with those from simulations.

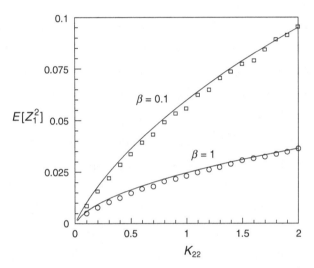

Figure 7.1.3 Stationary mean-square values of $Z_1(t)$ of system (7.1.55) (from Cai, 2004).

7.2 Cumulant-Neglect Closure

7.2.1 *Response Moments*

Consider a two-dimensional nonlinear stochastic system which can be described by two first-order differential equations

$$
\begin{aligned}
\dot{X}_1 &= f_1(X_1, X_2) + \sum_{k=1}^{m} g_{1k}(X_1, X_2)W_k(t), \\
\dot{X}_2 &= f_2(X_1, X_2) + \sum_{k=1}^{m} g_{2k}(X_1, X_2)W_k(t),
\end{aligned}
\tag{7.2.1}
$$

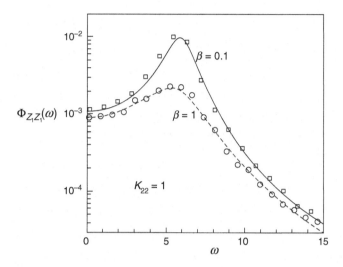

$\Phi_{Z_1 Z_1}(\omega)$

$\beta = 0.1$

$\beta = 1$

$K_{22} = 1$

Figure 7.1.4   Spectral densities of $Z_1(t)$ of system (7.1.55) (from Cai, 2004).

where $W_k(t)$ are Gaussian white noises. Equation set (7.2.1) is equivalent to the following set of Ito stochastic differential equations

$$
\begin{aligned}
dX_1 &= \left(f_1 + \sum_{l,s=1}^{m} \pi K_{ls} \sum_{j=1}^{2} g_{js} \frac{\partial g_{1l}}{\partial X_j} \right) dt \\
&\quad + \left(\sum_{l,s=1}^{m} 2\pi K_{ls} \sum_{j=1}^{2} g_{1l} g_{js} \right)^{\frac{1}{2}} dB_1(t), \\
dX_2 &= \left(f_2 + \sum_{l,s=1}^{m} \pi K_{ls} \sum_{j=1}^{2} g_{js} \frac{\partial g_{2l}}{\partial X_j} \right) dt \\
&\quad + \left(\sum_{l,s=1}^{m} 2\pi K_{ls} \sum_{j=1}^{2} g_{2l} g_{js} \right)^{\frac{1}{2}} dB_2(t),
\end{aligned}
$$

(7.2.2)

where $B_1(t)$ and $B_2(t)$ are independent unit Wiener processes and K_{ls} are spectral densities of $W_l(t)$ and $W_s(t)$. Denote

$M(X_1, X_2) = X_1^i X_2^j$ and use the Ito differential rule to obtain

$$
dM = \left[\left(f_1 + \sum_{l,s=1}^{m} \pi K_{ls} \sum_{k=1}^{2} g_{ks} \frac{\partial g_{1l}}{\partial X_k} \right) \frac{\partial M}{\partial X_1} \right.
$$

$$
+ \left(f_2 + \sum_{l,s=1}^{m} \pi K_{ls} \sum_{k=1}^{2} g_{ks} \frac{\partial g_{2l}}{\partial X_k} \right) \frac{\partial M}{\partial X_2}
$$

$$
\left. + \sum_{l,s=1}^{m} \pi K_{ls} \sum_{k,r=1}^{2} g_{kl} g_{rs} \frac{\partial^2 M}{\partial X_k \partial X_r} \right] dt
$$

$$
+ \left(\sum_{l,s=1}^{m} 2\pi K_{ls} \sum_{k=1}^{2} g_{1l} g_{ks} \right)^{\frac{1}{2}} \frac{\partial M}{\partial X_1} dB_1(t)
$$

$$
+ \left(\sum_{l,s=1}^{m} 2\pi K_{ls} \sum_{k=1}^{2} g_{2l} g_{ks} \right)^{\frac{1}{2}} \frac{\partial M}{\partial X_2} dB_2(t). \qquad (7.2.3)
$$

Taking the ensemble average, we have

$$
\frac{dE[M]}{dt} = E \left[\left(f_1 + \sum_{l,s=1}^{m} \pi K_{ls} \sum_{k=1}^{2} g_{ks} \frac{\partial g_{1l}}{\partial X_k} \right) \frac{\partial M}{\partial X_1} \right.
$$

$$
\left. + \left(f_2 + \sum_{l,s=1}^{m} \pi K_{ls} \sum_{k=1}^{2} g_{ks} \frac{\partial g_{2l}}{\partial X_k} \right) \frac{\partial M}{\partial X_2} \right]
$$

$$
+ E \left[\sum_{l,s=1}^{m} \pi K_{ls} \sum_{k,r=1}^{2} g_{kl} g_{rs} \frac{\partial^2 M}{\partial X_k \partial X_r} \right]. \qquad (7.2.4)
$$

The left-hand side of (7.2.4) is the time derivative of a statistical moment of order N, where $N = i + j$, whereas the right-hand side depends on the functional form of f_1, f_2 and g_{ls}. If these functions are linear, the right-hand side contains only the Nth- and lower-order statistical moments. In this case, equations for statistical moments can be solved recursively, beginning from $N = 1$. However, if at least one of the f_1, f_2 and g_{ls} functions is a nonlinear polynomial, then the right-hand side of (7.2.4) also contains moments of order higher than N. An exact solution is no longer obtainable since the moment equations for $N = 1, 2$, constitute an infinite hierarchy.

It is shown in Section 2.5.1 that for a Gaussian random variable with zero mean, all of its cumulants of orders higher than two are zero. For a non-Gaussian random variable, its cumulants of higher order are expected to be small if it deviates not far from Gaussian distribution. Based on such reasoning, the scheme of cumulant-neglect closure was proposed (Beran, 1968; Wu and Lin, 1984; Ibrahim *et al.*, 1985). Let all cumulants of orders higher than a given order N be set to zero. Then a statistical moment of an order higher than N can be expressed in terms of those moments of order N and lower than N, according to Section 2.6.4. For example, if all random variables X_1, X_2, have zero mean, and if the truncation order is set at $N = 2$, i.e., letting all cumulants of order higher than two be zero, we obtain from Eqs. (2.6.25) and (2.6.27)

$$E[X_j X_k X_l] = 0,$$

$$E[X_j X_k X_l X_m] = 3\{E[X_j X_k]E[X_l X_m]\}_s,$$

$$E[X_j X_k X_l X_m X_p] = 0,$$

$$E[X_j X_k X_l X_m X_p X_r] = 15\{E[X_j X_k]\{3E[X_l X_m]E[X_p X_r]\}_s\}_s$$
$$+ 15\{E[X_j X_k]E[X_l X_m]E[X_p X_r]\}_s,$$

$$\vdots$$
$$(7.2.5)$$

where symbol $\{\cdot\}_s$ denotes a symmetrizing operation with respect to all its arguments, as explained in Section 2.6.4.

Equation (7.2.5) are the same relations as those of Gaussian random variables, indicating the cumulant-neglect closure of the second order is the same as assuming that the variables are Gaussian distributed. Thus, it is the same as the method of Gaussian closure (Crandall, 1978). However, by letting $N = 4$, we have

$$E[X_j X_k X_l X_m X_p] = 10\{E[X_j X_k]E[X_l X_m X_p]\}_s,$$

$$E[X_j X_k X_l X_m X_p X_r] = 15\{E[X_j X_k](E[X_j X_k X_p X_r]$$

$$-3\{E[X_l X_m]E[X_p X_r]\}_s)\}_s$$

$$+15\{E[X_j X_k]E[X_l X_m]E[X_p X_r]\}_s$$

$$+10\{E[X_j X_k X_l]E[X_m X_p X_r]\}_s$$

$$\vdots \qquad (7.2.6)$$

By selecting a truncated level N, the total number of equations of the form (7.2.4) becomes finite in order to computing unknown statistical moments up to the Nth order. Although these equations may be nonlinear, they can be solved numerically. If the system tends to statistical stationarity, in which case the statistical moments are independent of time, then Eq. (7.2.4) reduces to an algebraic equation.

The cumulant-neglect closure procedure is illustrated in the following example.

Example 7.2.1 Consider the Duffing oscillator subjected to a Gaussian white noise excitation

$$\ddot{X} + \alpha \dot{X} + X + \varepsilon X^3 = W(t), \qquad (7.2.7)$$

where $\alpha > 0$, ε is a small positive parameter representing the degree of nonlinearity. The stationary probability density for $X_1 = X$ and $X_2 = \dot{X}$ is known as

$$p(x_1, x_2) = C \exp\left[-\frac{\alpha}{\pi K}\left(\frac{1}{2}x_2^2 + \frac{1}{2}x_1^2 + \frac{1}{4}\varepsilon x_1^4\right)\right]. \qquad (7.2.8)$$

Since the statistical moments of any order can be calculated from (7.2.8), this example is just used here for illustration purpose and for checking the accuracy of the cumulant-neglect closure. Equation (7.2.4) now reads

$$\frac{dE[M]}{dt} = E\left[X_2 \frac{\partial M}{\partial X_1}\right] - E\left[(\alpha X_2 + X_1 + \varepsilon X_1^3)\frac{\partial M}{\partial X_2}\right]$$

$$+\pi K E\left[\frac{\partial^2 M}{\partial X_2^2}\right]. \qquad (7.2.9)$$

Letting $M = X_1^i X_2^j$, and denoting $m_{ij} = E[X_1^i X_2^j]$, we obtain the equations of the statistical moments for $i + j = 1$

$$\frac{dm_{10}}{dt} = m_{01},$$

$$\frac{dm_{01}}{dt} = -m_{10} - \eta m_{01} - \varepsilon m_{30},$$

(7.2.10)

the equations for $i + j = 2$

$$\frac{dm_{20}}{dt} = 2m_{11},$$

$$\frac{dm_{11}}{dt} = m_{02} - m_{20} - \alpha m_{11} - \varepsilon m_{40},$$

(7.2.11)

$$\frac{dm_{02}}{dt} = -2m_{11} - 2\alpha m_{02} - 2\varepsilon m_{31} + 2\pi K,$$

the equations for $i + j = 3$

$$\frac{dm_{30}}{dt} = 3m_{21},$$

$$\frac{dm_{21}}{dt} = 2m_{12} - \alpha m_{21} - m_{30} - \varepsilon m_{50},$$

$$\frac{dm_{12}}{dt} = m_{03} - 2\alpha m_{12} - 2m_{21} - 2\varepsilon m_{41} + 2\pi K m_{10},$$

$$\frac{dm_{03}}{dt} = -3\alpha m_{03} - 3m_{12} - 3\varepsilon m_{32} + 6\pi K m_{01},$$

(7.2.12)

and the equations for $i + j = 4$

$$\frac{dm_{40}}{dt} = 4m_{31},$$

$$\frac{dm_{31}}{dt} = 3m_{22} - \alpha m_{31} - m_{40} - \varepsilon m_{60},$$

$$\frac{dm_{22}}{dt} = 2m_{13} - 2\alpha m_{22} - 2m_{31} - 2\varepsilon m_{51} + 2\pi K m_{20},$$

$$\frac{dm_{13}}{dt} = m_{04} - 3\alpha m_{13} - 3m_{22} - 3\varepsilon m_{42} + 6\pi K m_{11},$$

$$\frac{dm_{04}}{dt} = -4\alpha m_{04} - 4m_{13} - 4\varepsilon m_{33} + 12\pi K m_{02}. \qquad (7.2.13)$$

It is observed that in each set of equations for $N = i + j$, there are moments of order higher than N, and the equations cannot be solved. Now carrying the cumulant-neglect closure of truncation order of 2, i.e., letting all cumulants of order higher than two be zero, we have from (2.6.27)

$$m_{30} = 0, \quad m_{40} = 3m_{20}^2, \quad m_{31} = 3m_{20}m_{11}. \qquad (7.2.14)$$

Then the stationary non-zero second-order moments are solved from equation sets (7.2.10) and (7.2.11) as follows:

$$E[X_1^2] = m_{20} = \frac{1}{6\varepsilon}\left(-1 + \sqrt{1 + \frac{12\pi K\varepsilon}{\alpha}}\right), \quad E[X_2^2] = m_{02} = \frac{\pi K}{\alpha}.$$

$$(7.2.15)$$

If the truncated order is 4, we obtain

$$m_{60} = 15m_{20}m_{40} - 30m_{20}^3,$$
$$(7.2.16)$$
$$m_{42} = 6m_{20}m_{22} + m_{02}m_{40} - 6m_{20}^2 m_{02}.$$

Using (7.2.16), the non-zero second- and fourth-order moments can be solved from Eqs. (7.2.10) through (7.2.13), among which $E[X_1^2] = m_{20}$ is solved from the cubic equation

$$30\varepsilon^2 m_{20}^3 + 15m_{20}^2 + \left(1 - \frac{12\pi K\varepsilon}{\alpha}\right)m_{20} - \frac{\pi K}{\alpha} = 0. \qquad (7.2.17)$$

More tedious derivation is needed for higher-order cumulant-neglect closure. As shown in Fig. 7.2.1 for system (7.2.7) with $\alpha = 0.1$, $\pi K = 0.1$ (Wu and Lin, 1984), the results obtained from cumulant-neglect closure procedure have the tendency to converge to the exact solution for all values of ε, and a higher-order cumulant-neglect closure is more accurate.

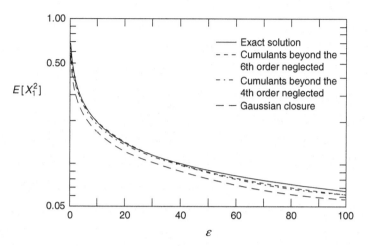

Figure 7.2.1   Stationary mean square values of $X_1(t)$ of system (7.2.7) (from Wu and Lin, 1984).

It should be pointed out that the cumulant-neglect closure is mainly applicable for a system subjected to external excitations. When parametric excitations are also present, it may result in erroneous results, especially when the system response is near a stability boundary (Sun and Hsu, 1987; Bruckner and Lin, 1987).

Now we try to apply the method to MDOF nonlinear stochastic systems governed by

$$\dot{X}_j = f_j(\mathbf{X}) + \sum_{l=1}^{m} g_{jl}(\mathbf{X}) W_l(t), \quad j = 1, 2, \ldots, n. \qquad (7.2.18)$$

Equation set (7.2.18) is equivalent to the following set of Ito stochastic differential equations

$$dX_j = \left(f_j + \sum_{k=1}^{n} \sum_{l,s=1}^{m} \pi K_{ls} g_{ks} \frac{\partial g_{jl}}{\partial X_k} \right) dt$$

$$+ \left(\sum_{k=1}^{n} \sum_{l,s=1}^{m} 2\pi K_{ls} g_{jl} g_{ks} \right)^{\frac{1}{2}} dB_j(t). \qquad (7.2.19)$$

Denote $M(\mathbf{X}) = X_1^{k_1} X_2^{k_2} \cdots X_n^{k_n}$ where $k_1, k_2, \ldots, k_n$ are non-negative integers. According to the Ito differential rule,

$$
dM = \left[\sum_{j=1}^{n} \left(f_j + \sum_{k=1}^{n} \sum_{l,s=1}^{m} \pi K_{ls} g_{ks} \frac{\partial g_{jl}}{\partial X_k} \right) \frac{\partial M}{\partial X_j} \right.
$$
$$
\left. + \sum_{l,s=1}^{m} \sum_{k,r=1}^{n} \pi K_{ls} g_{kl} g_{rs} \frac{\partial^2 M}{\partial X_k \partial X_r} \right] dt
$$
$$
+ \left(\sum_{j,k=1}^{n} \sum_{l,s=1}^{m} 2\pi K_{ls} g_{jl} g_{ks} \right)^{\frac{1}{2}} \frac{\partial M}{\partial X_j} dB_k(t). \qquad (7.2.20)
$$

Taking the ensemble average, we have

$$
\frac{d}{dt} E[M] = E \left[\sum_{j=1}^{n} \left(f_j + \sum_{k=1}^{n} \sum_{l,s=1}^{m} \pi K_{ls} g_{ks} \frac{\partial g_{jl}}{\partial X_k} \right) \frac{\partial M}{\partial X_j} \right]
$$
$$
+ E \left[\sum_{l,s=1}^{m} \sum_{k,r=1}^{n} \pi K_{ls} g_{kl} g_{rs} \frac{\partial^2 M}{\partial X_k \partial X_r} \right]. \qquad (7.2.21)
$$

Similar to the SDOF case, all equations of (7.2.21) for a certain order $N = k_1 + k_2 + \cdots + k_n$ constitute an infinity hierarchy due to the nonlinearity of functions f_i and/or g_{jk}. If these functions are polynomials of $\mathbf{X}$, cumulant-neglect closure scheme can be applied, and approximate results can be obtained.

7.2.2 *Response Correlation Functions and Spectral Densities*

The cumulant-neglect closure may also be used to obtain approximate correlation functions and spectral densities of responses for certain types of nonlinear stochastic systems. For illustration, consider a system with a nonlinear restoring force

$$
\ddot{X} + \alpha \dot{X} + u(X) = W(t), \qquad (7.2.22)
$$

where $\alpha > 0$, and $u(X)$ is assumed to be an odd polynomial of X. Letting $X_1 = X$ and $X_2 = \dot{X}$, the stochastic processes $X_1(t)$ and $X_2(t)$ possess stationary probability density

$$p(x_1, x_2) = C \exp \left\{ -\frac{\alpha}{\pi K} \left[\frac{1}{2} x_2^2 + \int u(x_1) dx_1 \right] \right\}. \qquad (7.2.23)$$

The joint statistical moments can be calculated from

$$m_{ij} = E[X_1^i X_2^j] = \int\limits_{-\infty}^{\infty} \int\limits_{-\infty}^{\infty} x_1^i x_2^j p(x_1, x_2) dx_1 dx_2. \qquad (7.2.24)$$

It follows from (7.2.23) and (7.2.24) that $m_{ij} = 0$, either i or j is odd

$$m_{i0} = C\pi \sqrt{\frac{2K}{\alpha}} \int\limits_{-\infty}^{\infty} x_1^i \exp\left[-\frac{\alpha}{\pi K} \int u(x_1) dx_1 \right] dx_1, \quad i \text{ is even,}$$

$$m_{0j} = \left(\frac{\pi K}{\alpha} \right)^{\frac{j}{2}} (j-1)(j-3)\cdots 1, \qquad\qquad j \text{ is even}$$

$$m_{ij} = m_{i0}m_{0j}, \qquad\qquad\qquad\qquad \text{both } i \text{ and}$$
$$j \text{ are even.}$$
$$(7.2.25)$$

Although the first-order properties can be determined, the second-order properties, such as correlation functions and spectral densities are not obtainable.

Using Ito differential rule, we have the following equation sets

$$dX_1 = X_2 dt,$$
$$dX_2 = [-\alpha X_2 - u(X_1)]dt + \sqrt{2\pi K} dB(t). \qquad (7.2.26)$$
$$dX_1^3 = 3X_1^2 X_2 dt,$$
$$dX_1^2 X_2 = [-\alpha X_1^2 X_2 - X_1^2 u(X_1) + 2X_1 X_2^2]dt$$
$$\qquad\quad + \sqrt{2\pi K} X_1^2 dB(t),$$

$$dX_1X_2^2 = [-2\alpha X_1X_2^2 - 2X_1X_2u(X_1) + 2\pi KX_1 + X_2^3]dt$$
$$+ 2\sqrt{2\pi K}X_1X_2dB(t),$$
$$dX_2^3 = [-3\alpha X_2^3 - 3X_2^2u(X_1) + 6\pi KX_2]dt + 3\sqrt{2\pi K}X_2^2dB(t).$$
$$(7.2.27)$$

Multiplying each equation in (7.2.26) and (7.2.27) by $X_1(t_0)$, and taking ensemble average, we have

$$\frac{dQ_{10}}{d\tau} = Q_{01},$$

$$\frac{dQ_{01}}{d\tau} = -E\{u[X_1(t)]X_1(t-\tau)\} - \alpha Q_{01}, \qquad (7.2.28)$$

$$\frac{dQ_{30}}{d\tau} = 3Q_{21},$$

$$\frac{dQ_{21}}{d\tau} = -E\{u[X_1(t)]X_1^2(t)X_1(t-\tau)\} - \alpha Q_{21} + 2Q_{12},$$

$$\frac{dQ_{12}}{d\tau} = 2\pi KQ_{10} - 2E\{u[X_1(t)]X_1(t)X_2(t)X_1(t-\tau)\}$$
$$- 2\alpha Q_{12} + Q_{03},$$

$$\frac{dQ_{03}}{d\tau} = 6\pi KQ_{01} - 3E\{u[X_1(t)]X_2^2(t)X_1(t-\tau)\} - 3\alpha Q_{03},$$
$$(7.2.29)$$

where $\tau = t - t_0$, and

$$Q_{ij} = E[X_1^i(t)X_2^j(t)X_1(t-\tau)]. \qquad (7.2.30)$$

Associated with (7.2.28) and (7.2.29) are the following initial conditions

$$Q_{ij}(0) = m_{i+1,j}, \qquad (7.2.31)$$

which can be computed from Eq. (7.2.25). It is clear that the auto-correlation function of $X_1(t)$ is

$$R_{11}(\tau) = E[X_1(t)X_1(t-\tau)] = Q_{10}(\tau), \qquad (7.2.32)$$

and the cross-correlation function of $X_1(t)$ and $X_2(t)$ is

$$R_{12}(\tau) = E[X_1(t-\tau)X_2(t)] = Q_{01}(\tau). \qquad (7.2.33)$$

Equation sets (7.2.28) and (7.2.29) cannot be solved since they constitute an infinite hierarchy. However, they may be solved approximately using a suitable closure scheme. The simplest one is the Gaussian closure, in which all cumulants of orders higher that two are set to zero. For more accurate results, the cumulant-neglect closure can be applied.

If only the spectral density functions are of interest, the transformation method described in Sections 4.5.2 and 5.5.2 can be employed to obtain algebraic equation set for the spectral densities and solve for the desired results.

Example 7.2.2 Consider a similar equation as that in Example 7.2.1,

$$\ddot{X} + \alpha \dot{X} + kX + \delta X^3 = W(t). \tag{7.2.34}$$

The nonlinear function in Eq. (7.2.22) is now $u(X) = kX + \delta X^3$. For this case, Eq. (7.2.28) becomes

$$\frac{dQ_{10}}{d\tau} = Q_{01},$$
$$\frac{dQ_{01}}{d\tau} = -kQ_{10} - \delta Q_{30} - \alpha Q_{01}. \tag{7.2.35}$$

Following the method of integral transformation described in Section 4.5.2, we obtain

$$i\omega \bar{Q}_{10} - \bar{Q}_{01} = \frac{1}{\pi} m_{20},$$
$$\delta \bar{Q}_{30} + k\bar{Q}_{10} + (i\omega + \alpha)\bar{Q}_{01} = 0, \tag{7.2.36}$$

where

$$\bar{Q}_{ij}(\omega) = \Im[Q_{ij}(\tau)] = \frac{1}{\pi} \int_0^\infty Q_{ij}(\tau) e^{-i\omega \tau} d\tau. \tag{7.2.37}$$

The spectral density of X_1 is then

$$\Phi_{XX}(\omega) = \text{Re}[\bar{Q}_{10}(\omega)]. \tag{7.2.38}$$

Gaussian closure (second-order cumulant-neglect closure)

Letting the fourth-order cumulants be zero, we obtain from Eq. (2.6.27)

$$Q_{30} = E[X_1^3(t)X_1(t-\tau)] = 3E[X_1^2(t)]E[X_1(t)X_1(t-\tau)] = 3m_{20}Q_{10}. \quad (7.2.39)$$

Substitution of (7.2.39) into (7.2.35) results in a set of closed equations

$$\frac{dQ_{10}}{d\tau} = Q_{01},$$

$$\frac{dQ_{01}}{d\tau} = -(k + 3\delta m_{20})Q_{10} - \alpha Q_{01}. \quad (7.2.40)$$

With the initial conditions

$$Q_{10}(0) = m_{20}, \quad Q_{01}(0) = m_{11} = 0, \quad (7.2.41)$$

the correlation function $R_{11}(\tau)$ can be solved as

$$R_{11}(\tau) = Q_{10}(\tau) = m_{20}e^{-\alpha\tau/2}\left(\cos \omega_d\tau + \frac{\alpha}{2\omega_d}\sin \omega_d\tau\right), \quad (7.2.42)$$

where

$$\omega_d = \sqrt{k + 3\delta m_{20} - \alpha^2/4}, \quad (7.2.43)$$

Applying the integral transformation (7.2.37) to (7.2.40) leads to

$$i\omega\bar{Q}_{10} - \bar{Q}_{01} = \frac{1}{\pi}m_{20},$$

$$(k + 3\delta m_{20})\bar{Q}_{10} + (i\omega + \alpha)\bar{Q}_{01} = 0. \quad (7.2.44)$$

The spectral density of X_1 is then solved from (7.2.44) as

$$\Phi_{XX}(\omega) = \text{Re}[\bar{Q}_{10}(\omega)] = \frac{\eta(k + 3\delta m_{20})m_{20}}{\pi[(k + 3\delta m_{20} - \omega^2)^2 + \alpha^2\omega^2]}. \quad (7.2.45)$$

Fourth-order cumulant-neglect closure

Employing (2.6.27), we obtain

$$Q_{50} = E[X_1^5(t)X_1(t-\tau)] = 10m_{20}Q_{30} + 5cQ_{10},$$

$$Q_{41} = E[X_1^4(t)X_2(t)X_1(t-\tau)] = 6m_{20}Q_{21} + cQ_{01},$$

$$Q_{32} = E[X_1^3(t)X_2^2(t)X_1(t-\tau)] = m_{02}Q_{30} + 3m_{20}Q_{12} - 3m_{22}Q_{10},$$

$$(7.2.46)$$

where

$$c = m_{40} - 6m_{20}^2. \qquad (7.2.47)$$

Substituting (7.2.46) into (7.2.29), we have

$$\frac{dQ_{30}}{d\tau} = 3Q_{21},$$

$$\frac{dQ_{21}}{d\tau} = -(k + 10\delta m_{20})Q_{30} - \alpha Q_{21} + 2Q_{12} - 5\delta cQ_{10},$$

$$\frac{dQ_{12}}{d\tau} = -(2k + 12\delta m_{20})Q_{21} - 2\alpha Q_{12} + Q_{03} + 2\pi K Q_{10} - 2\delta cQ_{01},$$

$$\frac{dQ_{03}}{d\tau} = -3\delta m_{02}Q_{30} - (3k + 9\delta m_{20})Q_{12} - 3\alpha Q_{03}$$

$$+ 9\delta m_{22}Q_{10} + 6\pi K Q_{01}. \qquad (7.2.48)$$

Equations in (7.2.35) and (7.2.48) constitute a set of linear ordinary differential equations with constant coefficients for $Q_{ij}(\tau)$, $i + j = 1$, 3. With the initial conditions given by (7.2.31), they can be solved at least numerically, and $Q_{10}(\tau)$ and $Q_{01}(\tau)$ are the autocorrelation function $R_{11}(\tau)$ and cross-correlation function $R_{12}(\tau)$, respectively, according to (7.2.32) and (7.2.33).

In frequency domain, equation set (7.2.48) is transformed to

$$i\omega \bar{Q}_{30} - 3\bar{Q}_{21} = \frac{m_{40}}{\pi},$$

$$(k + 10\delta m_{20})\bar{Q}_{30} + (i\omega + \alpha)\bar{Q}_{21} - 2\bar{Q}_{12} + 5\delta c\bar{Q}_{10} = 0,$$

$$(2k + 12\delta m_{20})\bar{Q}_{21} + (i\omega + 2\alpha)\bar{Q}_{12} - \bar{Q}_{03} - 2\pi K\bar{Q}_{10}$$

$$+ 2\delta c\bar{Q}_{01} = \frac{m_{22}}{\pi},$$

$$3\delta m_{02}\bar{Q}_{30} + (3k + 9\delta m_{20})\bar{Q}_{12}$$

$$+ (i\omega + 3\alpha)\bar{Q}_{03} - 9\delta m_{22}\bar{Q}_{10} - 6\pi K\bar{Q}_{01} = 0. \qquad (7.2.49)$$

Equations in (7.2.36) and (7.2.49) constitute a set of complex linear algebraic equations for $\bar{Q}_{ij}(\omega), i + j = 1, 3$. The spectral density of X_1 is then the real part of $\bar{Q}_{10}(\omega)$.

Numerical calculations were carried out for the above system using Gaussian closure, the fourth-order and sixth-order cumulant-neglect closures (Cai and Lin, 1997). Some results calculated with different order of cumulant-neglect closures are shown in Fig. 7.2.2 for system (7.2.34) with $k = 1$, $\alpha = 0.08$ and $K = 0.01$. Also shown in the figures are results from Monte Carlo simulations for comparison. It was shown that for a strong nonlinearity ($\delta = 0.5$ or 1), even the fourth-order cumulant-neglect closure cannot produce accurate results; while the sixth-order closure are required to obtain more accurate results.

In principle, the cumulant-neglect closure can be applied to MDOF nonlinear stochastic systems. However, the derivation procedure will be tedious, and may be not practical.

7.3 Method of Equivalent Nonlinear Systems

In the equivalent linearization and the partial linearization described in Sections 7.1.1 and 7.1.2 respectively, the original nonlinear systems are replaced by systems for which the exact probabilistic solutions are obtainable. Nevertheless, the replacement systems in these procedures are subclasses of the exactly solvable nonlinear systems of generalized stationary potential. If the selected pool is enlarged to include the entire solvable class of generalized stationary potential, the accuracy of the approximate solutions may be improved. Moreover, the equivalent linearization and partial linearization are applied to obtain approximate probability solutions only for systems under external excitations. The procedure in Section 7.1.3 allows the

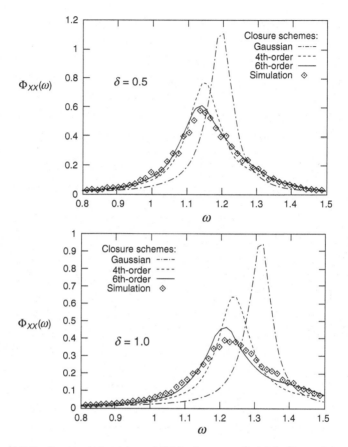

Figure 7.2.2   Spectral densities of $X(t)$ of system (7.2.34) (from Cai and Lin, 1997).

presence of parametric excitations, but it can only solve for statistical properties of responses, not the probabilistic properties. Thus, it is desirable to develop techniques for approximate probabilistic solutions of systems under both external and parametric excitations.

7.3.1 *Method of Weighted Residual*

Consider a nonlinear stochastic system

$$\frac{d}{dt}X_j = F_j(\mathbf{X}) + \sum_{l=1}^{m} g_{jl}(\mathbf{X})W_l(t), \quad j = 1, 2, \ldots, n, \qquad (7.3.1)$$

where $W_l(t)$ are Gaussian white noises. Assume that an exact solution for system (7.3.1) is not obtainable, and we wish to find a replacement system within the solvable class of generalized stationary potential in some statistical sense. Let the replacement system be

$$\frac{d}{dt}X_j = f_j(\mathbf{X}) + \sum_{l=1}^{m} g_{jl}(\mathbf{X})W_l(t), \quad j = 1, 2, \ldots, n. \tag{7.3.2}$$

Note that only functions $F_j(\mathbf{X})$ in the original system (7.3.1) is replaced by $f_j(\mathbf{X})$, and the excitation terms remain unchanged. Let $p(\mathbf{x})$ be the exact stationary probability density of response $\mathbf{X}$ for system (7.3.2), i.e., $p(\mathbf{x})$ satisfies the reduced FPK equation

$$\sum_{j=1}^{n} \frac{\partial}{\partial x_j} \left[\left(f_j(\mathbf{x}) + \pi \sum_{k=1}^{n} \sum_{l,s=1}^{m} K_{ls} g_{ks}(\mathbf{x}) \frac{\partial}{\partial x_k} g_{jl}(\mathbf{x}) \right) p \right]$$
$$-\pi \sum_{j,k=1}^{n} \sum_{l,s=1}^{m} K_{ls} g_{ks}(\mathbf{x}) \frac{\partial}{\partial x_k} g_{jl}(\mathbf{x}) = 0. \tag{7.3.3}$$

Our objective is to select a set of $f_j(\mathbf{X})$ functions such that system (7.3.2) is closest to system (7.3.1) in some statistical sense. Since $p(\mathbf{x})$ is not true solution for system (7.3.1), it will not satisfy its FPK; therefore, the following residual error occurs

$$\delta(\mathbf{x}) = \sum_{j=1}^{n} \frac{\partial}{\partial x_j} \left[\left(F_j(\mathbf{x}) + \pi \sum_{k=1}^{n} \sum_{l,s=1}^{m} K_{ls} g_{ks}(\mathbf{x}) \frac{\partial}{\partial x_k} g_{jl}(\mathbf{x}) \right) p \right]$$
$$-\pi \sum_{j,k=1}^{n} \sum_{l,s=1}^{m} K_{ls} g_{ks}(\mathbf{x}) \frac{\partial}{\partial x_k} g_{jl}(\mathbf{x}). \tag{7.3.4}$$

The difference of (7.3.3) and (7.3.4) is known as the residual error,

$$\delta(\mathbf{x}) = \sum_{j=1}^{n} \frac{\partial}{\partial x_j} \{ [F_j(\mathbf{x}) - f_j(\mathbf{x})] p(\mathbf{x}) \}, \tag{7.3.5}$$

$\delta(\mathbf{x})$ is a measure of error of the approximate solution $p(\mathbf{x})$ from the true solution of system (7.3.1); therefore, it needs to be minimized in some sense. For this purpose, the method of weighted residual (e.g.,

Finlayson, 1972) was proposed, in which function $f_j(\mathbf{X})$, hence $p(\mathbf{x})$, are chosen such that

$$\Delta_M = \int M(\mathbf{x})\delta(\mathbf{x})d\mathbf{x}$$

$$= \int M(\mathbf{x}) \sum_{j=1}^{n} \frac{\partial}{\partial x_j} \{[F_j(\mathbf{x}) - f_j(\mathbf{x})]\,p(\mathbf{x})\}d\mathbf{x} = 0, \quad (7.3.6)$$

where $M(\mathbf{x})$ is known as the weighting function to be selected. The subscript M of the integral Δ indicates that the value of Δ depends on the choice of the weighting function $M(\mathbf{x})$. To proceed further, we impose rather general restrictions that the M function is differentiable with respect to x_i and that, at the boundaries of the $\mathbf{x}$ domain,

$$M(\mathbf{x})[F_j(\mathbf{x}) - f_j(\mathbf{x})]p(\mathbf{x}) = 0, \quad \mathbf{x} \text{ at boundaries.} \quad (7.3.7)$$

Carrying out integration by parts on every x_i and applying (7.3.7), we obtain from (7.3.6)

$$\Delta_M = E\left\{ \sum_{j=1}^{n} [F_j(\mathbf{X}) - f_j(\mathbf{X})] \frac{\partial M(\mathbf{X})}{\partial X_j} \right\} = 0, \quad (7.3.8)$$

where the ensemble average is carried out with respect to the approximate probability density $p(\mathbf{x})$. By appropriately selecting a set of M functions, we obtain a corresponding set of constraints from (7.3.8). The $f_j(\mathbf{x})$ functions for the replacing system can then be determined from these constraints. The required number of constraints, hence the required number of weighting functions, depends on the number of parameters to be determined in functions $f_j(\mathbf{x})$.

7.3.2 *Dissipation Energy Balancing*

Now let the above method of weighted residual be applied to a SDOF nonlinear stochastic system governed by

$$\ddot{X} + H(X, \dot{X}) + v(X) = \sum_{l=1}^{m} g_l(X, \dot{X})W_l(t), \quad (7.3.9)$$

where $H(X, \dot{X})$ is the damping force and $v(X)$ is the restoring force. We assume that this system does not belong to the exactly solvable class of generalized stationary potential. Replace system (7.3.9) by a solvable one

$$\ddot{X} + h(X, \dot{X}) + v(X) = \sum_{l=1}^{m} g_l(X, \dot{X}) W_l(t). \qquad (7.3.10)$$

In the replacement system (7.3.10), the restoring force and the excitation terms remain unchanged. Thus, the essence of the replacement is to replace the damping force so that the exact solution is obtainable for the replacing system. Letting $X_1 = X$ and $X_2 = \dot{X}$, changing each of (7.3.9) and (7.3.10) into a set of two first-order differential equations, and comparing them with (7.3.1) and (7.3.2) respectively, we obtain from (7.3.8)

$$\Delta_M = E\left\{[H(X_1, X_2) - h(X_1, X_2)]\frac{\partial M(X_1, X_2)}{\partial X_2}\right\} = 0. \qquad (7.3.11)$$

By selecting appropriate weighting functions $M(X_1, X_2)$, constraints can be established to obtain an approximate solution.

First we need to identify the additional restoring force $u^*(X_1)$ and the additional damping force $h^*(X_1, X_2)$ according to Eq.' (6.3.16), i.e.,

$$\pi \sum_{l,s=1}^{m} K_{ls} g_l \frac{\partial g_s}{\partial X_2} = h^*(X_1, X_2) + u^*(X_1). \qquad (7.3.12)$$

Then the effective restoring force and the total effective energy are

$$u(X_1) = v(X_1) + u^*(X_1), \quad \Lambda = \frac{1}{2}X_2^2 + \int u(X_1)dX_1. \qquad (7.3.13)$$

Since an exact stationary solution is assumed to be obtainable for system (7.3.10), function $h(X_1, X_2)$ can be expressed as the form (6.3.22), i.e.,

$$h(X_1, X_2) = \pi X_2 \phi'(\Lambda) \sum_{l,s=1}^{m} K_{ls} g_l g_s - h^*(X_1, X_2). \qquad (7.3.14)$$

Substitute (7.3.14) into (7.3.11) to yield

$$\Delta_M = E\left\{\left[H(X_1, X_2) + h^*(X_1, X_2) - \pi X_2 \phi'(\Lambda) \sum_{l,s=1}^{m} K_{ls} g_l g_s\right]\right.$$

$$\left. \times \frac{\partial M}{\partial X_2}\right\} = 0. \tag{7.3.15}$$

It is noted that $H(X_1, X_2) + h^*(X_1, X_2)$ is the effective damping force of the original system taking into account of the contribution from the Wong–Zakai correction. The stationary probability density of the system response is then

$$p(x_1, x_2) = C\exp\{-\phi[\lambda(x_1, x_2)]\}, \tag{7.3.16}$$

where the probability potential function $\phi[\lambda(x_1, x_2)]$ can be determined from (7.3.15) by selecting an appropriate weighting function $M(X_1, X_2)$.

Letting $M = X_2^2$, we obtain from (7.3.15)

$$E\left\{X_2\left[H(X_2, X_2) + h^*(X_2, X_2) - \pi X_2 \phi'(\Lambda) \sum_{l,s=1}^{m} K_{ls} g_l g_s\right]\right\} = 0. \tag{7.3.17}$$

Use the approximate probability density in (7.3.16) to calculate the assemble average in (7.3.17) to obtain

$$\int_{-\infty}^{\infty}\int_{-\infty}^{\infty} e^{-\phi(\lambda)} x_2 \left[H(x_1, x_2) + h^*(x_1, x_2) - \pi x_2 \phi'(\lambda) \sum_{l,s=1}^{m} K_{ls} g_l g_s\right]$$

$$\times dx_1 dx_2 = 0. \tag{7.3.18}$$

Assume that the effective restoring force $u(x_1)$ is an odd function, and denote the effective potential energy as

$$U(x_1) = \int u(x_1) dx_1. \tag{7.3.19}$$

Combining (7.3.13) and (7.3.19), we have

$$x_2 = \begin{cases} \sqrt{2\lambda - 2U(x_1)} & x_2 \geq 0 \\ -\sqrt{2\lambda - 2U(x_1)} & x_2 < 0 \end{cases}. \tag{7.3.20}$$

The integration on x_1 and x_2 in Eq. (7.3.18) can be transformed to that on x_1 and λ as follows

$$\int_{-\infty}^{\infty} e^{-\phi(\lambda)} d\lambda \int_0^{a(\lambda)} \left[H(x_1, x_2) + h^*(x_1, x_2) \right.$$

$$\left. - \pi x_2 \phi'(\lambda) \sum_{l,s=1}^{m} K_{ls} g_l g_s \right]_{x_2 = \sqrt{2\lambda - 2U(x_1)}} dx_1 = 0,$$

$$\tag{7.3.21}$$

where $a(\lambda)$ is the amplitude corresponding to the energy level λ, namely, a is determined from $U(a) = \lambda$, and x_2 is treated as a function of x_1 and λ, as indicated in (7.3.20). Since $\phi(\lambda)$ is still unknown, we replace the constrain (7.3.21) by a more stringent condition, requiring that the integration on x_1 vanishes for every λ, i.e.,

$$\int_0^{a(\lambda)} \left[H(x_1, x_2) + h^*(x_1, x_2) \right.$$

$$\left. - \pi x_2 \phi'(\lambda) \sum_{l,s=1}^{m} K_{ls} g_l g_s \right]_{x_2 = \sqrt{2\lambda - 2U(x_1)}} dx_1 = 0,$$

$$\tag{7.3.22}$$

which leads to the following expression for $\phi'(\lambda)$

$$\phi'(\lambda) = \frac{\int_0^{a(\lambda)} [H(x_1, x_2) + h^*(x_1, x_2)]_{x_2 = \sqrt{2\lambda - 2U(x_1)}} \, dx_1}{\pi \int_0^{a(\lambda)} \left[x_2 \sum_{l,s=1}^{m} K_{ls} g_l g_s \right]_{x_2 = \sqrt{2\lambda - 2U(x_1)}} dx_1}. \tag{7.3.23}$$

In summary, the solution procedure consists of the following:

(i) To identify the additional restoring force $u^*(x_1)$ and additional damping force $h^*(x_1, x_2)$ from (7.3.12),
(ii) Construct the total effective energy λ according to (7.3.13),
(iii) Calculate $\phi'(\lambda)$ from (7.3.23), and integrate to obtain $\phi(\lambda)$, and
(iv) Find the stationary probability density $p(x_1, x_2)$ from (7.3.16).

An important advantage to use Eq. (7.3.23) to find the probability potential directly is that a prior knowledge of $h(X_1, X_2)$ function in the replacement system (7.3.10) is not needed. Thus, possible errors associated with the choice of a form for $h(X_1, X_2)$ are avoided. Moreover, taking into account that the restoring force and the excitation terms remain unchanged in the replacing system, it can be said that the approximation obtained is indeed the best within the class of generalized stationary potential.

The physical implication of constraint (7.3.11) with $M = X_2^2$, and hence (7.3.17), is that the average dissipated energy per unit time remains the same for the replacing and replaced systems, while the stronger condition (7.3.22) indicates that the average dissipated energies are identical at every energy level. Therefore, the method was referred to as dissipation energy balancing (Cai and Lin, 1988).

Equation (7.3.23) can be evaluated at least numerically. If the effective restoring force is linear, a closed-form analytical solution is often possible. In this case, we may write

$$u(x_1) = k_e x_1, \quad U(x_1) = \frac{1}{2} k_e x_1^2, \quad \lambda = \frac{1}{2} k_e x_1^2 + \frac{1}{2} x_2^2. \quad (7.3.24)$$

By letting

$$x_1 = \frac{\sqrt{2\lambda}}{\sqrt{k_e}} \cos\theta, \quad x_2 = \sqrt{2\lambda} \sin\theta. \quad (7.3.25)$$

Equation (7.3.23) is transformed to

$$\phi'(\lambda) = \frac{\int_0^{2\pi} \sin\theta \left[H(x_1, x_2) + h^*(x_1, x_2)\right]_{x_1, x_2 \to \lambda, \theta} d\theta}{\pi \sqrt{2\lambda} \int_0^{2\pi} \sin^2\theta \left[\sum_{l,s=1}^{m} K_{ls} g_l g_s\right]_{x_1, x_2 \to \lambda, \theta} d\theta}, \quad (7.3.26)$$

where the notation $x_1, x_2 \to \lambda, \theta$ indicates that x_1 and x_2 are expressed in terms of λ and θ according to transformation (7.3.25). The integration in (7.3.26) can be carried out analytically if $H(x_1, x_2)$ and $g_l(x_1, x_2)$ are polynomials of x_1 and x_2.

If parametric excitations are absent, i.e., if all g_l are constants, then Eq. (7.3.23) is reduced to

$$\phi'(\lambda) = \frac{\int_0^{a(\lambda)} \left[H\left(x_1, \sqrt{2\lambda - 2U(x_1)}\right) \right] dx_1}{\pi \sum\limits_{l,s=1}^{m} K_{ls} g_l g_s \int_0^{a(\lambda)} \sqrt{2\lambda - 2U(x_1)} dx_1} \tag{7.3.27}$$

and (7.3.26) is reduced to

$$\phi'(\lambda) = \frac{\int_0^{2\pi} H(\lambda, \theta) \sin \theta d\theta}{\pi^2 \sum\limits_{l,s=1}^{m} K_{ls} g_l g_s \sqrt{2\lambda}}, \tag{7.3.28}$$

where $H(\lambda, \theta)$ is obtained from $H(x_1, x_2)$ by changing the arguments from x_1 and x_2 to λ and θ according to (7.3.25).

Example 7.3.1 Consider the nonlinear stochastic system

$$\ddot{X} + \alpha \dot{X} + \beta \dot{X}^3 + \delta X^3 = W(t), \quad \beta, \delta > 0. \tag{7.3.29}$$

The restoring force, potential energy and total energy are, respectively,

$$u(x_1) = \delta x_1^3, \quad U(x_1) = \frac{1}{4} \delta x_1^4, \quad \lambda = \frac{1}{4} \delta x_1^4 + \frac{1}{2} x_2^2. \tag{7.3.30}$$

The equivalent linearization leads to an approximation of Gaussian distribution

$$p(x_1, x_2) = C \exp\left[-\frac{\alpha_e}{2\pi K} (k_e x_1^2 + x_2^2) \right], \tag{7.3.31}$$

where the linearization coefficients are calculated from (7.1.17) as

$$\alpha_e = \frac{1}{2} \left(\alpha + \sqrt{\alpha^2 + 12\pi \beta K} \right), \quad k_e = \sqrt{\frac{3\pi \delta K}{\alpha_e}}. \tag{7.3.32}$$

If the method of partial linearization is applied, we have

$$p(x_1, x_2) = C \exp \left[-\frac{\alpha_e}{2\pi K} \left(\frac{1}{2} \delta x_1^4 + x_2^2 \right) \right], \qquad (7.3.33)$$

with the same α_e given in (7.3.32). In the dissipation energy balancing method, Eq. (7.3.28) can be applied, leading to

$$p(x_1, x_2) = C \exp \left\{ -\frac{1}{2\pi K} \left[\alpha \left(\frac{1}{2} \delta x_1^4 + x_2^2 \right) + \frac{3\beta}{7} \left(\frac{1}{2} \delta x_1^4 + x_2^2 \right)^2 \right] \right\}. \qquad (7.3.34)$$

The derivation of (7.3.34) is left for readers as Exercise Problem 7.8.

Numerical calculations were carried out in (Lin and Cai, 1995) to obtain the mean-square values of displacement $X(t)$ for system (7.3.29) for $\alpha = 0.1$ and $K = 1$. In Fig. 7.3.1, the computed stationary mean-square values are plotted against the nonlinear damping parameter. Also shown in the figures are results calculated using the linearization in Section 7.1.1 and the partial linearization in Section 7.1.2. The figures show that the results obtained from the dissipation energy balancing are the most accurate, and those calculated from the linearization were the least accurate when compared with the Monte Carlo simulation results.

Example 7.3.2 Consider the following van der Pol-Rayleigh type oscillator,

$$\ddot{X} + \alpha \dot{X} + \beta X^2 \dot{X} + \gamma \dot{X}^3 + \omega_0^2 X = X W_1(t) + \dot{X} W_2(t) + W_3(t), \qquad (7.3.35)$$

where $W_1(t)$, $W_2(t)$ and $W_2(t)$ are Gaussian white noises with spectral densities K_{ij} ($i, j = 1, 2, 3$). It is assumed that $K_{23} = 0$. Comparing (7.3.35) and the standard form (7.3.9), we see that

$$\begin{aligned} H(X_1, X_2) &= \alpha X_2 + \beta X_1^2 X_2 + \gamma X_2^3, \quad v(X_1) = \omega_0^2 X_1, \\ g_1(X_1, X_2) &= X_1, \quad g_2(X_1, X_2) = X_2, \quad g_3(X_1, X_2) = 1. \end{aligned} \qquad (7.3.36)$$

To apply the dissipation energy balancing, we need to identify the additional restoring force $u^*(x_1)$ and the additional damping force

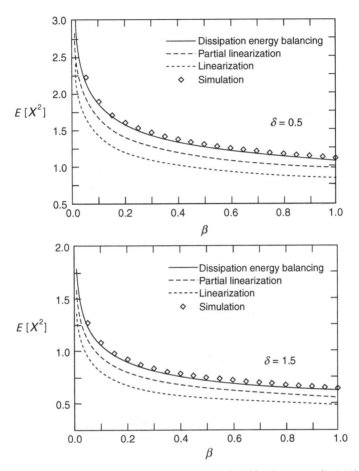

Figure 7.3.1   Stationary mean-square values of $X(t)$ of system (7.3.29) (from Lin and Cai, 1995).

$h^*(x_1, x_2)$ according to (7.3.12). Since

$$\pi \sum_{l,s=1}^{m} K_{ls} g_l(x_1, x_2) \frac{\partial}{\partial x_2} g_s(x_1, x_2) = \pi K_{12} x_1 + \pi K_{22} x_2, \quad (7.3.37)$$

we have

$$u^*(x_1) = \pi K_{12} x_1, \quad h^*(x_1, x_2) = \pi K_{22} x_2. \quad (7.3.38)$$

Therefore, the effective restoring force, the effective potential energy, and the total effective energy are given in (7.3.24), in which

$$k_e = \omega_0^2 + \pi K_{12}. \tag{7.3.39}$$

Since the restoring force is linear, Eq. (7.3.28) is applicable to obtain

$$\phi'(\lambda) = \frac{2k_e(\alpha + \pi K_{22}) + (\beta + 3k_e\gamma)\lambda}{\pi \left[2k_e K_{33} + (K_{11} + 3k_e K_{22})\lambda\right]}. \tag{7.3.40}$$

Different scenarios will be discussed separately. First consider the case in which the parametric excitations are absent. Letting $K_{11} = 0$, $K_{22} = 0$ and $K_{12} = 0$, Eq. (7.3.40) is simplified to

$$\phi'(\lambda) = \frac{2\alpha\omega_0^2 + (\beta + 3\omega_0^2\gamma)\lambda}{2\pi\omega_0^2 K_{33}}, \tag{7.3.41}$$

and the probability density is obtained as

$$p(x_1, x_2) = C \exp \left\{ -\frac{1}{2\pi K_{33}} \right.$$
$$\left. \left[\alpha(\omega_0^2 x_1^2 + x_2^2) + \frac{1}{8}\left(\frac{\beta}{\omega_0^2} + 3\gamma\right)(\omega_0^2 x_1^2 + x_2^2)^2\right] \right\}. \tag{7.3.42}$$

On the other hand, if at least one parametric excitation is present so that $K_{11} + 3k_e K_{22} \neq 0$, we obtain from (7.3.40)

$$\phi(\lambda) = \frac{\beta + 3k_e\gamma}{\pi(K_{11} + 3k_e K_{22})}\lambda - \ln\left[(K_{11} + 3k_e K_{22})\lambda + 2k_e K_{33}\right]^\delta$$
$$+ \text{Constant}, \tag{7.3.43}$$

where

$$\delta = \frac{2k_e[K_{33}(\beta + 3k_e\gamma) - (\alpha + \pi K_{22})(K_{11} + 3k_e K_{22})]}{\pi(K_{11} + 3k_e K_{22})^2}. \tag{7.3.44}$$

The probability density is then

$$p(x_1, x_2) = C[(K_{11} + 3k_e K_{22})\lambda + 2k_e K_{33}]^\delta$$
$$\times \exp\left[-\frac{\beta + 3k_e\gamma}{\pi(K_{11} + 3k_e K_{22})}\lambda\right]. \tag{7.3.45}$$

Another special case is that the system damping and restoring forces are linear, i.e., $\beta = 0$ and $\gamma = 0$. Then (7.3.45) reduces to

$$p(x_1, x_2) = C\left[(K_{11} + 3k_e K_{22})\lambda + 2k_e K_{33}\right]^\delta. \tag{7.3.46}$$

From (7.3.46), the second-order moments can be calculated as follows (see Exercise Problem 2.22)

$$E[X_1^2] = \frac{\pi K_{33}}{k_e(\alpha - 2\pi K_{22}) - \pi K_{11}}, \quad E[X_1 X_2] = 0,$$

$$E[X_2^2] = \frac{\pi k_e K_{33}}{k_e(\alpha - 2\pi K_{22}) - \pi K_{11}}. \tag{7.3.47}$$

Even the probability density (7.3.46) is approximate, the second-order moments in (7.3.47) are exact, as can be verified by deriving them from the original system (see Exercise Problem 6.12). The expressions in Eq. (7.3.47) implies that the conditions for existence of the second-order moments are

$$\alpha > \frac{\pi K_{11}}{k_e} + 2\pi K_{22}. \tag{7.3.48}$$

For the following special cases, in which the condition for generalized stationary potential is satisfied, the present procedure yields the exact results:

1. $K_{11} = 0$, $K_{22} = 0$, $K_{33} \neq 0$, $\beta = \gamma\omega_0^2$,

$$p(x_1, x_2) = C\exp\left\{-\frac{1}{2\pi K_{33}}\left[\alpha(\omega_0^2 x_1^2 + x_2^2) + \frac{1}{4}\gamma(\omega_0^2 x_1^2 + x_2^2)^2\right]\right\}. \tag{7.3.49}$$

2. $K_{12} = 0$, $K_{13} = 0$, $K_{33} \neq 0$, $K_{11} = \frac{\beta K_{33}}{\alpha + \pi K_{22}}$, $K_{22} = \frac{\gamma K_{33}}{\alpha + \pi K_{22}}$,

$$p(x_1, x_2) = C\exp\left[-\frac{\alpha + \pi K_{22}}{2\pi K_{33}}(\omega_0^2 x_1^2 + x_2^2)\right]. \tag{7.3.50}$$

3. $K_{12} = 0$, $K_{13} = 0$, $K_{11} = 0$, $K_{33} = 0$, $\beta = 0$, $\alpha = -\pi K_{22} \neq 0$,

$$p(x_1, x_2) = C\exp\left[-\frac{\gamma}{2\pi K_{22}}(\omega_0^2 x_1^2 + x_2^2)^2\right]. \tag{7.3.51}$$

4. $K_{12} = 0$, $K_{13} = 0$, $K_{22} = 0$, $K_{33} = 0$, $K_{11} \neq 0$, $\alpha = 0$, $\gamma = 0$,

$$p(x_1, x_2) = C \exp \left[-\frac{\beta}{2\pi K_{11}} (\omega_0^2 x_1^2 + x_2^2) \right]. \tag{7.3.52}$$

5. $K_{12} = 0$, $K_{13} = 0$, $K_{33} \neq 0$, $K_{11} = \omega_0^2 K_{22}$, $\beta = \gamma \omega_0^2$,

$$p(x_1, x_2) = C \left[K_{33} + K_{22}(\omega_0^2 x_1^2 + x_2^2) \right]^\delta$$

$$\times \exp \left[-\frac{\gamma}{2\pi K_{22}} (\omega_0^2 x_1^2 + x_2^2) \right], \tag{7.3.53}$$

where

$$\delta = \frac{\gamma K_{33} - K_{22}(\alpha + \pi K_{22})}{2\pi K_{22}^2}. \tag{7.3.54}$$

In this example, we assume that $K_{23} = 0$ so that the mean value of X_1 is zero. The more complicated case of $K_{23} \neq 0$ was discussed in the book of Lin and Cai (1995).

Now we try to apply the methods of weighted residual and dissipation energy balancing to the following MDOF nonlinear stochastic system

$$\ddot{Z}_j + H_j(\mathbf{Z}, \dot{\mathbf{Z}}) + u_j(\mathbf{Z}) = \sum_{l=1}^{m} g_{jl}(\mathbf{Z}, \dot{\mathbf{Z}}) W_l(t), \quad j = 1, 2, \ldots, n.$$

$$\tag{7.3.55}$$

We assume that this system does not belong to the exactly solvable class of generalized stationary potential. Now replace system (7.3.55) by a solvable one

$$\ddot{Z}_j + h_j(\mathbf{Z}, \dot{\mathbf{Z}}) + u_j(\mathbf{Z}) = \sum_{l=1}^{m} g_{jl}(\mathbf{Z}, \dot{\mathbf{Z}}) W_l(t), \quad j = 1, 2, \ldots, n.$$

$$\tag{7.3.56}$$

Letting $X_{2j-1} = Z_j$, $X_{2j} = \dot{Z}_j$, and $p(\mathbf{x})$ be the exact stationary probability density of its response; i.e., $p(\mathbf{x})$ satisfies the reduced

FPK equation

$$x_{2j} \sum_{j=1}^{n} \frac{\partial p}{\partial x_{2j-1}}$$

$$+ \sum_{j=1}^{n} \frac{\partial}{\partial x_{2j}} \left[\left(-h_j - u_j + \pi \sum_{k=1}^{n} \sum_{l,s=1}^{m} K_{ls} g_{ks} \frac{\partial}{\partial x_k} g_{jl} \right) p \right]$$

$$- \pi \sum_{j,k=1}^{n} \sum_{l,s=1}^{m} K_{ls} \frac{\partial^2}{\partial x_{2j} \partial x_{2k}} (g_{jl} g_{ks} p) = 0. \tag{7.3.57}$$

Our objective is to select a set of h_j functions such that system (7.3.56) is closest to system (7.3.55) in some statistical sense. Since $p(\mathbf{x})$ is not true solution for system (7.3.55), it will not satisfy its FPK equation; therefore, the following residual error occurs

$$\delta = x_{2j} \sum_{j=1}^{n} \frac{\partial p}{\partial x_{2j-1}}$$

$$+ \sum_{j=1}^{n} \frac{\partial}{\partial x_{2j}} \left[\left(-H_j - u_j + \pi \sum_{k=1}^{n} \sum_{l,s=1}^{m} K_{ls} g_{ks} \frac{\partial}{\partial x_k} g_{jl} \right) p \right]$$

$$- \pi \sum_{j,k=1}^{n} \sum_{l,s=1}^{m} K_{ls} \frac{\partial^2}{\partial x_{2j} \partial x_{2k}} (g_{jl} g_{ks} p). \tag{7.3.58}$$

Combining (7.3.57) and (7.3.58), we have

$$\delta(\mathbf{x}) = \sum_{j=1}^{n} \frac{\partial}{\partial x_{2j}} \{ [H_j(\mathbf{x}) - h_j(\mathbf{x})] p(\mathbf{x}) \}. \tag{7.3.59}$$

The residual error as a function of $\mathbf{x}$ is a measure of error of the approximate solution $p(\mathbf{x})$ from the true solution of system (7.3.55). To minimize $\delta(\mathbf{x})$ in some sense, the method of weighted residual is applied to yield

$$\Delta_M = E \left\{ \sum_{j=1}^{n} [H_j(\mathbf{X}) - h_j(\mathbf{X})] \frac{\partial M(\mathbf{X})}{\partial X_{2j}} \right\} = 0. \tag{7.3.60}$$

If we choose $M = X_2^2, X_4^2, \ldots, X_{2n}^2$, the following constraints are obtained

$$E\left\{X_{2j}[H_j(\mathbf{X}) - h_j(\mathbf{X})]\right\} = 0, \quad j = 1, 2, \ldots, n, \qquad (7.3.61)$$

which allow to solve n unknown parameters in h_j functions. Equation set (7.3.61) implies that the average energy dissipations are the same in each coordinate for the original and replacement systems.

Three important issues are pointed out here: (i) the ensemble averages in (7.3.61) would be calculated based on the probability density of the replacement system (7.3.56), (ii) the forms of h_j functions should be adopted so that system (7.3.56) possesses an exact stationary solution and (iii) the exact stationary probability density function $p(\mathbf{x})$ of the replacing system (7.3.56) will be formulated according to the principle of generalized stationary potential. Taking into consideration of these three issues, the approximate probability density of system (7.3.55) can be found directly through iteration procedure of equation set (7.3.61). The next example will illustrate the procedure.

Example 7.3.3 Consider a two-DOF nonlinear stochastic system with damping coupling, governed by

$$\begin{aligned}
&\ddot{Z}_1 + \alpha_1 \dot{Z}_1 + \beta_1 \dot{Z}_1(Z_1^2 + Z_2^2) + \omega_1^2(Z_1 + \gamma_1 Z_1^3) = W_1(t), \\
&\ddot{Z}_2 + \alpha_2 \dot{Z}_2 + \beta_2 \dot{Z}_2(Z_1^2 + Z_2^2) + \omega_2^2(Z_2 + \gamma_2 Z_2^3) = W_2(t),
\end{aligned} \qquad (7.3.62)$$

where $W_1(t)$ and $W_2(t)$ are correlated Gaussian white noises with the correlation functions

$$E[W_i(t)W_j(t + \tau)] = 2\pi K_{ij}\delta(\tau). \qquad (7.3.63)$$

Comparing (7.3.62) with the standard form (7.3.55), we have

$$\begin{aligned}
H_j &= \alpha_j X_{2j} + \beta_j X_{2j}(X_1^2 + X_3^2), \\
v_j &= \omega_j^2(X_{2j-1} + \gamma_j X_{2j-1}^3), \quad j = 1, 2.
\end{aligned} \qquad (7.3.64)$$

Since parametric excitations are absent, there is no Wong–Zakai correction. For the replacement system (7.1.56) belonging to the class of generalized stationary potential, we know from Eq. (6.3.49) that the probability potential has the form

$$\phi(\mathbf{x}) = \phi(\lambda_1, \lambda_2), \quad \lambda_j = \frac{1}{2}x_{2j}^2 + \frac{1}{2}\omega_j^2\left(x_{2j-1}^2 + \frac{1}{2}\gamma_j x_{2j-1}^4\right), \quad j = 1, 2$$

(7.3.65)

and that functions h_j are given by (6.3.48) as

$$h_j = \sum_{k=1}^{2}\left[b_{2j,2k}^{(2j)}\frac{\partial\phi}{\partial x_{2k}}\right], \quad j = 1, 2.$$

(7.3.66)

The second derivate moments b_{ij} can be obtained from (7.3.60), and they can be split as follows

$$b_{22}^{(2)} = \frac{1}{2}b_{22} = \pi K_{11}, \quad b_{44}^{(4)} = \frac{1}{2}b_{44} = \pi K_{22},$$

$$b_{24}^{(2)} = b_{42}^{(4)} = \frac{1}{2}b_{24} = \pi K_{12}.$$

(7.3.67)

Substituting (7.3.67) into (7.3.66), we have

$$\bar{h}_1 = \pi K_{11}x_2\frac{\partial\phi}{\partial\lambda_1} + \pi K_{12}x_4\frac{\partial\phi}{\partial\lambda_2},$$

$$\bar{h}_2 = \pi K_{12}x_2\frac{\partial\phi}{\partial\lambda_1} + \pi K_{22}x_4\frac{\partial\phi}{\partial\lambda_2}.$$

(7.3.68)

To proceed further, we must choose a function form for the probability potential ϕ. Consider a simple linear form

$$\phi = c_{10}\lambda_1 + c_{01}\lambda_2,$$

(7.3.69)

where c_{10} and c_{01} are two constants to be determined. This implies that the approximated probability density has the form of

$$
\begin{aligned}
p(\mathbf{x}) &= Ce^{-\phi(\mathbf{x})}\\
&= C\exp\left\{-c_{10}\left[\frac{1}{2}x_2^2 + \frac{1}{2}\omega_1^2\left(x_1^2 + \frac{1}{2}\gamma_1 x_1^4\right)\right]\right.\\
&\qquad\left. -c_{01}\left[\frac{1}{2}x_4^2 + \frac{1}{2}\omega_2^2\left(x_3^2 + \frac{1}{2}\gamma_2 x_3^4\right)\right]\right\}
\end{aligned}
$$

(7.3.70)

Substituting H_j in (7.3.64) and h_j in (7.3.66) into the constraints (7.3.61), and taking into account of (7.3.70), we obtain

$$E[\alpha_1 X_2^2 + \beta_1 X_2^2(X_1^2 + X_3^2) - \pi K_{11}c_{10}X_2^2 - \pi K_{12}c_{01}X_2X_4] = 0,$$
$$E[\alpha_2 X_4^2 + \beta_2 X_4^2(X_1^2 + X_3^2) - \pi K_{12}c_{10}X_2X_4 - \pi K_{22}c_{01}X_4^2] = 0.$$

$$(7.3.71)$$

In view of the form of the probability density given in (7.3.70), we obtain a set of formal solutions for c_{10} and c_{01} from (7.3.71) as follows:

$$c_{10} = \frac{1}{\pi K_{11}}\{\alpha_1 + \beta_1(E[X_1^2] + E[X_3^2])\},$$

$$c_{01} = \frac{1}{\pi K_{22}}\{\alpha_2 + \beta_2(E[X_1^2] + E[X_3^2])\}.$$

$$(7.3.72)$$

Since the ensemble averages in (7.3.72) depend on the probability density (7.3.70), which in turn is a function of c_{10} and c_{01}, equation set (7.3.72) must be solved by iteration. Numerical calculation was carried out in (Cai and Lin, 1996) for system (7.3.62) with parameters of $\alpha_1 = 0.02$, $\alpha_2 = 0.05$, $\beta_1 = 0.01$, $\beta_2 = 0.01$, $\omega_1 = 1$, $\omega_2 = 1.5$, $\gamma_2 = 1$, $K_{11} = 4$, $K_{12} = 2$, $K_{22} = 1$, and three different levels of nonlinearity γ_1. The calculated stationary probability densities of $Z_1(t)$ and $Z_2(t)$ are depicted in Figs. 7.3.2 and 7.3.3, respectively. It is seen that the accuracy of the approximate results are quite good compared with those obtained from simulations.

In passing, we note that the choice of the functional form of ϕ is made as a simple linear function in this example. The accuracy may be improved if a more complicated form with more parameters is adopted, in which case more M-functions are required in criterion (7.3.60) to generate more constraints. The choice of the ϕ-functional form is unnecessary when applying the procedure of dissipation energy balancing to a SDOF nonlinear stochastic system, as shown in (7.3.23) and Examples 7.3.1 and 7.3.2.

7.4 Stochastic Averaging Methods

As shown in Chapters 4, 5 and the previous sections of this chapter, Gaussian white noise excitation are required in most cases to acquire

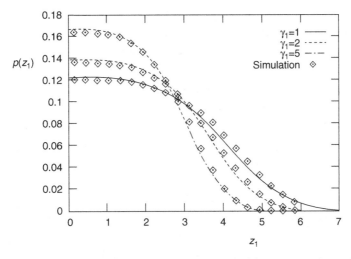

Figure 7.3.2 Stationary probability densities of $Z_1(t)$ of system (7.3.62) (from Cai and Lin, 1996).

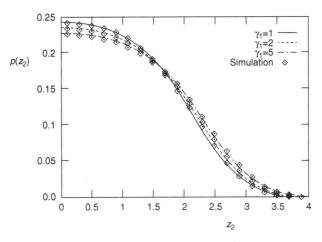

Figure 7.3.3 Stationary probability densities of $Z_2(t)$ of system (7.3.62) (from Cai and Lin, 1996).

exact or approximate solutions of system responses which are Markov diffusion processes. It is known that the concept of white noise is a mathematical idealization with its correlation function being a delta function. Such a requirement is not met for a real physical process. We would like to develop a procedure to approximate a non-white process by a white noise, and investigate the conditions for such an approximation to be feasible.

Let the system equations of motion be

$$\frac{d}{dt}X_j(t) = f_j(\mathbf{X}, t) + \sum_{l=1}^{m} g_{jl}(\mathbf{X}, t)\xi_l(t), \quad j = 1, 2, \ldots, n, \quad (7.4.1)$$

where $\xi_l(t)$ are jointly stationary processes with zero mean and the following correlation functions

$$E[\xi_l(u)\xi_s(v)] = R_{ls}(\tau), \quad \tau = v - u. \quad (7.4.2)$$

Define the correlation time of $\xi_l(t)$ and $\xi_s(t)$ as

$$\tau_{ls} = \frac{1}{\sqrt{R_{ll}(0)R_{ss}(0)}} \int_{-\infty}^{0^-} |R_{ls}(\tau)| \, d\tau, \quad (7.4.3)$$

which is a measure of "memory" of the present $\xi_s(t)$ with respect to the past $\xi_l(t - \tau)$. It is zero for a white noise. If all excitations $\xi_l(t)$ can be approximated as white noises, the probability density of the system response $\mathbf{X}$ satisfies the FPK equation. Thus, the main objective of the approximation is to determine the first and second derivate moments a_j and b_{jk} from the original equations (7.4.1). Denote the time interval as Δt between two consecutive measurements. To determine a_j according to Eq. (4.2.11), we calculate the increment as follows

$$X_j(t + \Delta t) - X_j(t) = \int_{t}^{t+\Delta t} f_j(\mathbf{X}_u, u)du + \sum_{l=1}^{m} \int_{t}^{t+\Delta t} g_{jl}(\mathbf{X}_u, u)\xi_l(u)du,$$

$$(7.4.4)$$

where $\mathbf{X}_u$ is an abbreviation of $\mathbf{X}(u)$. Expand f_j and g_{jl} in (7.4.4) at time instant t,

$$f_j(\mathbf{X}_u, u) = f_j(\mathbf{X}_t, t) + (u - t)\frac{\partial}{\partial t}f_j(\mathbf{X}_t, t)$$

$$+ \sum_{r=1}^{n} [X_r(u) - X_r(t)]\frac{\partial}{\partial X_r}f_j(\mathbf{X}_t, t) + \cdots, \quad (7.4.5)$$

$$g_{jl}(\mathbf{X}_u, u) = g_{jl}(\mathbf{X}_t, t) + (u - t)\frac{\partial}{\partial t}g_{jl}(\mathbf{X}_t, t)$$

$$+ \sum_{r=1}^{n}[X_r(u) - X_r(t)]\frac{\partial}{\partial X_r}g_{jl}(\mathbf{X}_t, t) + \cdots. \qquad (7.4.6)$$

In (7.4.5) and (7.4.6), we again substitute $X_r(u) - X_r(t)$ as

$$X_r(u) - X_r(t) = \int_t^u f_r(\mathbf{X}_v, v)dv + \sum_{s=1}^{m}\int_t^u g_{rs}(\mathbf{X}_v, v)\xi_s(v)dv.$$

$$(7.4.7)$$

Combining (7.4.4) through (7.4.7) and keeping the leading terms, we obtain

$$X_j(t + \Delta t) - X_j(t) = \int_t^{t+\Delta t} f_j(\mathbf{X}_t, t)du + \sum_{l=1}^{m}\int_t^{t+\Delta t} g_{jl}(\mathbf{X}_t, t)\xi_l(u)du$$

$$+ \sum_{l=1}^{m}\int_t^{t+\Delta t}(u - t)\left[\frac{\partial}{\partial t}g_{jl}(\mathbf{X}_t, t)\right]\xi_l(u)du$$

$$+ \sum_{l,s=1}^{m}\int_t^{t+\Delta t}\xi_l(u)du\sum_{r=1}^{n}\frac{\partial}{\partial X_r}g_{jl}(\mathbf{X}_t, t)$$

$$\times \int_t^u g_{rs}(\mathbf{X}_v, v)\xi_s(v)dv. \qquad (7.4.8)$$

Substitution of (7.4.8) into (4.2.11) results in

$$a_j(\mathbf{x}_t, t) = f_j(\mathbf{x}_t, t) + \frac{1}{\Delta t}\sum_{l,s=1}^{m}\sum_{r=1}^{n}\int_t^{t+\Delta t}du\int_t^u\left[\frac{\partial}{\partial x_r}g_{jl}(\mathbf{x}_t, t)\right]$$

$$\times g_{rs}(\mathbf{x}_v, v)E[\xi_l(u)\xi_s(v)]dv + O(\Delta t), \qquad (7.4.9)$$

where $O(\Delta t)$ indicates the remaining terms of order Δt. In deriving (7.4.9), we assume that the variations of f_j and g_{jl} functions are not affected significantly by the randomness in $\mathbf{X}(t)$ within the interval Δt; thus they are not included in the ensemble averaging. Substituting (7.4.2) into (7.4.9) and change integration variable from v to $\tau = v - u$, we obtain

$$a_j(\mathbf{x}_t, t) = f_j(\mathbf{x}_t, t) + \frac{1}{\Delta t} \sum_{l,s=1}^{m} \sum_{s=1}^{n} \int_{t}^{t+\Delta t} du \int_{t-u}^{0} \left[\frac{\partial}{\partial x_r} g_{jl}(\mathbf{x}_t, t) \right]$$

$$\times g_{rs}(\mathbf{x}_{u+\tau}, u + \tau) R_{ls}(\tau) d\tau + O(\Delta t). \tag{7.4.10}$$

If Δt is so adopted that it is much larger than the correlation time τ_{ls}, i.e., $R_{ls}(\tau) \approx 0$ for $\tau > \Delta t$, then by changing the integration order, i.e., integrating on u first, (7.4.10) can be written as

$$a_j(\mathbf{x}_t, t) = f_j(\mathbf{x}_t, t) + \sum_{l,s=1}^{m} \sum_{r=1}^{n} \int_{-\Delta t}^{0} \left[\frac{\partial}{\partial x_r} g_{jl}(x_t, t) \right]$$

$$\times g_{rs}(x_{t+\tau}, t + \tau) R_{ls}(\tau) d\tau. \tag{7.4.11}$$

For the second derivate moments, substituting (7.4.4) into (4.2.11) and following the similar procedure, we obtain

$$b_{jk}(\mathbf{x}_t, t) = \sum_{l,s=1}^{m} \int_{-\Delta t}^{\Delta t} g_{jl}(\mathbf{x}_t, t) g_{ks}(\mathbf{x}_{t+\tau}, t + \tau) R_{ls}(\tau) d\tau. \tag{7.4.12}$$

In above derivation, we assume that functions f_j and g_{jl} are varying slowly within the interval Δt. To meet this condition, we may require Δt not greater than the system relaxation time, which measures the change rate of the system motion without excitations. The relaxation time, denoted by τ_{rel}, has been defined for either an oscillatory or non-oscillatory system. For an oscillatory system, τ_{rel} is the time needed for the amplitude of oscillation to reduce by a factor of e^{-1}, or increase by a factor of e. In the non-oscillatory case, the amplitude is replaced by the motion itself. If Δt is larger than τ_{rel}, functions f_j and g_{jl} may have non-negligible changes and too many details may be lost.

From the derivation of (7.4.11) and (7.4.12), it is known that the time interval Δt should be much larger than the excitation correlation times and smaller than the system relaxation time.

Therefore, the condition for (7.4.11) and (7.4.12) to be valid is that the system relaxation time is much longer than the correlation times of all excitations. Under this condition, the system response can be approximated as a Markov diffusion process.

Since Δt is much larger than the excitation correlation times, the correlation functions are non-zero only within a small neighborhood around $\tau = 0$. Thus, the low limit of the integrations in (7.4.11) and (7.4.12) may be extended to $-\infty$, and the upper limit in (7.4.12) may be extended to ∞, i.e.,

$$a_j(\mathbf{x}_t, t) = f_j(\mathbf{x}_t, t) + \sum_{l,s=1}^{m} \sum_{s=1}^{n} \int_{-\infty}^{0} \left[\frac{\partial}{\partial x_r} g_{jl}(\mathbf{x}_t, t) \right]$$
$$\times g_{rs}(\mathbf{x}_{t+\tau}, t+\tau) R_{ls}(\tau) d\tau, \tag{7.4.13}$$

$$b_{jk}(\mathbf{x}_t, t) = \sum_{l,s=1}^{m} \int_{-\infty}^{\infty} g_{jl}(\mathbf{x}_t, t) g_{ks}(\mathbf{x}_{t+\tau}, t+\tau) R_{ls}(\tau) d\tau. \tag{7.4.14}$$

These two equations allow us to calculate the approximate first- and second-derivate moments in the FPK equation. They will be used for different versions of stochastic averaging to approximate the non-white excitations as white noises, and the system response as a (vector) Markov diffusion process.

In the special case in which the random excitations are white noises, i.e., delta correlated, that is,

$$E[\xi_l(u)\xi_s(v)] = R_{ls}(v - u) = 2\pi K_{ls}\delta(v - u), \tag{7.4.15}$$

where K_{ls} are constant spectral densities, Eqs. (7.4.13) and (7.4.14) reduce to Eqs. (4.3.18) and (4.3.19), i.e.,

$$a_j(\mathbf{x}_t, t) = f_j(\mathbf{x}_t, t)$$
$$+ \sum_{l,s=1}^{m} \sum_{r=1}^{n} \pi K_{ls} g_{rs}(\mathbf{x}_t, t) \frac{\partial}{\partial x_r} g_{jl}(\mathbf{x}_t, t), \tag{7.4.16}$$

$$b_{jk}(\mathbf{x}_t, t) = \sum_{l,s=1}^{m} 2\pi K_{ls} g_{jl}(\mathbf{x}_t, t) g_{ks}(\mathbf{x}_t, t). \tag{7.4.17}$$

It is known that the second term on the right-hand side of (7.4.16) is the Wong–Zakai correction term.

An additional scheme used in the stochastic averaging method is the time averaging which is carried out differently in two cases. One case is that functions f_j and g_{jl} in Eqs. (7.4.13) and (7.4.14) or (7.4.16) and (7.4.17) are periodic functions of time with a period T_p. In this case, the time averaging can be done in one period as

$$\langle [\cdot] \rangle_t = \lim_{T \to \infty} \frac{1}{2T} \int_{-T}^{T} [\cdot] dt = \frac{1}{T_p} \int_{0}^{T_p} [\cdot] dt. \qquad (7.4.18)$$

When an analysis involves a Markov diffusion process, it is often convenient to work with the Ito stochastic differential equations. The averaged Ito equations are given in (4.2.53) with the drift and diffusion coefficients obtained by performing the time averaging on Eqs. (7.4.13) and (7.4.14) as follows

$$m_j(\mathbf{X}) = \langle f_j(\mathbf{X}_t, t) \rangle_t$$

$$+ \sum_{l,s=1}^{m} \sum_{r=1}^{n_1} \int_{-\infty}^{0} \left\langle g_{rs}(\mathbf{X}_{t+\tau}, t+\tau) \frac{\partial}{\partial X_r} g_{jl}(\mathbf{X}_t, t) \right\rangle_t R_{ls}(\tau) d\tau,$$

$$(7.4.19)$$

$$(\boldsymbol{\sigma} \boldsymbol{\sigma}^T)_{jk} = \sum_{r=1}^{n_1} \sigma_{jr}(\mathbf{X}) \sigma_{kr}(\mathbf{X})$$

$$= \sum_{l,s=1}^{m} \int_{-\infty}^{\infty} \langle g_{jl}(\mathbf{X}_t, t) g_{ks}(\mathbf{X}_{t+\tau}, t+\tau) \rangle_t R_{ls}(\tau) d\tau. \qquad (7.4.20)$$

In (7.4.19) and (7.4.20), the averaged drift and diffusion coefficients are smoothed, and no longer depend on the time explicitly. After the time averaging, the total response vector of the averaged stochastic system is still a vector diffusion process of the same dimension as that of the original system. However, it may happen that one single response itself (or a subset of the response vector) constitutes a diffusion process (or diffusion vector process) after the time averaging. In this case, system dimension is reduced.

The procedure to obtain (7.4.19) and (7.4.20) was proposed by Stratonovich (1963), and it has been known as the stochastic averaging in literature. It is an extension of the well-known averaging method of Bogliubov and Mitropolski (1961) from deterministic problems to stochastic ones. Its rigorous mathematical proof was provided by Khasminskii (1966) in a limit theorem, which states that $\mathbf{X}$ approaches a Markov diffusion process in probability as the right-hand sides of Eqs. (7.4.13) and (7.4.14) approach zero.

For the case of Gaussian white-noise excitations, the averaged equations of (7.4.19) and (7.4.20) are

$$m_j(\mathbf{X}) = \langle f_j(\mathbf{X}_t, t)\rangle_t + \sum_{l,s=1}^{m}\sum_{r=1}^{n} \pi K_{ls}\left\langle g_{rs}(\mathbf{X}_t, t)\frac{\partial}{\partial x_r}g_{jl}(\mathbf{X}_t, t)\right\rangle_t,$$

$$(7.4.21)$$

$$(\boldsymbol{\sigma}\boldsymbol{\sigma}^T)_{jk} = \sum_{l,s=1}^{m} 2\pi K_{ls}\langle g_{jl}(\mathbf{X}_t, t)g_{ks}(\mathbf{X}_t, t)\rangle_t. \qquad (7.4.22)$$

In another case of carrying out the time averaging, the system state variables can be distinguished as two different types: fast varying and slowly varying. Without loss of generality, assume that the first $n_1(n_1 < n)$ state variables in system (7.4.1) are slowly varying, and all others are fast varying. Then we may write the first n_1 equations as

$$\frac{d}{dt}X_j(t) = \varepsilon f_j(\mathbf{X}, t) + \varepsilon^{\frac{1}{2}}\sum_{l=1}^{m} g_{jl}(\mathbf{X}, t)\xi_l(t), \quad j = 1, \ldots, n_1,$$

$$(7.4.23)$$

to replace the first n_1 equations in system (7.4.1). In (7.4.23), a small parameter $\varepsilon \ll 1$ is introduced to indicate that the first and second terms on the right-hand-side are assumed to be of order ε and $\varepsilon^{1/2}$, respectively, and $X_j(t)$, $j = 1, 2, \ldots, n_1$, are slowly varying. As it will soon become clear, this is equivalent to assuming the contributions of the same order from the two terms to the system response. Carrying the stochastic averaging and the time-averaging, we obtain Ito equations for $X_j(t)$ $(j = 1, 2, \ldots, n_1)$ from (7.4.23) and (7.4.1) as follows

$$dX_j(t) = m_j(\tilde{\mathbf{X}})dt + \sum_{l=1}^{m} \sigma_{jl}(\tilde{\mathbf{X}})dB_l(t), \qquad (7.4.24)$$

where $\tilde{\mathbf{X}} = [X_1, X_2, \ldots X_{n_1}]^T$, and

$$m_j(\tilde{\mathbf{X}}) = \varepsilon \left\{ \langle f_j(\mathbf{X}_t, t) \rangle_t + \sum_{l,s=1}^{m} \sum_{r=1}^{n_1} \int_{-\infty}^{0} \right.$$

$$\left. \times \left\langle g_{rs}(\mathbf{X}_{t+\tau}, t + \tau) \frac{\partial}{\partial X_r} g_{jl}(\mathbf{X}_t, t) \right\rangle_t R_{ls}(\tau) d\tau \right\},$$

$$(7.4.25)$$

$$(\boldsymbol{\sigma\sigma}^T)_{jk} = \sum_{r=1}^{n_1} \sigma_{jr}(\tilde{\mathbf{X}})\sigma_{kr}(\tilde{\mathbf{X}})$$

$$= \varepsilon \sum_{l,s=1}^{m} \int_{-\infty}^{\infty} \langle g_{jl}(\mathbf{X}_t, t) g_{ks}(\mathbf{X}_{t+\tau}, t + \tau) \rangle_t R_{ls}(\tau) d\tau.$$

$$(7.4.26)$$

When carrying out the time-averaging, the slowly varying variables are considered to be constants, and all fast varying variables are averaged out; therefore, only the slowly varying variables remain, and they also constitute a Markov diffusion process themselves. The slowly varying variables reflect the overall trend of the system motion. As a result, the system dimension is reduced to that of the slowly varying variables, and the problem is simplified.

The procedure to obtain (7.4.24) through (7.4.26) was known as the stochastic averaging theorem of Ito equations with fast and slowly varying variables, proposed by Khasminskii (1968).

It is clear that the derivation of Eqs. (7.4.19) through (7.4.22) or (7.4.24) through (7.4.26) includes two procedures. The first one is to approximate the excitation processes as Gaussian white noises and the system response as a Markov diffusion process. The obtained equations (7.4.13) and (7.4.14), known as the unsmoothed version, in which all state variables remain. The second procedure is to carry out the time averaging, eliminating periodic functions and keeping all state variables in the first case, and eliminating the fast varying variables and reducing the system dimension in the second case. The combined procedure is known as the smoothed version. The stochastic averaging may refer to either the smoothed version of (7.4.19) through (7.4.22) or (7.4.24) through (7.4.26), or the unsmoothed version of (7.4.13) and (7.4.14).

The stochastic averaging will be applied to different situations, as presented in the following sections.

7.4.1 *Stochastic Averaging of Amplitude Envelope*

As mentioned above, the time-averaging is applied to reduce the system dimension when the system equations contain both slowly and fast varying variables. This is usually not the case with the original set of state variables and associated equations of motion. However, if certain types of slowly varying variables can be identified, the system equations can be transformed to those including both the slowly and fast varying variables as state variables. Consider the following oscillator with linear stiffness, weak nonlinear damping and weak excitations,

$$\ddot{X} + \varepsilon h(X, \dot{X}) + \omega_0^2 X = \varepsilon^{\frac{1}{2}} \sum_{l=1}^{m} g_l(X, \dot{X}) \xi_l(t), \qquad (7.4.27)$$

where $\xi_l(t)$ are stationary broadband processes. It is clear that both X and $\dot{X}$ are not slowly varying. We assume that the Wong–Zakai correction does not give rise to an additional restoring force. Let

$$X = A(t) \cos \theta, \quad \dot{X} = -A(t) \omega_0 \sin \theta, \quad \theta = \omega_0 t + \phi(t),$$
$$(7.4.28)$$

where $A(t)$ is the amplitude process, and can be expressed as

$$A(t) = \sqrt{X^2 + \frac{\dot{X}^2}{\omega_0^2}}. \tag{7.4.29}$$

Using Eqs. (7.4.27) and (7.4.28), we obtain (see Exercise Problem 7.16)

$$\dot{A}\cos\theta - A\dot{\phi}\sin\theta = 0. \tag{7.4.30}$$

$$\dot{A}\sin\theta + A\dot{\phi}\cos\theta = \frac{1}{\omega_0}\varepsilon h(A\cos\theta, -A\omega_0\sin\theta)$$

$$-\frac{1}{\omega_0}\varepsilon^{\frac{1}{2}}\sum_{l=1}^{m}g_l(A\cos\theta, -A\omega_0\sin\theta)\xi_l(t). \tag{7.4.31}$$

Solve $\dot{A}$ and $\dot{\phi}$ from (7.4.30) and (7.4.31) to obtain

$$\dot{A} = \frac{\sin\theta}{\omega_0}\left[\varepsilon h(A\cos\theta, -A\omega_0\sin\theta)\right.$$

$$\left.-\frac{1}{\omega_0}\varepsilon^{\frac{1}{2}}\sum_{l=1}^{m}g_l(A\cos\theta, -A\omega_0\sin\theta)\xi_l(t)\right], \tag{7.4.32}$$

$$\dot{\phi} = \frac{\cos\theta}{\omega_0 A}\left[\varepsilon h(A\cos\theta, -A\omega_0\sin\theta)\right.$$

$$\left.-\frac{1}{\omega_0}\varepsilon^{\frac{1}{2}}\sum_{l=1}^{m}g_l(A\cos\theta, -A\omega_0\sin\theta)\xi_l(t)\right]. \tag{7.4.33}$$

The right-hand sides of (7.4.32) and (7.4.33) are small, indicating that both $A(t)$ and $\phi(t)$ are slowly varying. Physically, they represent the amplitude and phase of a nearly linear oscillator, perturbed by small damping nonlinearity and weak excitations.

Treating $A(t)$ as $X_1(t)$ and $\phi(t)$ as $X_2(t)$ and comparing (7.4.32) and (7.4.33) with the standard form (7.4.1), we know

$$
\begin{aligned}
f_1(A, \phi) &= \frac{\sin \theta}{\omega_0} h(A \cos \theta, -A\omega_0 \sin \theta), \\
f_2(A, \phi) &= \frac{\cos \theta}{\omega_0 A} h(A \cos \theta, -A\omega_0 \sin \theta),
\end{aligned}
\tag{7.4.34}
$$

$$
\begin{aligned}
g_{1l} &= -\frac{\sin \theta}{\omega_0^2} g_l(A \cos \theta, -A\omega_0 \sin \theta), \\
g_{2l} &= -\frac{\cos \theta}{\omega_0^2 A} g_l(A \cos \theta, -A\omega_0 \sin \theta).
\end{aligned}
\tag{7.4.35}
$$

After applying the stochastic averaging procedure, i.e., approximating the broad-band excitations by white noises and carrying out the time averaging, $A(t)$ and $\phi(t)$ constitute a Markov diffusion vector process, and the drift and diffusion coefficients can be found from Eqs. (7.4.19) and (7.4.20). Since the system is nearly linear, the motion is nearly periodic with a "quasi-period" $2\pi/\omega_0$. The time-averaging (7.4.18) can then be calculated over one quasi-period, i.e.,

$$
\langle [\cdot] \rangle_t = \frac{1}{2\pi} \int_0^{2\pi} [\cdot] d\theta.
\tag{7.4.36}
$$

It is found that the averaged equation for $A(t)$ does not contain $\phi(t)$, and the smoothed amplitude process $A(t)$ itself is a Markov diffusion process governed by an Ito differential equation

$$
dA = m(A)dt + \sigma(A)dB(t).
\tag{7.4.37}
$$

Equation (7.4.37) is a one-dimensional Ito equation, and its analysis becomes much simpler, as shown in Section 4.4.

Although the amplitude process $A(t)$ is an important feature of the motion, the joint stationary probability density, as well as the marginal probability densities of $X(t)$ and $\dot{X}(t)$ may be also needed for analysis, which will be derived below. The joint distribution function $F_{XA}(x, a)$ can be written as

$$F_{XA}(x, a) = \text{Prob}[(X \le x) \cap (A \le a)]$$

$$= \text{Prob}\left[(X \le x) \cap \left(\sqrt{X^2 + \dot{X}^2/\omega_0^2} \le a\right)\right]$$

$$= \text{Prob}[(X \le x) \cap (-y \le \dot{X} \le y)]$$

$$= \text{Prob}[(X \le x) \cap (\dot{X} \le y)] - \text{Prob}[(X \le x) \cap (\dot{X} \le -y)]$$

$$= F_{X\dot{X}}(x, y) - F_{X\dot{X}}(x, -y), \tag{7.4.38}$$

where

$$y = \omega_0 \sqrt{a^2 - x^2} \tag{7.4.39}$$

Thus, we have from (7.4.38)

$$p_{XA}(x, a) = \frac{\partial^2}{\partial x \partial a} F_{XA}(x, a)$$

$$= \left[\frac{\partial^2}{\partial x \partial y} F_{X\dot{X}}(x, y)\right] \frac{\partial y}{\partial a} - \left[\frac{\partial^2}{\partial x \partial y} F_{X\dot{X}}(x, -y)\right] \frac{\partial(-y)}{\partial a}$$

$$= p_{X\dot{X}}(x, \dot{x}) \frac{2\omega_0 a}{\sqrt{a^2 - x^2}}. \tag{7.4.40}$$

In deriving (7.4.40), the symmetry of $\dot{X}$ is used, i.e., $p_{X\dot{X}}(x, -\dot{x}) = p_{X\dot{X}}(x, \dot{x})$. By omitting the subscripts, (7.4.40) can be written as

$$p(x, \dot{x}) = \frac{\sqrt{a^2 - x^2}}{2\omega_0 a} p(x, a) = \frac{\sqrt{a^2 - x^2}}{2\omega_0 a} p(x|a)p(a), \tag{7.4.41}$$

where $p(x|a)$ is the conditional probability density. For a fixed a, the probability of $X(t)$ in a region near x is inversely proportional to the velocity; therefore, $p(x|a)$ can be expressed as

$$p(x|a) = \frac{C}{|\dot{x}|} = \frac{C}{\omega_0 \sqrt{a^2 - x^2}}. \tag{7.4.42}$$

Integrating both sides with respect to x from $-a$ to a, we obtain

$$C = \left[\int_{-a}^{a} \frac{dx}{\omega_0 \sqrt{a^2 - x^2}}\right]^{-1} = \frac{\omega_0}{\pi}. \tag{7.4.43}$$

Substituting (7.4.43) into (7.4.42), then into (7.4.41), we obtain

$$p(x, \dot{x}) = \frac{1}{2\pi\omega_0 a} p(a); \quad a = \sqrt{x^2 + \frac{\dot{x}^2}{\omega_0^2}}. \tag{7.4.44}$$

The marginal probability density function $p(x)$ or $p(\dot{x})$ can then be calculated from (7.4.44) by integration with respect to $\dot{x}$ or x, as shown in Eq. (2.6.6) or (2.6.7).

Example 7.4.1 Consider a system governed by

$$\ddot{X} + 2\zeta\omega_0\dot{X} + \omega_0^2[1 + \xi_1(t)]X = \xi_2(t), \tag{7.4.45}$$

where $\xi_1(t)$ and $\xi_2(t)$ are two stationary and independent broadband processes with spectral densities $\Phi_{11}(\omega)$ and $\Phi_{22}(\omega)$, respectively. Equation (7.4.45) can be used to describe the first-mode motion of a column subjected to both random axial and transverse loads (Lin and Cai, 1995). Using the transformation of (7.4.28), we obtain the equations of $A(t)$ and $\phi(t)$ according to (7.4.32) and (7.4.33),

$$\dot{A} = -2\zeta\omega_0 A \sin^2\theta + \omega_0 A \sin\theta\cos\theta\xi_1(t) - \frac{1}{\omega_0}\sin\theta\xi_2(t),$$
$$\tag{7.4.46}$$

$$\dot{\phi} = -2\zeta\omega_0\sin\theta\cos\theta + \omega_0\cos^2\theta\xi_1(t) - \frac{1}{A\omega_0}\cos\theta\xi_2(t).$$
$$\tag{7.4.47}$$

We assume that the correlation times of $\xi_1(t)$ and $\xi_2(t)$ are much shorter than the system relaxation time which is of the order $(\zeta\omega_0)^{-1}$. Furthermore, we assume that the damping is small and the excitations are weak. Under these assumptions, both $A(t)$ and $\phi(t)$ are slowly varying, and the stochastic averaging is applicable.

Compared with the standard form (7.4.1) in which $A(t)$ and $\phi(t)$ playing the roles of $X_1(t)$ and $X_2(t)$, respectively, we have

$$f_1 = -2\zeta\omega_0 A \sin^2\theta, \quad f_2 = -2\zeta\omega_0\sin\theta\cos\theta,$$

$$g_{11} = \omega_0 A \sin\theta\cos\theta, \quad g_{12} = -\frac{1}{\omega_0}\sin\theta,$$
$$\tag{7.4.48}$$

$$g_{21} = \omega_0\cos^2\theta, \quad g_{22} = -\frac{1}{A\omega_0}\cos\theta.$$

Carrying out the time-averaging in (7.4.21) and (7.4.22), we obtain

$$m_1 = -\zeta\omega_0 A$$

$$+ \int_{-\infty}^{0} \left[\frac{3}{8}\omega_0^2 A \cos(2\omega_0\tau)R_{11}(\tau) + \frac{1}{2\omega_0^2 A}\cos(\omega_0\tau)R_{22}(\tau) \right] d\tau$$

$$= - \left[\zeta\omega_0 - \frac{3\pi}{8}\omega_0^2\Phi_{11}(2\omega_0) \right] A + \frac{\pi}{2\omega_0^2 A}\Phi_{22}(\omega_0). \qquad (7.4.49)$$

$$\sigma_{11}\sigma_{11} = \int_{-\infty}^{\infty} \left[\frac{1}{8}\omega_0^2 A^2 \cos(2\omega_0\tau)R_{11}(\tau) + \frac{1}{2\omega_0^2}\cos(\omega_0\tau)R_{22}(\tau) \right] d\tau$$

$$= \frac{\pi}{4}\omega_0^2\Phi_{11}(2\omega_0)A^2 + \frac{\pi}{\omega_0^2}\Phi_{22}(\omega_0), \qquad (7.4.50)$$

$$\sigma_{12}\sigma_{12} = 0, \qquad (7.4.51)$$

where use has been made of the relationship between the correlation and spectral density functions according to Eq. (3.6.1), i.e.,

$$\int_{-\infty}^{\infty} R_{XX}(\tau)e^{-i\omega\tau}d\tau = \int_{-\infty}^{\infty} R_{XX}(\tau)\cos(\omega\tau)d\tau$$

$$= 2\int_{-\infty}^{0} R_{XX}(\tau)\cos(\omega\tau)d\tau = 2\pi\Phi_{XX}(\omega). \qquad (7.4.52)$$

In the above time-averaging over a quasi-period, the amplitude A is considered to be a constant, and the smoothed drift and diffusion coefficients are then functions of the amplitude. Now the amplitude process $A(t)$ itself is a Markov diffusion process, governed by the Ito equation

$$dA = \left\{ -\left[\zeta\omega_0 - \frac{3\pi}{8}\omega_0^2\Phi_{11}(2\omega_0) \right] A + \frac{\pi}{2\omega_0^2 A}\Phi_{22}(\omega_0) \right\} dt$$

$$+ \left[\frac{\pi}{4}\omega_0^2\Phi_{11}(2\omega_0)A^2 + \frac{\pi}{\omega_0^2}\Phi_{22}(\omega_0) \right]^{\frac{1}{2}} dB(t). \qquad (7.4.53)$$

Equation (7.4.53) shows that the system behavior depends on the spectrum value of the parametric excitation $\xi_1(t)$ at frequency $2\omega_0$ and the spectrum value of the external excitation $\xi_2(t)$ at frequency ω_0.

As discussed in Section 4.4, the nature of the one-dimensional Markov diffusion process $A(t)$ can be investigated in terms of its boundary behaviors, which depend on whether the external random excitation $\xi_2(t)$ is present. First, we consider the case where $\xi_2(t)$ exists. At the left boundary $a = 0$, the drift coefficient is unbounded, and it is singular of the second kind. The diffusion exponent, drift exponent, and character value are calculated from Eqs. (4.4.15), (4.4.16) and (4.4.17) as follows

$$\alpha = 0, \quad \beta = 1, \quad c = 1, \quad \text{at } a = 0. \tag{7.4.54}$$

According to Table 4.4.3, $a = 0$ is an entrance boundary. The right boundary at infinity is also singular of the second kind, and its diffusion exponent, drift exponent and character value are obtained from Eqs. (4.4.18), (4.4.19) and (4.4.20) as follows

$$\alpha = 2, \quad \beta = 1, \quad c = \frac{8\zeta - 3\pi\omega_0\Phi_{11}(2\omega_0)}{\pi\omega_0\Phi_{11}(2\omega_0)}, \quad \text{at } a = \infty. \tag{7.4.55}$$

From Table 4.4.4, it is identified as repulsively natural if $c > -1$, that is, if

$$\zeta > \frac{\pi}{4}\omega_0\Phi_{11}(2\omega_0). \tag{7.4.56}$$

Otherwise, it is strictly natural or attractively natural. A stationary probability density exists under condition (7.4.56), and is obtained according to Eq. (4.4.7) as

$$p(a) = Ca(a^2 + D)^{-\delta}, \tag{7.4.57}$$

where

$$D = \frac{4\Phi_{22}(\omega_0)}{\omega_0^4\Phi_{11}(2\omega_0)}, \quad \delta = \frac{4\zeta}{\pi\omega_0\Phi_{11}(2\omega_0)}, \quad C = 2(\delta - 1)D^{\delta-1}. \tag{7.4.58}$$

The integrability of the probability density $p(a)$ or the positivity of the normalization constant C requires $\delta > 1$, namely, the condition

(7.4.56). The joint stationary probability density of $X(t)$ and $\dot{X}(t)$ can be obtained according to Eq. (7.4.44).

It is noticed that the system response only depends on the spectral density of the external excitation $\xi_2(t)$ at the system natural frequency ω_0, and that of the parametric excitation $\xi_1(t)$ at $2\omega_0$. This can be understood according to the nature of the stochastic averaging. In the procedure, the correlation times of the excitations are required to be much shorter than the system relaxation time, and the spectra to be much broader than ω_0. Thus, the spectrum values remain almost unchanged near ω_0 and $2\omega_0$, and $\Phi_{11}(2\omega_0)$ and $\Phi_{22}(\omega_0)$ represent the spectrum values near $2\omega_0$ and ω_0, respectively. The significance of $\Phi_{22}(\omega_0)$ is obvious since the system response function has a sharp peak near ω_0 due to small damping. That $\Phi_{11}(2\omega_0)$ plays an important role also agrees with the deterministic analysis if $\xi_1(t)$ were to be replaced by a sinusoidal excitation, system equation would become a damped Mathieu equation, and $2\omega_0$ would be the primary resonant frequency.

We digress to remark that if the random parametric excitation were to be multiplied to the velocity $\dot{X}$ rather than the displacement X, the spectrum value at $\omega = 0$ is also involved, as shown in Exercise Problem 7.20.

If the parametric excitation is absent, the stationary probability densities are obtained as

$$p(a) = \frac{2\zeta\omega_0^3}{\pi\Phi_{22}(\omega_0)} a \exp\left[-\frac{\zeta\omega_0^3}{\pi\Phi_{22}(\omega_0)}a^2\right], \qquad (7.4.59)$$

$$p(x, \dot{x}) = \frac{\zeta\omega_0^2}{\pi^2\Phi_{22}(\omega_0)} \exp\left[-\frac{\zeta\omega_0^3}{\pi\Phi_{22}(\omega_0)}\left(x^2 + \frac{\dot{x}^2}{\omega_0^2}\right)\right], \quad (7.4.60)$$

indicating that, for a linear oscillator under an external broad-band excitation, the amplitude $A(t)$ is Rayleigh-distributed, and both displacement $X(t)$ and velocity $\dot{X}$ are Gaussian-distributed, which are the well-known results.

If the external excitation $\xi_2(t)$ is absent, it can be seen from Eq. (7.4.49) that, at the left boundary $a = 0$, both the drift and diffusion coefficients are zero, and the diffusion exponent, drift exponent, and character value are calculated from Eqs. (4.4.12), (4.4.13) and (4.4.14) as

$$\alpha = 2, \quad \beta = 1, \quad c = \frac{-8\zeta + 3\pi\omega_0\Phi_{11}(2\omega_0)}{\pi\omega_0\Phi_{11}(2\omega_0)}. \tag{7.4.61}$$

From Table 4.4.2, we know that

the left boundary $a = 0$ is

$$\begin{cases} \text{attractively natural} & \text{if } \zeta > \dfrac{\pi}{4}\omega_0\Phi_{11}(2\omega_0) \\[2mm] \text{strictly natural} & \text{if } \zeta = \dfrac{\pi}{4}\omega_0\Phi_{11}(2\omega_0) \\[2mm] \text{repulsively natural} & \text{if } \zeta < \dfrac{\pi}{4}\omega_0\Phi_{11}(2\omega_0). \end{cases} \tag{7.4.62}$$

The right boundary at infinity is the same as for the case when both external and parametric excitations are present, namely,

the right boundary $a = \infty$ is

$$\begin{cases} \text{repulsively natural} & \text{if } \zeta > \dfrac{\pi}{4}\omega_0\Phi_{11}(2\omega_0) \\[2mm] \text{strictly natural} & \text{if } \zeta = \dfrac{\pi}{4}\omega_0\Phi_{11}(2\omega_0) \\[2mm] \text{attractively natural} & \text{if } \zeta < \dfrac{\pi}{4}\omega_0\Phi_{11}(2\omega_0). \end{cases} \tag{7.4.63}$$

By examining behaviors of the two boundaries, it can be concluded that a non-trivial stationary probability density does not exist without the external excitation. It can also be deduced from the non-integrability of the probability density (7.4.57) if $D = 0$. Physically, if the damping is strong, the response will be attracted to zero. On the other hand, if the damping is not strong enough, the response will grow boundless.

Example 7.4.2 Consider the nonlinear stochastic system,

$$\ddot{X} + \alpha\dot{X} + \beta X^2\dot{X} + \gamma\dot{X}^3 + \omega_0^2 X = XW_1(t) + \dot{X}W_2(t) + W_3(t), \tag{7.4.64}$$

where $W_1(t)$, $W_2(t)$ and $W_3(t)$ are independent Gaussian white noises with spectral densities K_{ii} $(i = 1, 2, 3)$. Letting $X = X_1$ and $\dot{X} = X_2$,

Eq. (7.4.64) is replaced by two first-order equations

$$\dot{X}_1 = X_2, \tag{7.4.65}$$

$$\dot{X}_2 = -\omega_0^2 X_1 - \alpha X_2 - \beta X_1^2 X_2$$
$$-\gamma X_2^3 + X_1 W_1(t) + X_2 W_2(t) + W_3(t), \tag{7.4.66}$$

which can be converted to Ito differential equations

$$dX_1 = X_2 dt, \tag{7.4.67}$$

$$dX_2 = \left[-\omega_0^2 X_1 - \alpha X_2 - \beta X_1^2 X_2 - \gamma X_2^3 + \pi K_{22} X_2 \right] dt$$
$$+ [2\pi (K_{11} X_1^2 + K_{22} X_2^2 + K_{33}]^{\frac{1}{2}} dB(t). \tag{7.4.68}$$

Let

$$X_1 = A(t) \cos \theta, \quad X_2 = -A(t)\omega_0 \sin \theta, \quad \theta = \omega_0 t + \varphi(t). \tag{7.4.69}$$

The amplitude $A(t)$ can then be expressed as

$$A(t) = \sqrt{X_1^2 + \frac{X_2^2}{\omega_0^2}} \tag{7.4.70}$$

and the following partial derivatives are obtained

$$\frac{\partial A}{\partial X_1} = \cos \theta, \quad \frac{\partial A}{\partial X_2} = -\frac{\sin \theta}{\omega_0}, \quad \frac{\partial^2 A}{\partial X_2^2} = \frac{\cos^2 \theta}{A\omega_0^2}. \tag{7.4.71}$$

Using (7.4.71) and the Ito differential rule (4.2.56) to obtain an Ito equation for amplitude $A(t)$, and then applying the time-averaging for the drift and diffusion coefficients, we obtain a smoothed version of the Ito equation

$$dA = \frac{\partial A}{\partial X_1} dX_1 + \frac{\partial A}{\partial X_2} dX_2 + \frac{1}{2} \frac{\partial^2 A}{\partial X_2^2} (dX_2)^2 = m(A)dt + \sigma(A)dB(t), \tag{7.4.72}$$

where

$$m(A) = \left(-\frac{\alpha}{2} + \frac{3\pi}{8\omega_0^2}K_{11} + \frac{5\pi}{8}K_{22}\right)A$$
$$-\frac{1}{8}(\beta + 3\gamma\omega_0^2)A^3 + \frac{\pi}{2A\omega_0^2}K_{33}, \qquad (7.4.73)$$

$$\sigma^2(A) = \frac{\pi}{4\omega_0^2}(K_{11} + 3\omega_0^2 K_{22})A^2 + \frac{\pi}{\omega_0^2}K_{33}. \qquad (7.4.74)$$

The stationary probability density of $A(t)$ is calculated from (4.4.7) to yield

$$p(a) = Ca\left[(K_{11} + 3\omega_0^2 K_{22})a^2 + 4K_{33}\right]^{\delta}$$
$$\times \exp\left[-\frac{\omega_0^2(\beta + 3\omega_0^2\gamma)}{2\pi(K_{11} + 3\omega_0^2 K_{22})}a^2\right], \qquad (7.4.75)$$

where

$$\delta = \frac{2\omega_0^2[K_{33}(\beta + 3\omega_0^2\gamma) - (\alpha + \pi K_{22})(K_{11} + 3\omega_0^2 K_{22})]}{\pi(K_{11} + 3\omega_0^2 K_{22})^2}. \qquad (7.4.76)$$

It is of interest to note that the nonlinear damping is strong enough to bring the system back when the response is very large $(a \to \infty)$. If the external excitation is present, it will repulse the system from the left boundary $a = 0$, and a non-trivial stationary probability density exists. However, if the external excitation is absent, the linear damping plays the crucial role near $a = 0$. If it is weak to meet the condition

$$\alpha < \frac{\pi}{2}\left(\frac{K_{11}}{\omega_0^2} + K_{22}\right), \qquad (7.4.77)$$

the left boundary $a = 0$ is repulsively natural, and a non-trivial stationary probability density $p(a)$ exists. Otherwise, $a = 0$ is either strictly or attractively natural, and a non-trivial stationary probability density does not exist.

The joint probability density of $X_1(t)$ and $X_2(t)$ can be obtained from (7.4.75) according to (7.4.44) as

$$p(x_1, x_2) = C_1 \left[(K_{11} + 3\omega_0^2 K_{22}) \left(x_1^2 + \frac{x_2^2}{\omega_0^2} \right) + 4K_{33} \right]^\delta$$

$$\times \exp \left[-\frac{\omega_0^2 (\beta + 3\omega_0^2 \gamma)}{2\pi (K_{11} + 3K_{22})} \left(x_1^2 + \frac{x_2^2}{\omega_0^2} \right) \right]. \quad (7.4.78)$$

It can be verified that (7.4.78) obtained from the stochastic averaging is exactly the same as (7.3.51) derived from the dissipation energy balance method taking into account that $\lambda = \frac{1}{2}\omega_0^2 a^2$ and $\omega_0^2 = k_e$.

In this example, the excitations are Gaussian white noises; therefore, the approximation of non-white noises by white noises is not needed. The main task here is to identify the slowly varying variable and carry out the time-averaging. It is noted the assumption that the system linear restoring force is not changed due to the Wong–Zakai correction. If the Wong–Zakai correction produces an additional restoring force, the ω_0 in the transformation (7.4.28) needs to be modified accordingly.

7.4.2 *Stochastic Averaging of Energy Envelope*

If the restoring force is not linear, but strongly nonlinear, the transformation (7.4.28) is not appropriate since the frequency of the corresponding undamped free motion is changing with the amplitude. Consider the system governed by

$$\ddot{X} + \varepsilon h(X, \dot{X}) + u(X) = \varepsilon^{\frac{1}{2}} \sum_{l=1}^{m} g_l(X, \dot{X})\xi_l(t), \quad (7.4.79)$$

where $u(X)$ represents a strongly nonlinear restoring force, assumed to be an odd function of X. Also we assume that the Wong–Zakai correction does not cause the change of the system restoring force.

First, let us study the undamped free vibration of the system

$$\ddot{X} + u(X) = 0. \quad (7.4.80)$$

Since $\ddot{X} = \frac{d\dot{X}}{dt} = \frac{d\dot{X}}{dX}\frac{dX}{dt} = \dot{X}\frac{d\dot{X}}{dX}$, Eq. (7.4.80) can be rewritten as

$$\dot{X}\frac{d\dot{X}}{dX} + u(X) = 0. \tag{7.4.81}$$

Integrate (7.4.81) to obtain

$$\frac{1}{2}\dot{X}^2 + U(X) = \Lambda, \tag{7.4.82}$$

where

$$U(X) = \int\limits_0^X u(z)dz. \tag{7.4.83}$$

It is known that $U(X)$ is the potential energy, and Λ is the total energy, which is constant for an undamped free vibration. Assuming that $u(X)$ is a monotonic function, then the amplitude A of the oscillation is reached when the kinetic energy is zero, and the total energy is converted entirely to the potential energy, i.e.,

$$\Lambda = U(A) \quad \text{or} \quad A = U^{-1}(\Lambda). \tag{7.4.84}$$

The period of the free oscillation is obtained as

$$T = 4T_{\frac{1}{4}} = 4\int\limits_0^A \frac{1}{\sqrt{2\Lambda - 2U(X)}}dX. \tag{7.4.85}$$

It is clear that the period of the free motion depends on the energy level. Letting

$$\begin{aligned}\text{sgn}X\sqrt{U(X)} &= \sqrt{\Lambda}\cos\theta, \\ \dot{X} &= -\sqrt{2\Lambda}\sin\theta,\end{aligned} \tag{7.4.86}$$

Eq. (7.4.80) is transformed to

$$\begin{aligned}\dot{\Lambda} &= 0, \\ \dot{\theta} &= \frac{u}{\sqrt{2\Lambda}\cos\theta}.\end{aligned} \tag{7.4.87}$$

The first equation of (7.4.87) is a re-statement that the total energy is conserved. If the restoring force u is a linear function, the right-hand side of the second equation reduces to a constant, namely, the natural frequency of free motion. It is, therefore, reasonable to rewrite the equation as

$$\omega = \frac{u}{\sqrt{2\Lambda}\cos\theta}, \qquad (7.4.88)$$

which is the definition of the instantaneous frequency for the nonlinear system. The phase angle can be written as

$$\theta(t) = \int_0^t \omega(s)ds. \qquad (7.4.89)$$

The average $\omega(t)$ in one period, denoted by ω_Λ, is given by

$$\omega_\Lambda = \frac{1}{T}\int_0^T \omega(t)dt = \frac{1}{T}\int_0^T \frac{u}{\sqrt{2\Lambda}\cos\theta}dt = \frac{1}{T}\int_0^{2\pi} d\theta = \frac{2\pi}{T}. \qquad (7.4.90)$$

The above analysis shows that the undamped free motion is periodic with a period T depending on the energy level Λ, and the motion is not harmonic with a time-dependent frequency $\omega(t)$.

Now return to Eq. (7.4.79). Differentiating (7.4.82) and using (7.4.79), we obtain

$$\dot{\Lambda} = -\varepsilon\dot{X}h(X,\dot{X}) + \varepsilon^{\frac{1}{2}}\dot{X}\sum_{l=1}^m g_l(X,\dot{X})\xi_l(t). \qquad (7.4.91)$$

Another state variable chosen to associate with the energy process $\Lambda(t)$ may be the displacement process $X(t)$ with another equation of motion as

$$\dot{X} = \pm\sqrt{2\Lambda - 2U(X)}, \qquad (7.4.92)$$

where the positive and negative signs correspond to increasing and decreasing X, respectively. Equations (7.4.91) and (7.4.92) now constitute the governing equations for the system, instead of the original equation of motion (7.4.79). Equation (7.4.91) shows that the total energy $\Lambda(t)$ varies slowly if the damping and excitations are small. In addition, we assume that the correlation times of excitations $\xi_l(t)$ are short compared with the system relaxation time of order ε^{-1}, then the stochastic averaging is applicable, and energy process $\Lambda(t)$ is approximately Markov diffusion process governed by an Ito equation

$$d\Lambda = m(\Lambda)dt + \sigma(\Lambda)dB(t), \tag{7.4.93}$$

where the drift and diffusion coefficients are obtained from Eqs. (7.4.19) and (7.4.20) as follows

$$m(\Lambda) = -\varepsilon \left\langle \dot{X} h(X, \dot{X}) \right\rangle_t$$

$$+\varepsilon \int_{-\infty}^{0} \left\langle \sum_{l,s=1}^{m} [\dot{X} g_s(X, \dot{X})]_{t+\tau} \frac{\partial}{\partial \Lambda}[\dot{X} g_l(X, \dot{X})]_t \right\rangle_t R_{ls}(\tau)d\tau, \tag{7.4.94}$$

$$\sigma^2(\Lambda) = \varepsilon \int_{-\infty}^{\infty} \left\langle \sum_{l,s=1}^{m} [\dot{X} g_s(X, \dot{X})]_{t+\tau} [\dot{X} g_l(X, \dot{X})]_t \right\rangle_t R_{ls}(\tau)d\tau. \tag{7.4.95}$$

In (7.4.94) and (7.4.95), $\dot{X}$ is treated as function of Λ and X according to (7.4.92). The time-averaging is then calculated as

$$\langle [\cdot] \rangle_t = \frac{1}{T} \int_0^T [\cdot]dt = \frac{1}{T} \int_0^T \frac{[\cdot]}{\dot{X}}dX = \frac{1}{T\frac{1}{4}} \int_0^A \frac{[\cdot]_{\dot{X}=\sqrt{2\Lambda-2U(X)}}}{\sqrt{2\Lambda - 2U(X)}}dX. \tag{7.4.96}$$

When evaluating the integration, the integration limit A is a function of the energy level Λ according to Eq. (7.4.84).

If all the excitations are Gaussian white noises, Eqs. (7.4.94) and (7.4.95) are simplified as

$$m(\Lambda) = -\varepsilon \left\langle \dot{X} h(X, \dot{X}) \right\rangle_t$$

$$+\varepsilon\pi \left\langle \sum_{l,s=1}^{m} K_{ls}[\dot{X} g_s(X, \dot{X})] \frac{\partial}{\partial \Lambda}[\dot{X} g_l(X, \dot{X})] \right\rangle_t, \tag{7.4.97}$$

$$\sigma^2(\Lambda) = \varepsilon 2\pi \left\langle \sum_{l,s=1}^{m} K_{ls}\dot{X}^2 g_l(X, \dot{X})g_s(X, \dot{X}) \right\rangle_t. \tag{7.4.98}$$

The above procedure of stochastic averaging of energy envelope is also known as the quasi-conservative averaging (Landa and Stratonovich, 1962; Khasminskii, 1964).

The stationary probability density of $\Lambda(t)$ is obtained as

$$p(\lambda) = \frac{C}{\sigma^2(\lambda)} \exp\left[\int \frac{2m(\lambda)}{\sigma^2(\lambda)} d\lambda\right]. \qquad (7.4.99)$$

The joint stationary probability density of $X(t)$ and $\dot{X}(t)$ can be calculated following the similar procedure as in Section 7.4.1. For the present situation, Eq. (7.4.41) is changed to

$$p(x, \dot{x}) = \frac{1}{2}\sqrt{2\lambda - 2U(x)}p(x, \lambda) = \frac{1}{2}\sqrt{2\lambda - 2U(x)}p(x|\lambda)p(\lambda), \qquad (7.4.100)$$

where $p(x|\lambda)$ is the conditional probability density. For a fixed energy level λ, the probability of $X(t)$ in a region near x is inversely proportional to the velocity; therefore, $p(x|\lambda)$ can be expressed as

$$p(x|\lambda) = \frac{C}{|\dot{x}|} = \frac{C}{\sqrt{2\lambda - 2U(x)}}. \qquad (7.4.101)$$

Integrating both sides with respect to x from $-a$ to a, we obtain

$$C(\lambda) = \left[\int_{-a}^{a} \frac{dx}{\sqrt{2\lambda - 2U(x)}}\right]^{-1} = \frac{2}{T(\lambda)}. \qquad (7.4.102)$$

Substituting (7.4.102) into (7.4.101), and then into (7.4.100), we obtain

$$p(x, \dot{x}) = \frac{p(\lambda)}{T(\lambda)}; \quad \lambda = \frac{1}{2}\dot{x}^2 + U(x). \qquad (7.4.103)$$

Example 7.4.3 Consider the nonlinear system

$$\ddot{X} + \alpha\dot{X} + \beta\dot{X}^3 + \delta X^3 = W(t), \quad \beta, \delta > 0. \qquad (7.4.104)$$

The equation of the undamped free vibration corresponding to (7.4.104) is

$$\ddot{X} + \delta X^3 = 0, \tag{7.4.105}$$

with the potential energy and total energy

$$U = \frac{1}{4}\delta X^4, \quad \Lambda = \frac{1}{2}\dot{X}^2 + \frac{1}{4}\delta X^4 \tag{7.4.106}$$

and a period

$$T = 4T_{\frac{1}{4}} = 4 \int_0^A \frac{dX}{\sqrt{2\Lambda - \frac{1}{2}\delta X^4}}, \tag{7.4.107}$$

where the integration limit A is determined from $A = (4\Lambda/\delta)^{1/4}$. Changing the integration variable according to

$$X = \left(\frac{4\Lambda}{\delta}\sin^2\theta\right)^{\frac{1}{4}}, \tag{7.4.108}$$

we have

$$T_{\frac{1}{4}} = \frac{1}{4}\delta^{-\frac{1}{4}}\lambda^{-\frac{1}{4}}B\left(\frac{1}{4},\frac{1}{2}\right), \tag{7.4.109}$$

where $B(\cdot,\cdot)$ is the Beta function. In calculating the integral, the following formula has been used (Gradshteyn and Ryzhik, 1980):

$$\int_0^{\pi/2} \sin^{m-1}\theta\cos^{n-1}\theta d\theta = \frac{1}{2}B\left(\frac{m}{2},\frac{n}{2}\right). \tag{7.4.110}$$

Since the excitation is a Gaussian white noise, Eqs. (7.4.97) and (7.4.98) can be applied to obtain

$$m(\Lambda) = -\alpha\left\langle\dot{X}^2\right\rangle_t - \beta\left\langle\dot{X}^4\right\rangle_t + \pi K\left\langle\dot{X}\frac{\partial\dot{X}}{\partial\Lambda}\right\rangle_t, \tag{7.4.111}$$

$$\sigma^2(\Lambda) = 2\pi K\left\langle\dot{X}^2\right\rangle_t. \tag{7.4.112}$$

Evaluating the time-average terms according to (7.4.96) and noting that $\partial \dot{X}/\partial \Lambda = 1/\dot{X}$ due to (7.4.82),

$$\left\langle \dot{X}^2 \right\rangle_t = \frac{1}{T_{\frac{1}{4}}} \int_0^A \left[2\Lambda - \frac{1}{2}\delta X^4 \right]^{\frac{1}{2}} dX = \frac{4}{3}\Lambda, \qquad (7.4.113)$$

$$\left\langle \dot{X}^4 \right\rangle_t = \frac{1}{T_{\frac{1}{4}}} \int_0^A \left[2\Lambda - \frac{1}{2}\delta X^4 \right]^{\frac{3}{2}} dX = \frac{32}{7}\Lambda^2, \qquad (7.4.114)$$

$$\left\langle \dot{X}\frac{\partial \dot{X}}{\partial \Lambda} \right\rangle_t = \frac{1}{T_{\frac{1}{4}}} \int_0^A \left[2\Lambda - \frac{1}{2}\delta X^4 \right]^{-\frac{1}{2}} dX = 1. \qquad (7.4.115)$$

Thus the Ito equation for the energy process $\Lambda(t)$ is

$$d\Lambda = \left(-\frac{4}{3}\alpha\Lambda - \frac{32}{7}\beta\Lambda^2 + \pi K \right) dt + \sqrt{\frac{8\pi K}{3}}\Lambda dB(t), \qquad (7.4.116)$$

The stationary probability density of $\Lambda(t)$ can be calculated from Eq. (4.4.7) as

$$p(\lambda) = C\lambda^{-\frac{1}{4}} \exp\left[-\frac{1}{\pi K}\left(\alpha\lambda + \frac{12}{7}\beta\lambda^2 \right) \right]. \qquad (7.4.117)$$

Using Eq. (7.4.103), we obtain the stationary probability density for $p(x,\dot{x})$ as

$$p(x,\dot{x}) = \frac{p(\lambda)}{T(\lambda)}$$

$$= C\exp\left\{ -\frac{1}{2\pi K}\left[\alpha\left(\frac{1}{2}\delta x^4 + \dot{x}^2 \right) + \frac{6\beta}{7}\left(\frac{1}{2}\delta x^4 + \dot{x}^2 \right)^2 \right] \right\}. \qquad (7.4.118)$$

It can be shown that the result of (7.4.118) is the same as that obtained from the method of dissipation energy balancing (see Exercise Problem 7.8).

Example 7.4.4 As another example, consider the system

$$\ddot{X} + 2\zeta\omega_0\dot{X} + \omega_0^2 X + \delta X^3 = X\xi_1(t) + \xi_2(t), \qquad (7.4.119)$$

where the damping force and the excitations are assumed to be of orders ε and $\varepsilon^{1/2}$, respectively. The equation of undamped free vibration corresponding to (7.4.119) is

$$\ddot{X} + \omega_0^2 X + \delta X^3 = 0, \qquad (7.4.120)$$

with the total energy

$$\Lambda = \frac{1}{2}\dot{X}^2 + \frac{1}{2}\omega_0^2 X^2 + \frac{1}{4}\delta X^4 = \frac{1}{2}\omega_0^2 A^2 + \frac{1}{4}\delta A^4 \qquad (7.4.121)$$

and a period

$$T = 4T_{\frac{1}{4}} = 4\int_0^A \frac{dX}{\sqrt{2\Lambda - (\omega_0^2 X^2 + \delta X^4/2)}} = \frac{4K(k)}{\sqrt{\omega_0^2 + \delta A^2}}, \qquad (7.4.122)$$

where $K(k)$ is the complete elliptic integral with a modulus $k = \sqrt{\delta A^2/(2\omega_0^2 + 2\gamma A^2)}$.

Comparing (7.4.119) and the general form (7.4.79), we know that $g_1 = X$, and $g_2 = 1$. First consider a special case that the two excitations are Gaussian white noises. Applying (7.4.97) and (7.4.98), we obtain

$$m(\Lambda) = -2\zeta\omega_0 \left\langle \dot{X}^2 \right\rangle_t + \pi K_{11} \left\langle X^2 \right\rangle_t + \pi K_{22}$$

$$\sigma^2(\Lambda) = 2\pi K_{11} \left\langle X^2\dot{X}^2 \right\rangle_t + 2\pi K_{22} \left\langle \dot{X}^2 \right\rangle_t, \qquad (7.4.123)$$

where use has been made of $\partial\dot{X}/\partial\Lambda = 1/\dot{X}$, obtained from (7.4.82). The time-averaging terms in (7.4.123) can be calculated numerically according to (7.4.96). The stationary probability density $p(\lambda)$, as well as the joint stationary probability density $p(x, \dot{x})$, can then be calculated from (7.4.99) and (7.4.103), respectively.

If the excitations are stationary broadband process, and the correlation times are shorter than $(\zeta\omega_0)^{-1}$, the energy process is approximately Markov diffusion process, and Eqs. (7.4.94) and

(7.4.95) have to be employed. When carrying the averaging, (7.4.86) is used, and the following approximations can be made

$$\theta(t+\tau) = \int_0^{t+\tau} \omega(s)ds = \int_0^t \omega(s)ds + \int_t^{t+\tau} \omega(s)ds \approx \theta_t + \omega_\Lambda \tau,$$

(7.4.124)

$$X(t+\tau) = X(t) + \int_t^{t+\tau} \dot{X}(s)ds$$

$$\approx X(t) + \frac{\sqrt{2\Lambda}}{\omega_\Lambda}\{\cos[\theta(t) + \omega_\Lambda\tau] - \cos\theta(t)\}$$

$$= X_t + \frac{\text{sgn}(X_t)\sqrt{2U_t}}{\omega_\Lambda}(\cos\omega_\Lambda\tau - 1) + \frac{\dot{X}_t}{\omega_\Lambda}\sin\omega_\Lambda\tau,$$

(7.4.125)

$$\dot{X}(t+\tau) = -\sqrt{2\Lambda}\sin\theta(t+\tau) \approx -\sqrt{2\Lambda}\sin[\theta(t) + \omega_\Lambda\tau]$$

$$= \dot{X}_t\cos\omega_\Lambda\tau - \text{sgn}(X_t)\sqrt{2U_t}\sin\omega_\Lambda\tau,$$

(7.4.126)

where the subscript t associated with a variable indicates that it is evaluated at time t. Using approximations in Eqs. (7.4.124) through (7.4.126), the following terms in (7.4.97) and (7.4.98) are calculated,

$$\left\langle [\dot{X}g_1(X,\dot{X})]_{t+\tau}\frac{\partial}{\partial\Lambda}[\dot{X}g_1(X,\dot{X})]_t\right\rangle_t$$

$$= \left\langle X^2 - \frac{X\,\text{sgn}\,X\sqrt{2U}}{\omega_\Lambda}\right\rangle_t\cos\omega_\Lambda\tau$$

$$+ \left\langle \frac{X\,\text{sgn}\,X\sqrt{2U}}{\omega_\Lambda}\right\rangle_t\cos 2\omega_\Lambda\tau,$$

(7.4.127)

$$\left\langle [\dot{X}g_2(X,\dot{X})]_{t+\tau}\frac{\partial}{\partial\Lambda}[\dot{X}g_2(X,\dot{X})]_t\right\rangle_t = \cos\omega_\Lambda\tau,$$

(7.4.128)

$$\left\langle [\dot{X} g_1(X, \dot{X})]_{t+\tau} \frac{\partial}{\partial \Lambda} [\dot{X} g_2(X, \dot{X})]_t \right\rangle_t$$

$$= \left\langle \dot{X} [g_2(X, \dot{X})]_{t+\tau} \frac{\partial}{\partial \Lambda} [\dot{X} g_1(X, \dot{X})]_t \right\rangle_t = 0, \qquad (7.4.129)$$

$$\left\langle [\dot{X} g_1(X, \dot{X})]_{t+\tau} [\dot{X} g_1(X, \dot{X})]_t \right\rangle_t$$

$$= \left\langle X^2 \dot{X}^2 - \frac{\dot{X}^2 X \operatorname{sgn} X \sqrt{2U}}{\omega_\Lambda} \right\rangle_t \cos \omega_\Lambda \tau$$

$$+ \left\langle \frac{\dot{X}^2 X \operatorname{sgn} X \sqrt{2U}}{\omega_\Lambda} \right\rangle_t \cos 2\omega_\Lambda \tau, \qquad (7.4.130)$$

$$\left\langle [\dot{X} g_2(X, \dot{X})]_{t+\tau} [\dot{X} g_2(X, \dot{X})]_t \right\rangle_t = \left\langle \dot{X}^2 \right\rangle_t \cos \omega_\Lambda \tau, \qquad (7.4.131)$$

$$\left\langle [\dot{X} g_1(X, \dot{X})]_{t+\tau} [\dot{X} g_2(X, \dot{X})]_t \right\rangle_t$$

$$= \left\langle [\dot{X} g_2(X, \dot{X})]_{t+\tau} [\dot{X} g_1(X, \dot{X})]_t \right\rangle_t = 0. \qquad (7.4.132)$$

Substituting (7.4.127) through (7.4.129) into (7.4.97) and substituting (7.4.130) through (7.4.132) into (7.4.98), we obtain

$$m(\Lambda) = -2\zeta\omega_0 u_2 + \pi(u_1 - u_3)\Phi_{11}(\omega_\Lambda) + \pi u_3\Phi_{11}(2\omega_\Lambda) + \pi\Phi_{22}(\omega_\Lambda)$$

$$\sigma^2(\Lambda) = 2\pi(u_4 - u_5)\Phi_{11}(\omega_\Lambda) + 2\pi u_5\Phi_{11}(2\omega_\Lambda) + 2\pi u_2\Phi_{22}(\omega_\Lambda),$$

$$(7.4.133)$$

where

$$u_1 = \left\langle X^2 \right\rangle_t = \frac{1}{T_{\frac{1}{4}}} \int_0^A \frac{X^2}{\sqrt{2\Lambda - 2U(X)}} dX, \qquad (7.4.134)$$

$$u_2 = \left\langle \dot{X}^2 \right\rangle_t = \frac{1}{T_{\frac{1}{4}}} \int_0^A \sqrt{2\Lambda - 2U(X)} dX, \qquad (7.4.135)$$

$$u_3 = \left\langle \frac{X \, \text{sgn}(X) \sqrt{2U(X)}}{\omega_\Lambda} \right\rangle_t = \frac{2}{\pi} \int_0^A \frac{X \sqrt{2U(X)}}{\sqrt{2\Lambda - 2U(X)}} dX,$$

(7.4.136)

$$u_4 = \left\langle X^2 \dot{X}^2 \right\rangle_t = \frac{1}{T_{\frac{1}{4}}} \int_0^A X^2 \sqrt{2\Lambda - 2U(X)} \, dX,$$

(7.4.137)

$$u_5 = \left\langle \frac{X \dot{X}^2 \text{sgn}(X) \sqrt{2U(X)}}{\omega_\Lambda} \right\rangle_t = \frac{4}{\pi} \int_0^A X \sqrt{U(X)[\Lambda - U(X)]} \, dX.$$

(7.4.138)

In deriving (7.4.133), use has been made of

$$2 \int_{-\infty}^0 \cos(\omega\tau) R_{ii}(\tau) d\tau = \int_{-\infty}^\infty \cos(\omega\tau) R_{ii}(\tau) d\tau = 2\pi \Phi_{ii}(\omega)$$

$$\int_{-\infty}^0 \sin(\omega\tau) R_{ii}(\tau) d\tau \approx 0, \quad \int_{-\infty}^\infty \sin(\omega\tau) R_{ii}(\tau) d\tau \approx 0.$$

(7.4.139)

The approximations in (7.4.139) are justified on the basis of short correlation times of excitations.

Two special cases are noted below. First consider the case in which both $\xi_1(t)$ and $\xi_2(t)$ are Gaussian white noises. By letting $\Phi_{11}(\omega_\Lambda) = \Phi_{11}(2\omega_\Lambda) = K_{11}$ and $\Phi_{22}(\omega_\Lambda) = K_{22}$, Eq. (7.4.133) reduces to Eq. (7.4.123). Another case is the system with a linear stiffness. It is seen from (7.4.134) through (7.4.139) that $u_1 = u_3$ and $u_4 = u_5$, leading to

$$m(\Lambda) = -2\zeta\omega_0 u_2 + \pi u_3 \Phi_{11}(2\omega_\Lambda) + \pi \Phi_{22}(\omega_\Lambda),$$

$$\sigma^2(\Lambda) = 2\pi u_5 \Phi_{11}(2\omega_\Lambda) + 2\pi u_2 \Phi_{22}(\omega_\Lambda).$$

(7.4.140)

It can be proven that, in this case, the Ito equation (7.4.93) of the energy $\Lambda(t)$ with drift and diffusion coefficients given by (7.4.140) is equivalent to Ito equation (7.4.53) of the amplitude $A(t)$.

In this section, applications of the stochastic averaging method are mainly for SDOF nonlinear stochastic systems. It can also be applied to MDOF nonlinear stochastic systems; however, procedures are more complicated. Section 7.5 will introduce the stochastic averaging method of quasi Hamiltonian systems, which is applicable to MDOF strongly nonlinear stochastic systems.

The stochastic averaging method can be utilized not only for mechanical and structural systems, but also in other areas. In the next section, a type of nonlinear stochastic ecosystems will be investigated using the stochastic averaging method.

7.4.3 *Application to Nonlinear Stochastic Ecological System*

Evolution of species, either ecological or biological, is of great importance in various areas. Qualitatively, populations of species have been investigated for different situations. One typical case is prey-predator type, in which the prey species (or host) and predator species (or parasite) are interacted with each other, and are influenced by environments. Since uncertainties are always present in the environment changes, the systems are essentially stochastic. It is found that the stochastic averaging method is an effective tool to investigate the dynamics of prey-predator systems.

Deterministic Models

The earliest mathematical model describing the dynamics of predator-prey type ecosystems is the Lotka–Volterra model (Lotka, 1925; Volterra, 1926), governed by the following differential equations:

$$\begin{aligned}
\dot{x}_1 &= x_1(a - bx_2), \\
\dot{x}_2 &= x_2(-c + fx_1),
\end{aligned} \tag{7.4.141}$$

where x_1 and x_2 are the population densities of preys and predators, respectively, and a, b, c and f are positive constants. The term ax_1 on the right-hand side in the first equation indicates that the prey population would grow exponentially without the predators, while the term $-cx_2$ in the second equation indicates that the population of the predators would decrease exponentially without the preys.

The interactive terms $x_1 x_2$ in the two equations provide a balance between the two populations.

System (7.4.141) has an unstable equilibrium state $(0,0)$, and a stable, but non-asymptotic stable equilibrium state

$$x_{10} = \frac{c}{f}, \quad x_{20} = \frac{a}{b}. \tag{7.4.142}$$

Moreover, the system has a first integral

$$r(x_1, x_2) = f x_1 - c - c \ln \frac{f x_1}{c} + b x_2 - a - a \ln \frac{b x_2}{a}. \tag{7.4.143}$$

It can be shown that $r(x_{10}, x_{20}) = 0$, and $r(x_1, x_2) \geq 0$ for any positive x_1 and x_2. For any positive constant R, $r(x_1, x_2) = R$ represents a periodic trajectory. Its period can be determined from

$$T(R) = \oint dt = \oint \frac{dx_2}{x_2(f x_1 - c)} = \oint \frac{dx_1}{x_1(a - b x_2)}, \tag{7.4.144}$$

where x_1 and x_2 are related by $r(x_1, x_2) = R$.

Figure 7.4.1 depicts the equilibrium point O corresponding to $R = 0$, and three periodic trajectories corresponding to three R values, determined with system parameters $a = 0.9$, $b = 1$, $c = 0.5$ and $f = 0.5$. It shows that the prey and predator populations are changing periodically with time, along a path in the phase plane (the x_1, x_2 plane), which depends only on the initial states of x_1 and x_2. It also shows that an initial high level of prey density and/or predator density can lead to very low levels for both, even in an invariant environment.

System (7.4.141) shows that the prey population will grow without limit in the absence of the predators, contrary to what is expected of a realistic predator-prey ecosystem. To improve the classical Lotka–Volterra model, a self-competition term $-s x_1^2$ may be added to the prey equation in (7.4.141) (Volterra, 1931; May and Verga, 1973); namely,

$$\begin{aligned}
\dot{x}_1 &= x_1(a_1 - s x_1 - b x_2), \\
\dot{x}_2 &= x_2(-c + f x_1).
\end{aligned} \tag{7.4.145}$$

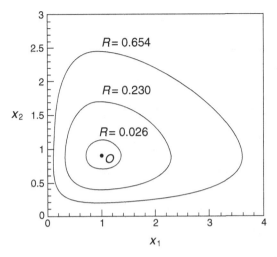

Figure 7.4.1 Equilibrium point and periodic solutions of system (7.4.141) (from Cai and Lin, 2004).

System (7.4.145) has an asymptotic stable equilibrium at $x_1 = c/f$ and $x_2 = (a_1 - sc/f)/b$. Note that systems (7.4.141) and (7.4.145) have the same equilibrium state $(c/f, a/b)$ if

$$a = a_1 - \frac{sc}{f}. \tag{7.4.146}$$

Therefore, equation set (7.4.145) may be rewritten as

$$\dot{x}_1 = x_1 \left[a - bx_2 - \frac{s}{f}(-c + fx_1) \right],$$
$$\dot{x}_2 = x_2(-c + fx_1). \tag{7.4.147}$$

In the absence of the predators, the prey density reaches its equilibrium state of a_1/s, which is inversely proportional to s, as expected. When the predators are present, however, the interaction between the prey and the predator populations is the more important factor, whereas the value of s affects only the density of the predators at the equilibrium state.

Figure 7.4.2 depicts two trajectories of system (7.4.145), corresponding to two different values of $s = 0.1$ and 0.02, respectively, and with the same $a_1 = 1$, $b = 1$, $c = 0.5$, $f = 0.5$. The motion of

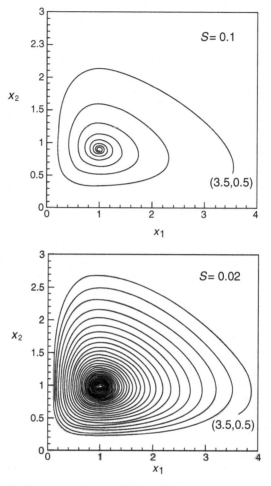

Figure 7.4.2 Trajectories of system (7.4.145) for two different values of s. (from Cai and Lin, 2004).

the system begins from point (3.5, 0.5). The system moves around the stable equilibrium state with a decreasing amplitude, and finally reaches the equilibrium state. The term $-sx_1^2$ models the effect of inter-species competition. With the larger $s = 0.1$, the system reaches its equilibrium faster, while with the smaller $s = 0.02$, the system moves slowly around the stable equilibrium. By comparing Figs. 7.4.1 and 7.4.2, it can be seen that the prey self-competition plays a dissipation role to the original system (7.4.141), and it changes a periodic trajectory to the final state of a single stable point.

Stochastic Model

The ecosystem model described by equation set (7.4.147) fails to describe a basic phenomenon of a natural ecosystem, namely, the changing environment may cause random variations in the prey growth rate and the predator death rate. Analogous to the deterministic model (7.4.147), a stochastic model is proposed as follows

$$\dot{X}_1 = X_1 \left[a - bX_2 - \frac{s}{f}(-c + fX_1) + W_1(t) \right],$$
$$\dot{X}_2 = X_2[-c + fX_1 + W_2(t)], \qquad (7.4.148)$$

where $X_1(t)$ and $X_2(t)$ are two stochastic processes, representing the prey and the predator population densities, respectively, and where $W_1(t)$ and $W_2(t)$ are two independent Gaussian white noises. The noises $W_1(t)$ and $W_2(t)$ are introduced to model the random variations in the prey birth rate and the predator death rate, respectively.

Equation set (7.4.148) can be re-written in the form of Ito stochastic differential equations as follows

$$dX_1 = X_1 \left[a - bX_2 - \frac{s}{f}(-c + fX_1) + \pi K_1 \right] dt + \sqrt{2\pi K_1} X_1 dB_1(t),$$
$$dX_2 = X_2[-c + fX_1 + \pi K_2]dt + \sqrt{2\pi K_2} X_2 dB_2(t),$$
$$(7.4.149)$$

where $B_1(t)$ and $B_2(t)$ are two independent unit Wiener processes, and K_1 and K_2 are spectral densities of $W_1(t)$ and $W_2(t)$, respectively. Compared with (7.4.148), the additional terms $\pi K_1 X_1 dt$ and $\pi K_2 X_2 dt$ are the Wong–Zakai correction terms.

Consider the following stochastic process

$$R(X_1, X_2) = fX_1 - c - c\ln\frac{fX_1}{c} + bX_2 - a - a\ln\frac{bX_2}{a}, \qquad (7.4.150)$$

which is the random counterpart of the first integral (7.4.143). The Ito equation for $R(X_1, X_2)$ can be obtained as follows, upon applying the Ito differential rule,

$$dR = \left[-\frac{s}{f}(fX_1 - c)^2 + \pi f K_1 X_1 + \pi b K_2 X_2 \right] dt$$

$$+ \sqrt{2\pi K_1}(fX_1 - c)dB_1(t) + \sqrt{2\pi K_2}(bX_2 - a)dB_2(t).$$

$$(7.4.151)$$

Assume that the coefficient s of the self-competition term is small, indicating that the term has a small influence when the prey density is small. Assume also that K_1 and K_2 are small, namely, the random disturbances are small. Under these assumptions, the right-hand side of (7.4.151) is small, and $R(t)$ is a slowly varying stochastic process. In this case, the stochastic averaging method can be applied to obtain an averaged Ito stochastic differential equation for R

$$dR = m(R)dt + \sigma(R)dB(t), \qquad (7.4.152)$$

where the drift coefficient m and the diffusion coefficient σ are obtained as follows

$$m(R) = \pi f K_1 \langle X_1 \rangle_t + \pi b K_2 \langle X_2 \rangle_t - \frac{s}{f} \langle (fX_1 - c)^2 \rangle_t,$$

$$(7.4.153)$$

$$\sigma^2(R) = 2\pi K_1 \langle (fX_1 - c)^2 \rangle_t + 2\pi K_2 \langle (bX_2 - a_1)^2 \rangle_t,$$

$$(7.4.154)$$

and where $\langle [\cdot] \rangle_t$ denotes the time average in one quasi-period, defined as

$$\langle [\cdot] \rangle_t = \frac{1}{T} \oint [\cdot] \, dt = \frac{1}{T} \oint \frac{[\cdot] \, dX_2}{X_2(fX_1 - c)} = \frac{1}{T} \oint \frac{[\cdot] \, dX_1}{X_1(a - bX_2)}.$$

$$(7.4.155)$$

The quasi-period T in (7.4.155) is given in (7.4.144), with x_1 and x_2 replaced by their random counterparts X_1 and X_2, respectively. The result obtained from each time average is a function of R. The following time averages can be obtained directly from Eqs. (7.4.141) and (7.4.142)

$$\langle X_1 \rangle_t = \frac{c}{f}, \quad \langle X_2 \rangle_t = \frac{a}{b}, \quad \langle X_1 X_2 \rangle_t = \frac{ac}{bf}, \tag{7.4.156}$$

$$a \left\langle (fX_1 - c)^2 \right\rangle_t = c \left\langle (bX_2 - a)^2 \right\rangle_t. \tag{7.4.157}$$

Defining

$$g(R) = a \oint \frac{(fX_1 - c)\mathrm{d}X_2}{X_2}, \tag{7.4.158}$$

we have

$$\left\langle (fX_1 - c)^2 \right\rangle_t = \frac{g(R)}{aT(R)}, \quad \left\langle (bX_2 - a)^2 \right\rangle_t = \frac{g(R)}{cT(R)}. \tag{7.4.159}$$

It follows from Eqs. (7.4.156) through (7.4.159) that

$$m(R) = \pi c K_1 + \pi a K_2 - \frac{s}{af} \frac{g(R)}{T(R)}, \tag{7.4.160}$$

$$\sigma^2(R) = \frac{2\pi}{ac}(cK_1 + aK_2)\frac{g(R)}{T(R)}. \tag{7.4.161}$$

Equations (7.4.152), (7.4.160) and (7.4.161) constitute the governing law for the one-dimensional Markov process $R(t)$.

The stochastic process $R(t)$, as defined in (7.4.151), can be considered as a representation of the system state. It is a function of two stochastic processes; namely, the prey density X_1 and the predator density X_2.

Asymptotic Behaviors

Governed by the Ito stochastic differential equation (7.4.152), the behaviors of the one dimensional Markov diffusion process $R(t)$ at the two boundaries at $R = 0$ and $R = \infty$ can be investigated based on a theory described in Section 4.4.

As can be deduced from Eq. (7.4.148), R approaches zero, when X_1 and X_2 approach c/f and a/b, respectively. Thus, we may write

$$\ln \frac{fX_1}{c} = \ln \left(1 + \frac{fX_1 - c}{c}\right) \approx 1 + \frac{fX_1 - c}{c} - \frac{1}{2}\left(\frac{fX_1 - c}{c}\right)^2,$$
(7.4.162)

$$\ln \frac{bX_2}{a} = \ln \left(1 + \frac{bX_2 - a}{a}\right) \approx 1 + \frac{bX_2 - a}{a} - \frac{1}{2}\left(\frac{bX_2 - a}{a}\right)^2.$$
(7.4.163)

Substituting (7.4.162) and (7.4.163) into (7.4.150),

$$a(fX_1 - c)^2 + c(bX_2 - a)^2 \approx 2acR.$$
(7.4.164)

Using Eqs. (7.4.157) and (7.4.164), we have

$$\langle (fX_1 - c)^2 \rangle_t \approx cR, \quad \langle (bx_2 - a)^2 \rangle_t \approx aR.$$
(7.4.165)

It follows from (7.4.160) and (7.4.161) that

$$m(R) \to \pi(cK_1 + aK_2), \quad \sigma^2(R) \to 2\pi(cK_1 + aK_2)R, \quad \text{as } R \to 0.$$
(7.4.166)

Since $\sigma(0) = 0$, the left boundary at $R = 0$ is singular of the first kind. The diffusion exponent, the drift exponent, and the character value are respectively,

$$\alpha = 1, \quad \beta = 0, \quad c = 1.$$
(7.4.167)

According to Table 4.4.2, the left boundary is an entrance, as long as $a > 0$ and $c > 0$. As the probability flow approaches this boundary, the repulsive force becomes larger, and it forces the system back to its defining interval. This indicates that, with random variations in the birth rate of the preys and/or the death rate of the predators, the system will not end up at a static single state, as in the deterministic model; instead, the system will continue to evolve dynamically.

As $R \to \infty$, it was shown in (Khasminskii and Klebaner, 2001) that

$$\langle (fX_1 - c)^2 \rangle_t \approx \frac{1}{2}cR, \quad \langle (bX_2 - a)^2 \rangle_t \approx \frac{1}{2}aR.$$
(7.4.168)

According to (7.4.160) and (7.4.161),

$$m(R) \to -\frac{sc}{2f}R, \quad \sigma^2(R) \to \pi(cK_1 + aK_2)R, \quad \text{as } R \to \infty,$$

$$(7.4.169)$$

indicating that the right boundary at $R = \infty$ is singular of the second kind. The diffusion exponent, the drift exponent and the character value are

$$\alpha = 1, \quad \beta = 1, \quad c = \frac{sc}{2\pi f(cK_1 + aK_2)}. \qquad (7.4.170)$$

Thus, under the condition of $s > 0$, the right boundary at $R = \infty$ is repulsively natural according to Table 4.4.4. The self-competition mechanism results in a negative drift term for large R, which guarantees that neither the prey population nor the predator population can grow without restraint.

It is of interest to note that, without the $-sX_1^2$ term in (7.4.148), the right boundary $R = \infty$ would be attractively natural, implying that the prey population could grow without limit, an outcome contrary to what is expected of nature. Therefore, the inclusion of the self-competition term $-sX_1^2$ in the model is necessary.

Since the left boundary at $R = 0$ is an entrance, and the right boundary at $R = \infty$ is repulsively natural, the stationary probability density of $R(t)$ exists.

Stationary Probability Density

The reduced FPK equation is given by

$$\frac{d}{dr}[m(r)p(r)] - \frac{1}{2}\frac{d^2}{dr^2}[\sigma^2(r)p(r)] = 0, \qquad (7.4.171)$$

which is solved to obtain

$$p(r) = \frac{C_1}{\sigma^2(r)} \exp \int \frac{2m(r)}{\sigma^2(r)} dr$$

$$= C\frac{T(r)}{g(r)} \exp \int \frac{\pi ac(cK_1 + aK_2)T(r) - \frac{sc}{f}g(r)}{\pi(cK_1 + aK_2)g(r)} dr,$$

$$(7.4.172)$$

where C and C_1 are two normalization constants. By noticing that

$$\frac{dg(r)}{dr} = a \oint \frac{f}{x_2} \frac{\partial x_1}{\partial r} dx_2 = a \oint \frac{f x_1 dx_2}{x_2(f x_1 - c)}$$

$$= a f T(r) \langle x_1 \rangle_t = a c T(r). \tag{7.4.173}$$

Equation (7.4.172) is simplified to

$$p(r) = CT(r) \exp(-\beta r), \tag{7.4.174}$$

where β is a constant given by

$$\beta = \frac{sc}{\pi f(cK_1 + aK_2)}. \tag{7.4.175}$$

The joint probability density of $R(t)$ and $X_1(t)$ can be written as

$$p(r, x_1) = p(r)p(x_1|r), \tag{7.4.176}$$

where $p(x_1|r)$ is the conditional probability density of $X_1(t)$ given $R(t) = r$. It can be obtained as follows

$$p(x_1|r)dx_1 = \frac{dt}{T(r)} = \frac{dx_1}{|\dot{x}_1| T(r)} = \frac{dx_1}{|x_1(a - bx_2)| T(r)}. \tag{7.4.177}$$

Substituting (7.4.177) into (7.4.176),

$$p(r, x_1) = \frac{p(r)}{|x_1(a - bx_2)|T(r)}, \tag{7.4.178}$$

in which x_2 is treated as a function of x_1 and r. Thus, the joint probability density $p(x_1, x_2)$ follows as

$$p(x_1, x_2) = p(r, x_1) \left| \frac{\partial(r, x_1)}{\partial(x_1, x_2)} \right| = \frac{p(r)}{x_1 x_2 T(r)}$$

$$= \frac{C}{x_1 x_2} \exp\left[-\beta r(x_1, x_2)\right], \tag{7.4.179}$$

where $\frac{\partial(r, x_1)}{\partial(x_1, x_2)}$ is the Jacobian of transformation. Upon substituting (7.4.143) into (7.4.179), we obtain

$$p(x_1, x_2) = p(x_1)p(x_2), \tag{7.4.180}$$

where

$$p(x_1) = \frac{(\beta f)^{\beta c}}{\Gamma(\beta c)} x_1^{\beta c - 1} \exp(-\beta f x_1), \qquad (7.4.181)$$

$$p(x_2) = \frac{(\beta b)^{\beta a}}{\Gamma(\beta a)} x_2^{\beta a - 1} \exp(-\beta b x_2), \qquad (7.4.182)$$

and $\Gamma(\cdot)$ is the gamma function. Equation (7.4.180) implies that $X_1(t)$ and $X_2(t)$ are independent when they reach the stationary state, which is an unexpected result for system (7.4.148) with nonlinear coupling between X_1 and X_2. Equations. (7.4.181) and (7.4.182) show that non-trivial $p(x_1)$ and $p(x_2)$ exist only if $\beta > 0$ and $a > 0$. These conditions lead to

$$0 < s < \frac{f a_1}{c}. \qquad (7.4.183)$$

The system will diverge without the self-competition term, namely, if $s = 0$, since no restriction will be imposed on the growth of the prey population. On the other hand, if $s > \frac{f a_1}{c}$, the growth of the prey population will be over restricted, leading to extinction of the predators.

The existence of the probability densities of the preys and predators under condition (7.4.183) indicates that (a) the ecosystem is dynamic, it will not approach a single state of fixed predator and prey populations; (b) neither the prey nor the predator will be extinct, except some unexpected events occur, which is not considered in the model; and (c) it is not possible to predict the exact prey and predator populations, only their probability densities can be estimated.

Numerical calculations were carried out in (Cai and Lin, 2004). Shown in Fig. 7.4.3 is the probability density $p(x_1)$ of the prey population X_1 for the stochastic system (7.4.148), calculated from (7.4.181), with $a_1 = 1$, $b = 1$, $c = 0.5$, $f = 0.5$, $\pi K_1 = \pi K_2 = 0.01$ and for two different s values of 0.1 and 0.02. Also depicted in Fig. 7.4.3 are results obtained from the Monte Carlo simulation. The theoretical and simulation results agree very well in both cases. With a large self-competition coefficient $s = 0.1$, the prey density is nearly

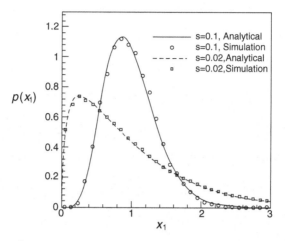

Figure 7.4.3 Stationary probability densities of the prey population $X_1(t)$ (from Cai and Lin, 2004).

centered around $x_1 = c/f = 1$, which is the equilibrium point for the deterministic counterpart without random variations in the prey birth rate and the predator death rate. With a small $s = 0.02$, the peak of the prey probability density is shifted to a value smaller than the equilibrium point $x_1 = 1$, and the probability for a high prey population becomes higher, indicating that the system is less stable.

More investigations on the topic of stochastic prey-predator problems have been conducted (Dimentberg, 2002, 2003; Cai and Lin, 2004, 2007; Wu and Zhu, 2008; Qi and Cai, 2013).

7.5 Stochastically Excited and Dissipated Hamiltonian Systems

The stochastically excited and dissipated Hamiltonian systems, governed by (6.4.4), have exact stationary solutions if certain conditions are satisfied, as described in Section 6.4. Such conditions are very restrictive, and not met in general for practical systems. Therefore, procedures are required to obtain approximate solutions. As described in Section 6.4, for stochastically excited and dissipated Hamiltonian systems, exact stationary solutions, if exist, are obtained differently for the five different types: completely non-integrable, completely integrable and non-resonant, completely integrable and resonant, partially integrable and non-resonant, and partially integrable and resonant. Similarly, procedures to obtain approximate solutions are also different for the five different types. For clear illustration, only the completely non-integrable case is considered here, and the method of equivalent nonlinear systems and the stochastic averaging method of quasi Hamiltonian systems will be introduced to obtain approximate solutions for MDOF nonlinear stochastic dynamical systems.

7.5.1 *Method of Equivalent Nonlinear Systems*

Consider a Gaussian-white-noise excited and dissipated Hamiltonian system of the completely non-integrable type. Assume that an exact stationary solution does not exist for such a system. The method of equivalent nonlinear systems for such system is developed to obtain the approximate stationary solution. To illustrate the method, consider the following system

$$
\dot{Q}_j = \frac{\partial H}{\partial P_j},
$$
$$
\dot{P}_j = -\frac{\partial H}{\partial Q_j} - \sum_{k=1}^{n} c_{jk}(\mathbf{Q}, \mathbf{P})\frac{\partial H}{\partial P_k} + \sum_{l=1}^{m} g_{jl}(\mathbf{Q})W_l(t).
$$

(7.5.1)

Equation (7.5.1) indicates that the parametric excitation terms are associated only with the general displacements $\mathbf{Q}$. Assume the equivalent nonlinear system with an exact stationary solution has the equations of motion of the form

$$
\dot{Q}_j = \frac{\partial H}{\partial P_j},
$$
$$
\dot{P}_j = -\frac{\partial H}{\partial Q_j} - \sum_{k=1}^{n} c'_{jk}(\mathbf{Q}, \mathbf{P})\frac{\partial H}{\partial P_k} + \sum_{l=1}^{m} g_{jl}(\mathbf{Q})W_l(t).
$$

(7.5.2)

Comparison of the original system (7.5.1) with the replacement system (7.5.2) shows that only the damping terms are different. As shown in Section 6.4.3, an exact stationary solution exists for system

(7.5.2) if the following conditions are satisfied

$$c'_{jk}(\mathbf{Q}, \mathbf{P}) = \lambda(H) B_{jk}(\mathbf{Q}), \quad B_{jk}(\mathbf{Q}) = \pi \sum_{l,s=1}^{m} K_{ls} g_{jl}(\mathbf{Q}) g_{ks}(\mathbf{Q}).$$

$$(7.5.3)$$

In this case, the stationary probability density is of the form

$$p(\mathbf{q}, \mathbf{p}) = C \exp[-\phi(H)] = C \exp\left[-\int_0^{H(\mathbf{q},\mathbf{p})} \lambda(u) du\right], \quad (7.5.4)$$

where $\phi(H) = d\lambda(H)/dH$, known as the probability potential function. The residual errors, i.e., the differences between systems (7.5.1) and (7.5.2), are

$$\delta_j = \sum_{k=1}^{n} [c_{jk}(\mathbf{Q}, \mathbf{P}) - \lambda(H) B_{jk}(\mathbf{Q})] \frac{\partial H}{\partial P_k}, \quad j = 1, 2, \ldots, n. \quad (7.5.5)$$

It is clear that each δ_j is the difference of damping forces in the jth degree between the replacing system (7.5.2) and the original system (7.5.1). For accurate result, all $\delta_j(\mathbf{Q}, \mathbf{P})$ need to be minimized in some statistical sense. Several criteria were proposed to minimize the difference $\delta_j(\mathbf{Q}, \mathbf{P})$ (Zhu, Soong and Lei, 1994). One of them is to let the average energy dissipation rate remain the same for the replacing and replaced systems, leading to

$$E\left(\sum_{j=1}^{n} \delta_j \frac{\partial H}{\partial P_j}\right)$$

$$= E\left\{\sum_{j,k=1}^{n} [c_{jk}(\mathbf{Q}, \mathbf{P}) - \lambda(H) B_{jk}(\mathbf{Q})] \frac{\partial H}{\partial P_k} \frac{\partial H}{\partial P_j}\right\} = 0. \quad (7.5.6)$$

Use the approximate stationary probability density (7.5.4) to rewrite (7.5.6) as

$$\int e^{-\phi(H)} \sum_{j,k=1}^{n} [c_{jk}(\mathbf{q}, \mathbf{p}) - \lambda(H) B_{jk}(\mathbf{q})] \frac{\partial H}{\partial p_k} \frac{\partial H}{\partial p_j} d\mathbf{q} d\mathbf{p} = 0. \quad (7.5.7)$$

The left-hand side of (7.5.7) is a $2n$-fold integration with $\mathbf{q} = (q_1, q_2, \ldots, q_n)$ and $\mathbf{p} = (p_1, p_2, \ldots, p_n)$ as the integration variable. Now change the integration variable p_1 to H, and transform (7.5.7) to

$$\int_0^\infty e^{-\phi(H)} dH \int_\Omega \sum_{j,k=1}^n [c_{jk}(\mathbf{q}, \mathbf{p}) - \lambda(H) B_{jk}(\mathbf{q})] \frac{\partial H}{\partial p_k} \frac{\partial H}{\partial p_j} \left(\frac{\partial H}{\partial p_1}\right)^{-1}$$

$$\times d\mathbf{q} dp_2 \cdots dp_n = 0. \tag{7.5.8}$$

In (7.5.8), p_1 is replaced by $\mathbf{q}, p_2, \ldots, p_n$ and H according to equation $H = H(\mathbf{q}, \mathbf{p})$, and the integration domain Ω is determined by inequality $H(\mathbf{q}, p_1 = 0, p_2, \ldots, p_n) \leq H$. Using the same idea as in Section 7.3.2, we require a more restrictive criterion that the energy dissipation rate remains the same for every H leading to

$$\int_\Omega \sum_{j,k=1}^n [c_{jk}(\mathbf{q}, \mathbf{p}) - \lambda(H) B_{jk}(\mathbf{q})] \frac{\partial H}{\partial p_k} \frac{\partial H}{\partial p_j} \left(\frac{\partial H}{\partial p_1}\right)^{-1} d\mathbf{q} dp_2 \cdots dp_n = 0$$

$$\tag{7.5.9}$$

and an expression for $\lambda(H)$ follows as

$$\lambda(H) = \frac{\int_\Omega \sum_{j,k=1}^n c_{jk}(\mathbf{q}, \mathbf{p}) \frac{\partial H}{\partial p_k} \frac{\partial H}{\partial p_j} \left(\frac{\partial H}{\partial p_1}\right)^{-1} d\mathbf{q} dp_2 \cdots dp_n}{\int_\Omega \sum_{j,k=1}^n B_{jk}(\mathbf{q}) \frac{\partial H}{\partial p_k} \frac{\partial H}{\partial p_j} \left(\frac{\partial H}{\partial p_1}\right)^{-1} d\mathbf{q} dp_2 \cdots dp_n},$$

$$\tag{7.5.10}$$

Equation (7.5.10) can be used to find function $\lambda(H)$, and hence $\phi(H)$ and $p(\mathbf{q}, \mathbf{p})$ according to (7.5.4), numerically, if not analytically. The above procedure is illustrated in the following example.

Example 7.5.1 Consider the two-DOF nonlinear stochastic system governed by

$$\ddot{X} - \alpha_1 \dot{X} + \beta_1 X^2 \dot{X} + \omega_1^2 X + aY + b(X - Y)^3 = c_1 X W_1(t),$$
$$\ddot{Y} - (\alpha_1 - \alpha_2)\dot{Y} + \beta_2 Y^2 \dot{Y} + \omega_2^2 Y + aX + b(Y - X)^3 = c_2 Y W_2(t),$$

$$\tag{7.5.11}$$

where $W_1(t)$ and $W_2(t)$ are two independent Gaussian white noises with spectral density K_1 and K_2, respectively. Treat $\mathbf{Q} = (X, Y)$ and

$\mathbf{P} = (\dot{X}, \dot{Y})$, and find the Hamiltonian of the system as

$$H = \frac{1}{2}\dot{X}^2 + \frac{1}{2}\dot{Y}^2 + U(X, Y), \tag{7.5.12}$$

where $U(X, Y)$ is the potential energy given by

$$U(X, Y) = \frac{1}{2}\omega_1^2 X^2 + \frac{1}{2}\omega_2^2 Y^2 + aXY + \frac{1}{4}b(X - Y)^4. \tag{7.5.13}$$

According to (7.5.3), the equivalent nonlinear system is of the form

$$\ddot{X} + b_{11}\lambda(H)\dot{X} + \frac{\partial U}{\partial X} = c_1 X W_1(t),$$

$$\ddot{Y} + b_{22}\lambda(H)\dot{Y} + \frac{\partial U}{\partial Y} = c_2 Y W_2(t), \tag{7.5.14}$$

where

$$b_{11} = 2\pi K_1 c_1^2 X^2, \, b_{22} = 2\pi K_2 c_2^2 Y^2. \tag{7.5.15}$$

System (7.5.14) possesses an exact stationary probability density of the form of (7.5.3) where function $\lambda(H)$ is to be determined from (7.5.10), leading to

$$\lambda(H) = \frac{2\int_\Omega \left[-\alpha_1 \dot{x} + \beta_1 x^2 \dot{x} - (\alpha_1 - \alpha_2)\frac{\dot{y}^2}{\dot{x}} + \frac{\beta_2 y^2 \dot{y}^2}{\dot{x}} \right]_{\dot{x}=\pm\sqrt{2H-2U(x,y)-\dot{y}^2}} dxdyd\dot{y}}{\int_\Omega \left(b_{11}\dot{x} + b_{22}\frac{\dot{y}^2}{\dot{x}} \right)_{\dot{x}=\pm\sqrt{2H-2U(x,y)-\dot{y}^2}} dxdyd\dot{y}}. \tag{7.5.16}$$

For each H, $\lambda(H)$ can be calculate numerically from (7.5.16), and hence $p(x, y, \dot{x}, \dot{y})$ are obtained from (7.5.4). By letting $x = r\cos\theta$ and $y = r\sin\theta$, the triple integrations in (7.5.16) can be simplified to single integrations with respect to θ, and $\lambda(H)$ can be calculated.

Some numerical results were given in (Zhu, Soong, and Lei, 1994), and shown in Fig. 7.5.1. The system parameters were chosen as: $\beta_1 = 0.01$, $\beta_2 = 0.02$, $\omega_1 = 1$, $\omega_2 = 2$, $a = 0.01$, $b = 1$ and $c_1 = c_2 = 0.2$. The spectral densities of the two excitations were assumed to be equal, denoted by $K_1 = K_2 = K$, and varying in the calculations. In the figures, the solid lines represent the results obtained by using the method of dissipation energy balancing proposed above, and the

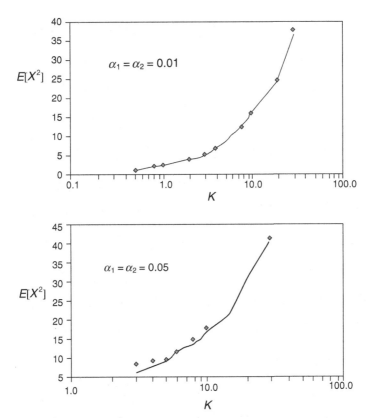

Figure 7.5.1 Stationary mean square values of displacement $X(t)$ of system (7.5.11) (from Zhu, Soong, and Lei, 1994).

symbols are the simulation results. The numerical results show that the proposed method yields quite accurate results.

Besides the criterion of equal average energy dissipation rate, other criteria can also be applied (Zhu, Soong, and Lei, 1994).

7.5.2 *Stochastic Averaging Method of Quasi-Hamiltonian Systems*

Consider a stochastically excited and dissipated Hamiltonian system governed by

$$\dot{Q}_j = \frac{\partial H'}{\partial P_j},$$

$$\dot{P}_j = -\frac{\partial H'}{\partial Q_j} - \varepsilon \sum_{k=1}^{n} c_{jk}(\mathbf{Q}, \mathbf{P})\frac{\partial H'}{\partial P_k} + \varepsilon^{\frac{1}{2}} \sum_{l=1}^{m} g_{jl}(\mathbf{Q}, \mathbf{P})W_l(t),$$

$$(7.5.17)$$

where ε is a small parameter, and $W_l(t)$ are Gaussian white noises with correlation functions

$$E[W_l(t)W_s(t+\tau)] = 2\pi K_{ls}\delta(\tau), \quad l, s = 1, 2, \ldots, m. \qquad (7.5.18)$$

As shown in (7.5.17), both the damping forces and excitations are weak. We name such a system as quasi-Hamiltonian system. Also only the type of completely quasi non-integrable Hamiltonian systems is considered.

Motion equations (7.5.17) can be modeled as Stratonovich stochastic differential equations and then converted into Ito stochastic differential equations by adding Wong–Zakai correction terms as follows

$$dQ_j = \frac{\partial H'}{\partial P_j}dt,$$

$$dP_j = \left[-\frac{\partial H'}{\partial Q_j} - \varepsilon \sum_{k=1}^{n} c_{jk}\frac{\partial H'}{\partial P_k} + \varepsilon\pi \sum_{k=1}^{n}\sum_{l,s=1}^{m} K_{js}g_{ks}\frac{\partial g_{jl}}{\partial P_k} \right] dt$$

$$+ \varepsilon^{\frac{1}{2}} \sum_{l=1}^{m} g_{jl}dB_l(t), \qquad (7.5.19)$$

where $B_l(t)$ are Wiener processes with

$$E[dB_l(t)dB_s(t+\tau)] = 2\pi K_{ls}\delta(\tau)dt, \quad l, s = 1, 2, \ldots, m.$$

$$(7.5.20)$$

The triple summation terms on the right-hand side of (7.5.19) are the Wong–Zakai correction terms. As shown in Section 7.3.2, these terms may affect the system stiffness and damping. Taking the effects on the system restoring and damping forces, the original Hamiltonian H' may be modified as H, and the damping coefficients are changed from

c_{jk} to m_{jk}. Then, the system equations of motion are modified as

$$dQ_j = \frac{\partial H}{\partial P_j} dt,$$

$$dP_j = - \left(\frac{\partial H}{\partial Q_j} - \varepsilon \sum_{k=1}^{n} m_{jk} \frac{\partial H}{\partial P_k} \right) dt$$

$$+ \varepsilon^{\frac{1}{2}} \sum_{l=1}^{m} g_{jl} dB_l(t). \qquad (7.5.21)$$

Since the system Hamiltonian H is a function of $\mathbf{Q}$ and $\mathbf{P}$, an Ito differential equation for H can be derived from (7.5.21) by using the Ito differential rule,

$$dH = \sum_{j=1}^{n} \frac{\partial H}{\partial Q_j} dQ_j + \sum_{j=1}^{n} \frac{\partial H}{\partial P_j} dP_j + \frac{1}{2} \sum_{j,k=1}^{n} \frac{\partial^2 H}{\partial P_j \partial P_k} (dP_j)(dP_k)$$

$$= \varepsilon \left(- \sum_{j,k=1}^{n} m_{jk} \frac{\partial H}{\partial P_j} \frac{\partial H}{\partial P_k} + \sum_{j,k=1}^{n} \sum_{j,s=1}^{m} \pi K_{ls} g_{jl} g_{ks} \frac{\partial^2 H}{\partial P_j \partial P_k} \right) dt$$

$$+ \varepsilon^{\frac{1}{2}} \sum_{j=1}^{n} \sum_{l=1}^{m} g_{jl} \frac{\partial H}{\partial P_j} dB_l(t). \qquad (7.5.22)$$

The right-hand side of (7.5.22) is small; therefore, the Hamiltonian H is a slowly varying process. Now we replace the state variable P_1 by H, and reformulate the system equations of motion as Eq. (7.5.22) and all equations in (7.5.21) except the one for P_1. Since H is slowly varying and $Q_1, Q_2, \ldots, Q_n, P_2, \ldots, P_n$ are rapidly varying processes. The stochastic averaging theorem due to Khasminskii (1968) can be applied to approximate H as a Markov diffusion process governed by the Ito differential equation

$$dH = m(H)dt + \sigma(H)dB(t), \qquad (7.5.23)$$

where $B(t)$ is a unit Weiner process, and the drift coefficient $m(H)$ and diffusion coefficient $\sigma(H)$ can be obtained from (7.5.22) by carrying out the time averaging, i.e.,

$$m(H) = \varepsilon \left\langle -\sum_{j,k=1}^{n} m_{jk} \frac{\partial H}{\partial P_j} \frac{\partial H}{\partial P_k} + \sum_{j,k=1}^{n} \sum_{j,s=1}^{m} \pi K_{ls} g_{jl} g_{ks} \frac{\partial^2 H}{\partial P_j \partial P_k} \right\rangle_t,$$

(7.5.24)

$$\sigma^2(H) = \varepsilon 2\pi \left\langle \sum_{j,k=1}^{n} \sum_{l,s=1}^{m} K_{ls} g_{jl} g_{ks} \frac{\partial H}{\partial P_j} \frac{\partial H}{\partial P_k} \right\rangle_t.$$

(7.5.25)

Note that the state-space response of a non-integrable Hamiltonian system is ergodic on the $(2n-1)$-dimensional equal energy surface. Then the time averaging can be replaced by space averaging with respect to all the independent fast varying variables $Q_1, Q_2, \ldots, Q_n$ and $P_2, \ldots, P_n$. Specifically, the time averaging of a function $F(\mathbf{q}, \mathbf{p})$ can be calculated as follows:

$$\langle F(\mathbf{q}, \mathbf{p}) \rangle_t = \frac{1}{T(H)} \int_\Omega F(\mathbf{q}, \mathbf{p}) \left(\frac{\partial H}{\partial p_1} \right)^{-1} dq_1 dq_2 \cdots dq_n dp_2 \cdots dp_n,$$

(7.5.26)

where

$$T(H) = \int_\Omega \left(\frac{\partial H}{\partial p_1} \right)^{-1} dq_1 dq_2 \cdots dq_n dp_2 \cdots dp_n$$

(7.5.27)

and domain Ω is defined as $H(q_1, q_2, \ldots, q_n, p_1 = 0, p_2, \ldots, p_n) \leq H$.

After the averaging, the drift coefficient $m(H)$ and diffusion coefficient $\sigma(H)$ are obtained, and the stationary probability density $p(h)$ of the Hamiltonian process H can be calculated from

$$p(h) = \frac{C}{\sigma^2(h)} \exp \left[\int \frac{2m(h)}{\sigma^2(h)} dh \right].$$

(7.5.28)

The joint stationary probability density of the generalized displacement vector $\mathbf{Q}$ and the generalized momentum vector $\mathbf{P}$ can be calculated from

$$p(\mathbf{q}, \mathbf{p}) = p(\mathbf{q}, p_2, \ldots, p_n, h) \left| \frac{\partial h}{\partial p_1} \right| = p(\mathbf{q}, p_2, \ldots, p_n | h) p(h) \left| \frac{\partial h}{\partial p_1} \right|.$$

(7.5.29)

For a fixed Hamiltonian $H(\mathbf{q}, \mathbf{p}) = h$, the conditional probability $p(\mathbf{q}, p_2, \ldots, p_n | h)$ is inversely proportional to $\left| \frac{\partial h}{\partial p_1} \right|$; therefore, $p(\mathbf{q}, p_2, \ldots, p_n | h)$ can be expressed as

$$p(\mathbf{q}, p_2, \ldots, p_n | h) = C_1 \left| \frac{\partial h}{\partial p_1} \right|^{-1}.$$

(7.5.30)

Integrating both sides over domain Ω, and using (7.5.27), we obtain

$$C_1 = \left[\int_\Omega \left(\frac{\partial H}{\partial p_1} \right)^{-1} dq_1 dq_2 \cdots dq_n dp_2 \cdots dp_n \right]^{-1} = [T(h)]^{-1}.$$

(7.5.31)

Substituting (7.5.30) and (7.5.31) into (7.5.29), we obtain

$$p(\mathbf{q}, \mathbf{p}) = \left[\frac{p(h)}{T(h)} \right]_{h=H(\mathbf{q}, \mathbf{p})}.$$

(7.5.32)

Example 7.5.2 In system (7.5.11) in Example 7.5.1, the damping coefficients multiplied by ε and the excitations multiplied by $\varepsilon^{1/2}$, then (7.5.11) is a quasi non-integrable Hamiltonian system and the stochastic averaging method described above can be applied. The Hamiltonian H and the potential energy U are given in Eqs. (7.5.12) and (7.5.13), respectively. The drift and diffusion coefficients are obtained from Eqs. (7.5.24) and (7.5.25) as follows

$$m(H) = \varepsilon \langle (\alpha_1 - \beta_1 x^2) \dot{x}^2 + (\alpha_1 - \alpha_2 - \beta_2 y^2) \dot{y}^2$$
$$+ \pi K_1 c_1^2 x^2 + \pi K_2 c_2^2 y^2 \rangle_t,$$

(7.5.33)

$$\sigma^2(H) = \varepsilon 2\pi \langle K_1 c_1^2 x^2 \dot{x}^2 + K_2 c_2^2 y^2 \dot{y}^2 \rangle_t.$$

(7.5.34)

Carrying the time averaging procedure given in Eqs. (7.5.26) and (7.5.27), we obtain

$$m(H) = \frac{\varepsilon}{T(H)} \int_{\Omega} \left[(\alpha_1 - \beta_1 x^2)\dot{x} + (\alpha_1 - \alpha_2 - \beta_2 y^2)\frac{\dot{y}^2}{\dot{x}} \right.$$

$$\left. + \pi K_1 c_1^2 \frac{x^2}{\dot{x}} + \pi K_2 c_2^2 \frac{y^2}{\dot{x}} \right] dxdyd\dot{y}, \tag{7.5.35}$$

$$\sigma^2(H) = \frac{\varepsilon 2\pi}{T(H)} \int_{\Omega} \left(K_1 c_1^2 x^2 \dot{x} + K_2 c_2^2 \frac{y^2 \dot{y}^2}{\dot{x}} \right) dxdyd\dot{y}, \tag{7.5.36}$$

where

$$T(H) = \int_{\Omega} \frac{1}{\dot{x}} dxdyd\dot{y}, \tag{7.5.37}$$

where domain Ω is defined as $H(x, y, \dot{x} = 0, \dot{y}) = \frac{1}{2}\dot{y}^2 + U(x, y) \leq H$.

Similar to Example 7.5.1, the multi-fold integrations in Eqs. (7.5.35) through (7.5.37) can be simplified to single integrations with respect to θ by letting $x = r\cos\theta$ and $y = r\sin\theta$. Some numerical results were given in (Zhu and Yang, 1997), and are shown in Figs. 7.5.2 and 7.5.3. The stationary probability density of the system Hamiltonian $H(t)$ was calculated by the stochastic averaging for system (7.5.11) with parameters as: $\alpha_1 = \alpha_2 = 0.01$, $\beta_1 = 0.01$, $\beta_2 = 0.02$, $\omega_1 = 1$, $\omega_2 = 2$, $a = 0.01$, $b = 1$, $c_1 = c_2 = 0.2$ and $K_1 = K_2 = 3$. The result is shown in Fig. 7.5.2 by the solid line. The stationary mean square values of $X(t)$, also calculated from the stochastic averaging, are shown in Fig. 7.5.3 by the solid line. The system parameters are $\alpha_1 = \alpha_2 = 0.03$, $\beta_1 = 0.03$, $\beta_2 = 0.04$, $\omega_1 = 1$, $\omega_2 = 2$, $a = 0.01$, $b = 1$ and $c_1 = c_2 = 0.2$. The spectral densities $K_1 = K_2 = K$ are changing. Also shown in the figure by the dashed line is the results calculated according to the equal average energy dissipation rate introduced in Section 7.5.1. In both figures, results from the Monte Carlo simulations are also shown for comparison. It is seen that the stochastic averaging produces quite accurate results.

The procedures of stochastic averaging method for different types of quasi-Hamiltonian systems are quite different. For more

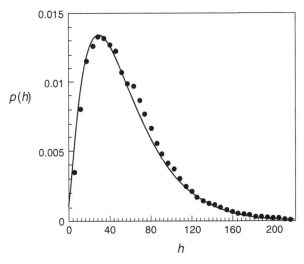

Figure 7.5.2 Stationary probability density of Hamiltonian $H(t)$ of system (7.5.11). Solid line — stochastic averaging, symbols — simulation (from Zhu and Yang, 1997).

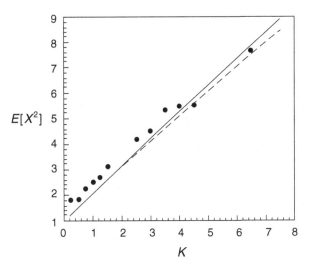

Figure 7.5.3 Stationary mean square values of displacement $X(t)$ of system (7.5.11). Solid line — stochastic averaging, dashed line — equal average energy dissipation rate, symbols — simulation (from Zhu and Yang, 1997).

information, readers can refer to (Zhu and Yang, 1997; Zhu, Huang and Yang, 1997; Zhu, Huang and Suzuki, 2002). The stochastic averaging method of quasi Hamiltonian systems has also been extended to more kinds of systems and more kinds of stochastic excitations (Zhu, 2006).

Exercise Problems

7.1 Given a nonlinear system

$$\ddot{X} + \alpha \dot{X} + \beta \dot{X}^3 + X = W(t),$$

where $W(t)$ is a Gaussian white noise with a spectral density K, calculate the approximate stationary mean-square value of $X(t)$ by using (a) the linearization, and (b) the Gaussian closure.

7.2 For the system of Problem 7.1, derive a set of equations for the stationary moments on the basis of the fourth-order cumulant-neglect closure.

7.3 Consider a nonlinear system

$$\ddot{X} + \alpha \dot{X} + \gamma X^2 \dot{X} + \beta \dot{X}^3 + \omega_0^2 X = W(t),$$

where $W(t)$ is a Gaussian white noise with a spectral density K. Use the linearization procedure to calculate the approximate stationary mean-square values of $X(t)$ and $\dot{X}(t)$.

7.4 Obtain and compare the approximate stationary solutions for each of the following systems using the linearization and partial linearization procedures.

(a) $\ddot{X} + \alpha \dot{X} + \gamma X^2 \dot{X} + X^3 = W(t),$
(b) $\ddot{X} + \alpha \dot{X} + \beta \dot{X}^3 + \omega_0^2 X + \delta X^3 = W(t),$

where $W(t)$ is a Gaussian white noise with a spectral density K.

7.5 Consider a nonlinear SDOF system under a random excitation

$$\ddot{X} + 2\zeta \omega_0 \dot{X} + \beta \dot{X}^3 + \omega_0^2 X + \delta X^3 = F(t),$$

where $F(t)$ is generated from the following one-dimensional linear filter

$$\dot{F} + \alpha F = W(t),$$

in which $W(t)$ is a Gaussian white noise. Derive the procedures for the approximate stationary second-order moments by (a) applying the equivalent linearization directly to the SDOF system, and (b) using the linearization procedure to the augmented 3-dimensional system.

7.6 Consider the following nonlinear system under both external and parametric excitations

$$\ddot{X} + \alpha\dot{X} + \beta\dot{X}^3 + kX + \delta X^3 = XW_1(t) + W_2(t),$$

where Gaussian white noises $W_1(t)$ and $W_2(t)$ are assumed to be independent with spectral densities K_{11} and K_{22}, respectively. Use the quasi-linearization method to solve for (a) the stationary second — and fourth-order moments, and (b) the stationary correlation function and the power spectral density.

7.7 For the same system in Problem 7.1,

$$\ddot{X} + \alpha\dot{X} + \beta\dot{X}^3 + X = W(t),$$

find the approximate stationary probability density of $X(t)$ by using the method of dissipation energy balancing.

7.8 Use the method of dissipation energy balancing to solve the nonlinear system

$$\ddot{X} + \alpha\dot{X} + \beta\dot{X}^3 + \delta X^3 = W(t), \quad \beta, \delta > 0.$$

7.9 Consider the same nonlinear system in Problem 7.3, i.e.,

$$\ddot{X} + \alpha\dot{X} + \gamma X^2\dot{X} + \beta\dot{X}^3 + \omega_0^2 X = W(t).$$

(a) Use the method of dissipation energy balancing to find the approximate stationary probability density $p(x, \dot{x})$ of $X(t)$ and $\dot{X}(t)$.

(b) From $p(x, \dot{x})$ obtained from (a), derive the stationary probability density of the amplitude process $A(t) = [X^2(t) + \dot{X}^2(t)/\omega_0^2]^{1/2}$.

7.10 Obtain approximate stationary probability densities for the two systems in Problem 7.4 using the procedure of dissipation energy balancing.

(a) $\ddot{X} + \alpha\dot{X} + \gamma X^2\dot{X} + X^3 = W(t)$,

(b) $\ddot{X} + \alpha\dot{X} + \beta\dot{X}^3 + \omega_0^2 X + \delta X^3 = W(t)$.

7.11 Consider the nonlinear system under a parametric white noise excitation

$$\ddot{X} + \alpha\dot{X} + \beta\dot{X}^3 + X = XW(t).$$

(a) Use the method of dissipation energy balancing to find the approximate stationary probability density $p(x, \dot{x})$ of $X(t)$ and $\dot{X}(t)$.

(b) Determine the condition for a normalizable $p(x, \dot{x})$.

7.12 Consider the system

$$\ddot{X} + f(X, \dot{X}) + X = XW(t),$$

where

$$f(X, \dot{X}) = \dot{X} \sum_{i+j=0}^{N} d_{ij} \left| X^i \dot{X}^j \right|.$$

(a) Use the method of dissipation energy balancing to find the approximate stationary probability density $p(x, \dot{x})$ of $X(t)$ and $\dot{X}(t)$.

(b) Determine the condition under which $p(x, \dot{x})$ is normalizable.

7.13 Use the method of dissipation energy balancing to find the stationary probability density $p(x, \dot{x})$ of $X(t)$ and $\dot{X}(t)$ for the system

$$\ddot{X} + f(X, \dot{X}) + X = XW(t),$$

where

$$f(X, \dot{X}) = \sum_{m=0}^{M} a_m \dot{X} X^2 (X^2 + \dot{X}^2)^m.$$

Show that the obtained $p(x, \dot{x})$ is the exact result.

7.14 Consider the system

$$\ddot{X} + f(X, \dot{X}) + X = XW_1(t) + W_2(t),$$

where $W_1(t)$ and $W_2(t)$ are Gaussian white noises with spectral densities K_{kl}, k, $l = 1, 2$, and

$$f(X, \dot{X}) = \dot{X} \sum_{i+j=0}^{N} d_{ij}|X^i \dot{X}^j|.$$

(a) Use the method of dissipation energy balancing to find the approximate stationary probability density $p(x, \dot{x})$ of $X(t)$ and $\dot{X}(t)$.

(b) Determine the condition under which $p(x, \dot{x})$ is normalizable.

(c) Show that in the special case, $d_{00} = \alpha$, $d_{20} = \alpha K_{11}/K_{22}$, other $d_{ij} = 0$, i.e.,

$$f(X, \dot{X}) = \alpha \dot{X} \left(1 + \frac{K_{11}}{K_{22}} X^2 \right).$$

The method of dissipation energy balancing leads to the exact result.

7.15 Consider the system

$$\ddot{X} + f(X, \dot{X}) + u(X) = XW_1(t) + W_2(t),$$

where $W_1(t)$ and $W_2(t)$ are independent Gaussian white noises with spectral densities K_{11} and K_{22}, respectively, and

$$f(X, \dot{X}) = \dot{X} \sum_{i+j=0}^{N} d_{ij}|X^i \dot{X}^j|.$$

(a) Use the method of dissipation energy balancing to find the approximate stationary probability density $p(x, \dot{x})$ of $X(t)$ and $\dot{X}(t)$.

(b) Show that in the special case, $d_{00} = \alpha$, $d_{20} = \alpha K_{11}/K_{22}$, other $d_{ij} = 0$, i.e.,

$$f(X, \dot{X}) = \alpha \dot{X} \left(1 + \frac{K_{11}}{K_{22}} X^2 \right).$$

The method of dissipation energy balancing leads to the exact result.

7.16 For system (7.4.27), i.e.,

$$\ddot{X} + \varepsilon h(X, \dot{X}) + \omega_0^2 X = \varepsilon^{\frac{1}{2}} \sum_{l=1}^{m} g_l(X, \dot{X}) \xi_l(t),$$

derive the differential equations for the amplitude process $A(t)$ and phase process $\phi(t)$, defined by the transformation

$$X = A(t) \cos \theta, \quad \dot{X} = -A(t)\omega_0 \sin \theta, \quad \theta = \omega_0 t + \phi(t).$$

7.17 Consider the same system as that in Problem 7.9, i.e.,

$$\ddot{X} + \alpha \dot{X} + \gamma X^2 \dot{X} + \beta \dot{X}^3 + \omega_0^2 X = W(t),$$

where $W(t)$ is a Gaussian white noises with a spectral density K. Apply the transformation

$$X = A(t) \cos \theta, \quad \dot{X} = -A(t)\omega_0 \sin \theta, \quad \theta = \omega_0 t + \phi(t).$$

(a) Derive the differential equations for the amplitude process $A(t)$ and phase process $\phi(t)$.
(b) Obtain the Ito stochastic differential equations and FPK equation for the Markov process $[A(t), \phi(t)]$, using the stochastic averaging method.
(c) Find the stationary probability density $p(a)$ of $A(t)$, and compare the result with that obtained in Problem (7.9).
(d) Obtain the stationary probability density $p(x, \dot{x})$ of the original variables $X(t)$ and $\dot{X}(t)$ according to Eq. (7.4.44).

7.18 Consider a nonlinearly damped oscillator under a parametric random excitation governed by

$$\ddot{X} + 2\zeta\omega_0\dot{X} + \eta\left|\dot{X}\right|^\delta \operatorname{sgn}\dot{X} + \omega_0^2[1 + W(t)]X = 0,$$

$$\zeta, \eta > 0, \quad 0 \leq \delta < 1,$$

where $W(t)$ is a Gaussian white noises with a spectral density K. Apply the transformation

$$X = A(t)\cos\theta, \quad \dot{X} = -A(t)\omega_0\sin\theta, \quad \theta = \omega_0 t + \phi(t).$$

(a) Derive the differential equations for the amplitude process $A(t)$ and phase process $\phi(t)$.

(b) Apply the stochastic averaging method to obtain the averaged Ito equation for $A(t)$.

(c) Analyze the boundary behaviors of averaged $A(t)$, and determine whether a non-trivial stationary probability density exists. If it exists, find it.

7.19 Redo Problem 7.18(c) for the case of $\delta > 1$.

7.20 Consider a system governed by

$$\ddot{X} + [2\zeta\omega_0\dot{X} + \xi_2(t)]\dot{X} + \omega_0^2 X = \xi_1(t),$$

where $\xi_1(t)$ and $\xi_2(t)$ are stationary and independent with broadband spectral densities $\Phi_{ii}(\omega)$, $i = 1, 2$.

(a) Derive the differential equations for the amplitude process $A(t)$ and phase process $\phi(t)$ using the transformation

$$X = A(t)\cos\theta, \quad \dot{X} = -A(t)\omega_0\sin\theta, \quad \theta = \omega_0 t + \phi(t).$$

(b) Apply the stochastic averaging method to obtain the averaged Ito equation for $A(t)$.

(c) Solve for the stationary probability density for the averaged amplitude process $A(t)$, and find its integrability condition.

(d) Using Ito differential rule to derive the Ito equation for $A^n(t)$, and find an iterative formula for the stationary moments $E[A^n(t)]$.

7.21 An oscillator with a linear damping and a nonlinear stiffness is governed by

$$\ddot{X} + \alpha\dot{X} + k|X^\rho|\mathrm{sgn}X = W(t), \quad \alpha, k, \rho > 0,$$

where $W(t)$ is a Gaussian white noise with a spectral density K.

(a) Use the stochastic averaging procedure to derive the averaged Ito equation for the energy process

$$\Lambda = \frac{1}{2}\dot{X}^2 + k\int |X^\rho|\mathrm{sgn}XdX.$$

(b) Find the stationary probability density of the energy process $\Lambda(t)$.

7.22 Obtain approximate stationary probability densities for the two systems in Problems 7.4 and 7.10 using the procedure of stochastic averaging.

(a) $\ddot{X} + \alpha\dot{X} + \gamma X^2\dot{X} + X^3 = W(t)$.
(b) $\ddot{X} + \alpha\dot{X} + \beta\dot{X}^3 + \omega_0^2 X + \delta X^3 = W(t)$.

7.23 Determine the approximate stationary solutions for the system

$$\ddot{X} + \alpha\dot{X} + X^3 = XW_1(t) + W_2(t),$$

where $W_1(t)$ and $W_2(t)$ are independent Gaussian white noises with spectral densities K_{11} and K_{22}, respectively.

(a) Use the energy dissipation balancing method.
(b) Use the stochastic averaging procedure.

7.24 Prove that methods of stochastic averaging and energy dissipation balancing result in the same results for a general nonlinear system

$$\ddot{X} + f(X, \dot{X}) + u(X) = \sum_{l=1}^{m} g_l(X, \dot{X})W_l(t),$$

where $W_l(t)$ are Gaussian white noises with correlation functions

$$R[W_l(t)W_s(t+\tau)] = 2\pi K_{ls}\delta(\tau).$$

7.25 Consider the following linear system subjected to both harmonic and random excitations

$$\ddot{X} + 2\zeta\omega_0\dot{X} + \omega_0^2 X = \lambda \sin \nu t + W(t),$$

where $W(t)$ is a Gaussian white noise with a spectral density K. For the resonant case of $\omega_0 \approx \nu$, make the transformation

$$X = A(t)\cos\theta, \quad \dot{X} = -A(t)\nu\sin\theta, \quad \theta = \nu t + \phi(t).$$

Derive the Ito equations for the amplitude process $A(t)$ and phase process $\phi(t)$, and the corresponding FPK equation.

7.26 Consider the following system subjected to both harmonic and random excitations

$$\ddot{X} + 2\zeta\omega_0\dot{X} + \gamma X^2\dot{X} + \omega_0^2 X = \lambda \sin \nu t + W(t),$$

where $W(t)$ is a Gaussian white noise with a spectral density K.

(a) Replace the original equation by two first-order equations of X_s and X_c, define by the transformation

$$X = X_s \sin \nu t + X_c \cos \nu t,$$
$$\dot{X} = \nu(-X_c \sin \nu t + X_s \cos \nu t).$$

(b) For the resonant case, $\nu \approx \omega_0$, apply the stochastic averaging method to obtain the Ito stochastic differential equations for X_s and X_c.

(c) Find the condition under which the averaged system belongs to the class of stationary potential, and obtain $p(x_s, x_c)$ under this condition.

CHAPTER 8

STABILITY AND BIFURCATION
OF STOCHASTICALLY
EXCITED SYSTEMS

Chapters 5, 6 and 7 describe techniques and methods to obtain probabilistic and statistical solutions of stochastically excited systems. In many situations, qualitative or topological behaviors of such systems can be acquired without solving the system equations. Nevertheless, the knowledge of the qualitative behaviors of dynamical systems is often important and useful. The concepts of stability and bifurcation are introduced for the purpose. Analogous to the deterministic systems, analysis of stability and bifurcation have been developed for stochastically excited systems. In this chapter, definitions of different stochastic stabilities are given first in terms of convergence of a stochastic process in a probabilistic or statistical sense. Then, the stabilities of parametrically excited linear systems, nonlinear systems and quasi-Hamiltonian systems are analyzed. Differently from the stability analysis, the bifurcation theory investigates the sudden change of the system qualitative behaviors when a system parameter passes a critical point. For stochastically excited nonlinear systems, the description of the qualitative behavior change is quite different from that for deterministic systems. In this chapter, the deterministic bifurcation theory will be introduced first so that readers understand the concept of bifurcation. Then extension of deterministic bifurcation to stochastic systems follows.

8.1 Stochastic Stability

Dynamic stability has always been an important topic in research. When a system is stochastic, the stability condition, if existing, must be given in terms of statistical or probabilistic form. In order to explain the fundamental concepts of stochastic stability, a brief review of dynamic stability of deterministic systems is necessary.

A typical system for investigating dynamic stability is the well-known Mathieu–Hill system, governed by

$$\ddot{x} + 2\zeta\omega_0\dot{x} + (\omega_0^2 + \varepsilon\cos\omega t)x = 0. \tag{8.1.1}$$

The term $\varepsilon\cos\omega t$ plays the role of a parametric excitation which makes the stiffness time-changing. For system (8.1.1), $x = 0$ and $\dot{x} = 0$ is the trivial solution, which represents an equilibrium state. In general, the purposes of stability analysis are to determine (i) whether the system will eventually approach the equilibrium state under an initial state very close to the equilibrium state, (ii) whether the system will be unbounded beginning from a non-equilibrium state and (iii) whether there are other equilibrium states and what are system behaviors near them. These are qualitative properties of the system, not the detailed quantitative solutions.

In general, the state of a dynamic system can be represented by a n-dimensional vector $\mathbf{x}(t)$. The boundedness and convergence of $\mathbf{x}(t)$ can be investigated in terms of a suitable norm of $\mathbf{x}(t)$, denoted by $\|\mathbf{x}(t)\|$. Examples of such norms are

$$\|\mathbf{x}(t)\| = \sum_{i=1}^{n} |x_i(t)|, \quad \|\mathbf{x}(t)\| = \left[\mathbf{x}^T(t)\mathbf{x}(t)\right]^{\frac{1}{2}} = \left[\sum_{i=1}^{n} x_i^2(t)\right]^{\frac{1}{2}}$$

$$\|\mathbf{x}(t)\| = \left[\mathbf{x}^T(t)\mathbf{A}\mathbf{x}(t)\right]^{\frac{1}{2}} = \left[\sum_{i,j=1}^{n} a_{ij}x_i(t)x_j(t)\right]^{\frac{1}{2}}, \tag{8.1.2}$$

where $\mathbf{A}$ is a positive definite square matrix, and a_{ij} are elements of $\mathbf{A}$. The norms in (8.1.2) is known as Euclidean norm. It is noticed that an Euclidean norm is non-negative, and of homogeneity of degree one. The following concepts, introduced by Lyapunov (1892), have been adopted universally in investigation of dynamical stability.

Lyapunov stability. The trivial solution is said to be stable if for every $\varepsilon > 0$, there exists a $\delta(\varepsilon, t_0) > 0$ such that

$$\sup_{t \geq t_0} \|\mathbf{x}(t; \mathbf{x}_0, t_0)\| < \varepsilon, \tag{8.1.3}$$

provided $\|\mathbf{x}_0\| \leq \delta$ where $\mathbf{x}_0 = \mathbf{x}(0)$, and the notation "sup" is an abbreviation of the term supremum (least upper bound).

Lyapunov asymptotic stability. The trivial solution is said to be asymptotically stable if it is stable, and if there exists a $\delta(t_0) > 0$ such that

$$\lim_{t \to \infty} \|\mathbf{x}(t; \mathbf{x}_0, t_0)\| = 0, \tag{8.1.4}$$

provided $\|\mathbf{x}_0\| \leq \delta$.

The Lyapunov stability indicates that the system state can be very close to the trivial solution, while the Lyapunov asymptotic stability says that the system will eventually approach the trivial solution. To illustrate these concepts, let consider a linear oscillator

$$\ddot{x} + 2\zeta\omega_0\dot{x} + \omega_0^2 x = 0. \tag{8.1.5}$$

Without damping, i.e., $\zeta = 0$, the trivial solution $x = 0, \dot{x} = 0$ is stable, but not asymptotically stable. However, it is asymptotically stable as long as $\zeta > 0$.

8.1.1 *Concepts and Classification of Stochastic Stability*

The response of a dynamical system under stochastic excitations is also a stochastic process, say $\mathbf{X}(t)$. Stability of such a system should also defined in terms of boundedness and convergence of a norm of $\mathbf{X}(t)$, $\|\mathbf{X}(t)\|$, which is a one-dimensional stochastic process. As described in Section 3.5.1, convergence of a stochastic process can be defined in different senses; therefore, stochastic stability can be also defined in different ways. A set of commonly used definitions were summarized in a survey article of Kozin (1969), and also will be used in this book.

Lyapunov stability with probability 1. The trivial solution is said to be stable in Lyapunov sense with probability 1 if, for every pair of ε_1, $\varepsilon_2 > 0$, there exists a $\delta(\varepsilon_1, \varepsilon_2, t_0) > 0$ such that

$$\text{Prob}\left\{ \bigcup_{\|\mathbf{x}_0\| \leq \delta} \left[\sup_{t \geq t_0} \|\mathbf{X}(t; \mathbf{x}_0, t_0)\| \geq \varepsilon_1 \right] \right\} \leq \varepsilon_2, \qquad (8.1.6)$$

where $\mathbf{x}_0 = \mathbf{x}(0)$ is deterministic. The definition (8.1.6) states that almost all sample functions are stable in Lyapunov sense except for those sample functions associated with an arbitrarily small probability ε_2. Since ε_1 and ε_2 can be arbitrarily small, this type of stability is also known as the almost sure sample stability, or simply sample stability. It is concerned with all sample functions in the time interval $[t_0, t]$.

Lyapunov asymptotic stability with probability 1. The trivial solution is said to be asymptotically stable in Lyapunov sense with probability 1 if (8.1.6) holds, and if, for every $\varepsilon > 0$, there exists a $\delta(\varepsilon, t_0) > 0$ such that

$$\lim_{t_1 \to \infty} \text{Prob}\left[\sup_{t \geq t_1} \|\mathbf{X}(t; \mathbf{x}_0, t_0)\| \geq \varepsilon \right] = 0, \qquad (8.1.7)$$

provided $\|\mathbf{x}_0\| \leq \delta$. Since ε is arbitrarily small, it is also known as the almost sure asymptotic sample stability, or simply asymptotic sample stability.

Stability in probability. The trivial solution is said to be stable in probability if, for every pair of ε_1, $\varepsilon_2 > 0$, there exists a $\delta(\varepsilon_1, \varepsilon_2, t_0) > 0$ such that

$$\text{Prob}\left[\|\mathbf{X}(t; \mathbf{x}_0, t_0)\| \geq \varepsilon_1\right] \leq \varepsilon_2, \qquad t \geq t_0, \qquad (8.1.8)$$

provided $\|\mathbf{x}_0\| \leq \delta$.

Asymptotic stability in probability. The trivial solution is said to be asymptotically stable in probability if (8.1.8) holds, and if, for every $\varepsilon > 0$, there exists a $\delta(\varepsilon, t_0) > 0$ such that

$$\lim_{t \to \infty} \text{Prob}\left[\|\mathbf{X}(t; \mathbf{x}_0, t_0)\| \geq \varepsilon\right] = 0, \qquad (8.1.9)$$

provided $\|\mathbf{x}_0\| \leq \delta$.

Stability in the mth moment. The trivial solution is said to be stable in the mth moment if, for every $\varepsilon > 0$, there exists a $\delta(\varepsilon, t_0) > 0$ such that

$$E\left[\|\mathbf{X}(t; \mathbf{x}_0, t_0)\|^m\right] \leq \varepsilon, \quad m > 0, t \geq t_0, \tag{8.1.10}$$

provided $\|\mathbf{x}_0\| \leq \delta$.

Asymptotic stability in the mth moment. The trivial solution is said to be asymptotically stable in the mth moment if (8.1.10) holds, and if, for every $\varepsilon > 0$, there exists a $\delta(\varepsilon, t_0) > 0$ such that

$$\lim_{t \to \infty} E\left[\|\mathbf{X}(t; \mathbf{x}_0, t_0)\|^m\right] = 0, \quad m > 0, \tag{8.1.11}$$

provided $\|\mathbf{x}_0\| \leq \delta$.

In principle, the sample stability and asymptotic sample stability describe the qualitative behavior of the stochastic system more precisely, since they reveal the properties of all sample functions on the entire time domain $t \geq t_0$. However, these types of stability conditions are often difficult to obtain for practical dynamical systems.

As shown in Fig. 3.5.1, the convergence with probability one implies the convergence in probability, indicating that (8.1.6) implies (8.1.8), and (8.1.7) implies (8.1.9). Therefore, the sample stability is more stringent than the stability in probability. However, the two types of stability are equivalent for linear systems under parametric excitations.

Also shown in Fig. 3.5.1, the convergence in mean square implies the convergence in probability. Thus, the stability in the second-order moment is more stringent than the stability in probability. For a parametrically excited linear system, the stability in the second-order moment is the most stringent, and the other two stabilities are equivalent.

8.1.2 *Asymptotic Sample Stability of Parametrically Excited Linear Systems*

In this section, we restrict the discussion on stability of parametrically excited linear systems, i.e., systems with linear properties under parametric excitations associated with linear functions of the state variables. It is assumed that the excitations can be modeled or approximated as Gaussian white noises so that the system responses are Markov diffusion processes. It turns out (Khasminskii, 1967) that a necessary and sufficient condition can be obtained for the asymptotic sample stability.

We begin with the following Ito stochastic differential equations for an n-dimensional Markov diffusion process $\mathbf{X}(t)$

$$dX_j = \sum_{k=1}^{n} a_{jk} X_k dt + \sum_{k=1}^{n} \sum_{r=1}^{m} \gamma_{jkr} X_k dB_r(t), \quad j = 1, 2, \ldots, n.$$

$$(8.1.12)$$

The sample stability of the system may be investigated in terms of the following norm

$$\|\mathbf{X}(t)\| = \left[\mathbf{X}^T(t)\mathbf{X}(t)\right]^{\frac{1}{2}} = \left[\sum_{i=1}^{n} X_i^2(t)\right]^{\frac{1}{2}}. \qquad (8.1.13)$$

Denote U_j as the normalized X_j, defined as

$$U_j(t) = \frac{X_j}{\|\mathbf{X}(t)\|}. \qquad (8.1.14)$$

Since the n components of vector $\mathbf{U}(t)$ are subjected to the constraint

$$\|\mathbf{U}(t)\| = 1. \qquad (8.1.15)$$

Equation (8.1.15) defines an n-dimensional unit sphere. Since $\mathbf{X}(t)$ is a n-dimensional vector Markov diffusion process, $\mathbf{U}(t)$ is a $(n-1)$-dimensional vector Markov diffusion process. The growth or decay of the norm $\|\mathbf{X}(t)\|$ can be characterized by its logarithm

$$\Upsilon(t) = \ln \|\mathbf{X}(t)\|. \qquad (8.1.16)$$

Using the Ito differential rule, the Ito equation for $\Upsilon(t)$ can be derived from (8.1.12) through (8.1.16) as

$$d\Upsilon = Q(\mathbf{U})dt + \sum_{r=1}^{m} P_r(\mathbf{U})dB_r(t), \qquad (8.1.17)$$

where $Q(\mathbf{U})$ and $P_r(\mathbf{U})$ can be obtained for a specific system. Their general expressions can be found in (Lin and Cai, 1995). Integration of (8.1.17) from 0 to t results in

$$\Upsilon(t) - \Upsilon(0) = \ln \|\mathbf{X}(t)\| - \ln \|\mathbf{X}(0)\|$$

$$= \int_0^t Q[\mathbf{U}(\tau)]d\tau + \sum_{r=1}^{m} \int_0^t P_r[\mathbf{U}(\tau)]dB_r(\tau). \quad (8.1.18)$$

Dividing (8.1.18) by t, we have

$$\frac{1}{t} \ln \|\mathbf{X}(t)\| = \frac{1}{t} \ln \|\mathbf{X}(0)\| + \frac{1}{t} \int_0^t Q[\mathbf{U}(\tau)]d\tau + \sum_{r=1}^{m} \frac{1}{t} \int_0^t P_r[\mathbf{U}(\tau)]dB_r(\tau).$$

$$(8.1.19)$$

As $t \to \infty$, the first term on the right-hand side vanishes. The last term also vanishes since $\mathbf{U}(\tau)$ is bounded and the Wiener processes $B_r(\tau)$ grow as $[\tau\ln(\ln\tau)]^{1/2}$ (Lin and Cai, 1995). Then (8.1.19) reduces to

$$\lim_{t\to\infty} \frac{1}{t} \ln \|\mathbf{X}(t)\| = \lim_{t\to\infty} \frac{1}{t} \int_0^t Q[\mathbf{U}(\tau)]d\tau. \qquad (8.1.20)$$

If $\mathbf{U}(t)$ is stationary and ergodic, the time average can be replaced by the ensemble average, and we obtain from (8.1.20),

$$\lambda = \lim_{t\to\infty} \frac{1}{t} \ln \|\mathbf{X}(t)\| = E[Q[\mathbf{U}(\tau)]. \qquad (8.1.21)$$

Equation (8.1.21) implies that

$$\|\mathbf{X}(t)\| \sim e^{\lambda t}, \quad t \to \infty, \qquad (8.1.22)$$

which indicates that parameter λ characterizes the exponential growth ($\lambda > 0$) or decay ($\lambda < 0$) of the dynamical system responses as time increases. The constant λ given in (8.1.21) is known as the Lyapunov exponent. In general, a Lyapunov exponent specifies the exponential growth or decay of a quantity, similar to Eq. (8.1.22) for the norm of $\mathbf{X}(t)$. For a n-dimensional stochastic dynamical system, there exist n Lyapunov exponents corresponding to n state variables. They may be arranged as $\lambda_1 \leq \lambda_2 \leq \cdots \leq \lambda_n = \lambda_T$. The stability behavior of the stochastic dynamical system is dependent on these Lyapunov exponents. However, for system (8.1.12), the Lyapnov exponent for the norm $\|\mathbf{X}(t)\|$ given by Eq. (8.1.21) converges to the largest Lyapunov exponent λ_T of the system, provided that the normalized process $\mathbf{U}(t)$ is ergodic on the entire unit sphere (Oseledec, 1968). The asymptotic sample stability of the stochastic dynamical system (8.1.12) depends only on the largest Lyapunov exponent λ_T. Therefore, the necessary and sufficient condition for the asymptotic sample stability is

$$\lambda_T = E[Q[\mathbf{U}(\tau)] < 0. \tag{8.1.23}$$

To determine the condition for the asymptotic sample stability, we need to find the stationary probability density of the normalized process $\mathbf{U}(t)$. If there is no singular boundaries over the entire unit sphere, $\mathbf{U}(t)$ may be ergodic and its stationary probability density exists over the entire unit sphere. The problem becomes more complicated if the unit sphere is divided into regions by singular boundaries. In this case, some regions may not admit ergodic solutions, while some others may not be communicative and admit different ergodic solutions. Such complexities are now illustrated in the following two-dimensional system.

Consider the system governed by the Ito differential equations

$$dX_1 = (\alpha_{11}X_1 + \alpha_{12}X_2)dt + \sum_{k=1}^{2}\sum_{r=1}^{m}\gamma_{1kr}X_k dB_r(t),$$

$$dX_2 = (\alpha_{21}X_1 + \alpha_{22}X_2)dt + \sum_{k=1}^{2}\sum_{r=1}^{m}\gamma_{2kr}X_k dB_r(t). \tag{8.1.24}$$

Let

$$U_1(t) = \frac{X_1}{\|\mathbf{X}(t)\|} = \cos \Theta, \quad U_2(t) = \frac{X_2}{\|\mathbf{X}(t)\|} = \sin \Theta. \quad (8.1.25)$$

Applying Ito differential rule to $\Theta = \tan^{-1}(X_2/X_1)$, we obtain an Ito equation for $\Theta(t)$ process

$$d\Theta = m(\Theta)dt + \sigma(\Theta)dB(t). \quad (8.1.26)$$

The stationary probability density $p(\theta)$ of one-dimensional process $\Theta(t)$, if exists, is governed by the reduced FPK equation

$$\frac{d}{d\theta}[m(\theta)p(\theta)] - \frac{1}{2}\frac{d^2}{d\theta^2}[\sigma^2(\theta)p(\theta)] = 0. \quad (8.1.27)$$

Integrating (8.1.27) with respect to θ, we obtain

$$\frac{1}{2}\frac{d}{d\theta}[\sigma^2(\theta)p(\theta)] - m(\theta)p(\theta) = G_c, \quad (8.1.28)$$

where G_c is a constant known as the probability flow (see Section 4.2.2). The solution $p(\theta)$ of Eq. (8.1.28) depends on whether there are singular points in $[0, 2\pi)$, and the natures of these singular points, if exist, according to the behaviors of the drift and diffusion coefficients (see Section 4.4).

The logarithm of the norm (8.1.16) used to investigate the asymptotic sample stability is now

$$\Upsilon(t) = \frac{1}{2}\ln(X_1^2 + X_2^2). \quad (8.1.29)$$

It can be shown that the drift coefficient of $\Upsilon(t)$ has the form of

$$m_\Upsilon = Q(\cos 2\theta, \sin 2\theta). \quad (8.1.30)$$

The condition for the asymptotic sample stability can be obtained from Eq. (8.1.21) after the stationary probability density $p(\theta)$ is obtained.

Example 8.1.1 Consider a system governed by

$$\ddot{Z} + 2\zeta\dot{Z} + [1 + W(t)]Z = 0. \tag{8.1.31}$$

Letting

$$Z = Xe^{-\zeta t}. \tag{8.1.32}$$

Equation (8.1.31) is simplified to

$$\ddot{X} + [1 - \zeta^2 + W(t)]X = 0. \tag{8.1.33}$$

The Lyapunov exponent λ_X of system (8.1.33) and λ_Z of system (8.1.31) are related as

$$\lambda_Z = \lambda_X - \zeta. \tag{8.1.34}$$

Denote $X_1 = X$ and $X_2 = \dot{X}$, convert Eq. (8.1.33) into two first-order equations, and obtain the corresponding Ito differential equations as follows

$$\begin{aligned}
dX_1 &= X_2 dt, \\
dX_2 &= -(1 - \zeta^2)X_1 dt + \sqrt{2\pi K}X_1 dB(t),
\end{aligned} \tag{8.1.35}$$

where K is the spectral density of $W(t)$. Now change the state variables from X_1 and X_2 to Υ and Θ, defined by

$$\Upsilon(t) = \frac{1}{2}\ln(X_1^2 + X_2^2), \qquad \Theta = \tan^{-1}\frac{X_2}{X_1}. \tag{8.1.36}$$

The Ito equations for $\Upsilon(t)$ and $\Theta(t)$ are obtained from (8.1.35) and (8.1.36) as follows

$$\begin{aligned}
d\Upsilon &= m_\Upsilon dt + \sigma_\Upsilon dB(t), \\
d\Theta &= m_\Theta dt + \sigma_\Theta dB(t),
\end{aligned} \tag{8.1.37}$$

where

$$m_\Upsilon = \frac{1}{2}\zeta^2\sin 2\Theta + \pi K\cos 2\Theta\cos^2\Theta,$$

$$\sigma_\Upsilon = \sqrt{\frac{\pi K}{2}}\sin 2\Theta, \tag{8.1.38}$$

$$m_\Theta = -1 + \zeta^2\cos^2\Theta - \pi K\sin 2\Theta\cos^2\Theta,$$

$$\sigma_\Theta = \sqrt{2\pi K}\cos^2\Theta.$$

It is noted that both m_Θ and σ_Θ are only functions of Θ, thus $\Theta(t)$ itself is a Markov diffusion process defined on a unit circle. To use Eq. (8.1.23) to investigate the asymptotic sample stability of the system, the stationary probability density $p(\theta)$ must be found first. For this purpose, the behaviors of m_Θ and σ_Θ need to be examined. At $\Theta = \pi/2$ and $3\pi/2$, $\sigma_\Theta = 0$ and $m_\Theta = -1$, the two points are singular of the first kind and are shunts. The probability flow can pass from one region of $(-\pi/2, \pi/2)$ to another of $(\pi/2, 3\pi/2)$, and the $\Theta(t)$ process is ergodic within the entire domain of $[0, 2\pi)$.

The stationary probability density $p(\theta)$ of process $\Theta(t)$ is governed by the reduced FPK equation (8.1.27). Taking into consideration the fact that both m_Θ and σ_Θ are function of $\sin 2\Theta$ and $\cos 2\Theta$ with a period of π, $p(\theta)$ is also a periodic function of π. Using this property, the solution is obtained as (Nishioka, 1976; Xie, 1990)

$$p(\theta) = \begin{cases} Cf(\theta), & -\dfrac{\pi}{2} < \theta < \dfrac{\pi}{2} \\[2ex] Cf(\theta - \pi), & \dfrac{\pi}{2} < \theta < \dfrac{3\pi}{2} \end{cases}, \qquad (8.1.39)$$

where C is a normalization constant and

$$f(\theta) = \frac{\psi(\theta)}{\sigma_\Theta^2(\theta)} \int\limits_{-\pi/2}^{\theta} \psi^{-1}(\phi) d\phi, \quad \psi(\theta) = \exp\left[\int \frac{2m_\Theta(\theta)}{\sigma_\Theta^2(\theta)} d\theta\right]. \tag{8.1.40}$$

Substituting m_Θ and σ_Θ in (8.1.38) into (8.1.40) and carrying out the integration, we obtain

$$f(\theta) = Ce^{-g(\theta)} \sec^2\theta \int\limits_{-\pi/2}^{\theta} e^{-g(\phi)} \sec^2\phi \, d\phi, \quad -\frac{\pi}{2} < \theta < \frac{\pi}{2}, \tag{8.1.41}$$

where

$$g(\theta) = \frac{1}{3\pi K}(3 - 3\zeta^2 + \tan^2\theta)\tan\theta. \tag{8.1.42}$$

The largest Lyapunov exponent for system (8.1.33) is then calculated from

$$\lambda_X = 2 \int\limits_{-\pi/2}^{\pi/2} \left(\frac{1}{2}\zeta^2 \sin 2\theta + \pi K \cos 2\theta \cos^2 \theta \right) p(\theta)d\theta. \quad (8.1.43)$$

The exact value of λ_X can be calculated numerically from (8.1.43). An approximate closed form expression were obtained (Xie, 1990) as

$$\lambda_X = \frac{\pi K}{\left(1 + \sqrt{1 - \zeta^2}\right)^2}. \quad (8.1.44)$$

The largest Lyapunov exponent for system (8.1.31) is then obtained from Eq. (8.1.34) as

$$\lambda_Z = -\zeta + \frac{\pi K}{\left(1 + \sqrt{1 - \zeta^2}\right)^2}. \quad (8.1.45)$$

The condition for the asymptotic sample stability can be derived from $\lambda_Z < 0$. For small damping $\zeta \ll 1$, an approximate condition is

$$\zeta > \frac{\pi K}{4}. \quad (8.1.46)$$

Example 8.1.2 Consider a similar system as (8.1.31), but subjected to a broadband stochastic excitation

$$\ddot{X} + 2\zeta\omega_0\dot{X} + \omega_0^2[1 + \xi(t)]X = 0. \quad (8.1.47)$$

Transforming the state variables from X and $\dot{X}$ to amplitude A and phase ϕ according to

$$X = A(t)\cos\theta, \quad \dot{X} = -A(t)\omega_0\sin\theta, \quad \theta = \omega_0 t + \phi(t), \quad (8.1.48)$$

we have

$$\begin{aligned}\dot{A} &= -2\zeta\omega_0 A \sin^2\theta + A\omega_0 \sin\theta\cos\theta\xi(t), \\ \dot{\phi} &= -2\zeta\omega_0 \sin\theta\cos\theta + \omega_0 \cos^2\theta\xi(t).\end{aligned} \quad (8.1.49)$$

Carrying the stochastic averaging, we obtain an Ito equation for the amplitude process

$$dA = \left[-\zeta\omega_0 + \frac{3}{8}\pi\omega_0^2\Phi(2\omega_0)\right]Adt + \sqrt{\frac{\pi}{4}\omega_0^2\Phi(2\omega_0)}AdB(t).$$
(8.1.50)

Using the Ito differential rule, we have

$$d(\ln A) = \left[-\zeta\omega_0 + \frac{1}{4}\pi\omega_0^2\Phi(2\omega_0)\right]dt + \sqrt{\frac{\pi}{4}\omega_0^2\Phi(2\omega_0)}dB(t).$$
(8.1.51)

It follows that the asymptotic sample stability condition is

$$\zeta > \frac{1}{4}\pi\omega_0\Phi(2\omega_0).$$
(8.1.52)

For the special case of white noise, (8.1.52) reduces to

$$\zeta > \frac{1}{4}\pi\omega_0 K.$$
(8.1.53)

Noting that $A = \sqrt{X^2 + \dot{X}^2/\omega_0^2}$, which is an suitable norm according to (8.1.2). Thus, the asymptotic sample stability condition for system (8.1.50) is approximate one for system (8.1.47).

It is of interest to study the sample behaviors of $A(t)$ at the two boundaries at zero and infinity. At the left boundary at zero, the drift exponent, diffusion exponent and character value are determined from Eqs. (4.4.18), (4.4.19) and (4.4.20) as

$$\beta = 1, \quad \alpha = 2, \quad c = \frac{-8\zeta + 3\pi\omega_0 K}{\pi\omega_0 K}.$$
(8.1.54)

According to Table 4.4.2, it is attractively natural if $c < 1$, strictly natural if $c = 1$, and repulsively natural if $c > 1$. At the right boundary at infinity, the drift exponent, diffusion exponent, and character value are determined as

$$\beta = 1, \quad \alpha = 2, \quad c = \frac{8\zeta - 3\pi\omega_0 K}{\pi\omega_0 K}.$$
(8.1.55)

According to Table 4.4.4, it is attractively natural if $c < -1$, strictly natural if $c = -1$, and repulsively natural if $c > -1$. The sample behaviors at the two boundaries are illustrated in Fig. 8.1.1.

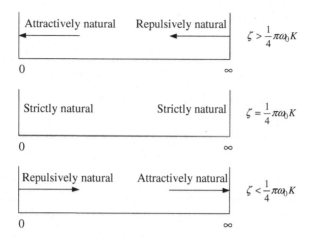

Figure 8.1.1 Sample behaviors of amplitude process $A(t)$ at the two boundaries.

From Fig. 8.1.1, the system behavior can be inferred as follows

The amplitude process $A(t)$ is

$$
\begin{cases}
\text{asymptotically stable in probability} & \text{if } \zeta > \dfrac{1}{4}\pi\omega_0 K \\[2ex]
\text{uncertain} & \text{if } \zeta = \dfrac{1}{4}\pi\omega_0 K \\[2ex]
\text{unbounded} & \text{if } \zeta < \dfrac{1}{4}\pi\omega_0 K
\end{cases}
.
$$

For the present linear system, the asymptotic stability in probability is equivalent to the asymptotic sample stability.

Comparing systems (8.1.31) and (8.1.47), it is found that the two approximate stability conditions, (8.1.46) and (8.1.53) are consistent. Of course, the stochastic averaging method is applicable to a broad band excitation in system (8.1.47), not necessarily a white noise.

8.1.3 *Asymptotic Moment Stability of Parametrically Excited Linear Systems*

Another type of stability, moment stability, is also very useful for practical stochastic systems. Compared with the sample stability, it is much easier to determine if the excitations can be modeled as Gaussian white noises or can be formulated in terms of Gaussian white noises. In this case, the system equations of motion can be converted to Ito stochastic differential equations.

Let the Ito equations for a n-dimensional system be

$$dX_j(t) = m_j(\mathbf{X}, t)dt + \sum_{l=1}^{m} \sigma_{jl}(\mathbf{X}, t)dB_l(t), \quad j = 1, 2, \ldots, n.$$

$$(8.1.56)$$

For a scalar function $F(\mathbf{X})$, its Ito equation can be obtained by applying the Ito differential rule (4.2.56) as follows

$$
\begin{aligned}
dF(\mathbf{X}) &= \sum_{j=1}^{n} \frac{\partial F}{\partial X_j} dX_j + \frac{1}{2} \sum_{j,k=1}^{n} \frac{\partial^2 F}{\partial X_j \partial X_k} dX_j dX_k \\
&= \left(\sum_{j=1}^{n} m_j \frac{\partial F}{\partial X_j} + \frac{1}{2} \sum_{l=1}^{m} \sum_{j,k=1}^{n} \sigma_{jl} \sigma_{kl} \frac{\partial^2 F}{\partial X_j \partial X_k} \right) dt \\
&\quad + \sum_{l=1}^{m} \sum_{j=1}^{n} \sigma_{jl} \frac{\partial F}{\partial X_j} dB_l(t).
\end{aligned}
$$

$$(8.1.57)$$

Taking the expectation of Eq. (8.1.57), we obtain

$$\frac{d}{dt} E[F(\mathbf{X})] = E \left(\sum_{j=1}^{n} m_j \frac{\partial F}{\partial X_j} + \frac{1}{2} \sum_{l=1}^{m} \sum_{j,k=1}^{n} \sigma_{jl} \sigma_{kl} \frac{\partial^2 F}{\partial X_j \partial X_k} \right).$$

$$(8.1.58)$$

To derive the moment equations, let

$$F(\mathbf{X}) = X_1^{k_1} X_2^{k_2} \cdots X_n^{k_n}, \tag{8.1.59}$$

where $k_1, k_2, \ldots, k_n$ are non-negative integers. Denote $N = k_1 + k_2 + \cdots + k_n$, and exclude the case of $N = 0$. By keeping the same N but changing $k_1, k_2, \ldots$ and k_n, we derive a set of deterministic ordinary differential equations (ODEs) for various Nth-order moments. As described previously, the set of moment equations for a given N is closed, namely, all equations contain only the moments of the Nth and lower orders. Therefore, by letting $N = 1, 2, \ldots$, the stability conditions for moments can be determined successively from lower order to higher order.

For cases of broadband excitations, the original equations of motion may be approximated by Ito equations using the stochastic averaging method if certain conditions are met. Since the dimension is reduced, the problems will be simpler. But there may be errors which should be within the ranges caused by the stochastic averaging.

It is noted that the stability in moments is in fact the deterministic stability problem.

Example 8.1.3 Now we investigate the moment stability for the same system in Example 8.1.1, i.e., for the system

$$\ddot{X} + 2\zeta\dot{X} + [1 + W(t)]X = 0. \tag{8.1.60}$$

Letting $X_1 = X$ and $X_2 = \dot{X}$, system (8.1.60) is converted to two first-order equations, and furthermore, to the corresponding Ito differential equations as follows

$$\begin{aligned}
dX_1 &= X_2 dt, \\
dX_2 &= (-2\zeta X_2 - X_1)dt - \sqrt{2\pi K}X_1 dB(t),
\end{aligned} \tag{8.1.61}$$

where K is the spectral density of $W(t)$. Denote $m_{ij} = E[X_1^i X_2^j]$. The equation for the first-order moments m_{10} and m_{01} can be derived by taking ensemble averages of (8.1.61),

$$\frac{d}{dt}\begin{Bmatrix} m_{10} \\ m_{01} \end{Bmatrix} = \begin{bmatrix} 0 & 1 \\ -1 & -2\zeta \end{bmatrix}\begin{Bmatrix} m_{10} \\ m_{01} \end{Bmatrix}. \tag{8.1.62}$$

The eigenvalues of the coefficient matrix on the right-hand side of (8.1.62) are found as

$$\lambda_{1,2} = -\zeta \pm i\sqrt{1 - \zeta^2}. \tag{8.1.63}$$

If the real parts of the two eigenvalues are negative, the trivial solution is asymptotically stable in the first-order moments. Thus the stability condition is

$$\zeta > 0. \tag{8.1.64}$$

Using the Ito differential rule, we obtain the Ito equations for X_1^2, X_1X_2 and X_2^2 as follows

$$dX_1^2 = 2X_1X_2dt,$$
$$d(X_1X_2) = (-X_1^2 - 2\zeta X_1X_2 + X_2^2)dt - \sqrt{2\pi K}X_1^2dB(t),$$
$$dX_2^2 = (-2\pi KX_1^2 - 2X_1X_2 - 4\zeta X_2^2)dt - 2\sqrt{2\pi K}X_1X_2dB(t).$$
$$(8.1.65)$$

Carrying out the ensemble averages of (8.1.65), we obtain the equations for the second-order moments m_{20}, m_{11} and m_{02},

$$\frac{d}{dt}\left\{\begin{array}{c} m_{20} \\ m_{11} \\ m_{02} \end{array}\right\} = \begin{bmatrix} 0 & 2 & 0 \\ -1 & -2\zeta & 1 \\ 2\pi K & -2 & -4\zeta \end{bmatrix}\left\{\begin{array}{c} m_{20} \\ m_{11} \\ m_{02} \end{array}\right\}. \qquad (8.1.66)$$

The characteristic equation of system (8.1.66) for the eigenvalues is

$$\begin{vmatrix} \lambda & -2 & 0 \\ 1 & \lambda + 2\zeta & -1 \\ -2\pi K & 2 & \lambda + 4\zeta \end{vmatrix} = \lambda^3 + 6\zeta\lambda^2 + 4\lambda(1+2\zeta^2) + 4(2\zeta - \pi K) = 0.$$
$$(8.1.67)$$

According to the well-known Routh–Hurwitz criteria (e.g., Chetayev, 1961), the real parts of all eigenvalues are negative if and only if the following conditions are met

$$\Delta_1 = 6\zeta > 0, \qquad (8.1.68)$$

$$\Delta_2 = \begin{bmatrix} 6\zeta & 1 \\ 4(2\zeta - \pi K) & 4(1 + 2\zeta^2) \end{bmatrix} > 0, \qquad (8.1.69)$$

$$\Delta_3 = \begin{bmatrix} 6\zeta & 1 & 0 \\ 4(2\zeta - \pi K) & 4(1 + 2\zeta^2) & 6\zeta \\ 0 & 0 & 4(2\zeta - \pi K) \end{bmatrix} > 0. \,(8.1.70)$$

Inequalities (8.1.68)–(8.1.70) are all satisfied if

$$\zeta > \frac{1}{2}\pi K. \qquad (8.1.71)$$

The condition (8.1.71) is the necessary and sufficient for the trivial solution to be asymptotically stable in the second-order moments. Compared with the sample stability condition (8.1.46), the stability condition for the second-order moments is more stringent.

Example 8.1.4 Consider the same system as that in Example 8.1.2. As given by Eq. (8.1.50), application of the stochastic averaging method yields an Ito equation for the amplitude process

$$dA = \left[-\zeta\omega_0 + \frac{3}{8}\pi\omega_0^2\Phi(2\omega_0) \right] A dt + \sqrt{\frac{\pi}{4}\omega_0^2\Phi(2\omega_0)} A dB(t),$$

$$(8.1.72)$$

where the amplitude process $A(t)$ is defined as

$$A = \sqrt{X^2 + \frac{\dot{X}^2}{\omega_0^2}}.$$

$$(8.1.73)$$

Applying the Ito differential rule, we obtain an Ito equation for function A^n,

$$dA^n = n\left[-\zeta\omega_0 + \frac{\pi}{8}(n+2)\omega_0^2\Phi(2\omega_0) \right] A^n dt$$

$$+ n\sqrt{\frac{\pi}{4}\omega_0^2\Phi(2\omega_0)} A^n dB(t).$$

$$(8.1.74)$$

Taking ensemble average, we have

$$\frac{d}{dt}E[A^n] = n\left[-\zeta\omega_0 + \frac{\pi}{8}(n+2)\omega_0^2\Phi(2\omega_0) \right] E[A^n].$$

$$(8.1.75)$$

Then the asymptotic stability condition in the nth-order moments for the trivial solution is

$$\zeta > \frac{\pi}{8}(n+2)\omega_0\Phi(2\omega_0),$$

$$(8.1.76)$$

indicating that the condition is more stringent with an increasing n. As expected due to the relationship (8.1.73), the case of $n = 2$ in (8.1.76) is consistent with (8.1.71) in the last example.

According to the definitions of stochastic stability in Section 8.1.1, only the convergence to the trivial solution is of concern. That excludes the presence of external excitations since any external excitation causes the trivial solution to become unstable. Nevertheless, adding an external excitation provides means for investigating stability in moments by examining the stationary probability density function, as illustrated in the following example.

Example 8.1.5 Add an external excitation to the system in Example 8.1.2, i.e.,

$$\ddot{X} + 2\zeta\omega_0\dot{X} + \omega_0^2[1 + \xi(t)]X = \xi_1(t), \qquad (8.1.77)$$

where $\xi_1(t)$ is a broadband stationary process, independent of $\xi(t)$, and with a spectral density $\Phi_{11}(\omega)$. Applying the stochastic averaging method, an Ito equation for the amplitude process is derived as

$$dA = m(A)dt + \sigma(A)dB(t), \qquad (8.1.78)$$

where the drift and diffusion coefficients are (see Example 7.4.1)

$$m(A) = -\left[\zeta\omega_0 - \frac{3\pi}{8}\omega_0^2\Phi(2\omega_0)\right]A + \frac{\pi}{2\omega_0^2 A}\Phi_{11}(\omega_0). \quad (8.1.79)$$

$$\sigma(A) = \left[\frac{\pi}{4}\omega_0^2\Phi(2\omega_0)A^2 + \frac{\pi}{\omega_0^2}\Phi_{11}(\omega_0)\right]^{\frac{1}{2}}. \qquad (8.1.80)$$

If the stationary probability density exists, it can be found from Eq. (4.4.7) as

$$p(a) = \frac{C_1}{\sigma^2(a)}\exp\left[\int\frac{2m(a)}{\sigma^2(a)}da\right] = Ca(a^2 + D)^{-\delta}, \qquad (8.1.81)$$

where

$$D = \frac{4\Phi_{11}(\omega_0)}{\omega_0^4\Phi(2\omega_0)}, \quad \delta = \frac{4\zeta}{\pi\omega_0\Phi(2\omega_0)}, \quad C = 2(\delta - 1)D^{\delta-1}. \quad (8.1.82)$$

The stationary nth moment of $A(t)$ is then calculated as

$$E[A^n] = \int_0^\infty a^n p(a)da = \frac{(\delta - 1)}{\Gamma(\delta)}\Gamma\left(\frac{n}{2} + 1\right)\Gamma\left(\delta - \frac{n}{2} - 1\right)D^{\frac{-n}{2}}.$$

$$(8.1.83)$$

The condition for $E[A^n]$ to be positive and meaningful is

$$\delta > \frac{n}{2} + 1, \quad \text{i.e., } \zeta > \frac{\pi}{8}(n + 2)\omega_0\Phi(2\omega_0). \qquad (8.1.84)$$

Under this condition, we have

$$\lim_{\Phi_{11}(\omega_0)\to 0} E[A^n] = 0. \qquad (8.1.85)$$

If condition (8.1.84) is met, the nth moment will not go unbounded. Taking into account of Eq. (8.1.85), (8.1.84) is the asymptotic stability condition of the nth-order moment for the trivial solution in the absence of the external excitation $\xi_1(t)$. As expected, (8.1.84) is the same as (8.1.76).

For the case of $n = 0$, the condition becomes

$$\zeta > \frac{\pi}{4}\omega_0\Phi(2\omega_0). \tag{8.1.86}$$

It is the condition for the trivial solution to be asymptotic stability in probability. Since the system property is linear, condition (8.1.86) for the asymptotic stability in probability is equivalent to the asymptotic sample stability condition given in Eq. (8.1.52) of Example 8.1.2.

8.1.4 *Asymptotic Stability of Nonlinear Stochastic Systems*

As shown in Section 8.1.2, the asymptotic sample stability of a parametrically excited linear system under Gaussian white noise excitations can be investigated by applying a procedure due to Khasminskii (1967). For nonlinear stochastic systems, the sample stability conditions are more difficult to determine. However, if the system is one-dimensional or can be reduced to one-dimensional, several methods are capable for stability analysis, including the sample stability, the stabilities in probability, and the stability in moments.

Consider a one-dimensional nonlinear stochastic system

$$\dot{X} = f(X) + g(X)W(t), \tag{8.1.87}$$

where $W(t)$ is a Gaussian white noise with a spectral density K. The corresponding Ito differential equation is

$$dX = m(X)dt + \sigma(X)dB(t), \tag{8.1.88}$$

where the drift and diffusion coefficients are

$$m(X) = f(X) + \pi K g(X)g'(X),$$
$$\sigma(X) = \sqrt{2\pi K}g(X). \tag{8.1.89}$$

Assume that the system possesses equilibrium points. For each equilibrium point x^*, the Lyapunov exponent can be obtained from the linearized equation at that point (Arnold, 1998)

$$dV = m'(x*)V\,dt + \sigma'(x*)V\,dB(t), \qquad (8.1.90)$$

where $V(t)$ is the linearized process. The Ito equation for $\ln[V(t)]$ is obtained by using the Ito differential rule as follows

$$d(\ln V) = \left\{ m'(x*) - \frac{1}{2}[\sigma'(x*)]^2 \right\} dt + \sigma'(x*)dB(t). \qquad (8.1.91)$$

The formal solution of (8.1.91) is

$$\ln V = \ln V_0 + \int_0^t \left\{ m'(x*) - \frac{1}{2}[\sigma'(x*)]^2 \right\} dt + \int_0^t \sigma'(x*)dB(t).$$
$$(8.1.92)$$

The Lyapunov exponent is then obtained as

$$\lambda = \lim_{t \to \infty} \frac{1}{t} \ln V$$
$$= \lim_{t \to \infty} \frac{1}{t} \int_0^t \left\{ m'(x*) - \frac{1}{2}[\sigma'(x*)]^2 \right\} dt = m'(x*) - \frac{1}{2}[\sigma'(x*)]^2.$$
$$(8.1.93)$$

In terms of the original system (8.1.87), Eq. (8.1.93) leads to

$$\lambda = f'(x*) + \pi K g(x*)g''(x*). \qquad (8.1.94)$$

The sign of λ determines the nature of the equilibrium, namely,

$$\begin{cases} \lambda < 0, & x* \text{ is locally asymptotically stable} \\ \lambda \geq 0, & x* \text{ is unstable.} \end{cases} \qquad (8.1.95)$$

The method of Lyapunov exponent can only identify the local stability which describes the system behavior very close to the equilibrium point. The system dynamic behaviors also include the nature at another boundary, such as an infinity boundary, as well as whether a non-trivial stationary probability distribution exists. These properties cannot be obtained only from the Lyapunov exponent of one boundary. A method capable to analyze the global stability is the method of boundary classification, as described next.

For one-dimensional nonlinear systems under Gaussian white noise excitations, the boundaries are generally singular, and they can be identified into different categories, as described in Section 4.4. If a boundary is a trivial solution, it is globally asymptotically stable in probability if and only if (i) it is an exit or attractively natural, and (ii) the other boundary is an entrance or repulsively natural.

Section 8.1.3 shows that the asymptotic stability in moments for parametrically excited linear systems can be studied in terms of the moment equations which are closed. For a nonlinear stochastic system, the moment equations form an infinite hierarchy, they cannot be used for stability analysis directly. Although a closure scheme can be employed, the stability analysis based on the truncated moment equations is often unreliable (Ariaratnam, 1980; Bruckner and Lin, 1987). If a non-trivial stationary probability density exists, then the trivial solution must be unstable in the sense of all order statistical moments. On the other hand, if a non-trivial stationary probability density does not exist, there are two possibilities: the system approaches to an equilibrium state or the system response is unbounded. Based on this knowledge, an approach as shown in Example 8.1.5 may be applicable. First add an external excitation in order to obtain a non-trivial stationary probability density, and hence, the statistical moments. Then the conditions for the boundedness of the statistical moments are determined. Finally, let the external excitation tend to zero, and obtain the limits of the moments to check if they are trivial.

The above stability analysis procedures for nonlinear stochastic systems are illustrated in the following examples.

Example 8.1.6 A nonlinearly damped oscillator under a parametric random excitation is governed by

$$\ddot{X} + 2\zeta\omega_0\dot{X} + \eta\left|\dot{X}\right|^{\delta} sgn\dot{X} + \omega_0^2[1 + W(t)]X = 0, \qquad \delta \geq 0, \delta \neq 1.$$
$$(8.1.96)$$

We assume that both the damping and excitation are weak so that the stochastic averaging method is applicable. Transforming the state variables from X and $\dot{X}$ to amplitude A and phase ϕ according to

$$X = A(t)\cos\theta, \quad \dot{X} = -A(t)\omega_0\sin\theta, \quad \theta = \omega_0 t + \phi(t), \quad (8.1.97)$$

we have

$$\dot{A} = -2\zeta\omega_0 A \sin^2\theta - \eta\omega_0^{\delta-1}|\sin\theta|^{\delta+1}A^\delta + A\omega_0\sin\theta\cos\theta W(t)$$

$$\dot{\phi} = -2\zeta\omega_0\sin\theta\cos\theta - \eta\omega_0^{\delta-1}|\sin\theta|^\delta\cos\theta\,\text{sgn}(\sin\theta)A^{\delta-1}$$

$$+ \omega_0\cos^2\theta\xi(t). \quad (8.1.98)$$

By applying the stochastic averaging method, the Ito equation for the averaged amplitude process is obtained as

$$dA = m(A)dt + \sigma(A)dB(t), \quad (8.1.99)$$

where

$$m(A) = \left(-\zeta\omega_0 + \frac{3}{8}\pi\omega_0^2 K\right)A - DA^\delta, \quad (8.1.100)$$

$$\sigma(A) = \sqrt{\frac{\pi}{4}\omega_0^2 K A}, \quad (8.1.101)$$

$$D = \frac{\eta\omega_0^{\delta-1}\Gamma\left(\frac{\delta}{2}+1\right)}{\sqrt{\pi}\Gamma\left(\frac{\delta+3}{2}\right)}. \quad (8.1.102)$$

As mentioned in Example 8.1.2, that the amplitude process $A(t)$ is a suitable norm for system (8.1.96), and the Lyapunov exponent at the equilibrium point $a = 0$ (the left boundary) is the approximate largest Lyapunov exponent of system (8.1.96), which dictates its asymptotic sample stability of the trivial solution. It is calculated from Eq. (8.1.93) as

$$\lambda = m'(a = 0) - \frac{1}{2}[\sigma'(a = 0)]^2 = -\zeta\omega_0 + \frac{1}{4}\pi\omega_0^2 K - D\delta\lim_{a\to 0}a^{\delta-1}. \quad (8.1.103)$$

Thus, it is found that (i) the equilibrium point $a = 0$ is locally asymptotically stable with probability one if $0 \leq \delta < 1$ or $\delta > 1$ and

$\zeta > \pi\omega_0 K/4$, and (ii) the equilibrium point $a = 0$ is unstable if $\delta > 1$ and $\zeta \le \pi\omega_0 K/4$. However, the system global qualitative behaviors cannot be deduced only from the Lyapunov exponent at $a = 0$.

Now we try to acquire the system global dynamical behaviors by using the method of boundary classification at $a = 0$ and $a = \infty$. Equations (8.1.100) and (8.1.102) show that the left boundary at zero is singular of the first kind since $\sigma(0) = 0$, and the right boundary at infinity is singular of the second kind due to $m(\infty) = \infty$. Two cases, $0 \le \delta < 1$, and $\delta > 1$, need to be discussed separately.

Case 1: $0 \le \delta < 1$. For the left boundary at zero, the drift and diffusion exponents are determined from Eqs. (4.4.12) and (4.4.13) as

$$\beta = \delta < 1, \quad \alpha = 2. \tag{8.1.104}$$

Since $m(0^+) < 0$, the left boundary is an exit according to Table 4.4.2.

At the right boundary at infinity, the drift exponent, diffusion exponent and character value are determined from Eqs. (4.4.18), (4.4.19) and (4.4.20) as

$$\beta = 1, \quad \alpha = 2, \quad c = \frac{8\zeta - 3\pi\omega_0 K}{\pi\omega_0 K}. \tag{8.1.105}$$

According to Table 4.4.4, it is attractively natural if $c < -1$, strictly natural if $c = -1$, and repulsively natural if $c > -1$.

Figure 8.1.2 shows the sample behaviors near the two boundaries schematically. The left boundary is an exit regardless of ζ value. Thus, the trivial solution is always locally asymptotically stable. However, it is globally asymptotically stable in probability if and only if the right boundary is repulsively natural. Therefore, the necessary and sufficient condition for the trivial solution to be globally asymptotically stable in probability is

$$\zeta > \frac{1}{4}\pi\omega_0 K. \tag{8.1.106}$$

Under condition (8.1.106), the stationary probability density is the delta function $\delta(a)$. If $\zeta < \pi\omega_0 K/4$, the right boundary at infinity is attractively natural, and a stationary probability density does not exist even the left boundary is an exit. In this case, a sample may

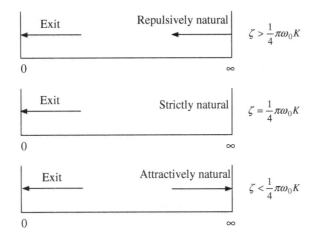

Figure 8.1.2 Sample behaviors at the two boundaries for $0 \leq \delta < 1$.

approach the left boundary, or may be unbounded. For the special case of $\zeta = \pi\omega_0 K/4$, the right boundary is strictly natural. When approaching the boundary, the sample behavior is uncertain; hence, the stationary probability density also does not exist.

It is noted that in the case of $\zeta > \pi\omega_0 K/4$, the Lyapunov exponent of the trivial solution is negative, indicating it is locally asymptotically stable. However, it is not globally asymptotically stable since the right boundary at infinity is attractively natural, and a sample initially not at $a = 0$ may also go unbounded.

For investigating the asymptotic stability in statistical moments, add an external excitation to system (8.1.96),

$$\ddot{X} + 2\zeta\omega_0\dot{X} + \eta|\dot{X}|^\delta \text{sgn}\dot{X} + \omega_0^2[1 + W(t)]X = W_1(t), \quad (8.1.107)$$

where $W_1(t)$ is another Gaussian White noise with a spectral density K_1, and independent of $W(t)$. The averaged drift and diffusion coefficients for the amplitude process are

$$m_1(A) = \left[-\zeta\omega_0 + \frac{3}{8}\pi\omega_0^2 K\right]A - DA^\delta + \frac{\pi K_1}{2\omega_0^2 A}, \quad (8.1.108)$$

$$\sigma_1(A) = \sqrt{\frac{\pi K_1}{\omega_0^2} + \frac{\pi}{4}\omega_0^2 K A^2}. \quad (8.1.109)$$

The stationary probability density are obtained from Eq. (4.4.7) as

$$p(a) = \frac{C}{4\pi K_1 + \pi\omega_0^4 K a^2}$$
$$\times \exp\left[-\int \frac{(8\zeta - 3\pi\omega_0 K)\omega_0^3 a + 8\omega_0^2 D a^\delta - 4\pi K_1/a}{4\pi K_1 + \pi\omega_0^4 K a^2} da\right].$$

$$(8.1.110)$$

The nth moment of $A(t)$ can be calculated from

$$E[A^n] = \int\limits_0^\infty a^n p(a) da. \qquad (8.1.111)$$

The integrability of (8.1.111) depends on the limiting behaviors of the integrand $a^n p(a)$ at the two boundaries, which can be deduced from the expression of $p(a)$, Eq. (8.1.100), as follows

$$a^n p(a) \rightarrow \begin{cases} \dfrac{C}{\pi\omega_0^4 K} a^{-(8\zeta/\pi\omega_0 K)+n+1}, & \text{as } a \to \infty \\[3mm] \dfrac{C}{4\pi K} a^{n+1}, & \text{as } a \to 0 \end{cases}. \qquad (8.1.112)$$

Equation (8.1.112) shows that the limiting behavior of $a^n p(a)$ at the right boundary controls the integrability of (8.1.111), resulting in the condition for the existence of the nth stationary moment $E[A^n]$ as

$$\zeta > \frac{\pi}{8}(n+2)\omega_0 K. \qquad (8.1.113)$$

If condition (8.1.113), which is stronger than (8.1.106) for $n > 0$, is met and the external excitation $W_1(t)$ vanishes, the stationary probability density is a delta function, and the statistical moments also vanish. Therefore, (8.1.113) is the condition for the trivial solution to be asymptotically stable in the nth moments for system (8.1.96).

It is noticed that (8.1.113) is the same as (8.1.76) for the linear system, indicating the extra nonlinear damping term $\eta|\dot{X}|^\delta \text{sgn}\dot{X}$ with $0 \le \delta < 1$ has no effect on the system stability in moments.

Case 2: $\delta > 1$. For the right boundary at infinity, the drift and diffusion exponents are determined as

$$\beta = \delta > 1, \quad \alpha = 2. \tag{8.1.114}$$

Since $m(\infty) < 0$, the right boundary is an entrance according to Table 4.4.4.

At the left boundary at zero, the drift exponent, diffusion exponent and character value are determined as

$$\beta = 1, \quad \alpha = 2, \quad c = \frac{-8\zeta + 3\pi\omega_0 K}{\pi\omega_0 K}. \tag{8.1.115}$$

According to Table 4.4.2, it is attractively natural if $c < 1$, strictly natural if $c = 1$, and repulsively natural if $c > 1$.

Figure 8.1.3 shows the sample behaviors near the two boundaries schematically. The right boundary is an entrance regardless of ζ value. The necessary and sufficient condition for the trivial solution to be asymptotically stable in probability is the left boundary is attractively natural, leading to the same condition given in (8.1.106), i.e., $\zeta > \pi\omega_0 K/4$.

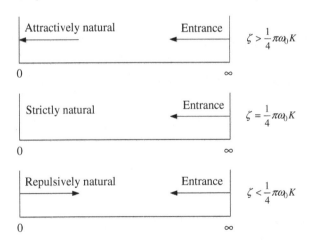

Figure 8.1.3 Sample behaviors at the two boundaries for $\delta > 1$.

If $\zeta < \pi\omega_0 K/4$, the left boundary at zero is repulsively natural, the trivial solution is unstable in probability, and a stationary probability density of non-delta type exists. With the drift and diffusion coefficients given by (8.1.100) and (8.1.101), the stationary probability density is found to be

$$p(a) = C a^{-(8\zeta/\pi\omega_0 K)+1} \exp\left[-\frac{8D}{\pi(\delta-1)\omega_0^2 K} a^{\delta-1}\right], \quad \delta > 1.$$

$$(8.1.116)$$

It can be shown that the condition for a valid probability density in (8.1.116), i.e., an integrable $p(a)$, is exactly $\zeta < \pi\omega_0 K/4$.

The nth stationary moment is calculated from the probability density as

$$E[A^n] = \int_0^\infty a^n p(a)\,da = \left[\frac{1}{8D}\pi(\delta-1)\omega_0^2 K\right]^{\frac{n}{\delta-1}} \frac{\Gamma\left(\gamma+\frac{n}{\delta-1}\right)}{\Gamma(\gamma)},$$

$$(8.1.117)$$

where

$$\gamma = \frac{2}{\delta-1}\left(-\frac{4\zeta}{\pi\omega_0 K}+1\right). \qquad (8.1.118)$$

The condition for a non-trivial $E[A^n]$ is $\gamma > 0$, namely, $\zeta < \pi\omega_0 K/4$, the same as the condition for a non-trivial probability density to exist. As $\zeta \to \pi\omega_0 K/4$, i.e., $\gamma \to 0^+$, we have

$$\lim_{\gamma\to 0^+} \Gamma(\gamma) = \infty, \qquad \lim_{\zeta\to\frac{1}{4}\pi\omega_0 K} E[A^n] = 0. \qquad (8.1.119)$$

Therefore, the condition for the asymptotic stability in probability, $\zeta > \pi\omega_0 K/4$, is the same as that in the statistical moments, regardless of the order of the moments.

Comparison of Figs. 8.1.2 and 8.1.3 shows that (i) the extra nonlinear damping term $\eta|\dot{X}|^\delta \mathrm{sgn}\dot{X}$ with $0 \le \delta < 1$ affects the sample behaviors at the left boundary since it increases the damping when $A \to 0$, and (ii) the extra nonlinear damping term with $\delta > 1$ increases the damping when $A \to \infty$, and affects the sample behaviors at the right boundary. These effects change the topological behaviors of the system, as described above.

8.1.5 *Asymptotic Stability of Quasi-Hamiltonian Systems*

Consider a quasi completely non-integrable Hamiltonian system governed by (7.5.17). Also assume that external excitations are absent and parametric excitations are present so that the stability analysis for the trivial solution is feasible. As shown in Section 7.5.2, the system Hamiltonian can be approximated as a one-dimensional Markov diffusion process by applying the stochastic averaging method of quasi Hamiltonian systems. The Hamiltonian is governed by the averaged Ito differential equation

$$dH = m(H)dt + \sigma(H)dB(t), \qquad (8.1.120)$$

where the drift coefficient $m(H)$ and diffusion coefficient $\sigma(H)$ can be obtained from (7.5.24) and (7.5.25) by carrying out the averaging defined in (7.5.26). In general, $m(H)$ and $\sigma(H)$ are nonlinear function, and Eq. (8.1.120) is a nonlinear Ito equation. According to (8.1.93), the Lyapunov exponent of $H(t)$ at the trivial solution $H = 0$ is

$$\lambda_H = m'(0) - \frac{1}{2}[\sigma'(0)]^2. \qquad (8.1.121)$$

The method of Lyapunov exponent is the simplest way for asymptotic sample stability analysis of the trivial solution (Zhu, 2004). The method of identifying the sample behavior at two boundaries is another way for stability analysis of the trivial solution (Zhu and Huang, 1998). The advantage of the latter method is that it can acquire the global qualitative behavior of system (8.1.120), as mentioned previously.

It is noted that the Hamiltonian function $H(t)$, although always non-negative, is not a Euclidean norm defined in (8.1.2). The question arises as whether the stability analysis of system (8.1.120) is applicable to the original quasi-Hamiltonian system (7.5.17). To surmount the problem, a new process $H^{1/2}$ is considered (Zhu and Huang, 1998; Zhu, 2004). If the conservative Hamiltonian system corresponding to the quasi-Hamiltonian system (7.5.17) is linear, $H^{1/2}$ is an Euclidean norm defined in (8.1.2). However, if the conservative Hamiltonian system is nonlinear, $H^{1/2}$ is not an Euclidean norm of homogeneity of degree one. Nevertheless, H usually represents the system total energy (may be in wide sense), and $H^{1/2}$ may be used to measure the distance of system state to the trivial solution in state space. Furthermore, the quadratic part dominates the Hamiltonian in vicinity of the trivial solution. Thus, it is reasonable to use $H^{1/2}$ as a norm in stability analysis. The Lyapunov exponent for process $Y(t) = H^{1/2}(t)$ at the trivial solution $Y = 0$ is obtained as

$$\lambda_Y = \lim_{t\to\infty} \frac{1}{t} \ln H^{1/2} == \frac{1}{2} \lim_{t\to\infty} \frac{1}{t} \ln H$$

$$= \frac{1}{2}\lambda_H = \frac{1}{2}\left\{ m'(0) - \frac{1}{2}[\sigma'(0)]^2 \right\}. \qquad (8.1.122)$$

Comparison of Eqs. (8.1.121) and (8.1.122) shows that $\lambda_H = 2\lambda_Y$. Since only the sign of the Lyapunov exponent is of concern in stability analysis, use of either λ_H or λ_Y leads to the same results.

The two methods for stability analysis will be illustrated in the following example.

Example 8.1.7 Consider the following quasi non-integrable Hamiltonian system

$$\ddot{X} + \alpha_1 \dot{X} + \beta_1 X^2 \dot{X} + \omega_1^2 X + aY + b|X - Y|^\delta \operatorname{sgn}(X - Y)$$
$$= C_1 X W_1(t),$$
$$\ddot{Y} + \alpha_2 \dot{Y} + \beta_2 Y^2 \dot{Y} + \omega_2^2 Y + aX + b|X - Y|^\delta \operatorname{sgn}(Y - X)$$
$$= C_2 Y W_2(t), \qquad (8.1.123)$$

where $\alpha_1 \neq 0$, $\alpha_2 \neq 0$, β_1, β_2, a, b and $\delta > 0$ are constants, and $W_1(t)$ and $W_2(t)$ are two independent Gaussian white noises with spectral densities K_1 and K_2, respectively. The system Hamiltonian is the total energy

$$H = \frac{1}{2}(\dot{X}^2 + \dot{Y}^2) + U(X, Y), \qquad (8.1.124)$$

where

$$U(X, Y) = \frac{1}{2}(\omega_1^2 X^2 + \omega_2^2 Y^2) + aXY + \frac{b}{1 + \delta}|X - Y|^{\delta+1}. \qquad (8.1.125)$$

Assuming that the damping forces and excitations are weak and the stochastic averaging method is applicable, we obtain an Ito equation for H as given in Eq. (8.1.120). The drift and diffusion coefficients are obtained as (Zhu, 2004)

$$m(H) = \frac{1}{T(H)} \int_\Omega \frac{1}{\dot{x}} [(-\alpha_1 - \beta_1 x^2)\dot{x}^2 + (-\alpha_2 - \beta_2 y^2)\dot{y}^2$$

$$+\pi K_1 C_1^2 x^2 + \pi K_2 C_2^2 y^2] dx dy d\dot{y}, \qquad (8.1.126)$$

$$\sigma^2(H) = \frac{2\pi}{T(H)} \int_\Omega \frac{1}{\dot{x}} \left(K_1 C_1^2 x^2 \dot{x}^2 + K_2 c_2^2 y^2 \dot{y}^2 \right) dx dy d\dot{y}, \qquad (8.1.127)$$

where

$$T(H) = \int_\Omega \frac{1}{\dot{x}} dx dy d\dot{y}, \qquad (8.1.128)$$

where domain Ω is defined as $H(x, y, \dot{x} = 0, \dot{y}) = \frac{1}{2}\dot{y}^2 + U(x,y) \leq H$.

The stability of the trivial solution $H = 0$ will be determined separately for two cases of $0 < \delta < 1$ and $\delta > 1$.

Case 1: $0 < \delta < 1$. First we apply the method of Lyapunov exponent. The linearized Ito equation of $H(t)$ at trivial solution $H = 0$ is

$$dH = \mu_1 H dt + \sqrt{\mu_2 H} dB(t), \qquad (8.1.129)$$

where (Zhu and Huang, 1998)

$$\mu_1 = -\frac{1}{3}(\alpha_1 + \alpha_2)\eta_1 + \frac{7\pi(C_1^2 K_1 + C_2^2 K_2)}{9(2a + \omega_1^2 + \omega_2^2)}, \qquad (8.1.130)$$

$$\mu_2 = \frac{2\pi(C_1^2 K_1 + C_2^2 K_2)}{3(2a + \omega_1^2 + \omega_2^2)}\eta_2, \qquad (8.1.131)$$

$$\eta_1 = \frac{6}{11} - \frac{1+\delta}{4(4+\delta)} - \frac{2(2-\delta)}{(2+\delta)(3+\delta)} + \frac{\delta^2 - 1}{48(5+\delta)}, \qquad (8.1.132)$$

$$\eta_2 = \frac{13}{12} - \frac{16}{(2+\delta)(5+\delta)} + \frac{4}{3+\delta} - \frac{3+\delta}{2(4+\delta)} + \frac{(1+\delta)(3+\delta)}{24(5+\delta)}.$$
$$(8.1.133)$$

Applying (8.1.93), we obtain the Lyapunov exponent as

$$\lambda = \frac{1}{2}\mu_1 - \frac{1}{4}\mu_2 = -\frac{1}{6}(\alpha_1 + \alpha_2)\eta_1 + \frac{1}{18}(7 - 3\eta_2)\frac{\pi(C_1^2 K_1 + C_2^2 K_2)}{(2a + \omega_1^2 + \omega_2^2)}.$$
$$(8.1.134)$$

Thus, the condition for the trivial solution to be locally asymptotically stable with probability one is $\lambda < 0$, i.e.,

$$\alpha_1 + \alpha_2 > D = \frac{1}{3\eta_1}(7 - 3\eta_2)\frac{\pi(C_1^2 K_1 + C_2^2 K_2)}{(2a + \omega_1^2 + \omega_2^2)}. \qquad (8.1.135)$$

Next, the stability analysis is carried out by identifying the sample behavior at two boundaries. It is found (Zhu, 2004) that the sample behaviors at the two boundaries are shown in Fig. 8.1.4.

It is seen from Fig. 8.1.4 that the left boundary $h = 0$ is attractively natural and the right boundary $h = \infty$ is an entrance if

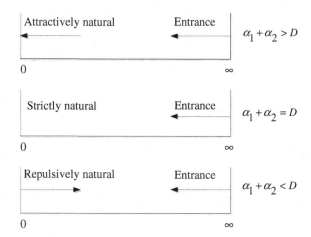

Figure 8.1.4   Sample behaviors at the two boundaries of system (8.1.123) with $0 < \delta < 1$.

condition (8.1.135) is satisfied, i.e., $\alpha_1 + \alpha_2 > D$. Under this condition, the trivial solution $H = 0$ is asymptotically stable in probability. On the other hand, if $\alpha_1 + \alpha_2 < D$, the trivial solution $H = 0$ is unstable, and a non-trivial stationary probability density exists.

Case 2: $\delta > 1$. In this case, the linearized Ito equation of H at trivial solution $H = 0$ is

$$dH = \mu_3 H dt + \sqrt{\mu_4} H dB(t), \tag{8.1.136}$$

where (Zhu, 2004)

$$\mu_3 = -\frac{1}{2}(\alpha_1 + \alpha_2) + \pi\eta\left(\frac{C_1^2 K_1}{\omega_1^2} + \frac{C_2^2 K_2}{\omega_2^2}\right), \tag{8.1.137}$$

$$\mu_4 = \frac{1}{3}\pi\eta\left(\frac{C_1^2 K_1}{\omega_1^2} + \frac{C_2^2 K_2}{\omega_2^2}\right), \tag{8.1.138}$$

$$\eta = \int_0^\pi \left(1 + \frac{a}{\omega_1\omega_2 \sin 2\theta}\right)^{-2} d\theta \Big/ \int_0^\pi \left(1 + \frac{a}{\omega_1\omega_2 \sin 2\theta}\right)^{-1} d\theta. \tag{8.1.139}$$

The Lyapunov exponent is then

$$\lambda = \frac{1}{2}\mu_3 - \frac{1}{4}\mu_4 = -\frac{1}{4}(\alpha_1 + \alpha_2) + \frac{1}{6}\pi\eta\left(\frac{C_1^2 K_1}{\omega_1^2} + \frac{C_2^2 K_2}{\omega_2^2}\right). \tag{8.1.140}$$

Thus, the trivial solution is locally asymptotically stable with probability one if $\lambda < 0$, i.e.,

$$\alpha_1 + \alpha_2 > G = \frac{2}{3}\pi\eta\left(\frac{C_1^2 K_1}{\omega_1^2} + \frac{C_2^2 K_2}{\omega_2^2}\right). \tag{8.1.141}$$

The boundary behaviors are the same as those in Case 1 of $0 < \delta < 1$ as long as constant D in Case 1 is replaced by constant G for the present case.

It is shown that the two methods lead to the same result for the stability of the trivial solution.

For more information about stochastic stability, please refer to (Kozin, 1969; Ariaratnam and Tam, 1979; Khasminskii, 1980; Arnold, 1974, 1998; Lin and Cai, 1995; Zhu, 2004).

8.2　Stochastic Bifurcation

Bifurcation theory is the study of changes in the qualitative or topological structure of systems dynamical behaviors in the steady state when system properties change smoothly and pass critical thresholds. For a system described by differential equations, a bifurcation occurs when a small smooth change of system parameter values (the bifurcation parameters) causes a sudden "qualitative" change in the system behavior. The name "bifurcation" was first introduced by Poincaré (Poincare, 1885). The bifurcation phenomena occur only in nonlinear systems.

Bifurcations are usually divided into two principal classes: (i) local bifurcation, which can be analyzed entirely through changes in the local stability properties of equilibriums (fixed points), periodic orbits or other invariant sets, as parameters cross through critical thresholds; and (ii) global bifurcations, which often occur when larger invariant sets of the system 'collide' with each other, or with equilibriums of the system. They cannot be detected purely by a stability analysis of the equilibriums. In this book, only local bifurcation is considered.

The bifurcation theory has been extended from the deterministic systems to stochastic systems in the past several decades. (Meunier and Verga, 1988; Sri Namachchivaya, 1990; Ariaratnam, 1980, 1993; Arnold, 1998; Zhu and Huang, 1999). In this section, basic knowledge of deterministic bifurcations is introduced first, and then classification and analysis procedures of stochastic bifurcations are described.

8.2.1　*Deterministic Bifurcations*

First consider one-dimensional systems. A typical one-dimensional ordinary differential equation (ODE) is written as

$$\dot{x} = f(x), \tag{8.2.1}$$

where x is a function of time t. We assume that Eq. (8.2.1) is an autonomous system, i.e., $f(x)$ does not depend on t explicitly. If the initial value of $x(t)$ is known, Eq. (8.2.1) can be solved, and we know exactly how the system is evolving.

A fixed point (or equilibrium point) x^* of system (8.2.1) is a root of equation $f(x) = 0$, i.e., $f(x^*) = 0$. If both, function f and its derivative $f' = df/dt$ are continuous at point x^*, the stability of equilibrium x^* may be identified by linearizing system (8.2.1) at $x = x^*$, i.e.,

$$\dot{z} = f'(x*)z, \tag{8.2.2}$$

where $z = x - x^*$. The solution of Eq. (8.2.2) is

$$z = z(0)\exp[f'(x*)t]. \tag{8.2.3}$$

We assume that $z(0) \neq 0$. Equation (8.2.3) indicates that (i) if $f'(x*) < 0$, $z \to 0$ and $x \to x^*$ as $t \to \infty$, then x^* is a stable equilibrium; (ii) if $f'(x*) > 0$, z, $x \to \infty$ as $t \to \infty$, then x^* is an unstable equilibrium; and (iii) if $f'(x*) = 0$, nothing can be said about the stability of x^*. Therefore, the stabilities of the fixed points are critical in describing the system topological behavior. Now assume that there is a system parameter α in the one-dimensional ODE. The characteristics of the fixed points, such as their number and stabilities, may depend on parameter α. The qualitative behavior of the system may or may not change when α changes. However, if a sudden change takes place when parameter α passes through a certain value, we say that a bifurcation occurs. Depending on the nature of the changes, bifurcations are classified into the following categories.

Transcritical bifurcation. A transcritical bifurcation is one in which a fixed point changes its stability, from unstable to stable, or vice versa, when the parameter passes through the bifurcation point. The normal form of a transcritical bifurcation is

$$\dot{x} = \alpha x - x^2. \tag{8.2.4}$$

There is either one ($\alpha = 0$) or two ($\alpha \neq 0$) fixed points. When $\alpha = 0$, the only fixed point is $x = 0$ which is semi-stable, i.e., stable from the right and unstable from the left. For $\alpha \neq 0$, there are two fixed points, $x = 0$ and $x = \alpha$. At $x = 0$, the linearized system is

$$\dot{z} = \alpha z, \qquad (8.2.5)$$

where $z = x$. Thus, $x = 0$ is stable if $\alpha < 0$ and unstable if $\alpha > 0$. For the fixed point $x = \alpha$, the linearized system is

$$\dot{z} = -\alpha z, \qquad (8.2.6)$$

where $z = x - \alpha$. Thus, $x = \alpha$ is unstable if $\alpha < 0$ and stable if $\alpha > 0$. Since the stabilities of both fixed point have switched when parameter α passes through zero, the bifurcation occurs at $\alpha = 0$. The stability characteristics of the system are shown in Fig. 8.2.1.

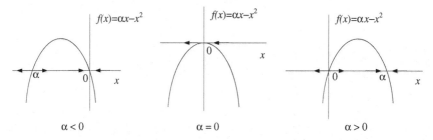

$$\alpha < 0 \qquad\qquad \alpha = 0 \qquad\qquad \alpha > 0$$

Figure 8.2.1   Stabilities of fixed points in transcritical bifurcation of system (8.2.4).

Shown in Fig. 8.2.2 is the so-called bifurcation diagram, in which the stable and unstable fixed points are represented by solid lines and dashed lines, respectively, and the arrows describe the moving trends of the system state.

Saddle-node bifurcation. In this type of bifurcation, fixed points are created or destructed when the parameter passes through the bifurcation point. The normal form of a saddle-node bifurcation is

$$\dot{x} = \alpha - x^2. \qquad (8.2.7)$$

For $\alpha > 0$, there are two fixed points given by $x = \pm\sqrt{\alpha}$. It is easy to show that the equilibrium $x = -\sqrt{\alpha}$ is stable, while $x = \sqrt{\alpha}$ is unstable. When $\alpha = 0$, there is a single fixed point at $x = 0$ which is semi-stable, and called saddle-node fixed point. Finally, if $\alpha < 0$,

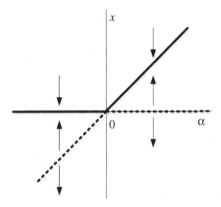

Figure 8.2.2   Diagram of transcritical bifurcation of system (8.2.4).

there are no fixed points at all. Thus, when parameter α decreases from a positive value and passes through zero, two fixed points merge to one, and then disappear. The stability of the system is illustrated in Fig. 8.2.3, while the bifurcation diagram is given in Fig. 8.2.4.

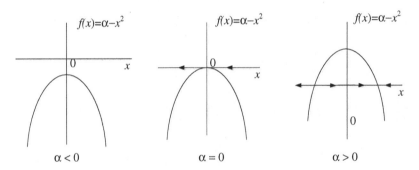

Figure 8.2.3   Stabilities of fixed points in saddle-node bifurcation of system (8.2.7).

Pitchfork bifurcation. Pitchfork bifurcations have two types — supercritical or subcritical. The normal form for supercritical pitchfork bifurcation is

$$\dot{x} = \alpha x - x^3. \tag{8.2.8}$$

The cases of $\alpha \leq 0$ and $\alpha > 0$, once again, give very different dynamics. If $\alpha \leq 0$, only a single fixed point, $x = 0$, exists, and it

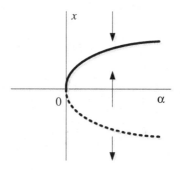

Figure 8.2.4 Diagram of saddle node bifurcation of system (8.2.7).

is stable. For the case of $\alpha > 0$, three fixed points appear, among which $x = 0$ becomes unstable and $x = \pm\sqrt{\alpha}$ are stable. The stability diagram is shown Fig. 8.2.5. The bifurcation diagram is shown in Fig. 8.2.6. The name of "pitchfork" is from the shape of the bifurcation diagram. The subjective "supercritical" indicates that the presence of the nonlinear term $-x^3$ stabilizes the system.

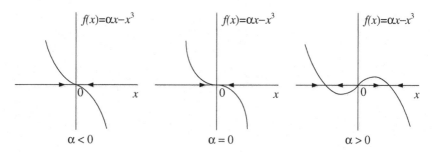

Figure 8.2.5 Stabilities of fixed points in supercritical pitchfork bifurcation of system (8.2.8).

The normal form for subcritical pitchfork bifurcation is

$$\dot{x} = \alpha x + x^3. \tag{8.2.9}$$

Figures 8.2.7 and 8.2.8 show the stability of the fixed points and the bifurcation diagram. In contrary to the supercritical case, the nonlinear term x^3 destabilizes the system.

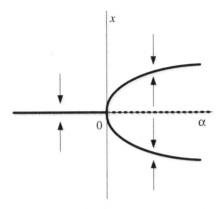

Figure 8.2.6 Diagram of supercritical pitchfork bifurcation of system (8.2.8).

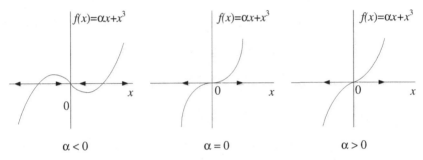

Figure 8.2.7 Stabilities of fixed points in subcritical pitchfork bifurcation of system (8.2.9).

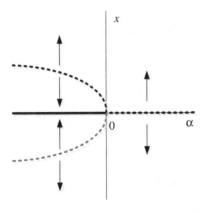

Figure 8.2.8 Diagram of subcritical pitchfork bifurcation of system (8.2.9).

Bifurcations of two-dimensional systems. The above three types of bifurcations take place for one-dimensional systems. Now consider two-dimensional autonomous systems, governed by

$$\dot{x} = f(x, y),$$
$$\dot{y} = g(x, y). \qquad (8.2.10)$$

Equilibrium points can also be found by solving equations

$$f(x, y) = 0,$$
$$g(x, y) = 0. \qquad (8.2.11)$$

The two equations in (8.2.11) define two lines on the $x - y$ phase plane, called nullclines. The fixed pints are the intersections of the two nullclines. To determine the stability of an equilibrium point (x^*, y^*), linear analysis may also be applied. First calculate the Jacobian matrix of the system, defined as

$$\mathbf{J} = \begin{bmatrix} \dfrac{\partial f}{\partial x} & \dfrac{\partial f}{\partial y} \\ \dfrac{\partial g}{\partial x} & \dfrac{\partial g}{\partial y} \end{bmatrix}_{x=x*, y=y*} . \qquad (8.2.12)$$

From this matrix, we can determine the eigenvalues from the characteristic equation

$$\det(\mathbf{J} - \lambda \mathbf{I}) = 0. \qquad (8.2.13)$$

In the case of a two-dimensional system, we have two eigenvalues which may be real or complex numbers. If the real parts of the two eigenvalues are both negative, the fixed point is stable and attracting the motion in two directions. If both real parts are positive, it is unstable and repelling in two directions. If one is positive and one is negative, it is a saddle point, and attracting in one direction and repelling in the other direction. A saddle fixed point is essentially unstable.

The linear analysis works well in general when both the real parts of two eigenvalues are not zero. If at least one of the real parts vanishes, it may be inappropriate.

For two-dimensional nonlinear systems, equilibrium points are one form of the steady states. For conservative systems, periodic solutions are also possible. Mathematically, a periodic solution is a first integral of the system equations corresponding to an energy (in wide sense) level. Each periodic solution depends on the initial conditions. For dissipative systems, another possible form besides the equilibrium points is the so-called limit cycle. A stable limit cycle is a closed trajectory in phase plane regardless of initial conditions. All the neighboring trajectories, outside or inside, approach it as time approaches infinity. The limit cycle, as a closed trajectory, also describes the periodic behavior of the system, and is an important characteristic of dissipative nonlinear systems. It is noted that the terms of "periodic solution" and "limit cycle" are used here with different meanings, although a limit cycle is also a periodic trajectory. Both cannot coexist in the same system. The number of periodic solutions may be infinite for a system, while there is only finite number of limit cycles in a system if they exist.

Example 8.2.1 The deterministic prey-predator system is described by the classical Lotka–Volterra model (7.4.141) in Section 7.4.3,

$$\begin{aligned} \dot{x}_1 &= x_1(a - bx_2), \\ \dot{x}_2 &= x_2(-c + fx_1), \end{aligned} \qquad (8.2.14)$$

where x_1 and x_2 are the population densities of preys and predators, respectively, and a, b, c and f are positive parameters. For system (8.2.14), there exist two fixed points, $(0, 0)$ and $(c/f, a/b)$, and periodic solutions

$$r(x_1, x_2) = fx_1 - c - c\ln\frac{fx_1}{c} + bx_2 - a - a\ln\frac{bx_2}{a} = R, \quad (8.2.15)$$

where R is a constant depending on the initial state. It can be shown that the fixed point $(0, 0)$ is a saddle point with one positive and one negative eigenvalues, and both the real parts of the eigenvalues at fixed point $(c/f, a/b)$ are zeros. For any positive constant R, (8.2.15) represents a periodic trajectory. With any non-trivial initial state $x_1 \neq c/f$ and $x_2 \neq a/b$, the system will follow a periodic trajectory, as shown in Fig. 7.4.1.

Example 8.2.2 System (8.2.14) shows that the prey population will grow without limit in the absence of the predators, contrary to what is expected of a realistic predator-prey ecosystem. The improved Lotka–Volterra model by adding a self-competition term $-sx_1^2$ is given by Eq. (7.4.145), i.e.,

$$\begin{aligned}
\dot{x}_1 &= x_1(a_1 - sx_1 - bx_2), \\
\dot{x}_2 &= x_2(-c + fx_1).
\end{aligned} \tag{8.2.16}$$

By letting the right-hand-sides of (8.2.16) be zero, three equilibrium points can be found for the system:

1. the trivial equilibrium point E_0, $x_1 = 0$ and $x_2 = 0$,
2. the predator-free equilibrium point E_1, $x_1 = a_1/s$ and $x_2 = 0$, and
3. the coexistence equilibrium point E^*,

$$x_1^* = c/f, \quad x_2^* = \left(a_1 - \frac{sc}{f}\right)/b. \tag{8.2.17}$$

Equation (8.2.17) shows that the existence of E^* requires

$$a_1 f - sc > 0. \tag{8.2.18}$$

We assume that (8.2.18) is satisfied.

Denote a fixed point as $(\tilde{x}_1, \tilde{x}_2)$. To linearize the system about the fixed point, let

$$z_1 = x_1 - \tilde{x}_1, \quad z_2 = x_2 - \tilde{x}_2. \tag{8.2.19}$$

Then the linearized equations for z_1 and z_2 are derived from Eq. (8.2.16) as follows

$$\begin{aligned}
\dot{z}_1 &= Az_1 + Bz_2, \\
\dot{z}_2 &= Cz_1 + Dz_2,
\end{aligned} \tag{8.2.20}$$

where

$$\begin{aligned}
A &= a_1 - 2s\tilde{x}_1 - b\tilde{x}_2, \quad B = -b\tilde{x}_1, \\
C &= f\tilde{x}_2, \quad D = -c + f\tilde{x}_1.
\end{aligned} \tag{8.2.21}$$

The characteristic equation associated with system (8.2.20) is

$$
\begin{vmatrix} A - \lambda & B \\ C & D - \lambda \end{vmatrix} = 0. \tag{8.2.22}
$$

The two eigenvalues are obtained as

$$
\lambda_1 = \frac{A + D}{2} + \frac{\sqrt{(A - D)^2 + 4BC}}{2}
$$

$$
\lambda_2 = \frac{A + D}{2} - \frac{\sqrt{(A - D)^2 + 4BC}}{2}. \tag{8.2.23}
$$

The stability of each equilibrium point can be determined according to the signs of the real parts of λ_1 and λ_2.

First consider the trivial equilibrium point E_0. Substitution of $\tilde{x}_1 = 0$ and $\tilde{x}_2 = 0$ into (8.2.21) and (8.2.23) results in $\lambda_1 = a_1$ and $\lambda_2 = -c$. Thus, it is a saddle point. For the predator-free equilibrium point E_1, we have

$$
\lambda_1 = \frac{1}{s}(a_1 f - cs) > 0, \quad \lambda_2 = -a_1 < 0. \tag{8.2.24}
$$

Therefore, E_1 it is also a saddle point. For the coexistence equilibrium point E^*,

$$
A = -\frac{sc}{f}, \quad B = -\frac{bc}{f}, \quad C = \frac{a_1 f - sc}{b}, \quad D = 0. \tag{8.2.25}
$$

According to (8.2.23), both eigenvalues possess negative real parts; thus, E^* is stable. If the system initial state (x_1, x_2) is inside the first quadrant, i.e., $x_1 > 0$ and $x_2 > 0$, the system will be approaching the stable fixed point E^*, as shown in Fig. 7.4.2.

Example 8.2.3 Consider another nonlinear system

$$
\dot{x}_1 = -x_2 + x_1(1 - x_1^2 - x_2^2),
$$

$$
\dot{x}_2 = x_1 + x_2(1 - x_1^2 - x_2^2). \tag{8.2.26}
$$

To investigate the stability of system (8.2.26), transform the variables to polar coordinates, i.e., let

$$
x_1 = \rho \cos \phi, \quad x_2 = \rho \sin \phi \tag{8.2.27}
$$

Taking differentiation of (8.2.27) with respect to t, and combining with (8.2.26), we obtain

$$\dot{\rho}\cos\phi - \rho\dot{\phi}\sin\phi = -\rho\sin\phi + \rho\cos\phi(1 - \rho^2),$$
$$\dot{\rho}\sin\phi + \rho\dot{\phi}\cos\phi = \rho\cos\phi + \rho\sin\phi(1 - \rho^2).$$

$$(8.2.28)$$

From (8.2.28), we have the system equations in polar form

$$\dot{\rho} = \rho(1 - \rho^2)$$
$$\dot{\phi} = 1.$$

$$(8.2.29)$$

Since the two functions $\rho(t)$ and $\phi(t)$ are uncoupled, the analysis is straightforward. The system (8.2.29) has an unstable fixed point $\rho = 0$ corresponding to $(x_1 = 0, x_2 = 0)$, and a stable equilibrium $\rho = 1$, which corresponds to a stable limit cycle

$$x_1^2 + x_2^2 = 1.$$

$$(8.2.30)$$

As long as the system begins from a point not at origin $(0, 0)$, it will eventually move along the circular limit cycle, as illustrated in Fig. 8.2.9. The second equation in (8.2.29) shows that the system circular motion has a constant speed.

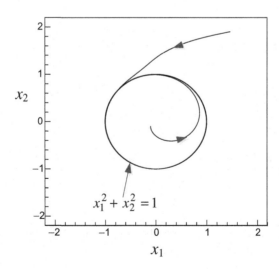

Figure 8.2.9 A limit cycle of system (8.2.26).

Similar to the one-dimensional case, the bifurcation of 2D systems is defined as any qualitative or topological change of system steady-state behaviors when the system parameters change. The qualitative change may be transit from one type to another type, creation or destruction of one type, or change of stability (from stable to unstable, or vice versa). Excluding the global bifurcation, there are two types of bifurcations in 2D systems: 1D-like bifurcations and Hopf bifurcation. The Hopf bifurcation is a new type that can only occur in a two-dimensional system. In order for a Hopf bifurcation to. occur, the eigenvalues at a fixed point must form a complex conjugate pair. Then, when the real part of the eigenvalues changes its sign, the stability of the involved equilibrium point changes, and the so-called Hopf bifurcation occurs. In the case when a stable fixed point becomes unstable and a stable limit cycle is born, we have a supercritical Hopf bifurcation. On the other hand, when a limit cycle disappear and the equilibrium point becomes stable, we have a subcritical Hopf bifurcation.

Example 8.2.4 Introduce a parameter α in system (8.2.26) of Example 8.2.3 to yield

$$\begin{aligned} \dot{x}_1 &= -x_2 + x_1(\alpha - x_1^2 - x_2^2), \\ \dot{x}_2 &= x_1 + x_2(\alpha - x_1^2 - x_2^2). \end{aligned} \tag{8.2.31}$$

Applying the same transformation (8.2.27), we obtain the system equations in polar form

$$\begin{aligned} \dot{\rho} &= \rho(\alpha - \rho^2), \\ \dot{\phi} &= 1. \end{aligned} \tag{8.2.32}$$

If $\alpha < 0$, the system has only one stable equilibrium point $\rho = 0$ corresponding to the origin ($x_1 = 0$, $x_2 = 0$) on the phase plane. It remains stable when $\alpha = 0$ but nonlinearly. When $\alpha > 0$, the equilibrium point $\rho = 0$ becomes unstable, and a stable fixed point $\rho = \sqrt{\alpha}$, corresponding to a stable limit cycle $x_1^2 + x_2^2 = \alpha$, appears. All orbits starting outside or inside the limit cycle except at the origin tend to the cycle as $t \to +\infty$. Therefore, $\alpha = 0$ is a bifurcation point at which the system undergoes a supercritical Hopf bifurcation. To verify the Hopf bifurcation phenomenon, the two eigenvalues for the equilibrium point $(0, 0)$ are found to be $\lambda_{1,2} = \alpha \pm i$. The real part indeed changes the sign when α passes zero. The system moving trajectories are illustrated in Fig. 8.2.10 on phase plane, and the bifurcation diagram is drawn in Fig. 8.2.11 in (x, y, α)-space. The family of limit cycles forms a paraboloid surface.

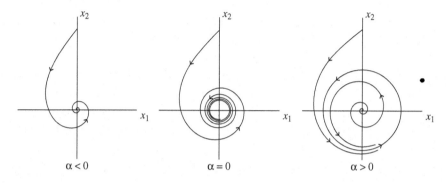

Figure 8.2.10   Bifurcation diagram on phase plane.

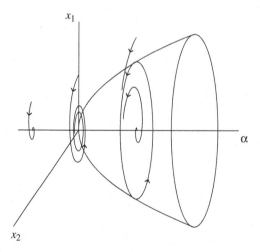

Figure 8.2.11   Diagram of supercritical Hopf bifurcation.

8.2.2 *Stochastic Bifurcations*

Similar to the deterministic case, stochastic bifurcation is the phenomenon of sudden qualitative change in the behavior for a stochastically excited system when a small smooth change of a system parameter value takes place. For deterministic systems, only steady state is considered, and the qualitative behavior for a two-dimensional system can be characterized by (i) equilibrium points (fixed points), stable or unstable, (ii) periodic trajectories, and (iii) limit cycles. Analogous to the steady state of deterministic systems, only stationary state is considered for stochastic systems. The periodic trajectories and limit cycles in deterministic systems are no longer present, and the system qualitative behavior is analyzed according to the so-called invariant measures, including stable or unstable fixed points and stationary probability density. A fixed point is in fact corresponding to a probability density of Dirac-delta function form. Therefore, the term of invariant measure is equivalent to the stationary probability density for the present situation. The qualitative change in stationary probability density is then used to define the bifurcation phenomenon.

Two classes of stochastic bifurcation. There are two classes of stochastic bifurcation: dynamical bifurcation (D-bifurcation) and phenomenological bifurcation (P-bifurcation). D-bifurcation specifies the change in nature of the probability density when a parameter passes through the bifurcation point, such as transition from a fixed point (probability density of Dirac-delta function type) to a non-trivial probability density, or from a non-trivial probability density to unbounded response, etc. On the other hand, P-bifurcation describes the shape change of the stationary probability density, such as transition from descendent shape (no peak) to unimodal (one peak), from unimodal to bimodal (two peaks), etc. The two classes are not related, but a single bifurcation point may be D-type, P-type or both. The precise mathematical definition for the D-bifurcation was given in (Arnold, 1998).

Determination of D- and P-bifurcations for one-dimensional systems is quite straightforward. Several methods can be applied, as described below.

Method of Lyapunov exponent. It is known that the sign of the Lyapunov exponent for an equilibrium point determines the local stability of the point. When a parameter value renders a zero Lyapunov exponent, the system qualitative behavior changes, from stable to unstable, or vice versa. Thus, at the point of zero Lyapunov exponent, a D-bifurcation takes place.

Method of boundary classification. There are generally two boundaries for a one-dimensional response process. Without loss of generality, assume the two boundaries as zero and infinity, and both are singular. According to Section 4.4, a boundary can be classified into different types based on the drift and diffusion coefficients. Roughly, the boundary types can be divided into two categories: (i) stable boundary including exit and attractively natural, and (ii) unstable boundary including strictly natural, repulsively natural and entrance. When a boundary changes its type from stable to unstable, or vice versa, the system qualitative behavior changes and D-bifurcation occurs.

Method of stationary probability density. For one-dimensional system governed by an Ito equation,

$$dX = m(X)dt + \sigma(X)dB(t), \qquad (8.2.33)$$

an expression of the stationary probability density is Eq. (4.4.7), i.e.,

$$p(x) = \frac{C}{\sigma^2(x)} \exp\left[\int \frac{2m(x)}{\sigma^2(x)} dx\right]. \qquad (8.2.34)$$

At the point where the integrability of (8.2.34) changes, D-bifurcation occurs. When the shape of $p(x)$ changes, P-bifurcation happens. If an analytical expression can be found for $p(x)$, this method is quite simple. On the other hand, the method may be not applicable if $p(x)$ must be calculated numerically.

Method of three indices. Assume that the boundary at infinity is an entrance or repulsively natural, then the system behavior solely depends on the nature of left boundary at zero. The method of three indices was designed for this case (Zhu and Huang, 1999). The three indices are diffusion exponent, drift exponent and character value, defined in Section 4.4.

Case 1. The boundary $x = 0$ is singular of the first kind, i.e., $\sigma(x) = 0$. The three indices are obtained from Eqs. (4.4.12), (4.4.13) and (4.4.14) as

$$\sigma^2(x) = \mathrm{O}|x|^\alpha, \quad m(x) = \mathrm{O}|x|^\beta,$$

$$c = \lim_{x \to 0} \frac{2m(x)x^{\alpha-\beta}}{\sigma^2(x)}, \quad \text{as } x \to 0, \tag{8.2.35}$$

(8.2.35) shows that $\alpha > 0$ and $\beta \geq 0$. Substituting (8.2.35) into (8.2.34), we have

$$p(x) = \mathrm{O}\left\{ x^{-\alpha} \exp\left[c \int x^{\beta-\alpha} dx \right] \right\}, \quad x \to 0. \tag{8.2.36}$$

Two scenarios need to be analyzed separately. If $\beta - \alpha = -1$, (8.2.36) is simplified as

$$p(x) = \mathrm{O}(x^{c-\alpha}), \quad x \to 0. \tag{8.2.37}$$

Then it is deduced that

$$\begin{cases} p(x) \text{ is not integrable,} & c \leq \alpha - 1 \\ p(x) \text{ exists with } p(0) = \infty, & \alpha - 1 < c < \alpha \\ p(x) \text{ exists with a finite maximum at } x = 0, & c = \alpha \\ p(x) \text{ exists with } p(0) = 0, & c > \alpha. \end{cases} \tag{8.2.38}$$

From (8.2.38), it is concluded that the D-bifurcation occurs at $c = \alpha - 1$, and P-bifurcation takes place at $c = \alpha$.

If $\beta - \alpha \neq -1$, (8.2.36) is specified as

$$p(x) = \mathrm{O}\left[x^{-\alpha} \exp\left(\frac{c}{\beta - \alpha + 1} x^{\beta - \alpha + 1} \right) \right], \quad x \to 0. \tag{8.2.39}$$

Since $\alpha > 0$ and $\beta \geq 0$, we know from (8.2.39) that

$$\begin{cases} p(x) \text{ exists with } p(0) = \infty, & 0 < \alpha < 1 \\ p(x) \text{ is not integrable,} & 1 \leq \alpha < \beta + 1 \\ p(x) \text{ exists with } p(0) = 0, & \alpha > \beta + 1. \end{cases} \tag{8.2.40}$$

Therefore, the D-bifurcation occurs at $\alpha = 1$ and $\alpha = \beta + 1$, and no P-bifurcation occurs.

Case 2. The boundary $x = 0$ is singular of the second kind, i.e., $|m(x)| = \infty$. The three indices are obtained from Eqs. (4.4.15), (4.4.16) and (4.4.17) as

$$\sigma^2(x) = \mathrm{O}|x|^{-\alpha}, \quad m(x) = \mathrm{O}|x|^{-\beta},$$

$$c = \lim_{x \to 0} \frac{2m(x)x^{\beta - \alpha}}{\sigma^2(x)}, \quad \text{as } x \to 0, \tag{8.2.41}$$

(8.2.41) shows that $\alpha > 0$ and $\beta > 0$. Substituting (8.2.41) into (8.2.34), we have

$$p(x) = \mathrm{O}\left\{x^\alpha \exp\left[c \int x^{\alpha - \beta} dx\right]\right\}, \quad x \to 0. \tag{8.2.42}$$

Similar analysis as that in Case 1 can be carried out.

Each of the above four methods has its advantages and limitations of applicability. For a specific system, one or more methods may be applicable. But the results obtained from different methods should be the same.

In the following, the deterministic transcritical and pitchfork bifurcations are extended by adding parametric noise excitations to the systems, shown in Examples 8.2.5 and 8.2.6. For a system with two dimensions under stochastic excitations, it may be reduced to one-dimensional system approximately by using the stochastic averaging method or other techniques. Then one-dimensional bifurcation analysis can be applied by using the above described methods. Such a procedure is shown in Examples 8.2.7 and 8.2.8. For bifurcation of two-dimensional systems, the analysis becomes more complicated. Example 8.2.8 gives an approximate bifurcation analysis for such a system. Finally, the Hamiltonian formulation is the best way to treat bifurcation of multi-dimensional systems, as illustrated in Example 8.2.9.

Example 8.2.5 Stochastic transcritical bifurcation. By adding a parametric excitation to Eq. (8.2.4), we have

$$\dot{X} = \alpha X - X^2 + XW(t), \tag{8.2.43}$$

where $W(t)$ is a Gaussian white noise with a spectral density K. For the problem to be meaningful, we restrict the stochastic process

$X(t)$ in $[0, \infty)$. The boundary $x = 0$ is a fixed point. The Lyapunov exponent for $x = 0$ is calculated from Eq. (8.1.93) as

$$\lambda = \alpha. \tag{8.2.44}$$

Then if $\alpha < 0$, the fixed point $x = 0$ is stable. If $\alpha > 0$, $x = 0$ is unstable and a non-trivial stationary probability density exists, that can be obtained from Eq. (4.4.7) as

$$p(x) = C x^{\frac{\alpha}{\pi K} - 1} \exp\left(-\frac{x}{\pi K}\right). \tag{8.2.45}$$

Therefore, $\alpha = 0$ is a D-bifurcation point, denoted as $\alpha_D = 0$. By analyzing the stationary probability density (8.2.45), it is known that (i) if $0 < \alpha < \pi K$, $p(x)$ is infinite at $x = 0$ and decreasing, (ii) if $\alpha = \pi K$, $p(x)$ is finite at $x = 0$ and decreasing, and (iii) if $\alpha > \pi K$, $p(x)$ is a unimodal function with $p(0) = 0$. It is concluded that $p(x)$ undergoes a P-bifurcation at point $\alpha_P = \pi K$. The bifurcation diagram of system (8.2.43) is shown in Fig. 8.2.12, where the solid and dashed lines indicate the stable and unstable trivial solution $x = 0$, respectively.

The fixed point $x = 0$ is the left boundary, and its nature can be analyzed using the method described in Section 4.4. It is found that

$$x = 0 \text{ is } \begin{cases} \text{attractively natural,} & \alpha < 0 \\ \text{strictly natural,} & \alpha = 0 \\ \text{repulsively natural,} & \alpha > 0. \end{cases}$$

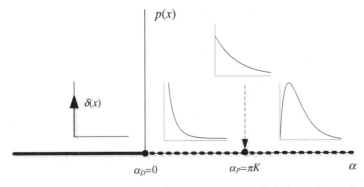

Figure 8.2.12 Diagram of stochastic transcritical bifurcation of system (8.2.42).

The right boundary at infinite is an entrance. The qualitative sample behaviors change at $\alpha = 0$ which is a D-bifurcation point, consistent with the method of Lyapunov exponent.

Example 8.2.6 Stochastic pitchfork bifurcation. The stochastic system corresponding to the pitchfork bifurcation problem is given by

$$\dot{X} = \alpha X - X^3 + XW(t). \tag{8.2.46}$$

We also restrict the stochastic process $X(t)$ in $[0, \infty)$. Similar to the transcritical bifurcation case, the Lyapunov exponent for $x = 0$ is $\lambda = \alpha$, the trivial solution is stable if $\alpha < 0$, unstable if $\alpha > 0$, and $\alpha = 0$ is the D-bifurcation point. In the case of $\alpha > 0$, a non-trivial stationary probability density can be obtained as

$$p(x) = C x^{\frac{\alpha}{\pi K} - 1} \exp\left(-\frac{x^2}{2\pi K}\right). \tag{8.2.47}$$

Although Eq. (8.2.47) is different from Eq. (8.2.45), the P-bifurcation phenomena are exact the same, namely (i) if $0 < \alpha < \pi K$, $p(x)$ is infinite at $x = 0$ and decreasing, (ii) if $\alpha = \pi K$, $p(x)$ is finite at $x = 0$ and decreasing, and (iii) if $\alpha > \pi K$, $p(x)$ is a unimodal function. The bifurcation diagram of system (8.2.46) is exact the same, given in Fig. 8.2.12.

Example 8.2.7 Stochastic Hopf Bifurcation of an oscillator with nonlinear damping and subjected to a parametric Gaussian white noise. The system is governed by

$$\ddot{X} + \alpha\dot{X} + \gamma X^2\dot{X} + \beta\dot{X}^3 + \omega_0^2 X = XW(t). \tag{8.2.48}$$

The linear damping coefficient α will be treated as the bifurcation parameter. Letting $X = X_1$ and $\dot{X} = X_2$, Eq. (8.2.48) is replaced by two first-order equations

$$\begin{aligned}
\dot{X}_1 &= X_2, \\
\dot{X}_2 &= -\omega_0^2 X_1 - \alpha X_2 - \gamma X_1^2 X_2 - \beta X_2^3 + X_1 W(t).
\end{aligned} \tag{8.2.49}$$

Without the excitation, the two eigenvalues can be found from the characteristic equation as $\lambda = (-\alpha \pm \sqrt{\alpha^2 - 4\omega_0^2})/2$. We assume that the system is underdamped, i.e., $\alpha^2 - 4\omega_0^2 < 0$. Then, a subcritical

Hopf bifurcation occurs when α passes zero, resulting in the sign change of the real parts of the two eigenvalues. If $\alpha < 0$, the fixed point $(0, 0)$ is unstable and a limit cycle exists, while $\alpha > 0$, the fixed point is stable.

For the case of a stochastic excitation present, carrying out the transformation

$$X_1 = A(t)\cos\theta, \quad X_2 = -A(t)\omega_0\sin\theta, \quad \theta = \omega_0 t + \phi(t) \quad (8.2.50)$$

and applying the stochastic averaging method, we obtain the averaged Ito equation for the amplitude process $A(t)$,

$$dA = m(A)dt + \sigma(A)dB(t), \quad (8.2.51)$$

where

$$m(A) = \left(-\frac{\alpha}{2} + \frac{3\pi}{8\omega_0^2}K\right)A - \frac{1}{8}(\gamma + 3\beta\omega_0^2)A^3, \quad (8.2.52)$$

$$\sigma(A) = \sqrt{\frac{\pi K}{4\omega_0^2}}A. \quad (8.2.53)$$

The left boundary $a = 0$ is a fixed point with a Lyapunov exponent calculated from (8.1.93) as

$$\lambda = -\frac{\alpha}{2} + \frac{\pi K}{4\omega_0^2}. \quad (8.2.54)$$

Therefore the D-bifurcation point occurs at $\lambda = 0$, i.e.,

$$\alpha_D = \frac{\pi K}{2\omega_0^2}. \quad (8.2.55)$$

If $\alpha > \alpha_D$, the fixed point $a = 0$ is stable. If $\alpha < \alpha_D$, $a = 0$ is unstable. By analyzing the sample behaviors at the two boundaries, it is found that a non-trivial stationary probability density exists

$$p(a) = Ca^{2\delta+1}\exp\left[-\frac{\omega_0^2(\gamma + 3\omega_0^2\beta)}{2\pi K}a^2\right], \quad (8.2.56)$$

where

$$\delta = -\frac{2\alpha\omega_0^2}{\pi K}. \quad (8.2.57)$$

The form of (8.2.56) reveals that the P-bifurcation point is

$$\alpha_P = \frac{\pi K}{4\omega_0^2}. \tag{8.2.58}$$

If $\alpha_P < \alpha < \alpha_D$, $p(a)$ is infinite at $a = 0$ and decreasing. At $\alpha = \alpha_P$, $p(a)$ is finite at $a = 0$ and decreasing, while $p(a)$ is a unimodal function if $\alpha < \alpha_P$.

Now we try to use the method of three indices. The right boundary at $a = \infty$ is an entrance, and the left boundary at $a = 0$ if singular of the first kind. The three indices are found to be

$$\alpha* = 2, \quad \beta = 1, \quad c = \frac{-4\alpha\omega_0^2 + 3\pi K}{\pi K}. \tag{8.2.59}$$

According to (8.2.38),

$$\begin{cases} p(a) \text{ is not integrable,} & c \leq 2 \\ p(a) \text{ exists with } p(0) = \infty, & 1 < c < 2 \\ p(a) \text{ exists with a peak at } a = 0, & c = 2 \\ p(a) \text{ exists with } p(0) = 0, & c > 2, \end{cases} \tag{8.2.60}$$

(8.2.60) yields the same results, but is simpler. The bifurcation diagram is shown in Fig. 8.2.13.

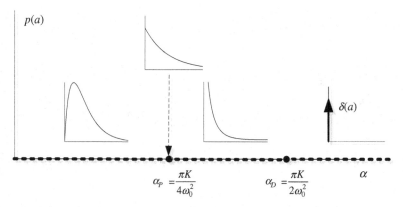

Figure 8.2.13 Diagram of stochastic bifurcation for the amplitude process of system (8.2.48).

Now we try to use the energy process defined as

$$\Lambda(t) = \frac{1}{2}\omega_0^2 X_1^2 + \frac{1}{2}X_2^2 = \frac{1}{2}\omega_0^2 A^2. \qquad (8.2.61)$$

Treating Λ as a function of A and using Eq. (2.7.4), the stationary probability density of Λ is obtained from (8.2.56) as follows

$$p(\lambda) = C\lambda^\delta \exp\left[-\frac{\gamma + 3\omega_0^2\beta}{\pi K}\lambda\right]. \qquad (8.2.62)$$

From (8.2.62), it is concluded that (i) the D-bifurcation point is $\alpha_D = \pi K/(2\omega_0^2)$, (ii) the P-bifurcation point is $\alpha_P = 0$. Figure 8.2.14 shows the bifurcation diagram for the energy process. Therefore, while the D-bifurcation point is the same for the amplitude and energy processes, the P-bifurcation points are different for the two processes.

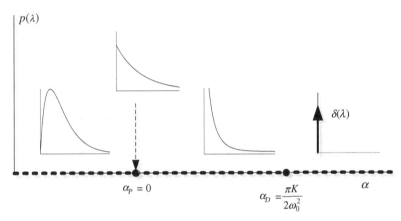

Figure 8.2.14 Diagram of stochastic bifurcation for the energy process of system (8.2.48).

The above analysis shows that the stochastic Hopf bifurcation consists of one D-bifurcation and one P-bifurcation, which is different from that of the deterministic system. When the excitation level represented by K decreases, the D-bifurcation point approaches the Hopf bifurcation point $\alpha = 0$ of the deterministic system.

Example 8.2.8 Stochastic Hopf bifurcation of a van der Pol system. A van der Pol oscillator parametrically excited by a Gaussian white noise is governed by

$$\ddot{X} + \alpha \dot{X} + X^2 \dot{X} + \omega_0^2 X = XW(t). \qquad (8.2.63)$$

The linear damping coefficient α will be treated as the bifurcation parameter. Letting $X = X_1$ and $\dot{X} = X_2$, Eq. (8.2.63) is replaced by two first-order equations

$$\begin{aligned} \dot{X}_1 &= X_2 \\ \dot{X}_2 &= -\omega_0^2 X_1 - \alpha X_2 - X_1^2 X_2 + X_1 W(t). \end{aligned} \qquad (8.2.64)$$

Without the noise, the deterministic system has a fixed point $(0, 0)$. The linearized system about the fixed point has two eigenvalues

$$\lambda_{1,2} = \frac{1}{2} \left(-\alpha \pm \sqrt{\alpha^2 - 4\omega_0^2} \right). \qquad (8.2.65)$$

When $\alpha > 0$, both eigenvalues have negative real parts; thus, the fixed point $(0, 0)$ is stable. On the other hand, if $\alpha < 0$, both eigenvalues have positive real parts. Then the fixed point is unstable, and a limit cycle appears. Therefore, $\alpha = 0$ is a Hopf bifurcation point. When α passes through zero, a supercritical Hopf bifurcation takes place. Figures 8.2.15(a) and 8.2.15(b) show the system motions for cases of $\alpha > 0$ and $\alpha < 0$, respectively.

For the stochastic system (8.2.63), the two Lyapunov exponents cannot be found exactly. If the noise is weak (small K), the two

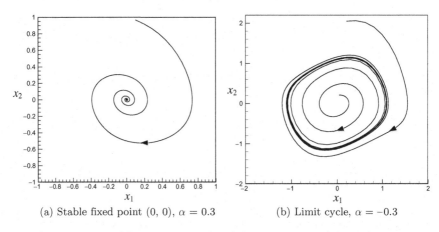

(a) Stable fixed point $(0, 0)$, $\alpha = 0.3$ (b) Limit cycle, $\alpha = -0.3$

Figure 8.2.15 Motion trajectories of deterministic van der Pol system.

Lyapunov exponents were found approximately as (Pardoux and Wihstutz, 1988)

$$\lambda_{1,2} = -\frac{\alpha}{2} \pm \frac{\pi K}{4\omega_0^2 - \alpha^2}. \tag{8.2.66}$$

Assume that α_1 is the solution of the equation

$$\lambda_1 = -\frac{\alpha}{2} + \frac{\pi K}{4\omega_0^2 - \alpha^2} = 0 \tag{8.2.67}$$

and that α_2 is the solution of the equation

$$\lambda_2 = -\frac{\alpha}{2} - \frac{\pi K}{4\omega_0^2 - \alpha^2} = 0. \tag{8.2.68}$$

If the linear damping is small, i.e., $\alpha^2 \ll 4\omega_0^2$, α_1 and α_2 can be expressed approximately as

$$\alpha_1 = \frac{\pi K}{2\omega_0^2}\left(1 + \frac{\alpha_1^2}{4\omega_0^2}\right), \quad \alpha_2 = -\frac{\pi K}{2\omega_0^2}\left(1 + \frac{\alpha_2^2}{4\omega_0^2}\right). \tag{8.2.69}$$

It can be shown that the top Lyapunov exponent λ_1 changes its sign at α_1, and another Lyapunov exponent λ_2 changes its sign at α_2. Then we have

$$\begin{cases} \lambda_1 < 0 \quad \text{and} \quad \lambda_2 < 0, \quad \text{if } \alpha > \alpha_1 \\ \lambda_1 > 0 \quad \text{and} \quad \lambda_2 < 0, \quad \text{if } \alpha_2 < \alpha < \alpha_1 \\ \lambda_1 > 0 \quad \text{and} \quad \lambda_2 > 0, \quad \text{if } \alpha < \alpha_2. \end{cases} \tag{8.2.70}$$

In the deterministic system, the two eigenvalues change their signs simultaneously at bifurcation point $\alpha = 0$. But in the present of a noise, the two Lyapunov exponents change their signs at different α values. It is clear that $\alpha_D = \alpha_1$ is a D-bifurcation point. When $\alpha > \alpha_D$, the fixed point $(0, 0)$ is asymptotically stable with probability one, and the invariant measure is $\delta(x)$. When $\alpha < \alpha_D$, the fixed point $(0, 0)$ is unstable, and a non-trivial stationary probability distribution exists. For the point at $\alpha = \alpha_2$, the instability nature of the fixed point and the system dynamical behavior do not change; therefore, it is not a D-bifurcation point. It is noted that the D-bifurcation point moves its location from the deterministic case to the stochastic case due to the noise $W(t)$.

For P-bifurcation, the shape of the non-trivial stationary probability density needs to be analyzed. Since the exact solution is not available, simulations are carried out to obtain the stationary probability density functions of system response $X(t)$ for $\omega_0 = 1$ and four different α values, as depicted in Fig. 8.2.16. The D-bifurcation point $\alpha_D = 0.0786$ for this case. In cases of $\alpha = -0.1$ and -0.01, the probability densities appear as crater shape although one is small and one is large. The crater shape disappears and the probability density has a peak at $x = 0$ for the case of $\alpha = 0.01$. With a large $\alpha = 0.1$, it is essentially a delta function. It is then concluded that the D-bifurcation point is $\alpha = \alpha_D$ and the P-bifurcation point is $\alpha = \alpha_P = 0$. The bifurcation diagram is shown schematically in Fig. 8.2.17.

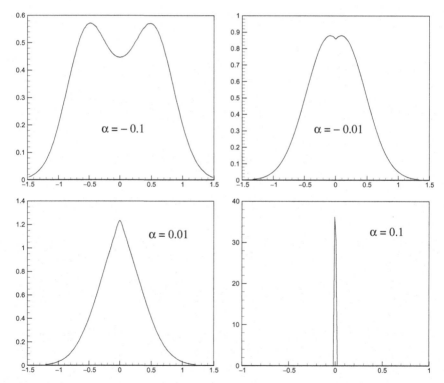

Figure 8.2.16 Stationary probability density functions of response process $X(t)$ of system (8.2.63) for different values of parameter α.

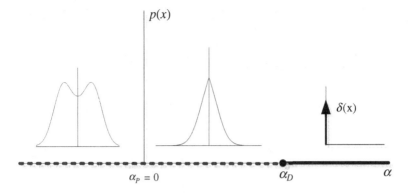

Figure 8.2.17   Diagram of stochastic Hopf bifurcation of system (8.2.63).

The stochastic bifurcation of system (8.2.63) can also be analyzed using the stochastic averaging method if the noise is weak and the damping is small. Comparing Eqs. (8.2.63) and (8.2.48), it is found that system (8.2.63) is a special case of system (8.2.48) with $\gamma = 1$ and $\beta = 0$. Thus, the results derived for system (8.2.48) can be applied to system (8.2.63). They are (i) for the amplitude process $A(t)$, the bifurcation diagram is shown in Fig. 8.2.13, and

$$\alpha_D = \frac{\pi K}{2\omega_0^2}, \quad \alpha_P = \frac{\pi K}{4\omega_0^2} \tag{8.2.71}$$

and (ii) for the energy process $\Lambda(t)$, the bifurcation diagram is shown in Fig. 8.2.14, and

$$\alpha_D = \frac{\pi K}{2\omega_0^2}, \quad \alpha_P = 0. \tag{8.2.72}$$

Comparing (8.2.71), (8.2.72) and the (8.2.69), it is found that the D-bifurcation points, which are the most important for the bifurcation phenomenon are almost the same with a small difference within the accuracy of the stochastic averaging method. As for the P-bifurcation, the two-dimensional analysis and the stochastic averaging of energy envelope obtain the same P-bifurcation point, but the stochastic averaging of amplitude envelope acquires a different P-bifurcation point.

Example 8.2.9 Stochastic Hopf bifurcation of MDOF quasi non-integrable Hamiltonian systems. Consider a quasi non-integrable Hamiltonian system governed by (7.5.17). The Hamiltonian is governed by the averaged Ito differential equation

$$dH = m(H)dt + \sigma(H)dB(t), \qquad (8.2.73)$$

where the drift coefficient $m(H)$ and diffusion coefficient $\sigma(H)$ can be obtained from (7.5.24) and (7.5.25) by carrying out the time averaging. In general, $m(H)$ and $\sigma(H)$ are nonlinear functions, and can be only calculated numerically. While the D-bifurcation can be identified by analyzing the boundary behaviors, the P-bifurcation is difficult to determine since an analytical expression for the stationary probability density, if exists, is not obtainable. To overcome the difficulty and simplify the analysis, the method of three indices is proposed to apply (Zhu and Huang, 1999).

Consider the same system as Eq. (8.1.123) in Example 8.1.7. After the stochastic averaging, the Hamiltonian is approximated as a Markov diffusion process, with its drift and diffusion coefficients given in Eqs. (8.1.126) and (8.1.127), respectively. While the right boundary at infinity is an entrance regardless of δ values, the left boundary at $h = 0$ have different behaviors depending on δ values.

For the case of $0 < \delta < 1$, we have

$$m(H) \to \mu_1 H, \sigma^2(H) \to \mu_2 H^2, \quad H \to 0, \qquad (8.2.74)$$

where μ_1 and μ_2 are given by Eqs. (8.1.130) and (8.1.131), respectively. According to (8.2.35), the three indices at the boundary $h = 0$ are

$$\alpha = 2, \quad \beta = 1, \quad c = \frac{2\mu_1}{\mu_2}. \qquad (8.2.75)$$

Using three indies method according to (8.2.38), the D-bifurcation occurs at $c = \alpha - 1 = 1$, leading to $\alpha_1 + \alpha_2 = D$, where D is given in (8.1.135), i.e.,

$$\alpha_1 + \alpha_2 = D = \frac{1}{3\eta_1}(7 - 3\eta_2)\frac{\pi(C_1^2 K_1 + C_2^2 K_2)}{(2a + \omega_1^2 + \omega_2^2)}, \qquad (8.2.76)$$

where η_1 and η_2 are given in Eqs. (8.1.132) and (8.1.133), respectively. The P-bifurcation takes place at $c = \alpha = 2$, leading to

$$\alpha_1 + \alpha_2 = P = \frac{1}{3\eta_1}(7 - 6\eta_2)\frac{\pi(C_1^2 K_1 + C_2^2 K_2)}{(2a + \omega_1^2 + \omega_2^2)}. \qquad (8.2.77)$$

The bifurcation diagram is shown in Fig. 8.2.18.

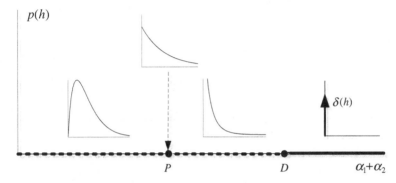

Figure 8.2.18 Stochastic Hopf bifurcation diagram of quasi non-integrable Hamiltonian system (8.1.123).

The analysis for the case of $\delta > 1$ is similar. The D-bifurcation occurs when

$$\alpha_1 + \alpha_2 = G = \frac{2}{3}\pi\eta\left(\frac{C_1^2 K_1}{\omega_1^2} + \frac{C_2^2 K_2}{\omega_2^2}\right), \qquad (8.2.78)$$

where η is given in Eq. (8.1.139). The P-bifurcation takes place when

$$\alpha_1 + \alpha_2 = \frac{1}{3}\pi\eta\left(\frac{C_1^2 K_1}{\omega_1^2} + \frac{C_2^2 K_2}{\omega_2^2}\right). \qquad (8.2.79)$$

The three indies method for determining stochastic Hopf bifurcation of quasi non-integrable Hamiltonian systems has been extended to quasi integrable Hamiltonian systems (Liu and Zhu, 2008).

More topics and more detailed analysis about stochastic bifurcation can be found in (Meunier and Verga, 1988; Sri Namachchivaya, 1990; Arnold, 1998; Zhu and Huang, 1998; Liu and Zhu, 2008).

Exercise Problems

8.1. Consider the system

$$\ddot{X} + [2\zeta + W(t)]\dot{X} + X = 0,$$

where $W(t)$ is a Gaussian white noise with a spectral density K.

(a) Derive the Ito differential equations for $\Upsilon(t)$ and $\Theta(t)$ defined by the transformation

$$\Upsilon(t) = \frac{1}{2}\ln(X^2 + \dot{X}^2), \qquad \Theta = \tan^{-1}\frac{\dot{X}}{X}.$$

(b) Show that $\Theta(t)$ itself is a Markov diffusion process, and obtain its FPK equation.

(c) Identify the singular points for the $\Theta(t)$ process in the interval of $[0, 2\pi)$, and find its stationary probability density.

(d) Obtain an expression for the largest Lyapunov exponent for the $X(t)$ process.

8.2 For the system

$$\ddot{X} + [2\zeta\omega_0 + W(t)]\dot{X} + \omega_0^2 X = 0,$$

where $W(t)$ is a Gaussian white noise with a spectral density K.

(a) Derive equations for the first- and second- order moments of $X(t)$ and $\dot{X}(t)$.

(b) Determine the stability conditions for the first- and second-order moments.

8.3 Consider the system

$$\ddot{X} + [2\zeta\omega_0 + \xi_2(t)]\dot{X} + \omega_0^2 X = \xi_1(t),$$

where $\xi_1(t)$ and $\xi_2(t)$ are two correlated stationary excitations with broad-band spectral density $\Phi_{ij}(\omega)$, $i, j = 1, 2$.

(a) Use the stochastic averaging method to derive the Ito equation for the averaged amplitude process $A(t)$ defined by the transformation

$$X = A(t)\cos\theta, \quad \dot{X} = -A(t)\omega_0\sin\theta, \quad \theta = \omega_0 t + \phi(t).$$

(b) Find the stationary probability density of $A(t)$, and determine the integrability condition for $A^n(t)$.

(c) Derive the Ito equation for $A^n(t)$, the equation for $E[A^n(t)]$, and find the stability condition for $E[A^n(t)]$.

(d) Compare the results obtained in (b) and (c).

8.4 For the following system

$$\ddot{X} + [2\zeta\omega_0 + \xi_2(t)]\dot{X} + \omega_0^2[1 + \xi_1(t)]X = 0,$$

where $\xi_1(t)$ and $\xi_2(t)$ are two correlated stationary excitations with broad-band spectral density $\Phi_{ij}(\omega)$, $i, j = 1, 2$.

(a) Use the stochastic averaging method to derive the Ito equation for the averaged amplitude process $A(t)$.

(b) Determine the stability conditions for the first- and second-order moments.

8.5 For the nonlinearly damped system,

$$\ddot{X} + \eta|\dot{X}|^\delta \mathrm{sgn}\dot{X} + \omega_0^2[1 + W(t)]X = 0, \quad \delta > 0.$$

(a) Use the results in Example 8.1.6 to write the Ito equation for averaged amplitude process $A(t)$.

(b) Investigate the stability behaviors of $A(t)$.

(c) Find the stability condition of the nth moment for the trivial solution $A(t) = 0$.

8.6 For the system with nonlinear stiffness governed by

$$\ddot{X} + 2\eta\dot{X} + k|X|^\delta \mathrm{sgn}X = XW(t), \quad k, \delta > 0.$$

(a) Derive the Ito equation for averaged energy process $\Lambda(t)$.

(b) Investigate the stability behaviors of $\Lambda(t)$.

8.7 Consider the nonlinearly damped system

$$\ddot{X} + \alpha\dot{X} + \beta\dot{X}^3 + X = \dot{X}W(t), \quad \beta > 0,$$

where $W(t)$ is a Gaussian white noise with a spectral density K.

(a) Use the stochastic averaging method to derive the Ito equation for the averaged amplitude process $A(t)$.

(b) Find the stationary probability density of $A(t)$, and determine its integrability condition.

(c) Analyze the system stability in terms of the boundary behaviors.

(d) Compare the results obtained in (b) and (c).

8.8 Another nonlinearly damped system is governed by

$$\ddot{X} + \eta|\dot{X}|^{\delta}\text{sgn}\dot{X} + \omega_0^2 X = \dot{X}W(t), \quad \eta, \delta > 0,$$

where $W(t)$ is a Gaussian white noise with a spectral density K.

(a) Use the stochastic averaging method to derive the Ito equation for the averaged amplitude process $A(t)$ defined by the same transformation as in Problem 8.3.

(b) Investigate the sample behaviors of $A(t)$ at the two boundaries.

(c) Find the stability condition for the trivial solution $A(t) = 0$.

(d) Find the stability condition of the nth moment for the trivial solution $A(t) = 0$.

8.9 A system with nonlinearly stiffness and subjected to a parametric white noise excitation is governed by

$$\ddot{X} + 2\eta\dot{X} + k|X|^{\delta}\text{sgn}X = \dot{X}W(t), \quad k, \delta > 0,$$

where $W(t)$ is a Gaussian white noise with a spectral density K.

(a) Use the quasi-conservative averaging procedure to derive the Ito equation for the averaged energy process $\Lambda(t)$

defined as

$$\Lambda(t) = \frac{1}{2}\dot{X}^2 + \frac{k}{\delta+1}|X|^{\delta+1}.$$

(b) Investigate the sample behaviors of $\Lambda(t)$ at the two boundaries.

(c) Find the stability condition for the trivial solution $\Lambda(t) = 0$.

(d) Find the stability condition of the nth moment for the trivial solution $\Lambda(t) = 0$.

8.10 Consider the nonlinear system in Problem 6.10, i.e.,

$$\ddot{X} + X^2\left[\beta + \frac{4\alpha}{X^2 + \dot{X}^2}\right]\dot{X} + [1 + W(t)]X = 0, \quad \beta > 0.$$

(a) According to the exact stationary probability density found in Problem 6.10, determine the condition for asymptotic stabilities in probability and in statistical moments of the trivial solution $(0, 0)$.

(b) Find the Ito differential equation for the amplitude process $A = \sqrt{X^2 + \dot{X}^2}$. Examine the sample behaviors at the two boundaries, and determine the condition for asymptotic stabilities in probability and in statistical moments of the trivial solution $A = 0$.

8.11 A typical example in real life of transcritical bifurcation could be the consumer–producer problem where the consumption is proportional to the quantity of resource. Denote $x(t)$ as the quantity of the resource, the system is represented by

$$\dot{x} = ax - bx^2 - \alpha x, \quad x \geq 0,$$

where a, b and α are positive parameters, a is the resource growth rate, α is the consumption rate, and the term $-bx^2$ indicates that there is a limit to the resource growth. Treat α as the bifurcation parameter.

(a) Determine the fixed points and their stability.

(b) Analyze the bifurcation of the system.

8.12 Add a stochastic noise in the assumption rate in the system in Problem 8.11, i.e.,

$$\dot{X} = aX - bX^2 - [\alpha + W(t)]X, \quad X \geq 0.$$

Carry out (a) the stability analysis, and (b) the bifurcation analysis.

8.13 As an example of pitchfork bifurcation, consider the system

$$\dot{x} = (\alpha - a)x - bx^3,$$

where a and b are positive parameters. Treat α as the bifurcation parameter.

(a) Determine the fixed points and their stability.

(b) Analyze the bifurcation of the system.

8.14 Carry out (a) the stability analysis, and (b) the bifurcation analysis for the system.

$$\dot{X} = [\alpha - a + W(t)]X - bX^3,$$

which is the stochastic version of the system in Problem 8.13 by adding a noise term.

8.15 Consider the same system in Problem 8.10, i.e.,

$$\ddot{X} + X^2 \left[\beta + \frac{4\alpha}{X^2 + \dot{X}^2} \right] \dot{X} + [1 + W(t)]X = 0, \quad \beta > 0.$$

Using α as the bifurcation parameter, analyze the system bifurcation phenomenon in terms of (a) the amplitude process defined as $A = \sqrt{X^2 + \dot{X}^2}$, and (b) the energy process defined as $\Lambda = (X^2 + \dot{X}^2)/2$.

8.16 Taking the damping ratio ζ as the bifurcation parameter, analyze the bifurcation picture for the amplitude process $A(t)$ in (a) the system in Problem 8.3,

$$\ddot{X} + [2\zeta\omega_0 + \xi_2(t)]\dot{X} + \omega_0^2 X = \xi_1(t)$$

and (b) the system in Problem 8.4,

$$\ddot{X} + [2\zeta\omega_0 + \xi_2(t)]\dot{X} + \omega_0^2[1 + \xi_1(t)]X = 0.$$

8.17 Taking δ and η as the bifurcation parameters, analyze the bifurcation picture for the energy process $\Lambda(t)$ in two systems: (a) the system in Problem 8.6,

$$\ddot{X} + 2\eta\dot{X} + k\,|X|^{\delta}\,\mathrm{sgn}X = XW(t), \quad k,\delta > 0$$

and (b) the system in Problem 8.9,

$$\ddot{X} + 2\eta\dot{X} + k\,|X|^{\delta}\,\mathrm{sgn}X = \dot{X}W(t), \quad k,\delta > 0.$$

8.18 Investigate the bifurcations for the amplitude process and energy process.

(a) Use α as the bifurcation parameter for the system in Problems 8.7,

$$\ddot{X} + \alpha\dot{X} + \beta\dot{X}^3 + X = \dot{X}W(t), \quad \beta > 0.$$

(b) Use both δ and η as bifurcation parameters for the system in Problem 8.8

$$\ddot{X} + \eta|\dot{X}|^{\delta}\mathrm{sgn}\dot{X} + \omega_0^2 X = \dot{X}W(t), \quad \eta,\delta > 0.$$

8.19 Analyze the bifurcation of the stochastic prey-predator ecosystem in Section 7.4.3 using self-competition coefficient s as the bifurcation parameter. The system is governed by

$$\dot{X}_1 = X_1[a - bX_2 - \tfrac{s}{f}(-c + fX_1) + W_1(t)],$$
$$\dot{X}_2 = X_2[-c + fX_1 + W_2(t)],$$

where

$$a = a_1 - \frac{sc}{f},$$

and a_1, b, c and f are positive constants.

8.20 Use the stochastic averaging of energy envelope to analyze the stability and bifurcation of the nonlinear system

$$\ddot{X} + 2\zeta\omega_0\dot{X} + \omega_0^2[1 + W(t)]X + \eta|X|^{\delta}\mathrm{sgn}X = 0, \quad \delta \neq 1,$$

(a) $0 \leq \delta < 1$. (b) $\delta > 1$.

CHAPTER 9

FIRST-PASSAGE PROBLEM OF STOCHASTICALLY EXCITED SYSTEMS

For dynamical systems subjected to stochastic excitations, there are two major failure modes. One is the so-called first-passage failure. It occurs as soon as a critical physical quantity of the system exceeds a prescribed safety boundary for the first time. In probabilistic words, it can be stated that a stochastic process exceeds, for the first time, a prescribed critical value. The time is called the first-passage time, and its probability density and its moments, as well as the reliability (probability of no exceeding) are of interest.

Another type of failure is due to accumulation of fatigue damage represented by certain physical quantities (Zhu and Lei, 1991). At the presence of uncertainties in loading and environments, the accumulation of fatigue damage is a random phenomenon, influenced by various factors. When it exceeds a critical threshold, failure takes place. Thus, it can be treated as the first-passage failure. In this chapter, one mode of the fatigue failure will be investigated when a dominant crack in material exceeds a critical size. It is an extension of the deterministic fracture mechanics to the case of random loading, where the propagation of a dominant crack is a random phenomenon. As a representation, the crack length at a critical location signifies the accumulated material damage.

461

Besides in investigating system failures, the first-passage analysis can also be applied when it is of interest that a system response reaches certain important thresholds. The first-passage problem is one of the difficult topics in the theory of stochastic dynamics. At the present time, analytical solutions are possible only if the stochastic processes in question can be treated as Markov diffusion processes. Still, known exact solutions are limited to one-dimensional cases. Since the state space of a practical problem is generally two-dimensional or higher, the reduction in dimension is necessary so that analytical solutions can be obtained for the first-passage problem. The stochastic averaging method provides an effective way to reduce the system dimension, as shown in Sections 7.4 and 7.5.

9.1 Reliability Function

Let a one-dimensional diffusion process $X(t)$ be governed by an Ito stochastic differential equation

$$dX(t) = m(X)dt + \sigma(X)dB(t). \tag{9.1.1}$$

We consider here only the cases in which the drift and diffusion coefficients m and σ do no depend on time t explicitly. When process $X(t)$ first reaches a critical value x_c, on the condition that $X(t_0) = x_0$ where $x_0 < x_c$, we say that the event of the first-passage occurs. The first-passage time T is clearly a random variable depending on the system, the excitation and the initial state x_0. The reliability function, denoted by $R(t, t_0, x_0)$, can be defined as the probability of $x_l \leq X(t) < x_c$ at time instant t with given initial state x_0 in $[x_l, x_c)$ i.e.,

$$R(t, t_0, x_0) = \int_{x_l}^{x_c} p(x, t|x_0, t_0)dx, \tag{9.1.2}$$

where x_l is the left boundary of the process $X(t)$. It is noted that $p(x, t|x_0, t_0)$ in (9.1.2) is the transition probability density of process $X(t)$ before it exceeds x_c. An important issue in the first-passage problem is that the critical state x_c corresponds to an absorbing boundary since a sample function is removed from the population of sample functions once it reaches the boundary. With an absorbing boundary, the process $X(t)$ can never become stationary. Furthermore, the total probability within the region $[x_l, x_c]$ is no longer conserved; otherwise, the integration in (9.1.2) would be unity.

It is known in Section 4.2.3 that the transition probability density $p(x, t|x_0, t_0)$ satisfies the Kolmogorov backward equation (4.2.21). Integration of (4.2.21) with respect to x over the range $[x_l, x_c)$ results in

$$\frac{\partial R}{\partial t_0} + m(x_0)\frac{\partial R}{\partial x_0} + \frac{1}{2}\sigma^2(x_0)\frac{\partial^2 R}{\partial x_0^2} = 0. \qquad (9.1.3)$$

In Eq. (9.1.3), R is treated as a function of t_0 and x_0, while t and x_c are treated as implicit parameters. By denoting $\tau = t - t_0$, Eq. (9.1.3) can be rewritten as

$$-\frac{\partial R}{\partial \tau} + m(x_0)\frac{\partial R}{\partial x_0} + \frac{1}{2}\sigma^2(x_0)\frac{\partial^2 R}{\partial x_0^2} = 0. \qquad (9.1.4)$$

Therefore, the reliability function can be re-denoted as $R(\tau, x_0)$. Associated with Eq. (9.1.4), we have the following initial and boundary conditions

$$R(\tau, x_0)|_{\tau=0} = 1, \quad x_l \le x_0 < x_c, \qquad (9.1.5)$$

$$R(\tau, x_0)|_{x_0=x_c} = 0. \qquad (9.1.6)$$

The physical meanings of (9.1.5) and (9.1.6) are obvious. Since Eq. (9.1.4) is of the second-order with respect to the spatial variable x_0, another boundary condition is required, which will be analyzed according to the nature of x_l as described below.

As discussed in Section 4.4 about one-dimensional Markov diffusion process, a boundary can be classified into singular or non-singular. If x_l is non-singular, i.e., $m(x_l)$ is finite and $\sigma(x_l)$ is non-zero, then it is a regular boundary. In principle, any conditions may be imposed to a regular boundary according the underlined physical problem. We assume that x_l is not critical, and the passage to the critical value x_c will occur eventually given $x_0 = x_l$. Then for the purpose of the first-passage analysis, a conservative assumption is that the left boundary is reflective, in which case, we have the second quantitative boundary condition

$$\left.\frac{\partial R}{\partial x_0}\right|_{x_0=x_l} = 0. \tag{9.1.7}$$

On the other hand, a singular boundary can be one of the following types: regular, entrance, exit, repulsively natural, strictly natural and attractively natural. If x_l is singular and is regular, entrance, or repulsively natural, then a sample path beginning at or near x_l can travel inward, and is capable to reach the critical state on the right-hand side. In this case, we may also use (9.1.7) as a boundary condition. However, if it is an exit, attractively natural, or strictly natural, then not every sample path that begins at or near x_l is to reach the critical state. Then the first-passage problem becomes meaningless.

Equations (9.1.4) through (9.1.7) governing the reliability function define an initial-boundary-value problem. While numerical solution algorithms are available, closed form analytical solutions are known only for very few cases as shown in the following examples.

Example 9.1.1 Consider a linear system under an external excitation, governed by

$$\ddot{X} + 2\zeta\omega_0\dot{X} + \omega_0^2 X = \xi(t), \tag{9.1.8}$$

where $\xi(t)$ is a broadband stationary process with a zero mean and a spectral density $\Phi(\omega)$. It is assumed that the damping is small and the excitation is weak. Define the amplitude process as $A(t) = \sqrt{X^2(t) + \dot{X}^2(t)/\omega_0^2}$. Taking into account that (i) the amplitude is the largest displacement and (ii) the square of the amplitude represent the system total energy, the first-passage failure may be posted as the reach of the amplitude process $A(t)$ to a critical value a_c for the first time. By applying the stochastic averaging method, the amplitude process is approximated as a Markov diffusion process governed by

$$dA(t) = m(A)dt + \sigma(A)dB(t), \tag{9.1.9}$$

with the averaged drift and diffusion coefficients

$$m(A) = \frac{\delta}{2A} - \xi\omega_0 A, \quad \sigma^2(A) = \delta, \quad \delta = \frac{\pi}{\omega_0^2}\Phi(\omega_0). \quad (9.1.10)$$

The equation for the reliability $R(\tau, a_0)$ is obtained from Eq. (9.1.4) as

$$-\frac{\partial R}{\partial \tau} + \left(\frac{\delta}{2a_0} - \xi\omega_0 a_0\right)\frac{\partial R}{\partial a_0} + \frac{1}{2}\delta\frac{\partial^2 R}{\partial a_0^2} = 0. \quad (9.1.11)$$

The initial and boundary conditions in (9.1.5) and (9.1.6) are posted as

$$R(\tau, a_0)|_{\tau=0} = 1, \quad a_0 < a_c, \quad (9.1.12)$$

$$R(\tau, a_0)|_{a_0=a_c} = 0. \quad (9.1.13)$$

Another boundary condition can be derived from Eq. (9.1.11) directly. Note that each of the three terms in Eq. (9.1.11) must be finite for the equation to be meaningful. Then from the second term, we have

$$\frac{\partial R(\tau, a_0)}{\partial a_0} \sim 0(a_0), \quad \text{as } a_0 \to 0. \quad (9.1.14)$$

It is noted that condition (9.1.14) is consistent with, but more precise than (9.1.7). Assume that the solution of Eq. (9.1.11) is of the form of

$$R(\tau, a_0) = e^{-\lambda\tau}P_\lambda(a_0), \quad (9.1.15)$$

where the subscript λ of $P_\lambda(a_0)$ indicates that it is a parameter in function P. Substituting (9.1.15) into (9.1.11), we obtain an ordinary differential equation for function P

$$\frac{1}{2}\delta\frac{d^2 P}{da_0^2} + \left(\frac{\delta}{2a_0} - \xi\omega_0 a_0\right)\frac{dP}{da_0} + \lambda P = 0. \quad (9.1.16)$$

Equation (9.1.16) is solved subjected to the boundary conditions

$$P|_{a_0=a_c} = 0; \quad \frac{dP}{da_0} \sim 0(a_0), \quad \text{as } a_0 \to 0. \quad (9.1.17)$$

To simply Eq. (9.1.16), change variable from a_0 to z

$$z = \frac{\xi \omega_0 a_0^2}{\delta} = \frac{\zeta \omega_0^3 a_0^2}{\pi \Phi(\omega_0)}, \tag{9.1.18}$$

transform Eq. (9.1.16) to

$$z \frac{d^2 P}{dz^2} + (1 - z) \frac{dP}{dz} + \frac{\lambda}{2\zeta \omega_0} P = 0 \tag{9.1.19}$$

and also transform the boundary conditions in (9.1.17) to

$$P|_{z=z_c} = 0, \tag{9.1.20}$$

$$\frac{dP}{dz}\bigg|_{z=0} = \left[\frac{dP}{da_0} \frac{da_0}{dz}\right]_{a_0=0} = \text{finite}, \tag{9.1.21}$$

where $z_c = \xi \omega_0 a_c^2/\delta$. Equation (9.1.19) is a degenerate hypergeometric equation (Whittaker and Watson, 1952), which has only one solution (eigenfunction) satisfying the boundary conditions (9.1.21)

$$P = M\left(-\frac{\lambda}{2\zeta \omega_0}, 1; z\right), \tag{9.1.22}$$

where M is the confluent or degenerate hypergeometric function. Using the boundary condition (9.1.20), the eigenvalues λ_n are the solutions of the equation

$$M\left(-\frac{\lambda_n}{2\zeta \omega_0}, 1; z_c\right) = 0. \tag{9.1.23}$$

Then the reliability function R can be expressed as the series

$$R(\tau, a_0) = \sum_{n=1}^{\infty} C_n M\left(-\frac{\lambda_n}{2\zeta \omega_0}, 1; \frac{\zeta \omega_0}{\delta} a_0^2\right) e^{-\lambda_n \tau}, \tag{9.1.24}$$

where constants C_n are to be determined from the initial condition (9.1.12). Substituting (9.1.24) into (9.1.12), we have

$$\sum_{n=1}^{\infty} C_n M\left(-\frac{\lambda_n}{2\zeta \omega_0}, 1; \frac{\zeta \omega_0}{\delta} a_0^2\right) = 1. \tag{9.1.25}$$

Applying the orthogonality property of the eigenfunctions

$$
\int_0^{z_c} e^{-z} M\left(-\frac{\lambda_m}{2\zeta\omega_0},\ 1; z\right) M\left(-\frac{\lambda_n}{2\zeta\omega_0},\ 1; z\right) dz
\begin{cases} = 0, & m \neq n \\ \neq 0, & m = n \end{cases}
$$

$$(9.1.26)$$

we obtain

$$
C_n = \frac{\int_0^{z_c} e^{-z} M\left(-\frac{\lambda_n}{2\zeta\omega_0}, 1; z\right) dz}{\int_0^{z_c} e^{-z} M^2\left(-\frac{\lambda_n}{2\zeta\omega_0}, 1; z\right) dz}.
\tag{9.1.27}
$$

This result was obtained by Lennox and Fraser (1974).

Example 9.1.2 Expand the problem of Example 9.1.1 by adding parametric excitations to have the following system

$$
\ddot{X} + \omega_0[2\zeta + \xi_2(t)]\dot{X} + \omega_0^2[1 + \xi_1(t)]X = \xi_3(t),
\tag{9.1.28}
$$

where $\xi_1(t)$, $\xi_2(t)$ and $\xi_3(t)$ are independent broadband stationary processes with zero mean values and spectral densities $\Phi_{ii}(\omega)$, $i = 1, 2, 3$, respectively. It is assumed that the damping is small and the excitations are weak. By applying the stochastic averaging method of amplitude envelope, the averaged equation is given by (9.1.9) with the drift and diffusion coefficients

$$
m(A) = \frac{\delta}{2A} - \alpha A, \quad \sigma^2(A) = (\delta + \gamma A^2),
\tag{9.1.29}
$$

where

$$
\alpha = \zeta\omega_0 - \frac{\pi\omega_0^2}{8}[2\Phi_{22}(0) + 3\Phi_{22}(2\omega_0) + 3\Phi_{11}(2\omega_0)],
$$

$$
\delta = \frac{\pi}{\omega_0^2}\Phi_{33}(\omega_0),
\tag{9.1.30}
$$

$$
\gamma = \frac{\pi\omega_0^2}{4}[2\Phi_{22}(0) + \Phi_{22}(2\omega_0) + \Phi_{11}(2\omega_0)].
$$

The reliability function $R(\tau, a_0)$ is governed by

$$-\frac{\partial R}{\partial \tau} + \left(\frac{\delta}{2a_0} - \alpha a_0\right)\frac{\partial R}{\partial a_0} + \frac{1}{2}(\delta + \gamma a_0^2)\frac{\partial^2 R}{\partial a_0^2} = 0. \quad (9.1.31)$$

The associated initial and boundary conditions are

$$R(\tau, a_0)|_{\tau=0} = 1, \quad a_0 < a_c, \quad (9.1.32)$$

$$R(\tau, a_0)|_{a_0=a_c} = 0, \quad (9.1.33)$$

$$\frac{\partial R(\tau, a_0)}{\partial a_0} \sim 0(a_0), \quad \text{as } a_0 \to 0. \quad (9.1.34)$$

Similar to the last example, letting the solution of Eq. (9.1.31) have the form of

$$R(\tau, a_0) = e^{-\lambda\tau}P_\lambda(a_0), \quad (9.1.35)$$

we obtain

$$\frac{1}{2}(\delta + \gamma a_0^2)\frac{d^2P}{da_0^2} + \left(\frac{\delta}{2a_0} - \alpha a_0\right)\frac{dP}{da_0} + \lambda P = 0. \quad (9.1.36)$$

The boundary conditions for Eq. (9.1.36) are

$$P|_{a_0=a_c} = 0; \quad \frac{dP}{da_0} \sim 0(a_0), \quad \text{as } a_0 \to 0. \quad (9.1.37)$$

Change variable a_0 to z as follows

$$z = -\frac{\gamma}{\delta}a_0^2. \quad (9.1.38)$$

Equation (9.1.36) and boundary conditions in (9.1.37) are transformed to

$$z(1-z)\frac{d^2P}{dz^2} + \left[1 - \left(\frac{1}{2} - \frac{\alpha}{\gamma}\right)z\right]\frac{dP}{dz} - \frac{\lambda}{2\gamma}P = 0, \quad (9.1.39)$$

$$P|_{z=z_c} = 0, \quad \frac{dP}{dz}\bigg|_{z=0} = \text{finite}, \quad (9.1.40)$$

where $z_c = -\gamma a_c^2/\delta$. Equation (9.1.39) is a hypergeometric equation (Whittaker and Watson, 1952). The one solution satisfying the boundary conditions in (9.1.40) is

$$
P(\lambda, z) =
\begin{cases}
F(c, d, 1, z), & -1 \leq z \leq 0 \\[2mm]
\dfrac{\Gamma(c-d)}{(-z)^c \Gamma(d)\Gamma(c-1)} F(c, c, c+1-d, z^{-1}) \\[2mm]
\quad + \dfrac{\Gamma(d-c)}{(-z)^d \Gamma(c)\Gamma(d-1)} \\[2mm]
\quad \times F(d, d, d+1-c, z^{-1}), & z < -1
\end{cases}
$$

(9.1.41)

where F is a hypergeometric function, and parameter c and d are given by

$$
c = -\frac{1}{2}\eta + \frac{1}{2}\sqrt{\eta^2 - \frac{2\lambda}{\gamma}}, \quad d = -\frac{1}{2}\eta - \frac{1}{2}\sqrt{\eta^2 - \frac{2\lambda}{\gamma}}, \quad \eta = \frac{1}{2} + \frac{\alpha}{\gamma}.
$$

(9.1.42)

Imposing the first boundary condition in (9.1.40), we obtain the eigenvalues λ_n as the solutions of the equation

$$
P(\lambda_n, z_c) = 0.
$$

(9.1.43)

Then the reliability function R is obtained as

$$
R(\tau, a_0) = \sum_{n=1}^{\infty} C_n P\left(\lambda_n, -\frac{\gamma}{\delta} a_0^2\right) e^{-\lambda_n \tau},
$$

(9.1.44)

where constants C_n are to be determined from the initial condition (9.1.32) and the orthogonality property of the eigenfunctions P

$$
\int_0^{z_c} (1-z)^{\eta-1} P(\lambda_m, z) P(\lambda_n, z)dz
\begin{cases}
= 0, & m \neq n \\
\neq 0, & m = n
\end{cases},
$$

(9.1.45)

to result in

$$
C_n = \frac{\int_0^{z_c} (1-z)^{\eta-1} P(\lambda_n, z)dz}{\int_0^{z_c} (1-z)^{\eta-1} P^2(\lambda_n, z)dz}.
$$

(9.1.46)

This result was obtained by Ariaratnam and Tam (1979).

9.2 The Generalized Pontryagin Equation

While solution of the reliability function is not easy, analysis of the first-passage time is simpler and more practical. The first-passage time T, as a random variable, depends on the system property, the excitation and the initial state. Its probability distribution and probability density functions are given respectively by

$$F_T(\tau, x_0) = \text{Prob}\,[T < \tau | X(t_0) = x_0] = 1 - R(\tau, x_0) \qquad (9.2.1)$$

and

$$p_T(\tau, x_0) = \frac{\partial F_T(\tau, x_0)}{\partial \tau} = -\frac{\partial R(\tau, x_0)}{\partial \tau}, \qquad (9.2.2)$$

where τ is the state variable of random variable T. The nth moment of T can then be calculated from

$$\mu_n(x_0) = E[T^n] = -\int_0^\infty \tau^n \frac{\partial R(\tau, x_0)}{\partial \tau} d\tau = n \int_0^\infty \tau^{n-1} R(\tau, x_0) d\tau.$$
$$(9.2.3)$$

In deriving (9.2.3), it is assumed reasonably that

$$\lim_{\tau \to \infty} \tau^n R(\tau, x_0) = 0. \qquad (9.2.4)$$

Multiplying (9.1.4) by τ^n, integrating on τ, and using (9.2.3), we obtain

$$(n+1)\mu_n + m(x_0)\frac{d}{dx_0}\mu_{n+1} + \frac{1}{2}\sigma^2(x_0)\frac{d^2}{dx_0^2}\mu_{n+1} = 0. \qquad (9.2.5)$$

The associated boundary conditions are

$$\mu_{n+1}(x_0)|_{x_0=x_c} = 0, \qquad (9.2.6)$$

$$\mu_{n+1}(x_0)|_{x_0=x_l} = \text{finite}. \qquad (9.2.7)$$

Condition (9.2.7) is based on the assumption that a first-passage will occur eventually, which is necessary for analysis of the first-passage problem. Equation (9.2.5) can be solved recursively from $n = 0$.

For the case of $n = 0$, Eq. (9.2.5) is an equation for the mean value μ_1 of T, given by

$$1 + m(x_0)\frac{d}{dx_0}\mu_1 + \frac{1}{2}\sigma^2(x_0)\frac{d^2}{dx_0^2}\mu_1 = 0. \qquad (9.2.8)$$

Equation (9.2.8) is the well-known classical Pontryagin equation (Andrnov, Pontryagin and Witt, 1933), and Eq. (9.2.5) is referred as the generalized Pontryagin equation (Ariaratnam and Tam, 1979). Since the first-passage time is non-negative, its moments of different orders have similar trends; therefore, the first moment μ_1 is representative and the most important.

The boundary condition (9.2.7) is qualitative, and may be useful to obtain closed-form solutions if possible. However, in many practical nonlinear systems, numerical procedure must be devised to solve Eqs. (9.2.5) or (9.2.8), and a quantitative boundary condition is required. As discussed in the last section, a boundary can be classified into singular or non-singular. If x_l is non-singular, we may assume that the left boundary x_l is reflective. From Eqs. (9.2.3) and (9.1.7), we have

$$\left.\frac{d\mu_{n+1}}{dx_0}\right|_{x_0=x_l} = n\int_0^\infty \tau^{n-1}\left[\frac{\partial R(\tau, x_0)}{\partial x_0}\right]_{x_0=x_l} d\tau = 0. \qquad (9.2.9)$$

On the other hand, if x_l is singular, it must be regular, entrance, or repulsively natural for all samples to reach the critical state eventually and to render the first-passage problem meaningful. Three scenarios are possible. If x_l is a regular shunt or an entrance shunt, i.e., $\sigma(x_l) = 0$ and $m(x_l) > 0$, we obtain the second boundary condition directly from Eq. (9.2.5) as (Zhu and Lei, 1989)

$$\left.\frac{d\mu_{n+1}}{dx_0}\right|_{x_0=x_l} = -\left.\frac{(n+1)\mu_n}{m(x_0)}\right|_{x_0=x_l}. \qquad (9.2.10)$$

If x_l is a regular trap, an entrance trap, or a repulsively natural trap, i.e., $\sigma(x_l) = 0$ and $m(x_l) = 0$, we may impose the second boundary condition as

$$0 \left| m(x_0) \frac{d\mu_{n+1}}{dx_0} \right| \sim 0 \left| \mu_n \right|, \quad x_0 \to x_l. \tag{9.2.11}$$

In particular, for $n = 0$,

$$m(x_0) \frac{d\mu_1(x_0)}{dx_0} = \text{finite}, \quad x_0 \to x_l. \tag{9.2.12}$$

Finally, If x_l is a singular boundary of the second kind, i.e., $m(x_l) = \infty$, we obtain from Eq. (9.2.5)

$$\left. \frac{d\mu_{n+1}}{dx_0} \right|_{x_0 = x_l} = 0, \tag{9.2.13}$$

since each term in (9.2.5) must be finite. Equation (9.2.13) is the same as (9.2.9), indicating that x_l is a reflective boundary.

To solve Eq. (9.2.5), let

$$y(x_0) = \frac{d}{dx_0} \mu_{n+1}. \tag{9.2.14}$$

Equation (9.2.5) is rewritten as

$$y' + \frac{2m(x_0)}{\sigma^2(x_0)} y = -\frac{2(n+1)\mu_n(x_0)}{\sigma^2(x_0)}. \tag{9.2.15}$$

Equation (9.2.15) is a first-order linear ODE, and can be solved to obtain

$$y = \frac{d\mu_{n+1}}{dx_0} = -2(n+1)\psi^{-1}(x_0) \left[\int_{x_l}^{x_0} \frac{\mu_n(v)}{\sigma^2(v)} \psi(v) dv + C_{n+1} \right], \tag{9.2.16}$$

where C_{n+1} is an integration constant, and function ψ is defined as

$$\psi(v) = \exp \left[\int \frac{2m(v)}{\sigma^2(v)} dv \right]. \tag{9.2.17}$$

Constant C_{n+1} can be determined from boundary condition (9.2.9), (9.2.10) or (9.2.11). By integrating (9.2.16) and imposing boundary condition (9.2.6), we obtained

$$
\mu_{n+1}(x_0) = - \int_{x_0}^{x_c} \mu'_{n+1}(u) du
$$

$$
= 2(n+1) \int_{x_0}^{x_c} \psi^{-1}(u) \left[\int_{x_l}^{u} \frac{\mu_n(v)}{\sigma^2(v)} \psi(v) dv + C_{n+1} \right] du.
$$

(9.2.18)

Therefore, closed-form solutions of the generalized Pontryagin equation are possible, and illustrated in the following sections.

9.3 Moments of First-Passage Time

9.3.1 *Moments of First-Passage Time of Response Amplitude*

Consider an oscillator with linear stiffness, weak damping and excitations, governed by

$$
\ddot{X} + \varepsilon h(X, \dot{X}) + \omega_0^2 X = \varepsilon^{\frac{1}{2}} \sum_{l=1}^{m} g_l(X, \dot{X}) \xi_l(t),
$$

(9.3.1)

where functions h and g_l may be nonlinear, and $\xi_l(t)$ are stationary random processes with broadband spectra. It is known in Section 7.4.1 that the stochastic averaging method can be applied to approximate the amplitude process $A = \sqrt{X^2 + \dot{X}^2/\omega_0^2}$ as a Markov diffusion process, which is governed by an averaged Ito stochastic differential equation

$$
dA = m(A)dt + \sigma(A)dB(t),
$$

(9.3.2)

where the drift coefficient $m(A)$ and diffusion coefficient $\sigma(A)$ can be calculated according to the procedure described in Section 7.4.

In terms of the amplitude process, the generalized Pontryagin equation now reads

$$(n+1)\mu_n + m(a_0)\frac{d}{da_0}\mu_{n+1} + \frac{1}{2}\sigma^2(a_0)\frac{d^2}{da_0^2}\mu_{n+1} = 0. \qquad (9.3.3)$$

With appropriate boundary conditions discussed in the last section, the moments of the first-passage time can be solved either analytically or numerically.

Example 9.3.1 Consider the same system of Example 9.1.2, i.e., a linear oscillator under both additive and parametric random excitations.

$$\ddot{X} + \omega_0[2\zeta + \xi_2(t)]\dot{X} + \omega_0^2[1 + \xi_1(t)]X = \xi_3(t), \qquad (9.3.4)$$

where $\xi_1(t)$, $\xi_2(t)$ and $\xi_3(t)$ are independent broadband stationary processes with zero mean values and spectral densities $\Phi_{ii}(\omega)$, $i = 1, 2, 3$, respectively. It is assumed that the damping is small and the excitations are weak. By applying the stochastic averaging method of amplitude envelope, the drift and diffusion coefficients for the amplitude process are given by Eq. (9.1.29), and the generalized Pontryagin equation is

$$\frac{1}{2}(\delta + \gamma a_0^2)\frac{d^2}{da_0^2}\mu_{n+1} + \left(\frac{\delta}{2a_0} - \alpha a_0\right)\frac{d}{da_0}\mu_{n+1} = -(n+1)\mu_n,$$
$$(9.3.5)$$

where parameters α, δ and γ are given in Eq. (9.1.30). Due to existence of the external excitation, the left boundary $a_0 = 0$ is unstable, and can be treated as a reflective boundary. The two boundary conditions are then

$$\mu_{n+1}\big|_{a_0=a_c} = 0, \qquad (9.3.6)$$

$$\frac{d\mu_{n+1}}{da_0}\bigg|_{a_0=0} = 0. \qquad (9.3.7)$$

Using Eq. (9.2.16), we have

$$\mu'_{n+1}(a_0) = -2(n+1)\frac{(\delta + \gamma a_0^2)^\eta}{a_0}\left[\int_0^{a_0}\frac{v\mu_n(v)}{(\delta+\gamma v^2)^{\eta+1}}dv + C_{n+1}\right].$$

$$(9.3.8)$$

According to boundary condition (9.3.7), $C_{n+1} = 0$. Integrate (9.3.8) and imposing boundary condition (9.3.6) to result in

$$\mu_{n+1}(a_0) = 2(n+1)\int_{a_0}^{a_c}\frac{(\delta+\gamma u^2)^\eta}{u}\left[\int_0^u\frac{v\mu_n(v)}{(\delta+\gamma v^2)^{\eta+1}}dv\right]du.$$

$$(9.3.9)$$

Letting $n = 0$ and $\mu_0(v) = 1$, the mean value of the first-passage time can then be derived from (9.3.9)

$$\mu_1(a_0) = \begin{cases} \dfrac{1}{\eta\gamma}\displaystyle\int_{a_0}^{a_c}\dfrac{1}{u}\left[\left(1+\dfrac{\gamma}{\delta}u^2\right)^\eta - 1\right]du, & \eta = \dfrac{\alpha}{\gamma}+\dfrac{1}{2}\neq 0 \\[3mm] \dfrac{1}{\gamma}\displaystyle\int_{a_0}^{a_c}\dfrac{1}{u}\ln\left(1+\dfrac{\gamma}{\delta}u^2\right)du, & \eta = \dfrac{\alpha}{\gamma}+\dfrac{1}{2}= 0 \end{cases}.$$

$$(9.3.10)$$

Results in (9.3.10) were obtained in (Ariaratnam and Tam, 1979).

Example 9.3.2 Consider also a linear system, but only subjected to an external excitation, i.e., let the parametric excitations $\xi_1(t)$ and $\xi_2(t)$ in Eq. (9.3.4) be zero. In this case, $\gamma = 0$ according to Eq. (9.1.30), and the results (9.3.9) and (9.3.10) are no longer applicable. The generalized Pontryagin equation is now

$$\frac{\delta}{2}\frac{d^2}{da_0^2}\mu_{n+1} + \left(\frac{\delta}{2a_0} - \zeta\omega_0 a_0\right)\frac{d}{da_0}\mu_{n+1} = -(n+1)\mu_n, \quad (9.3.11)$$

subjected to the same boundary conditions given in (9.3.6) and (9.3.7). Similar to Example 9.3.1, $C_{n+1} = 0$, and Eq. (9.2.18) leads to

$$\mu_{n+1}(a_0) = \frac{2(n+1)}{\delta} \int\limits_{a_0}^{a_c} \frac{1}{u} \exp\left(\frac{\zeta\omega_0}{\delta} u^2\right)$$

$$\times \left[\int\limits_0^u v\mu_n(v) \exp\left(-\frac{\zeta\omega_0}{\delta} v^2\right) dv\right] du. \quad (9.3.12)$$

The mean value is then

$$\mu_1(a_0) = \frac{1}{2\zeta\omega_0} \left[\bar{E}i\left(\frac{\zeta\omega_0}{\delta} a_c^2\right) - \bar{E}i\left(\frac{\zeta\omega_0}{\delta} a_0^2\right) - 2\ln\left(\frac{a_c}{a_0}\right)\right],$$
$$(9.3.13)$$

where $\bar{E}i$ is an exponential integral function defined as

$$\bar{E}i(x) = -\int\limits_{-x}^{\infty} \frac{1}{t} e^{-t} dt, \quad x > 0. \quad (9.3.14)$$

Result (9.3.13) was obtained by Ariaratnam and Pi (1973).

Example 9.3.3 Return to the linear system (9.3.4) with the external excitation $\xi_3(t)$ absent. Then $\delta = 0$ and the generalized Pontryagin equation is given by

$$\frac{1}{2}\gamma a_0^2 \frac{d^2}{da_0^2}\mu_{n+1} - \alpha a_0 \frac{d}{da_0}\mu_{n+1} = -(n+1)\mu_n. \quad (9.3.15)$$

The first boundary condition in (9.3.6) is still valid, but another boundary condition depends on the system parameters. At the left boundary $a_0 = 0$, both the drift coefficient $-\alpha a_0$ and diffusion coefficient γa_0^2 vanish; thus, it is singular of the first kind and is a trap. It is found from Eqs. (4.4.12) through (4.4.14) that at $a_0 = 0$ the drift and diffusion exponents are 1 and 2, respectively, and the character value is $c = -2\alpha/\gamma$. According to Table 4.4.2, it is attractively natural if $c < 1$. In this case, the trivial solution $A(t) = 0$ of the system equation is asymptotic stable with probability one, all samples will approach the left boundary, and the first-passage may not happen for some samples. In the case of $c = 1$, $a_0 = 0$ is strictly natural, the trivial solution $A(t) = 0$ is neither stable nor unstable. In both cases, it is not an admissible boundary for the first-passage problem. On the other hand, if $c = -2\alpha/\gamma > 1$, or according to (9.1.30),

$$\zeta < \frac{\pi\omega_0}{4}[\Phi_{22}(2\omega_0) + \Phi_{11}(2\omega_0)], \qquad (9.3.16)$$

the left boundary $a_0 = 0$ is repulsively natural and the system is unstable at $a_0 = 0$ in probability; therefore, a sample path will eventually reach the critical state a_c, and the first-passage problem is meaningful. In this case, the two boundary conditions are given in (9.2.6) and (9.2.11), i.e.,

$$\mu_{n+1}\big|_{a_0=a_c} = 0, \qquad (9.3.17)$$

$$0\left|a_0 \frac{d\mu_{n+1}}{da_0}\right| \sim 0\,|\mu_n|\,, \quad a_0 \to 0. \qquad (9.3.18)$$

According to Eq. (9.2.16), we obtain first from Eq. (9.3.15)

$$\frac{d\mu_{n+1}(a_0)}{da_0} = a_0^{2\eta-1}\left[-\frac{2(n+1)}{\gamma}\int_0^{a_0}\mu_n(v)v^{-(2\eta+1)}dv + C_{n+1}\right], \qquad (9.3.19)$$

where

$$\eta = \frac{1}{2} + \frac{\alpha}{\gamma}. \qquad (9.3.20)$$

Since $c = -2\alpha/\gamma > 1$, $\eta < 0$. The boundary condition (9.3.18) requires $C_{n+1} = 0$, and (9.3.19) results in

$$\mu_{n+1}(a_0) = \frac{2(n+1)}{\gamma}\int_{a_0}^{a_c} u^{2\eta-1}\left[\int_0^u \mu_n(v)v^{-(2\eta+1)}dv\right]du. \qquad (9.3.21)$$

For $n = 0$ and 1, we have from Eq. (9.3.21)

$$\mu_1(a_0) = -\frac{1}{\eta\gamma}\ln\frac{a_c}{a_0}, \qquad (9.3.22)$$

$$\mu_2(a_0) = \frac{1}{\eta^2\gamma^2}\left(-\frac{1}{\eta} + \ln\frac{a_c}{a_0}\right)\ln\frac{a_c}{a_0}. \qquad (9.3.23)$$

Equations (9.3.21) through (9.3.23) show that the statistical moments of the first-passage time are very large near $a_0 = 0$. This is because both the drift and diffusion near $a_0 = 0$ are very small and the motion is very slow.

Example 9.3.4 Consider a nonlinear system of van der Pol type,

$$\ddot{X} + 2\zeta\omega_0(1 + \beta X^2)\dot{X} + \omega_0^2 X = \xi(t), \qquad (9.3.24)$$

where $\xi(t)$ is a broad-band excitation with a spectral density $\Phi(\omega)$. It is assumed that the damping is small and the excitation is weak. Applying the stochastic averaging, the drift and diffusion coefficients for the amplitude process $A(t)$ are obtained as

$$m(A) = -\xi\omega_0 A - \frac{1}{4}\beta\xi\omega_0 A^3 + \frac{\delta}{2A}, \quad \sigma^2(A) = \delta, \quad \delta = \frac{\pi}{\omega_0^2}\Phi(\omega_0). \qquad (9.3.25)$$

The generalized Pontryagin equation is

$$\frac{1}{2}\delta\frac{d^2}{da_0^2}\mu_{n+1} + \left(-\zeta\omega_0 a_0 - \frac{1}{4}\beta\zeta\omega_0 a_0^3 + \frac{\delta}{2a_0}\right)\frac{d}{da_0}\mu_{n+1}$$
$$= -(n+1)\mu_n, \qquad (9.3.26)$$

subjected to the boundary conditions given in (9.3.6) and (9.3.7). The solution of (9.3.26) is obtained from Eq. (9.2.18) as

$$\mu_{n+1}(a_0) = \frac{2(n+1)}{\delta}\int\limits_{a_0}^{a_c}\frac{1}{u}\exp\left[\frac{\zeta\omega_0}{\delta}\left(u^2 + \frac{\beta}{8}u^4\right)\right]$$

$$\times \left\{\int\limits_0^u v\mu_n(v)\exp\left[-\frac{\zeta\omega_0}{\delta}(v^2 + \frac{\beta}{8}v^4)\right]dv\right\}du. \qquad (9.3.27)$$

It is noted that, with the existence of an external excitation, the left boundary at zero for the amplitude process $A(t)$ is singular of the second kind since the drift coefficient is positive and unbounded. Therefore, the boundary is reflective, indicating that samples near the boundary will travel inward, regardless of the damping mechanism and the existence of parametric excitations. In this case, the first-passage failure will occur eventually.

In the above several examples, closed forms of analytical solutions are obtained for the moments of the first-passage time. In fact, the generalized Pontryagin equation, as a second-order ordinary differential equation, is easy to solve numerically with appropriate boundary conditions.

9.3.2 *Moments of First-Passage Time of Response Energy*

If the system restoring force is strongly nonlinear, the stochastic averaging of energy envelope described in Section 7.4.2 should be applied. Consider the system governed by

$$\ddot{X} + \varepsilon h(X, \dot{X}) + u(X) = \varepsilon^{\frac{1}{2}} \sum_{l=1}^{m} g_l(X, \dot{X})\xi_l(t), \qquad (9.3.28)$$

where $u(X)$ represents a strongly nonlinear restoring force, assumed to be an odd function of X, and $\xi_l(t)$ are broad-band excitations with correlation functions

$$R_{ls}(\tau) = E[\xi_l(t)\xi_s(t + \tau)]. \qquad (9.3.29)$$

The potential energy and total energy are

$$U(X) = \int_0^X u(z)dz, \quad \Lambda = \frac{1}{2}\dot{X}^2 + U(X). \qquad (9.3.30)$$

It is shown in Section 7.4.2 that the energy process $\Lambda(t)$ is approximately a Markov diffusion process governed by an Ito equation

$$d\Lambda = m(\Lambda)dt + \sigma(\Lambda)dB(t), \qquad (9.3.31)$$

where the averaged drift and diffusion coefficients are calculated from (7.4.94) and (7.4.95). In term of the energy process, the generalized Pontryagin equation for the moments of first-passage time is

$$m(\lambda_0)\frac{d}{d\lambda_0}\mu_{n+1} + \frac{1}{2}\sigma^2(\lambda_0)\frac{d^2}{d\lambda_0^2}\mu_{n+1} = -(n + 1)\mu_n, \qquad (9.3.32)$$

with the boundary conditions (9.2.6) and (9.2.7) in general. The qualitative condition (9.2.7) may be replaced by one of the more specific quantitative conditions (9.2.9), (9.2.10) and (9.2.11).

If the broad-band excitations are not white noises, the drift and diffusion coefficients are usually calculated numerically, and the Pontryagin equation (9.3.32) is required to be also solved numerically. However, if all the excitations are Gaussian white noises, it may be possible to obtain $m(\Lambda)$ and $\sigma^2(\Lambda)$ analytically in closed form from Eqs. (7.4.97) and (7.4.98), and hence, the moments of the first-passage time from Eq. (9.3.32) also in closed form. Some examples are given below.

Example 9.3.5 Consider the oscillator with a nonlinear restoring force

$$\ddot{X} + 2\zeta \dot{X} + k\,|X|^\rho \operatorname{sgn}(X) = W(t), \qquad (9.3.33)$$

where k, $\rho > 0$, and $W(t)$ is a Gaussian white noise with a spectral density K. We assume that both the damping and excitation are small so that the stochastic averaging method is applicable. The total energy is given by

$$\Lambda = \frac{1}{2}\dot{X}^2 + \frac{k}{\rho+1}\,|X|^{\rho+1}. \qquad (9.3.34)$$

Using Eqs. (7.4.97) and (7.4.98), we obtain

$$m(\Lambda) = -2\zeta\delta\Lambda + \pi K, \quad \sigma^2(\Lambda) = 2\pi K\delta\Lambda, \quad \delta = \frac{2(\rho+1)}{(\rho+3)}. \qquad (9.3.35)$$

The generalized Pontryagin equation now reads

$$\pi K\delta\lambda_0\frac{d^2}{d\lambda_0^2}\mu_{n+1} + (\pi K - 2\zeta\delta\lambda_0)\frac{d}{d\lambda_0}\mu_{n+1} = -(n+1)\mu_n. \qquad (9.3.36)$$

One boundary condition is Eq. (9.2.6), i.e.,

$$\mu_{n+1}(\lambda_0)|_{\lambda_0=\lambda_c} = 0. \qquad (9.3.37)$$

Another one, if quantitative required, must be determined from the nature of the left boundary $\lambda_0 = 0$. Since $m(0) = \pi K$ and $\sigma^2(0) = 0$, $\lambda_0 = 0$ is singular of the first kind, and is either an entrance shunt or a regular shunt, according to Table 4.4.2. Therefore, the second boundary condition is Eq. (9.2.10), i.e.,

$$\frac{d\mu_{n+1}}{d\lambda_0}\bigg|_{\lambda_0=0} = -\frac{(n+1)\mu_n}{\pi K}\bigg|_{\lambda_0=0}. \tag{9.3.38}$$

With these two boundary conditions, Eq. (9.3.36) is solved according to (9.2.18) to obtain

$$\mu_{n+1}(\lambda_0) = \frac{n+1}{\pi K\delta} \int_{\lambda_0}^{\lambda_c} u^{-1/\delta} \exp\left(\frac{2\zeta}{\pi K}u\right)$$

$$\times \left[\int_0^u \mu_n(v)v^{1/\delta-1} \exp\left(-\frac{2\zeta}{\pi K}v\right) dv\right] du. \tag{9.3.39}$$

For the special case of linear system, i.e., $\rho = 1$ and $\delta = 1$, we have from (9.3.39)

$$\mu_1(\lambda_0) = \frac{1}{2\zeta}\left[\bar{E}i\left(\frac{2\zeta}{\pi K}\lambda_c\right) - \bar{E}i\left(\frac{2\zeta}{\pi K}\lambda_0\right) - \ln\left(\frac{\lambda_c}{\lambda_0}\right)\right]. \tag{9.3.40}$$

With $\rho = 1$, system (9.3.33) is linear, which is the same as that in Example 9.3.2 if replacing 2ζ by $2\zeta\omega_0$ and k by ω_0^2. As expected, the result given in (9.3.40) is the same as that in (9.3.13) by noting that $\Lambda = \frac{1}{2}kA^2$.

Example 9.3.6 Consider a Duffing oscillator under a parametric excitation of Gaussian white noise

$$\ddot{X} + 2\zeta\dot{X} + X + \delta X^3 = XW(t). \tag{9.3.41}$$

The total energy is given by

$$\Lambda = \frac{1}{2}\dot{X}^2 + \frac{1}{2}X^2 + \frac{1}{4}\delta X^4. \tag{9.3.42}$$

Applying the stochastic averaging described in Section 7.4.2, the averaged drift and diffusion coefficients for the energy process, obtained from Eqs. (7.4.97) and (7.4.98), are given by

$$m(\Lambda) = \frac{1}{\Delta} A^2$$

$$\times \left[\pi K \int_0^{\pi/2} \frac{\cos^2 \varphi}{k(A, \varphi)} d\varphi - 2\zeta(1 + \delta A^2) \int_0^{\pi/2} k(A, \varphi) \sin^2 \varphi d\varphi \right],$$

$$(9.3.43)$$

$$\sigma^2(\Lambda) = \frac{2}{\Delta} \pi K A^4 (1 + \delta A^2) \int_0^{\pi/2} k(A, \varphi) \sin^2 \varphi \cos^2 \varphi d\varphi, \qquad (9.3.44)$$

where

$$k(A, \varphi) = \sqrt{1 - \frac{\delta A^2 \sin^2 \varphi}{2(1 + \delta A^2)}}, \quad \Delta = \int_0^{\pi/2} \frac{d\varphi}{k(A, \varphi)} \qquad (9.3.45)$$

and A is a function of Λ, defined as

$$\Lambda = \frac{1}{4} A^2 (2 + \delta A^2), \quad A = \sqrt{\frac{1}{\delta} \left(\sqrt{1 + 4\delta\Lambda} - 1 \right)}. \qquad (9.3.46)$$

First we need to determine the condition for the first-passage problem to be meaningful by examining the left boundary $\lambda_0 = 0$. It is found that, as $\lambda_0 \to 0$, $A \to \sqrt{2\Lambda}$, $k(A, \phi) \to 1$, and

$$m(\Lambda) \to (\pi K - 2\zeta)\Lambda, \quad \sigma^2(\Lambda) \to \pi K \Lambda^2. \qquad (9.3.47)$$

According to (4.4.12) through (4.4.14), the diffusion exponent $\alpha = 2$, drift exponent $\beta = 1$, and the character value $c = 2 - 4\zeta/\pi K$. Similar to the analysis in Example 9.3.2, the first-passage problem is meaningful only if $c > 1$, in which case the left boundary is repulsively natural. Therefore, the condition for the admissible left boundary is

$$\zeta < \frac{1}{4} \pi K. \qquad (9.3.48)$$

Under this condition, the two boundary conditions for the generalized Pontryagin equation are

$$\mu_{n+1}|_{\lambda_0=\lambda_c} = 0, \tag{9.3.49}$$

$$0|m(\lambda_0)\frac{d\mu_{n+1}(\lambda_0)}{d\lambda_0}| \sim 0|\mu_n(\lambda_0)|, \quad \lambda_0 \to 0. \tag{9.3.50}$$

Due to the rather complex expressions of the drift and diffusion coefficients in Eqs. (9.3.43) and (9.3.44), an analytical solution for the generalized Pontryagin equation is difficult to obtain, and numerical solution procedure is required. For this purpose, the boundary condition in (9.3.50) is not adequate, and a quantitative one is needed. Consider the case of a small λ_0, and the drift and diffusion coefficients may be approximated as

$$m(\lambda_0) \approx (\pi K - 2\zeta)\lambda_0, \quad \sigma^2(\lambda_0) \approx K\lambda_0^2. \tag{9.3.51}$$

Thus, we obtain from the generalized Pontryagin equation

$$\mu'_{n+1}(\lambda_0) = -\frac{n+1}{\pi K}\lambda_0^{[(4\zeta/\pi K)-2]}\left[\int_0^{\lambda_0}\mu_n(v)v^{-4\zeta/\pi K}dv + C_{n+1}\right],$$

$$\lambda_0 \ll 1. \tag{9.3.52}$$

Using boundary condition (9.3.50) and taking into account the condition (9.3.48), we have $C_{n+1} = 0$. Specifically, for $n = 0$,

$$\mu'_1(\lambda_0) = \frac{2}{(4\zeta - \pi K)\lambda_0}, \quad \lambda_0 \ll 1. \tag{9.3.53}$$

Consider a small value $\lambda_l \ll 1$ as the left boundary for numerical solution of the Pontryagin equation. The quantitative boundary condition to replace (9.3.50) is

$$\mu'_1(\lambda_l) = \frac{2}{(4\zeta - \pi K)\lambda_l}. \tag{9.3.54}$$

After $\mu_1(\lambda_0)$ is solved numerically for the range $\lambda_l \leq \lambda_0 \leq \lambda_c$, higher-order moment $\mu_n(\lambda_0)$ can be solved sequentially for the same range.

Numerical calculations were carried out in (Cai and Lin, 1994) to obtain the mean first-passage time μ_1 for system (9.3.41) with

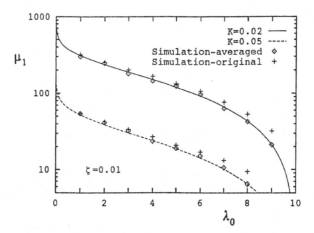

Figure 9.3.1 Mean first-passage time of system (9.3.41) with two different excitation levels. (from Cai and Lin, 1994).

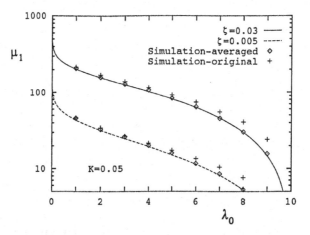

Figure 9.3.2 Mean first-passage time of system (9.3.41) with two different damping ratios (from Cai and Lin, 1994).

$\delta = 0.1$, $\lambda_c = 10$ and several combinations of ξ and K values. When applying boundary condition (9.3.53), two trial values of 0.01 and 0.005 were used for λ_0 to begin the computation, and the results were found to be almost the same. The calculated results are shown in Figs. 9.3.1 and 9.3.2 as solid and dashed lines. Monte Carlo simulations were carried out for the original system (9.3.41) and the averaged Ito equation of the energy process, and the results are shown as symbols in the figures. As expected, the results from solving the Pontryagin equation numerically are in excellent agreement with the simulation results for the averaged system. The difference between the simulation results for the original and averaged systems is quite minimal, indicating the error arising from stochastic averaging method is small.

Example 9.3.7 Consider a simplified model of ship rolling in random sea, governed by

$$\ddot{X} + \alpha \dot{X} + \beta \left| \dot{X} \right| \dot{X} + \gamma X - \delta X^3 = X \xi_1(t) + \xi_2(t), \qquad (9.3.55)$$

where X is the rolling angle, α, β, γ and δ are positive parameters, and $\xi_1(t)$ and $\xi_2(t)$ are stationary broad-band processes with correlation functions

$$E[\xi_j(t)\xi_k(t+\tau)] = R_{jk}(\tau), \quad j, k = 1, 2. \qquad (9.3.56)$$

The combined linear and quadratic damping in (9.3.55) was shown to be a good damping model (Dalzell, 1973; Roberts, 1985). The parametric excitation on the right-hand side is due to the fact that the restoring moment depends on the water level relative to the ship. The cubic softening nonlinearity in the restoring moment captures the essential feature that, as the rolling angle increases, the restoring moment also increases until it reaches a critical value, beyond which capsizing occurs.

First consider the undamped free rolling motion

$$\ddot{x} + \gamma x - \delta x^3 = 0. \qquad (9.3.57)$$

The restoring moment, the potential energy, and the total energy are given by, respectively,

$$u(x) = \gamma x - \delta x^3, \quad U(x) = \frac{1}{2}\gamma x - \frac{1}{4}\delta x^4, \quad \lambda = \frac{1}{2}\dot{x}^2 + U(x). \qquad (9.3.58)$$

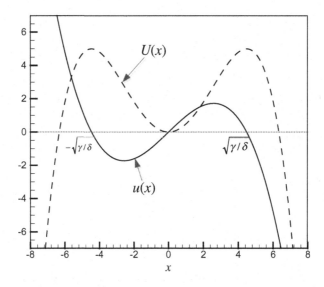

Figure 9.3.3   Restoring moment and potential energy of ship rolling motion.

Figure 9.3.3 shows schematically the restoring moment and the potential energy for the free rolling motion. The potential energy constitutes a potential well. At $x = \pm\sqrt{\gamma/\delta}$, it reaches the maximum

$$U_{\max} = \frac{\gamma^2}{4\delta}. \tag{9.3.59}$$

When the rolling angle is less than $\sqrt{\gamma/\delta}$, or equivalently, the total energy is less than $U_{\max} = \gamma^2/(4\delta)$, the rolling motion is periodic and confined within the potential well. The period is determined from

$$T = 4T_{\frac{1}{4}} = 4 \int_{0}^{a} \frac{dx}{\sqrt{2\lambda - 2U(x)}}, \tag{9.3.60}$$

where the upper limit a of integration is the amplitude determined from $U(a) = \lambda$. It is noted that (9.3.60) is valid under the condition $U(a) < U_{\max}$, i.e., $a < \sqrt{\gamma/\delta}$.

Once the rolling angle exceeds $\sqrt{\gamma/\delta}$, the restoring moment becomes negative, causing the angle larger and larger until capsizing of the ship. If the first-passage problem is investigated in term of the total energy, U_{max} is the critical value λ_c.

Return to the stochastic system (9.3.55), and apply the stochastic averaging method for the energy process $\Lambda(t)$. Following the procedure in Section 7.4.2 and in Example 7.4.3, we obtain the drift and diffusion coefficients as follows

$$m(\Lambda) = -\alpha u_2 - \beta u_6 + \pi(u_1 - u_3)\Phi_{11}(\omega_\Lambda)$$
$$+ \pi u_3 \Phi_{11}(2\omega_\Lambda) + \pi\Phi_{22}(\omega_\Lambda), \qquad (9.3.61)$$

$$\sigma^2(\Lambda) = 2\pi(u_4 - u_5)\Phi_{11}(\omega_\Lambda) + 2\pi u_5 \Phi_{11}(2\omega_\Lambda)$$
$$+ 2\pi u_2 \Phi_{22}(\omega_\Lambda), \qquad (9.3.62)$$

where $\Phi_{11}(\omega)$ and $\Phi_{22}(\omega)$ are spectral densities of $\xi_1(t)$ and $\xi_2(t)$, respectively, u_i, $i = 1, 2, \ldots, 5$ are given in (7.4.134) through (7.4.138), and u_6 is given by

$$u_6 = \left\langle \left| \dot{X} \right|^2 \right\rangle_t = \frac{1}{T_{\frac{1}{4}}} \int_0^A [2\Lambda - 2U(X)]^{\frac{3}{2}} \, dX. \qquad (9.3.63)$$

In the above equations, A and ω_Λ are amplitude and frequency corresponding to energy Λ, i.e., $U(A) = \Lambda$, $\omega_\Lambda = 2\pi/T$.

To acquire the boundary conditions for the generalized Pontryagin equation, the nature of the left boundary $\lambda_0 = 0$ must be investigated. It can be shown from Eqs. (9.3.61) and (9.3.62) that $m(0) = \pi\Phi_{22}(\sqrt{\gamma})$ and $\sigma(0) = 0$. Then the two boundary conditions are

$$\mu_{n+1}(\lambda_c) = 0, \quad \mu'_{n+1}(0) = -\frac{(n+1)\mu_n(0)}{\pi\Phi_{22}(\sqrt{\gamma})}. \qquad (9.3.64)$$

The second condition can be derived from the generalized Pontryagin equation (9.3.32) directly. With the two boundary conditions in (9.3.64), Eq. (9.3.32) can be solved numerically.

9.4 First-Passage Problem of Quasi-Hamiltonian Systems

Consider a quasi non-integrable Hamiltonian system governed by the Ito equations, Eq. (7.5.21), i.e.,

$$
dQ_j = \frac{\partial H}{\partial P_j} dt,
$$

$$
dP_j = -\left(\frac{\partial H}{\partial Q_j} - \varepsilon \sum_{k=1}^{n} m_{jk} \frac{\partial H}{\partial P_k} \right) dt + \varepsilon^{\frac{1}{2}} \sum_{l=1}^{m} g_{jl} dB_l(t).
$$

$$(9.4.1)$$

If the system damping terms are small and the excitations are weak, the Hamiltonian process $H(t)$ can be approximated by a Markov diffusion process by applying the stochastic averaging method of quasi non-integrable Hamiltonian systems, as described in Section 7.5.2. Thus, the Hamiltonian process $H(t)$ is governed by the averaged Ito differential equation

$$
dH = m(H)dt + \sigma(H)dB(t), \tag{9.4.2}
$$

where the drift coefficient $m(H)$ and diffusion coefficient $\sigma(H)$ are calculated from Eqs. (7.5.24) and (7.5.25), i.e.,

$$
m(H) = \varepsilon \left\langle -\sum_{j,k=1}^{n} m_{jk} \frac{\partial H}{\partial P_j} \frac{\partial H}{\partial P_k} + \sum_{j,k=1}^{n} \sum_{j,s=1}^{m} \pi K_{ls} g_{jl} g_{ks} \frac{\partial^2 H}{\partial P_j \partial P_k} \right\rangle_t,
$$

$$(9.4.3)$$

$$
\sigma^2(H) = \varepsilon^{\frac{1}{2}} 2\pi \left\langle \sum_{j,k=1}^{n} \sum_{l,s=1}^{m} K_{ls} g_{jl} g_{ks} \frac{\partial H}{\partial P_j} \frac{\partial H}{\partial P_k} \right\rangle_t, \tag{9.4.4}
$$

where the time-averaging can be replaced by the following space averaging

$$
\langle [\cdot] \rangle_t = \frac{1}{T(H)} \int_\Omega [\cdot] \left(\frac{\partial H}{\partial p_1} \right)^{-1} dq_1 dq_2 \cdots dq_n \ dp_2 \cdots dp_n, \tag{9.4.5}
$$

$$
T(H) = \int_\Omega \left(\frac{\partial H}{\partial p_1} \right)^{-1} dq_1 dq_2 \cdots dq_n \ dp_2 \cdots dp_n \tag{9.4.6}
$$

and the integration domain Ω is defined as $H(q_1, q_2, \ldots, q_n, p_1 = 0,$
$p_2, \ldots, p_n) \leq H$.

For most dynamical systems, H represents the total energy either in a narrow or a wide sense; therefore, it may be used for failure analysis as a critical physical quantity of the system. Assume that $H \geq 0$, and the system begins with $H = h_0$ initially at time instant t_0, and fails as soon as H reaches a critical value h_c. The reliability function, denoted by $R(\tau, h_0)$, where $\tau = t - t_0$, is then governed by Eq. (9.1.4), i.e.,

$$-\frac{\partial R}{\partial \tau} + m(h_0)\frac{\partial R}{\partial h_0} + \frac{1}{2}\sigma^2(h_0)\frac{\partial^2 R}{\partial h_0^2} = 0. \tag{9.4.7}$$

The initial condition and boundary conditions are

$$R(\tau, h_0)|_{\tau=0} = 1, \quad h_0 < h_c, \tag{9.4.8}$$

$$R(\tau, h_0)|_{h_0=h_c} = 0, \tag{9.4.9}$$

$$R(\tau, h_0)|_{h_0=0} = \text{finite}. \tag{9.4.10}$$

The generalized Pontryagin equation governing the moments of the first-passage time is given by Eq. (9.2.5), i.e.,

$$(n+1)\mu_n + m(h_0)\frac{d}{dh_0}\mu_{n+1} + \frac{1}{2}\sigma^2(h_0)\frac{d^2}{dh_0^2}\mu_{n+1} = 0. \tag{9.4.11}$$

The associated boundary conditions are

$$\mu_{n+1}(h_0)|_{h_0=h_c} = 0, \tag{9.4.12}$$

$$\mu_{n+1}(h_0)|_{h_0=0} = \text{finite}. \tag{9.4.13}$$

Condition (9.4.10) for the reliability and condition (9.4.13) for the moments are based on an assumption that the left boundary at $h_0 = 0$ is unstable and the first-passage will occur eventually. However, these conditions are qualitative, and may be useful to obtain closed-form solutions if possible, as shown in Sections 9.1 and 9.3. However, in many practical nonlinear systems, numerical procedure must be applied to solve Eqs. (9.4.7) or (9.4.11), and a quantitative boundary condition is required. For this purpose, the left boundary at $h_0 = 0$ must be analyzed. It is found that, for the present quasi-Hamiltonian system, the left boundary $h_0 = 0$ is singular of either first kind or second kind. In order for the first-passage problem to be meaningful, it must be regular, entrance, or repulsively natural nature. If it is singular of the first kind, i.e., $m(0) > 0$ and $\sigma(0) = 0$, then we have directly from Eqs. (9.4.7) and (9.4.11),

$$\frac{\partial R}{\partial \tau}\bigg|_{h_0=0} = m(0)\,\frac{\partial R}{\partial h_0}\bigg|_{h_0=0}, \tag{9.4.14}$$

$$\mu'_{n+1}(0) = -\frac{(n+1)\mu_n(0)}{m(0)}. \tag{9.4.15}$$

Equations (9.4.7) and (9.4.11) can then be solved numerically using the quantitative initial and boundary conditions. On the other hand, if it is singular of the second kind, i.e., $m(0) = 0$ and $\sigma(0) = 0$, we have

$$\frac{\partial R}{\partial \tau} = 0, \quad h_0 \to 0, \tag{9.4.16}$$

$$0[m(h_0)\mu'_{n+1}(h_0)] \sim 0[\mu_n(h_0)], n \neq 0, h_0 \to 0$$

$$m(0)\mu'_1(0) = \text{finite}. \tag{9.4.17}$$

Condition (9.4.16) is quantitative, but condition (9.4.17) is still qualitatively. Further approximation procedure may be needed to acquire a quantitatively boundary condition at or near $h_0 = 0$, as shown in the example below.

Example 9.4.1 (Gan and Zhu, 2001) Consider the following system

$$\ddot{X} + \lambda_1 \dot{X} + b\omega_1^2 X(\omega_1^2 X^2 + \omega_2^2 Y^2) + \omega_1^2 X + (a_1 + b_1 X)W_1(t) = 0,$$

$$\ddot{Y} + \lambda_2 \dot{Y} + b\omega_2^2 Y(\omega_1^2 X^2 + \omega_2^2 Y^2) + \omega_2^2 Y + (a_2 + b_2 Y)W_2(t) = 0,$$

$$\tag{9.4.18}$$

where λ_1, λ_2, a_1, a_2, b_1, b_2, b, ω_1 and ω_2 ($\neq \omega_1$) are constants, $W_1(t)$ and $W_2(t)$ are uncorrelated Gaussian white noises with spectral densities K_1 and K_2, respectively. The Hamiltonian of system (9.4.18) is

$$H = \frac{1}{2}\dot{X}^2 + \frac{1}{2}\dot{Y}^2 + U(X,Y), \tag{9.4.19}$$

where $U(X,Y)$ is the potential energy given by

$$U(X,Y) = \frac{1}{2}\omega_1^2 X^2 + \frac{1}{2}\omega_2^2 Y^2 + \frac{1}{4}b(\omega_1^2 X^2 + \omega_2^2 Y^2)^2. \qquad (9.4.20)$$

Assume that the damping forces are small and the excitations are weak. Application of the stochastic averaging method of quasi-Hamiltonian systems results in the Hamiltonian function as a one-dimensional diffusion process governed by Eq. (9.4.2). The averaged drift and diffusion coefficients are found to be (Gan and Zhu, 2001)

$$m(H) = 2\pi(K_1 a_1^2 + K_2 a_2^2) + \frac{\pi}{2}\left(\frac{K_1 b_1^2}{\omega_1^2} + \frac{K_2 b_2^2}{\omega_2^2}\right)$$

$$\times R^2(H) - (\lambda_1 + \lambda_2)H + \frac{1}{2}(\lambda_1 + \lambda_2)bR^4(H), \qquad (9.4.21)$$

$$\sigma^2(H) = 4\pi(K_1 a_1^2 + K_2 a_2^2)H - \pi(K_1 a_1^2 + K_2 a_2^2)R^2(H)$$

$$- \frac{\pi}{3}\left[b(K_1 a_1^2 + K_2 a_2^2) + \left(\frac{K_1 b_1^2}{\omega_1^2} + \frac{K_2 b_2^2}{\omega_2^2}\right)\right]R^4(H)$$

$$+ \pi\left(\frac{K_1 b_1^2}{\omega_1^2} + \frac{K_2 b_2^2}{\omega_2^2}\right)HR^2(H)$$

$$- \frac{\pi b}{8}\left(\frac{K_1 b_1^2}{\omega_1^2} + \frac{K_2 b_2^2}{\omega_2^2}\right)R^6(H), \qquad (9.4.22)$$

where $R(H)$ is the positive root of the equation

$$\frac{1}{4}R^4 + \frac{1}{2}R^2 = H. \qquad (9.4.23)$$

Since the drift and diffusion coefficients given in Eqs. (9.4.21) and (9.4.22) are rather complicated, analytical solutions for the reliability function from Eq. (9.4.7) and for the moments of the first-passage time from Eq. (9.4.11) are difficult to obtain, and numerical solution procedures are required. For this purpose, all initial and boundary conditions must be quantitative. Two cases will be discussed below.

First consider the case in which $a_1 \neq 0$ and $a_2 \neq 0$, namely, both the external excitations are present. Then as $H \to 0$, $R^2 \to 2H$, we obtain from Eqs. (9.4.21) and (9.4.22)

$$m(H) \rightarrow 2\pi(K_1 a_1^2 + K_2 a_2^2), \tag{9.4.24}$$

$$\sigma^2(H) \rightarrow 2\pi(K_1 a_1^2 + K_2 a_2^2)H. \tag{9.4.25}$$

Since $m(0) > 0$ and $\sigma^2(0) \sim 0$, $h_0 = 0$ is singular boundary of the first kind. The diffusion exponent, drift exponent and character value are calculated from Eqs. (4.4.12) through (4.4.14) as

$$\beta = 0, \quad \alpha = 1, \quad c = 2. \tag{9.4.26}$$

According to Table 4.4.3, $h_0 = 0$ is an entrance, which is an admissible boundary for the first-passage problem. Thus, initial and boundary conditions (9.4.8), (9.4.9) and (9.4.14) can be used to solve the reliability function from Eq. (9.4.7), and boundary conditions (9.4.12) and (9.4.15) are used to solve Eq. (9.4.11) for the moments of the first-passage time. Since all the initial and boundary conditions are quantitative, numerical calculations can be carried out, as done in (Gan and Zhu, 2001).

If $a_1 = 0$ and $a_2 = 0$, namely, only parametric excitations exist. Then we have from Eqs. (9.4.21) and (9.4.22)

$$m(H) = \left[\pi \left(\frac{K_1 b_1^2}{\omega_1^2} + \frac{K_2 b_2^2}{\omega_2^2} \right) - \frac{1}{2}(\lambda_1 + \lambda_2) \right] H, \quad H \rightarrow 0, \tag{9.4.27}$$

$$\sigma^2(H) = \frac{2\pi}{3} \left(\frac{K_1 b_1^2}{\omega_1^2} + \frac{K_2 b_2^2}{\omega_2^2} \right) H^2, \quad H \rightarrow 0. \tag{9.4.28}$$

In this case, $h_0 = 0$ is singular of the second kind. The diffusion exponent, drift exponent and character value are calculated from Eqs. (4.4.15) through (4.4.17) as

$$\beta = 1, \quad \alpha = 2, \quad c = 3 - \frac{3(\lambda_1 + \lambda_2)}{2\pi \left(\frac{K_1 b_1^2}{\omega_1^2} + \frac{K_2 b_2^2}{\omega_2^2} \right)}. \tag{9.4.29}$$

According to Table 4.4.3, $h_0 = 0$ is repulsively natural if $c > \beta = 1$, namely,

$$\lambda_1 + \lambda_2 < \frac{4\pi}{3} \left(\frac{K_1 b_1^2}{\omega_1^2} + \frac{K_2 b_2^2}{\omega_2^2} \right). \tag{9.4.30}$$

The first-passage problem is meaningful only if condition (9.4.30) is met, indicating that the damping forces are insufficient to attract the system motion to the left boundary. In this case, initial and boundary conditions (9.4.8), (9.4.9) and (9.4.16) can be used to solve Eq. (9.4.7) for the reliability function. However, for the moments of the first-passage time, a quantitative boundary condition has to be found to replace (9.4.13). Consider a small h_0, and the drift and diffusion coefficients can be approximated by expressions (9.4.27) and (9.4.28). Thus, we obtain from the generalized Pontryagin equation

$$\mu_{n+1}'(h_0) = -\frac{3(n+1)h_0^{-c}}{\pi \left(\frac{K_1 b_1^2}{\omega_1^2} + \frac{K_2 b_2^2}{\omega_2^2} \right)} \int_0^{h_0} \mu_n(v) v^{c-2} dv, \quad h_0 \ll 1. \tag{9.4.31}$$

Specifically, for $n = 0$,

$$\mu_1'(h_0) = -\frac{3}{\pi(c-1)h_0 \left(\frac{K_1 b_1^2}{\omega_1^2} + \frac{K_2 b_2^2}{\omega_2^2} \right)}, \quad h_0 \ll 1, \tag{9.4.32}$$

where parameter c is given in (9.2.29). Similar to the procedure in Example 9.3.6, take a small value $h_l \ll 1$ as the left boundary, and replace the qualitative boundary condition (9.4.13) by

$$\mu_1'(h_l) = -\frac{3}{\pi(c-1)h_l \left(\frac{K_1 b_1^2}{\omega_1^2} + \frac{K_2 b_2^2}{\omega_2^2} \right)}. \tag{9.4.33}$$

With boundary conditions (9.4.12) and (9.4.33), $\mu_1(h_0)$ can be solved numerically for the range $h_l \leq h_0 \leq h_c$. Higher-order moment $\mu_n(h_0)$ can be solved sequentially for the same range if necessary.

Numerical calculations were carried out in (Gan and Zhu, 2001) to solve the Pontryagin equation for the Hamiltonian process with system parameters: $\omega_1 = 1$, $\omega_2 = 2$, $\lambda_1 = 0.05$, $\lambda_2 = 0.03$, $b = 2$, $\pi K_1 = 0.6$, $\pi K_2 = 0.55$, $h_c = 40$ and two sets of a_1, a_2, b_1 and

b_2 values. The results are shown in Figs. 9.4.1 and 9.4.2 as solid lines. Results obtained from Monte Carlo simulations for the original system (9.4.18) and for the averaged Ito equation of the Hamiltonian process are also shown in the figures as solid squares and hollow squares, respectively. It is seen that results from the three procedures agree quite well.

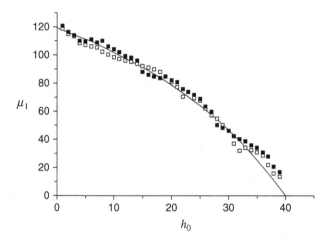

Figure 9.4.1   Mean first-passage time of system (9.4.18) with $a_1 = 0.6$, $a_2 = 0.8$, $b_1 = 0.6$ and $b_2 = 0.8$ (from Gan and Zhu, 2001).

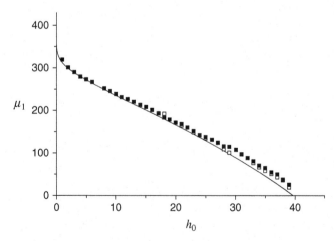

Figure 9.4.2   Mean first-passage time of system (9.4.18) with $a_1 = a_2 = 0$, $b_1 = 0.8$ and $b_2 = 0.85$ (from Gan and Zhu, 2001).

9.5 Fatigue Failure of Stochastically Excited Structures

As mentioned before, only one type of fatigue failure is investigated, that is, a dominant crack in material propagates and exceeds a critical size. Such type of fatigue failure for a structural or machine component due to dynamic loading always begins with an initial crack. The initial crack may be presented in the material during manufacturing process, or developed over time due to dynamic loading. During service operation, the initial crack propagates gradually, and failure occurs when the crack reaches a critical size. Even in a well-controlled laboratory environment, results of crack growth experiments exhibit uncertainty to various extents, as shown schematically in Fig. 9.5.1. Thus, it is reasonably to treat the growth of a fatigue crack as a random phenomenon, and the fatigue failure as a first-passage problem when the crack size exceeds the prescribed critical length for the first time.

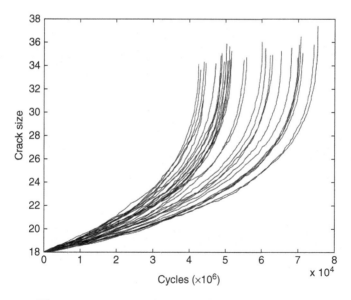

Figure 9.5.1   Time histories of crack propagation.

9.5.1 *Deterministic Model*

A typical deterministic model for fatigue crack growth under cyclic loading has been proposed (Hoeppner and Krupp, 1974; Miller and Gallagher, 1981)

$$\Delta a_n = a_{n+1} - a_n = f(k, \Delta k, a_n, R), \qquad (9.5.1)$$

where a_n is the crack length at the end of the nth stress cycle, n is the number of cycles, f is a non-negative function, k is the stress intensity factor, Δk is the range of the stress intensity factor, and R is the stress ratio. For ease of mathematical manipulation, transfer the discrete variable n to continuous time variable t, resulting in

$$\frac{da}{dt} = \eta f(k, \Delta k, a, R), \qquad (9.5.2)$$

where η is the number of cycles per unit time. It is generally recognized that the effect of the stress intensity factor range Δk is dominant. By neglecting the less important factors k, and R in Eq. (9.5.2), we have

$$\frac{da}{dt} = \eta g(a, \Delta k), \qquad (9.5.3)$$

where g is again a non-negative function. Noting the dependence of Δk on the crack length a and the stress range Δx, Eq. (9.5.3) can be written as

$$\frac{da}{dt} = \eta g(a, \Delta x). \qquad (9.5.4)$$

Experimental data show some general features of the crack growth. There exist two thresholds: (i) a lower one Δk_0, below which the crack growth is negligible, and (ii) a higher one Δk_{ft}, beyond which the crack growth rate is very high and fatigue failure occurs immediately. In the interval of the two thresholds, the relationship between a and Δk is approximately a straight line on a log–log scale, suggesting a power law function. Based on this argument, the well-known Paris–Erdogan model (Paris and Erdgan, 1963) was proposed as

$$\frac{da}{dt} = \eta \alpha (\Delta k)^\beta = \eta \alpha [h(a) \Delta x]^\beta, \qquad (9.5.5)$$

where α and β are non-negative material constants, and $h(a)$ is a non-negative function depending on the geometries of the crack and the component.

9.5.2 *Stochastic Model and Analysis*

Several stochastic model and solution techniques were proposed to treat the random growth of the fatigue crack. The following analysis is based on the papers (Zhu, Lin and Lei, 1992; Cai, Yu and Lin, 1995).

Denote $X(t)$ as the random stress process, which causes the crack growth. The usual definition of stress range in the cyclic loading case is no longer meaningful. For the random loading case, we define stress range ΔX as the difference between a local minimum and the following local maximum,

$$\Delta X(t) = |X(t_n) - X(t_{n+1})|, \quad t_n \le t < t_{n+1}, \qquad (9.5.6)$$

where t_n and t_{n+1} are the time instants at which the two consecutive extremes occur. Based on the deterministic model (9.5.4), the stochastic model is now proposed as

$$\frac{dA}{dt} = \eta g[A(t), \Delta X(t)]\psi(t), \qquad (9.5.7)$$

where $\psi(t)$ is a random environmental process introduced to account for such effects as temperature fluctuation, corrosion, etc. We assume that process $\psi(t)$ is independent of the stress process $X(t)$. The notations for the crack length and stress range are capitalized as A and ΔX to signify that they are now stochastic processes.

In general, the life of a structural or mechanical component, namely the time for the crack to growth to the critical size, is much longer than the correlation time of the stress range process $\Delta X(t)$. Then the fatigue crack length $A(t)$ is a slowly varying stochastic process compared with the stress range process $\Delta X(t)$. According to Section 7.4, the stochastic averaging method is applicable to approximate the slow process $A(t)$ as a Markov diffusion process governed by the Ito equation

$$dA = m(A)dt + \sigma(A)dB(t), \tag{9.5.8}$$

where

$$m(A) = \mu E[g]E[\psi] + \eta \int_{-\infty}^{0} \text{cov} \left[g(t+\tau), \frac{\partial g(t)}{\partial A} \right]$$

$$\times \text{cov} \left[\psi(t+\tau), \psi(t) \right] d\tau, \tag{9.5.9}$$

$$\sigma^2(A) = \eta^2 \int_{-\infty}^{\infty} \text{cov} \left[g(t+\tau), g(t) \right] \text{cov} \left[\psi(t+\tau), \psi(t) \right] d\tau.$$

$$\tag{9.5.10}$$

In deriving (9.5.9) and (9.5.10), considerations are taken of (i) both functions g and ψ have non-zero mean values, (ii) the time averaging in Eqs. (7.4.19) and (7.4.20) is not performed. The purpose of this type of stochastic averaging is only to approximate the underlying process as Markov diffusion process, and known as unsmoothed stochastic averaging, as described in Section 7.4.

It is noted that the Wiener process $B(t)$ in (9.5.8) has an unbounded variation; hence, the Markov model (9.5.8) does not guarantee the $A(t)$ is a non-decreasing process, as expected by the nature. However, if the drift coefficient $m(A)$ is positive and sufficiently large compared with the diffusion coefficient $\sigma(A)$, then the trend of an increasing $A(t)$ dominates. Denote a_i as the starting crack length under which the crack does not grow. We will impose a reflective boundary at the a_i to eliminate the possibility of $A(t)$ less than a_i.

As described in Section 9.1, the reliability function is defined as

$$R(\tau, a_0) = \int_{a_i}^{a_c} p(a, t|a_0, t_0)da, \tag{9.5.11}$$

where a_c is the critical crack length, and $\tau = t - t_0$. $R(\tau, a_0)$ is governed by

$$-\frac{\partial R}{\partial \tau} + m(a_0)\frac{\partial R}{\partial a_0} + \frac{1}{2}\sigma^2(a_0)\frac{\partial^2 R}{\partial a_0^2} = 0, \tag{9.5.12}$$

with the following initial and boundary conditions

$$[R(\tau, a_0)]_{\tau=0} = 1, \quad a_i \le a_0 < a_c, \tag{9.5.13}$$

$$[R(\tau, a_0)]_{a_0=a_c} = 0, \tag{9.5.14}$$

$$\left[\frac{\partial}{\partial a_0} R(\tau, a_0)\right]_{a_0=a_i} = 0. \tag{9.5.15}$$

The time T for the crack length to reach the critical length a_c, known as the first-passage time, is a random variable. As described in Section 9.2, the nth moment μ_n of the random time T is governed by the generalized Pontryagin equation

$$(n+1)\mu_n + m(a_0)\frac{d}{da_0}\mu_{n+1} + \frac{1}{2}\sigma^2(a_0)\frac{d^2}{da_0^2}\mu_{n+1} = 0. \tag{9.5.16}$$

The associated boundary conditions are

$$\mu_{n+1}(a_c) = 0, \tag{9.5.17}$$

$$\mu'_{n+1}(a_i) = 0. \tag{9.5.18}$$

Condition (9.5.18) is from the reflective nature of the left boundary at a_i. Equation (9.5.16) can be solved recursively from $n = 0$.

To solve the fatigue failure problem presented above, it is required to have the knowledge of (i) the functional form of $g[A(t), \Delta X(t)]$, (ii) the statistical properties of the random process $\psi(t)$ and the stress intensity factor range $\Delta X(t)$, and (iii) the starting crack length a_i and the critical crack length a_c.

To simplify the analysis, we adopt the randomized Paris–Erdogan fatigue model given by

$$\frac{dA}{dt} = \eta Q(A)(\Delta x)^\beta \psi(t). \tag{9.5.19}$$

Comparing (9.5.19) with (9.5.7) and (9.5.5), we know that

$$g(A, \Delta X) = Q(A)(\Delta X)^\beta, \quad Q(A) = \alpha[h(A)]^\beta. \tag{9.5.20}$$

The analysis can be simplified further by introducing another stochastic process

$$Z[A(t)] = \int_{a_i}^{A} \frac{du}{Q(u)}. \tag{9.5.21}$$

Since $Q(A)$ is a non-negative function, $Z[A(t)]$ is a monotonically increasing function with $A(t)$. With transformation (9.5.21), Eq. (9.5.19) becomes

$$\frac{dZ}{dt} = \eta(\Delta x)^{\beta}\psi(t), \tag{9.5.22}$$

in which the right-hand side is not a function of A, hence, is an external excitation by nature.

Similar to the crack length $A(t)$, process $Z(t)$ can also be approximated as a Markov diffusion process. Following the same procedure as that in deriving Eqs. (9.5.9) and (9.5.10), the drift and diffusion coefficients of process $Z(t)$ can be obtained as

$$m(Z) = \eta E[(\Delta X)^{\beta}]E[\psi], \tag{9.5.23}$$

$$\sigma^2(Z) = \eta^2 \int_{-\infty}^{\infty} \mathrm{cov}\left[(\Delta X_{t+\tau})^{\beta}, (\Delta X_t)^{\beta}\right] \mathrm{cov}\left[\psi(t+\tau), \psi(t)\right] d\tau. \tag{9.5.24}$$

Assume that the stress process $X(t)$ and environmental process $\psi(t)$ are stationary, then both $m(Z)$ and $\sigma^2(Z)$ are constant, denoted by m_Z and σ_Z^2, respectively.

Now the fatigue failure can be investigated for the new process $Z(t)$ instead of the original crack length process $A(t)$. The reliability function $R(\tau, z_0)$ is governed by

$$-\frac{\partial R}{\partial \tau} + m_Z \frac{\partial R}{\partial z_0} + \frac{1}{2}\sigma_Z^2 \frac{\partial^2 R}{\partial z_0^2} = 0, \tag{9.5.25}$$

subjected to the following initial and boundary conditions

$$R(\tau, z_0)]_{\tau=0} = 1, \quad 0 \le z_0 < z_c, \tag{9.5.26}$$

$$[R(\tau, z_0)]_{z_0 = z_c} = 0, \tag{9.5.27}$$

$$\left[\frac{\partial}{\partial z_0} R(\tau, z_0)\right]_{z_0 = 0} = 0, \tag{9.5.28}$$

where, according to (9.5.21), $z_i = 0$, and

$$z_0 = \int\limits_{a_i}^{a_0} \frac{du}{Q(u)}, \quad z_c = \int\limits_{a_i}^{a_c} \frac{du}{Q(u)}. \tag{9.5.29}$$

Note that z_c depends on both a_i and a_c. Use the method of separation of variables to solve the partial differential equation (9.5.25), i.e., let

$$R(\tau, z_0) = T(\tau)P(z_0). \tag{9.5.30}$$

We obtain two sets of ordinary differential equations

$$\frac{1}{2}\sigma_Z^2 \frac{d^2 P}{dz_0^2} + m_Z \frac{dP}{dz_0} + \lambda P = 0, \quad P(z_c) = 0, \quad P'(0) = 0, \tag{9.5.31}$$

$$\frac{dT}{d\tau} + \lambda T = 0. \tag{9.5.32}$$

Equations (9.5.31) and (9.5.32) define an eigenvalue problem. It is found that the eigenvalues and eigenfunctions are

$$\lambda_n = \frac{1}{2\sigma_Z^2}(m_Z^2 + \sigma_Z^4 \omega_n^2), \quad n = 1, 2, \ldots, \tag{9.5.33}$$

$$P_n = e^{-m_Z z_0/\sigma_Z^2}\left(\sin \omega_n z_0 + \frac{\sigma_Z^2}{m_Z}\cos \omega_n z_0\right), \tag{9.5.34}$$

where ω_n are determined from the nonlinear equation

$$\tan \omega_n z_c = -\frac{\sigma_Z^2}{m_Z}\omega_n. \tag{9.5.35}$$

It can be shown the following orthogonality of the eigenfunctions

$$\int\limits_0^{z_c} e^{m_Z z_0/\sigma_Z^2} P_n(z_0)P_m(z_0)dz_0 = \begin{cases} \neq 0, & m = n \\ = 0, & m \neq n. \end{cases} \tag{9.5.36}$$

The general solution of Eq. (9.5.25) is then the following superposition

$$R(\tau, z_0) = e^{-m_Z z_0/\sigma_Z^2} \sum_{n=1}^{\infty} d_n \left(\sin \omega_n z_0 + \frac{\sigma_Z^2}{m_Z} \omega_n \cos \omega_n z_0 \right) e^{-\lambda_n \tau}.$$

(9.5.37)

Applying the initial condition (9.5.26) and the orthogonality condition (9.5.36), constants d_n in (9.5.37) are determined as

$$d_n = \frac{2 m_Z \sigma_Z^2 e^{m_Z z_c/\sigma_Z^2} \sin \omega_n z_c}{z_c(m_Z^2 + \sigma_Z^4 \omega_n^2) + m_Z \sigma_Z^2}.$$

(9.5.38)

The probability density of the first-passage time is then

$$p(\tau, z_0) = -\frac{d}{d\tau} R(\tau, z_0)$$

$$= e^{-m_Z z_0/\sigma_Z^2} \sum_{n=1}^{\infty} d_n \lambda_n \left(\sin \omega_n z_0 + \frac{\sigma_Z^2}{m_Z} \omega_n \cos \omega_n z_0 \right) e^{-\lambda_n \tau}.$$

(9.5.39)

The statistical moments of the first passage time, $\mu_n = E[T^n]$, can be calculated by using (9.5.39). But they can be obtained directly from the generalized Pontryagin equation

$$\frac{1}{2}\sigma_Z^2 \frac{d^2}{dz_0^2}\mu_{n+1} + m_Z \frac{d}{dz_0}\mu_{n+1} = -(n+1)\mu_n,$$

(9.5.40)

with the associated boundary conditions

$$\mu_{n+1}(z_c) = 0,$$

(9.5.41)

$$\mu'_{n+1}(0) = 0.$$

(9.5.42)

Treating μ_n as a known function, using the two boundary conditions, the solution of (9.5.42) is obtained from (9.2.18) as

$$\mu_{n+1}(z_0) = \frac{2(n+1)}{\sigma_Z^2} \int_{z_0}^{z_c} \exp\left(-\frac{2m_z}{\sigma_Z^2} u \right) du \int_0^u \mu_n(v) \exp\left(\frac{2m_z}{\sigma_Z^2} v \right) dv.$$

(9.5.43)

In particular, the mean of the first-passage time ($n = 0$) is obtained as

$$\mu_1(z_0) = \frac{1}{m_Z}(z_c - z_0) + \frac{\sigma_Z^2}{2m_Z^2}\left(e^{-2m_z z_c/\sigma_Z^2} - e^{-2m_z z_0/\sigma_Z^2}\right).$$

$$(9.5.44)$$

The above analysis is based on the randomized Paris–Erdogan model (9.5.19). Still, the number of cycles per unit time η, the mean values and covariance functions of the stress range process $\Delta X(t)$ and environmental process $\psi(t)$ are required to calculate m_Z and σ_Z^2, as shown in Eqs. (9.5.23) and (9.5.24).

9.5.3 *The Case of Stationary Gaussian Stress Processes*

Now we consider a special type of stress process, stationary Gaussian process with zero mean. If it is a narrow-band process, almost all maxima are positive, all minima are negative, a zero crossing occurs between two consecutive extremes, and the number of cycles per unit time η in Eqs. (9.5.23) and (9.5.24) may be approximated as the average number of the up zero crossing. It is known that the average number of up zero crossing of a narrow-band stress process $X(t)$ is given by (Rice, 1944, 1945)

$$\eta = \frac{1}{2\pi}\left[\frac{\int_0^\infty \omega^2 \Phi_{XX}(\omega)d\omega}{\int_0^\infty \Phi_{XX}(\omega)d\omega}\right]^{\frac{1}{2}}, \qquad (9.5.45)$$

where $\Phi_{XX}(\omega)$ is the spectral density of $X(t)$. On the other hand, if the process is not of a narrow-band, it is reasonable to replace the parameter η by the average number of peaks per unit time, given by

$$\eta = \frac{1}{2\pi}\left[\frac{\int_0^\infty \omega^4 \Phi_{XX}(\omega)d\omega}{\int_0^\infty \omega^2 \Phi_{XX}(\omega)d\omega}\right]^{\frac{1}{2}}. \qquad (9.5.46)$$

It is noted that the use of (9.5.45) neglects the crack growth caused by those peak-to-next-trough or trough-to-next-peak excursions without an intermediate zero-crossing; thus, it is less conservative from an engineering design point of view.

The stress range process $\Delta X(t)$ may be approximated by twice of the envelope process. The replacement is more conservative since the stress range is always smaller than twice of the envelope. According to Section 3.9.2, the stationary stress process $X(t)$ can be represented by

$$X(t) = \int_{-\infty}^{\infty} e^{i\omega t} dZ(\omega), \qquad (9.5.47)$$

where $Z(\omega)$ is an orthogonal-increment process with the property

$$E[dZ(\omega)dZ*(\omega')] = \begin{cases} \Phi_{XX}(\omega)d\omega, & \omega = \omega' \\ 0, & \omega \neq \omega'. \end{cases} \qquad (9.5.48)$$

Define another stationary process $\hat{X}(t)$

$$\hat{X}(t) = \int_{-\infty}^{\infty} h(\omega)e^{i\omega t} dZ(\omega), \qquad (9.5.49)$$

where

$$h(\omega) = \begin{cases} i, & \omega > 0 \\ 0, & \omega = 0 \\ -i, & \omega < 0. \end{cases} \qquad (9.5.50)$$

It can be shown that

$$R_{XX}(\tau) = R_{\hat{X}\hat{X}}(\tau) = \int_{-\infty}^{\infty} e^{i\omega\tau} \Phi_{XX}(\omega)d\omega$$

$$= 2\int_{0}^{\infty} \Phi_{XX}(\omega)\cos(\omega\tau)d\omega, \qquad (9.5.51)$$

$$R_{X\hat{X}}(\tau) = -R_{\hat{X}X}(\tau) = \int_{-\infty}^{\infty} h*(\omega)e^{i\omega\tau} \Phi_{XX}(\omega)d\omega$$

$$= 2\int_{0}^{\infty} \Phi_{XX}(\omega)\sin(\omega\tau)d\omega. \qquad (9.5.52)$$

Carrying out the transformation

$$X(t) = A_e(t) \cos \Theta(t), \quad \hat{X}(t) = A_e(t) \sin \Theta(t), \qquad (9.5.53)$$

where $A_e(t)$ is the envelope process. The first-order and second-order probability density functions of $A(t)$ are given by (Rice, 1944, 1945)

$$p(a_e) = \frac{a_e}{\sigma_X^2} \exp\left(-\frac{a_e^2}{2\sigma_X^2}\right), \qquad (9.5.54)$$

$$p(a_{e1}, a_{e2}, \tau) = \frac{a_{e1} a_{e2}}{\sigma_X^2 (1 - \rho^2)} I_0\left[\frac{a_{e1} a_{e2} \rho}{\sigma_X^2 (1 - \rho^2)}\right] \exp\left[-\frac{a_{e1}^2 + a_{e2}^2}{2\sigma_X^2 (1 - \rho^2)}\right], \qquad (9.5.55)$$

where $I_0[\cdot]$ is the modified Bessel function of the zeroth order, σ_X^2 is the variance of $X(t)$, and $\rho = \rho(\tau)$ is the correlation coefficient of $A_e(t)$. σ_X^2 and $\rho(\tau)$ can be calculated from

$$\sigma_X^2 = \int_{-\infty}^{\infty} \Phi_{XX}(\omega) d\omega, \qquad (9.5.56)$$

$$\rho(\tau) = \frac{[R_{XX}^2(\tau) + R_{X\hat{X}}^2(\tau)]^{\frac{1}{2}}}{\sigma_X^2}. \qquad (9.5.57)$$

Given the spectral density $\Phi_{XX}(\omega)$ of the stress process $X(t)$, σ_X^2 and $\rho(\tau)$ can be calculated from Eqs. (9.5.51), (9.5.52), (9.5.56) and (9.5.57).

As mentioned above, the stress range process $\Delta X(t)$ can be approximated as twice of the envelope process, i.e., $\Delta X(t) = 2A_e(t)$, then for the stress range process we have

$$p(\Delta x) = \frac{\Delta x}{2\sigma_X^2} \exp\left(-\frac{\Delta x^2}{8\sigma_X^2}\right), \qquad (9.5.58)$$

$$p(\Delta x_1, \Delta x_2, \tau) = \frac{\Delta x_1 \Delta x_2}{16\sigma_X^2 (1 - \rho^2)} I_0\left[\frac{\Delta x_1 \Delta x_2 \rho}{4\sigma_X^2 (1 - \rho^2)}\right]$$

$$\times \exp\left[-\frac{\Delta x_1^2 + \Delta x_2^2}{8\sigma_X^2 (1 - \rho^2)}\right]. \qquad (9.5.59)$$

Using Eqs. (9.5.58) and (9.5.59), we obtain

$$E[(\Delta X)^\beta]$$
$$= \int_0^\infty (\Delta x)^\beta p(\Delta x) d(\Delta x) = (2\sqrt{2}\sigma_X)^\beta \Gamma\left(1 + \frac{\beta}{2}\right),$$

(9.5.60)

$$\text{cov}[(\Delta X_t)^\beta, (\Delta X_{t+\tau})^\beta]$$
$$= \int_0^\infty \int_0^\infty (\Delta x_1)^\beta (\Delta x_2)^\beta p(\Delta x_1, \Delta x_2, \tau) d(\Delta x_1)(\Delta x_1)$$

$$= (8\sigma_X^2)^\beta \sum_{n=1}^\infty c_n^2 \rho^{2n}(\tau),$$

(9.5.61)

where

$$c_n = \sum_{j=0}^n \frac{(-1)^j n!}{(n-j)!(j!)^2} \Gamma\left(1 + \frac{j}{2} + \frac{\beta}{2}\right).$$

(9.5.62)

With η calculated from (9.5.45) or (9.5.46), $E[(\Delta X)^\beta]$ from (9.5.59), and $\text{cov}[(\Delta X_t)^\beta, (\Delta X_{t+\tau})^\beta]$ from (9.5.60), the drift coefficient $m(Z)$ and the diffusion coefficient $\sigma(Z)$ can be obtained from (9.5.23) and (9.5.24) given the knowledge of the environmental process $\psi(t)$. Thus, the fatigue failure problem can be solved.

Example 9.5.1 Consider a thin square plate of size $l \times l$ with an initial crack length a_0, as shown in Fig. 9.5.2. The plate is supporting a rigid heavy mass m at its end, which is subjected to a Gaussian broad-band excitation $\xi(t)$. Denote the critical crack length as a_c, and assume that a_c is much smaller than l so that the plate stiffness degradation is negligible prior to failure. The randomness due to the environment, represented by function $\psi(t)$ in the above analysis, is also ignored.

Let $Y(t)$ be the displacement of the plate, governed by

$$m\ddot{Y} + c\dot{Y} + kY = \xi(t),$$

(9.5.63)

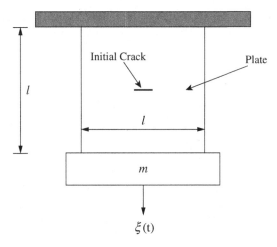

Figure 9.5.2 A plate with initial crack subjected to random load $\xi(t)$.

where the damping force and restoring force are assumed to be linear. If stress process $X(t)$ is a linear function of the displacement, then it is also governed by equation (9.5.63), provided that all terms are suitable rescaled, i.e.,

$$\ddot{X} + 2\zeta\omega_0\dot{X} + \omega_0^2 X = \xi_1(t). \tag{9.5.64}$$

The spectral density of $X(t)$ is

$$\Phi_{XX}(\omega) = \frac{\Phi_{\xi_1\xi_1}(\omega)}{(\omega^2 - \omega_0^2) + 4\zeta^2\omega^2\omega_0^2}. \tag{9.5.65}$$

Since the damping ratio ζ is usually small, $X(t)$ is a narrow-band stationary Gaussian process.

We use the randomized Paris–Erdogan model to treat the problem. Assume functions $h(A)$ and $Q(A)$ in Eqs. (9.5.19) and (9.5.20) to be

$$h(A) = \gamma\sqrt{A}, \quad Q(A) = \alpha[h(A)]^\beta = \alpha\gamma^\beta A^{\beta/2}. \tag{9.5.66}$$

Given (i) the material parameters α, β and γ, (ii) the starting, initial, and critical crack lengths, a_i, a_0 and a_c, and (iii) the spectral density

$\Phi_{XX}(\omega)$ of the stress process $X(t)$, the fatigue problem can be solved. The solution procedure is listed below.

(1) Use Eq. (9.5.29) to determine z_0 and z_c from a_i, a_0 and a_c.
(2) Use Eq. (9.5.45) to calculate η from $\Phi_{XX}(\omega)$.
(3) Use Eqs. (9.5.51), (9.5.52), (9.5.56), (9.5.57) to calculate σ_X^2 and $\rho(\tau)$.
(4) Use Eqs. (9.5.60) and (9.5.61) to find $E[(\Delta X)^\beta]$ and $\text{cov}[(\Delta X_t)^\beta, (\Delta X_{t+\tau})^\beta]$.
(5) Use Eqs. (9.5.23) and (9.5.24) to determine m_Z and σ_Z^2.
(6) Use Eqs. (9.5.33), (9.5.35) and (9.5.38) to calculate λ_n, ω_n and d_n.
(7) Use Eqs. (9.5.37) and (9.5.39) to determine $R(\tau, z_0)$ and $p(\tau, z_0)$.
(8) Use Eqs. (9.5.43) and (9.5.44) to calculate $\mu_{n+1}(z_0)$, $n = 0, 1, \ldots$.

Numerical examples were given in (Cai, Yu, and Lin, 1994).

Example 9.5.2 Another example is a linear SDOF structure under a random excitation whose spectral peak is located at a frequency not close to the natural frequency of the structure. As an example, let the excitation be a stationary Gaussian process with the so-called Jonswap spectrum (Hasselmann *et al.*, 1976)

$$\Phi(\omega) = \frac{a_1 g^2}{\omega^5} a_3^{\beta(\omega)} \exp\left(-\frac{a_2 \omega_m^4}{\omega^4}\right), \qquad (9.5.67)$$

where

$$\beta(\omega) = \exp\left[-\frac{(\omega - \omega_m)^2}{2\sigma^2 \omega_m^2}\right], \qquad \sigma = \begin{cases} \sigma_a, & |\omega| \le \omega_m \\ \sigma_b, & |\omega| > \omega_m \end{cases} \qquad (9.5.68)$$

and a_1, a_2, a_3, g, ω_m, σ_a and σ_b are constant. The Jonswap spectrum has been used often in analysis of ocean and naval structures. Figure 9.5.3 depicts the spectrum schematically.

The spectral density $X(t)$ of the stress process in a SDOF structure is then

$$\Phi_{XX}(\omega) = \frac{B}{(\omega^2 - \omega_0^2) + 4\zeta^2 \omega^2 \omega_0^2} \Phi(\omega), \qquad (9.5.69)$$

where B is a suitable positive constant. Figure 9.5.4 show a spectral density of $X(t)$ responding to the excitation of an ocean wave

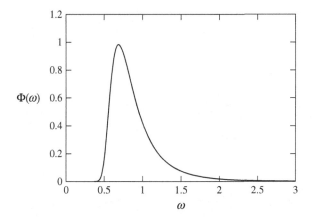

Figure 9.5.3 A Jonswap spectral density.

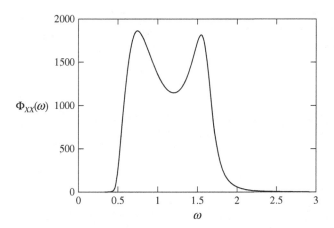

Figure 9.5.4 Spectral density of structure response.

field with a Jonswap spectrum. Two peaks appear in $\Phi_{XX}(\omega)$: one corresponds to the peak in the Jonswap spectrum, and another is located near the natural frequency of the structure.

Figure 9.5.5 is a simulated sample function of $X(t)$. It is seen that there are indeed some neighboring peak-trough pairs which are not associated with an intermediate zero-crossing. In this case, Eq. (9.5.46) should be used to calculate parameter η.

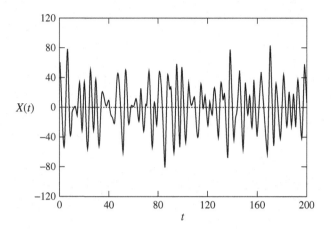

Figure 9.5.5   A sample function of structure response.

The same procedure as for Example 9.5.1 can be applied to solve the fatigue problem. Numerical examples were given in (Cai, Yu, and Lin, 1995).

Exercise Problems

9.1 Consider the linear system

$$\ddot{X} + 2\zeta\omega_0\dot{X} + \omega_0^2 X = W(t),$$

where $W(t)$ is a Gaussian white noise with a spectral density K. Using the energy process defined as

$$\Lambda = \frac{1}{2}\dot{X}^2 + \frac{1}{2}\omega_0^2 X^2,$$

to obtain the reliability function defined as

$$R(\tau, \lambda_0) = \text{Prob}[T > \tau | \Lambda(0) = \lambda_0],$$

where T is the first-passage time of energy process $\Lambda(t)$ to reach a critical value λ_c.

9.2 Do the same as that in Problem 9.1 for the linear system subjected to parametric excitations

$$\ddot{X} + \omega_0[2\zeta + W_2(t)]\dot{X} + \omega_0^2[1 + W_1(t)]X = W_3(t),$$

where $W_l(t)$ are Gaussian white noises with correlation functions

$$E[W_l(t)W_s(t+\tau)] = 2\pi K_{js}\delta(\tau), \quad l, s = 1, 2, 3.$$

9.3 A population genetic model is described by the Ito equation

$$dX = (1 + X)dt + X dB(t), \quad X \geq 0.$$

(a) Determine whether the first-passage problem is meaningful.

(b) If it is meaningful, find the statistical moments of the first-passage time.

9.4 For the system in Problem 9.1, use the energy process to find the statistical moments of the first-passage time.

9.5 Consider the system

$$\ddot{X} + \omega_0[2\zeta + W_2(t)]\dot{X} + \omega_0^2[1 + W_1(t)]X = 0,$$

where $W_l(t)$ are Gaussian white noises with correlation functions

$$E[W_l(t)W_s(t+\tau)] = 2\pi K_{js}\delta(\tau), \quad l, s = 1, 2, 3.$$

(a) Determine whether the first-passage problem is meaningful.

(b) If it is meaningful, find the statistical moments of the first-passage time.

9.6 For the system in Problem 9.2, use the energy process to find the statistical moments of the first-passage time.

9.7 Consider a nonlinear system of van der Pol type,

$$\ddot{X} + 2\zeta\omega_0(1 + \beta X^2)\dot{X} + \omega_0^2 X = \xi(t),$$

where $\xi(t)$ is an broadband excitation with a spectral density $\Phi(\omega)$. Use the stochastic averaging method to find the statistical moments of the first-passage time in terms of the energy process.

9.8 Consider a nonlinearly damped system

$$\ddot{X} + 2\zeta\omega_0\dot{X} + \beta\dot{X}^3 + \omega_0^2[1 + W_1(t)]X = W_2(t),$$

where $W_1(t)$ and $W_2(t)$ are Gaussian white noises with correlation functions

$$E[W_l(t)W_s(t+\tau)] = 2\pi K_{js}\delta(\tau), \quad l, s = 1, \ 2.$$

(a) Derive the Ito equation for the amplitude process using the stochastic averaging method.

(b) For the case of $W_1(t) = 0$ and $W_2(t) \neq 0$, find the statistical moments of the first-passage time.

(c) For the case of $W_2(t) = 0$ and $W_1(t) \neq 0$, determine the condition for the first-passage problem to be meaningful. When such a condition is satisfied, find the statistical moments of the first-passage time.

9.9 Give a nonlinear oscillator

$$\ddot{X} + \eta|\dot{X}|^\delta \mathrm{sgn}\dot{X} + \omega_0^2 X = \dot{X}W(t), \quad \eta > 0, \delta \geq 0.$$

(a) Derive the Ito equation for the amplitude process using the stochastic averaging method.

(b) Determine the condition for the first-passage problem to be meaningful.

(c) When the preceding condition is satisfied, find the statistical moments of the first-passage time.

9.10 Consider a system with a nonlinear restoring force

$$\ddot{X} + 2\eta\dot{X} + k\,|X|^\delta\,\mathrm{sgn}X = \dot{X}W(t), \quad k > 0, \delta \geq 0.$$

(a) Derive the Ito equation for the energy process using the stochastic averaging method.

(b) Determine the condition for the first-passage problem to be meaningful.

(c) When the preceding condition is satisfied, find the statistical moments of the first-passage time.

REFERENCES

Andrnov, A., Pontryagin, L., and Witt, A., 1933. "On the statistical investigation of dynamical systems", *Zh. Eksp. Teor. Fiz.* Vol. 3, 165–180 (in Russian).

Ariaratnam, S. T., 1980. "Bifurcation in nonlinear stochastic systems", in *New Approaches to Nonlinear Problems in Dynamics*, SIAM Publications, Philadelphia, 470–473.

Ariaratnam, S. T., 1993. "Stochastic bifurcation: An illustrate example", in *Proceedings of the 2nd International Conference of Nonlinear Mechanics*, Peking University Press, Beijing.

Ariaratnam, S. T. and Pi, H. N., 1973. "On the first-passage time for envelope crossing for a linear oscillator", *International Journal of Control*, Vol. 18, 89–96.

Ariaratnam, S. T. and Srikantaiah, T. K., 1978. "Parametric instability in elastic structures under stochastic loading", *Journal of Structural Mechanics*, Vol. 6, 349–365.

Ariaratnam, S. T. and Tam, D. S. F., 1979. "Random vibration and stability of a linear parametrically excited oscillator", *Journal of Applied Mathematics and Mechanics*, Vol. 59, 79–84.

Arnold, L., 1974. *Stochastic Differential Equations Theory and Application*, Wiley, New York.

Arnold, L., 1998. *Random Dynamical Systems*, Springer, Berlin.

Arnold, V. I., 1989. *Mathematical Methods of Classical Mechanics*, Spring-Verlag, New York.

Bochner, S., 1959. *Lectures on Fourier Integrals*, Princeton, New Jersey.

Bogliubov, N. N. and Mitropolski, Y. A., 1961. *Asymptotic Methods in the Theory of Nonlinear Oscillations*, Gorden and Breach, New York.

Bolotin, V. V., 1969. *Statistical Methods in Structural Mechanics*, Holden-Day, San Francisco.

Bruckner, A. and Lin, Y. K., 1987. "Generalization of equivalent linearization method for nonlinear random vibration problems", *International Journal of Non-Linear Mechanics*, Vol. 22, 227–235.

Cai, G. Q., 2003. "Response spectral densities of systems under both additive and multiplicative excitations", in *Proceedings of the IUTAM Symposium on Nonlinear Stochastic Dynamics*, Kluwer Academic Publishers, Norwell, MA, 299–306.

Cai, G. Q., 2004. "Nonlinear systems of multiple degrees of freedom under both additive and multiplicative random excitations", *Journal of Sound and Vibration*, Vol. 278, 889–901.

Cai, G. Q. and Lin, Y. K., 1988. "A new approximation solution technique for randomly excited nonlinear oscillators", *International Journal of Non-Linear Mechanics*, Vol. 23, 409–420.

Cai, G. Q. and Lin, Y. K., 1994. "On statistics of first-passage failure", *Journal of Applied Mechanics*, Vol. 61, 93–99.

Cai, G. Q. and Lin, Y. K., 1996. "Exact and approximate solutions for randomly excited MDOF non-linear systems", *International Journal of Non-Linear Mechanics*, Vol. 31, 647–655.

Cai, G. Q. and Lin, Y. K., 1997. "Response spectral densities of strongly nonlinear systems under random excitation", *Probabilistic Engineering Mechanics*, Vol. 12, 41–47.

Cai, G. Q. and Lin, Y. K., 2004. "Stochastic analysis of the Lotka-Volterra model for ecosystems", *Physical Review E*, Vol. 70, 041910.

Cai, G. Q. and Lin, Y. K., 2007. "Stochastic analysis of time-delayed ecosystems", *Physical Review E* , Vol. 76, 041913.

Cai, G. Q. and Suzuki, Y., 2005. "On statistical quasi-linearization", *International Journal of Non-Linear Mechanics*, Vol. 40, 1139–1147.

Cai, G. Q., Yu, J. S., and Lin, Y. K., 1994. "Fatigue crack growth of randomly excited structures", in Computational Stochastic Mechanics, *Proceedings of the 2nd International Conference on Computational Stochastic Mechanics*, Balkema, 673–679.

Cai, G. Q., Yu, J. S., and Lin, Y. K., 1995. "Fatigue life and reliability of randomly excited structures", *Journal of Ship Research*, Vol. 39, No. 1, 62–69.

Caughey, T. K. and Dienes, J. K., 1961. "Analysis of a nonlinear first-order system with a white noise input", *Journal of Applied Physics*, Vol. 23, 2476–2479.

Chetayev, N. G., 1961. *The Stability of Motion*, Pergamon Press, New York.

Crandall, S. H., 1958. *Random Vibration*, The MIT Press, MA.

Crandall, S. H., 1978. "Heuristic and equivalent linearization techniques for random vibration of non-linear oscillators", in *Proceedings of the 8th international Conference on Non-Linear Oscillation*, Vol. 1, Academia, Prague, 211–226.

Crandall, S. H. and Mark, W. D., 1963. *Random Vibration in Mechanical Systems*, Academic Press, New York.

Crandall, S. H. and Zhu, W. Q., 1983. "Random vibration: A survey of recent developments", *Journal of Applied Mechanics*, Vol. 50, No. 4b, 953–962.

Dalzell, J. F., 1973. "A note on the distribution of maxima of ship rolling", *Journal of Ship Research*, Vol. 17, 217–226.

Deodatis, G. and Micaletti, R. C., 2001. "Simulation of highly skewed non-Gaussian stochastic processes", *Journal of Engineering Mechanics*, Vol. 127, No. 12, 1284–1295.

Dimentberg, M. F., 1988. *Statistical Dynamics of Nonlinear and Time-Varying Systems*, Wiley, New York.

Dimentberg, M. F., 2002. "Lotka-Volterra system in a random environment", *Physical Review E*, Vol. 65, 036204.

Dimentberg, M. F., 2003. "Stochastic Lotka-Volterra system", in *Proceedings of IUTAM Symposium on Nonlinear Stochastic Dynamics*, Kluwer Academic Publishers, Dordrecht, The Netherlands, 307–317.

Einstein, A., 1956. "Investigation on the theory of Brownian movement", in *English Translation of Einstein Papers*, Dover Publications, New York.

Elishakoff, I., 1983. *Probabilistic Methods in Theory of Structures*, John Wiley and Sons, New York.

Elishakoff, I. and Cai, G. Q., 1992. "Approximate solution for nonlinear random vibration problems by partial stochastic linearization", in *Nonlinear Vibrations*, Vol. 144, ASME Winter Annual Meeting AMD, ASME, New York, 117–121.

Feller, W., 1952. "The parabolic differential equation and the associated semigroups of transformations", *Annals of Mathematics*, Vol. 55, No. 3, 468–519.

Feller, W., 1954. "Diffusion process in one dimension", *Transactions of American Mathematical Society*, Vol. 77, 1–31.

Finlayson, B. A., 1972. *The Method of Weighted Residuals and Variational Principles*, Academic Press, New York.

Gardiner, C. W., 1986, *Handbook of Stochastic Methods: For Physics, Chemistry, and the Natural Sciences*, Springer-Verlag, Berlin.

Gan C. B. and Zhu, W. Q., 2001. "First-passage failure of quasi-non-integrable-Hamiltonian systems", *International Journal of Non-Linear Mechanics*, Vol. 36, No. 2, 209–220.

Gradshteyn, I. S. and Ryzhik, I. M., 1980. *Table of Integrals, Series, and Products*, Academic Press, New York.

Graham, R. and Haken, H., 1971. "Generalized Thermo-dynamic potential for Markoff systems in detailed balance and far from thermal equilibrium", *Zeitschrift fur Physik*, Vol. 243, 289–302.

Grigoriu, M., 1998. "Simulation of stationary non-Gaussian translation processes", *Journal of Engineering Mechanics* , Vol. 124, No. 2, 121–126.

Hasselmann, K., Ross, D. B., Muller, P., and Sell, W., 1976. "A parametric wave prediction model", *Journal of Physical Oceanography*, Vol. 6, 200–228.

Hoeppner, D. W. and Krupp, W. E., 1974. "Prediction of component life by application of fatigue crack growth knowledge", *Engineering Fracture Mechanics*, Vol. 6, No. 1, 47–70.

Hou, Z., Zhou, M. F., Dimentberg, M., and Noori, M., 1996. "A stationary model for periodic excitation with uncorrelated random disturbance", *Probabilistic Engineering Mechanics*, Vol. 11, 191–203.

Huang, Z. L. and Zhu W. Q., 2000. "Exact stationary solutions of stochastically and harmonically excited and dissipated Hamiltonian systems", *Journal of Sound and Vibration*, Vol. 230, No. 3, 709–720.

Ibrahim, R. A., 1985. *Parametric Random Vibration*, John Wiley and Sons, New York.

Ibrahim, R. A., Soundararajan, A., and Heo, H., 1985. "Stochastic response of nonlinear dynamic systems based on non-Gaussian closure", *Journal of Applied Mechanics*, Vol. 52, No. 4, 965–970.

Ito, K., 1951a. "On stochastic differential equations", *Memoirs of the American Mathematical Society*, vol. 4, 289–302.

Ito, K., 1951b. "On a formula concerning stochastic differentials", *Nagoya Mathematical Journal*, Vol. 3, 55–65.

Ito, K. and McKean, H. P. Jr., 1965. *Diffusion Processes and Their Sample Paths*, Academic Press, New York.

Karlin, S. and Taylor, H. M., 1975. *A First Course in Stochastic Processes*, Academic Press, New York.

Karlin, S. and Taylor, H. M., 1981. *A Second Course in Stochastic Processes*, Academic Press, New York.

Khasminskii, R. Z., 1964. "On the behavior of a conservative system with small friction and small random noise", *Prikladnaya Mathematika (Applied Mathematics and Mechanics)*, Vol. 28, No. 5, 1126–1130 (in Russian).

Khasminskii, R. Z., 1966. "A limit theorem for the solution of differential equations with random right hand sides", *Theory of Probability and Application*, Vol. 11, No. 3, 390–405.

Khasminskii, R. Z., 1967. "Sufficient and necessary conditions of almost sure asymptotic stability of a linear stochastic system", *Theory of Probability and Application*, Vol. 12, No. 1, 144–147.

Khasminskii, R. Z., 1968. "On the averaging principle for Ito stochastic differential equations", *Kibernetika*, Vol. 3, No. 4, 260–279 (in Russian).

Khasminskii, R. Z., 1980. *Stochastic Stability of Differential Equations*, Kluwer Academic Publications, Norwell, MA.

Khasminskii, R. Z. and Klebaner, F. C., 2001. "Long term behavior of solutions of the Lotka-Volterra system under small random perturbations", *The Annals of Applied Probability*, Vol. 11, 952–963.

Kozin, F., 1969. "A survey of stability of stochastic systems", *Automatica*, Vol. 5, No. 1, 95–112.

Kozin, F. and Zhang, Z. Y., 1990. "On almost sure sample stability of nonlinear Ito differential equations", in *Stochastic Structural Dynamics 1 — New Theoretical Developments*, Spring-Verlag, Berlin, Heidelberg, 147–154.

Landa, P. S. and Stratonovoch, R. L., 1962. "Theory of stochastic transition in various systems between different states", *Vestnik MGU (Proceedings of Moscow University)*, Series III(1), 33–45 (in Russian).

Lennox, W. C. and Fraser, D. A., 1974. "On the first-passage distribution for the envelope of a non-stationary narrow-band stochastic process", *Journal of Applied Mechanics*, Vol. 41, No. 3, 793–797.

Lévy, P., 1948. *Processus Stochastiqueset Mouvement Brownien*, Gauthier-Villas, Paris.

Lin, Y. K., 1967. *Probabilistic Theory of Structural Dynamics*, McGraw-Hill, New York. Reprint R. E. Krieger, Melbourne, FL, 1976.

Lin, Y. K. and Cai, G. Q., 1988. "Exact stationary-response solutions for second-order nonlinear systems under parametric and external white-noise excitations: Part II", *Journal of Applied Mechanics*, Vol. 55, 702–705.

Lin, Y. K. and Cai, G. Q., 1995. *Probabilistic Structural Dynamics*, McGraw-Hill, New York. Reprint 2004.

Lotka, A. J., 1925, *Elements of Physical Biology*. William and Wilkins, Baltimore, New Jersey.

Liu, Z. H. and Zhu, W. Q., 2008. "Stochastic Hopf bifurcation of quasi-integrable Hamiltonian systems with time-delayed feedback control", *Journal of Theoretical and Applied Mechanics*, Vol. 46, 531–550.

Lyapunov, A. M., 1892. "problemegenerale de la stabilite du movement", *Comm. Soc. Math. Kharkov*, Vol. 2, 265–272. Reprinted in Annals of Mathematical Studies, Vol. 17, Princeton University Press, Princeton, 1947.

May, R. M. and Verga, A. D., 1973. *Stability and Complexity in Model Ecosystems*, Princeton University Press, Princeton, New Jersey.

Meunier, C. and Verga, A. D., 1988. "Noise and bifurcations", *Journal of Statistical Physics*, Vol. 50, No. 1/2, 345–375.

Miller, M. S. and Gallagher, J. P., 1981. "An analysis of several fatigue growth rate (FCGR) descriptions", in *Fatigue Crack Growth Measurement and Data Analysis*, ASTM-STP 738, 205–251.

Nishioka, K., 1976. "On the stability of two-dimensional linear stochastic systems", *Kodai Mathematics Seminar Reports*, Vol. 27, 211–230.

Oseledec, V. I., 1968. "A multiplicative ergodic theorem: Lyapunov characteristic number for dynamical systems", *Transactions of the Moscow Mathematical Society*, Vol. 19, 197–231.

Pardoux, E. and Wihstutz, V., 1988. "Lyapunov exponent and rotation number of two-dimensional linear stochastic system with small diffusion", *SIAM Journal of Applied Mathematics*, Vol. 48, 442–457.

Paris, P. C. and Erdogan, F., 1963. "A critical analysis of crack propagation laws", *Journal of Basic Engineering*, Vol. 85, 528–534.

Priestly, M. B., 1965. "Evolutionary spectra and non-stationary processes", *Journal of Royal Statistical Society*, Series B, Vol. 27, 204–237.

Poincaré, H., 1885. "L'Équilibred'une masse fluideanimée d'un mouvement de rotation", *ActaMathematica*, Vol. 7, 259–380.

Qi, L. and Cai, G. Q., 2013. "Dynamics of nonlinear ecosystems under colored noise disturbances", *Nonlinear Dynamics*, Vol. 73 No. 1, 463–474.

Rayleigh, 1919. "On the problem of random vibration and of random flights in one, two, or three dimensions", *Philosphical Magazine, VI, Ser. 37*, 321–347.

Rice, S. O., 1944, 1945. "Mathematical analysis of random noise", *Bell System Technical Journal*, Vol. 23, 282–332; Vol. 24, 46–156. Reprinted in *Selected Papers on Noise and Stochastic Processes*, Dover, New York, 1954.

Roberts, J. B., 1985. "Estimation of nonlinear ship roll damping from free-decay data", *Journal of Ship Research*, Vol. 29, 127–138.

Roberts, J. B. and Spanos, P. D., 1999. *Random Vibration and Statistical Linearization*, Dover Publications, New York.

Soong, T. T., 1973. *Random Differential Equations in Science and Engineering*, Academic Press, New York.

Soong, T. T. and Grogoriu M., 1993. *Random Vibration of Mechanical and Structural Systems*, Prentice Hall, New Jersey.

Sri Namachchivaya, 1990. "Stochastic bifurcation", *Applied Mathematics and Computation*, Vol. 38, 101–159.

Stratonovich, R. L., 1963. *Topics in the Theory of Random Noise*, Vol. 1, Gordon and Breach, New York.

Stratonovich, R. L., 1967. *Topics in the Theory of Random Noise*, Vol. 2, Gordon and Breach, New York.

Sun, J.-Q., 2006. *Stochastic Dynamics and Control*, Elsevier, Netherlands.

Sun, J.-Q. and Hsu, C. S., 1987. "Cumulant-neglect closure for nonlinear systems under random excitations", *Journal of Applied Mechanics*, Vol. 54, No. 3, 649–655.

Tabor, M., 1989. *Chaos and Integrability in Nonlinear Dynamics*, John Wiley and Sons, New York.

van Kampen, N. G., 1957. "Derivation of the phenomenological equations from the master equation. II: Even and odd variables", *Physica*, Vol. 23, No. 1, 41–57.

Volterra, V., 1926. "Variazioni e fluttuazioni del numerod'individui in specie d'animaniconviventi", *Mem. Acad. Lincei*, Vol. 2, 31–113.

Volterra, V., 1931. Leconssur la theoriemathematique de la lutte pour la vie, Gauthiers-Vilars, Paris, 1931.

Wedig, W. V., 1989. "Analysis and simulation of nonlinear stochastic systems", in *Nonlinear Dynamics in Engineering Systems*, Spring-Verlag, Berlin, 337–344.

Whittaker, E. T. and Watson, G. N., 1952. *A Course of Modern Analysis*, Cambridge University Press, Cambridge.

Winterstein, S. R., 1988. "Nonlinear vibration models for extremes and fatigue", *Journal of Engineering Mechanics*, Vol. 114, No. 10, 1772–1790.

Wong, E. and Zakai, M., 1965. "On the relation between ordinary and stochastic equations", *International Journal of Engineering Sciences*, Vol. 47, No. 1, 150–154.

Wu, C. and Cai, G. Q., 2004, "Effects of excitation probability distribution on system response", *International Journal of Nonlinear Mechanics*, Vol. 39, No. 9, 1463–1472.

Wu, W. F. and Lin, Y. K., 1984. "Cumulant-neglect closure for nonlinear oscillators under random parametric and external excitations", *International Journal of Non-Linear Mechanics*, Vol. 19, 349–362.

Wu, Y. and Zhu, W. Q., 2008. "Stochastic analysis of a pulse-type prey-predator model", *Physical Review E*, Vol. 77, 041911.

Xie, W. C., 1990. "Lyapunov exponents and their applications in structural dynamics", Ph. D. Thesis, University of Waterloo, Waterloo, Ontario, Canada.

Ying, Z. G. and Zhu W. Q., 2000. "Exact stationary solutions of stochastically excited and dissipated gyroscopic systems", *International Journal of Non-Linear Mechanics*, Vol. 35, 837–848.

Yong, Y. and Lin, Y. K., 1987. "Exact stationary-response solution for second-order nonlinear systems under parametric and external white-noise excitations", *Journal of Applied Mechanics*, Vol. 54, No. 2, 414–418.

Zhu, W. Q., 1992. *Random Vibration*, Scientific Publisher, Beijing (in Chinese).

Zhu, W. Q., 2003. *Nonlinear Stochastic Dynamics and Control — In Hamiltonian Formation*, Scientific Publisher, Beijing (in Chinese).

Zhu, W. Q., 2004. "Lyapunov exponent and stochastic stability of quasi non-integrable- Hamiltonian systems", *International Journal of Non-Linear Mechanics*, Vol. 39, No. 4, 569–579.

Zhu, W. Q., 2006. "Nonlinear stochastic dynamics and control in Hamiltonian formulation", *Applied Mechanics Reviews*, Vol. 59, No. 4, 230–248.

Zhu, W. Q. and Cai, G. Q., 2013. "On bounded stochastic processes", in *Bounded Noises in Physics, Biology and Engineering, Series: Modeling and Simulation in Science, Engineering and Technology*, Springer Science+Business Media, New York, 3–24.

Zhu, W. Q., Cai, G. Q., and Lin, Y. K., 1990. "On exact stationary solution of stochastically perturbed Hamiltonian systems", *Probabilistic Engineering Mechanics*, Vol. 5, No. 2, 84–87.

Zhu, W. Q. and Huang, Z. L., 1998. "Stochastic stability of quasi-non-integrable- Hamiltonian systems", *Journal of Sound and Vibration*, Vol. 218, No. 5, 769–789.

Zhu, W. Q. and Huang, Z. L., 1999. "Stochastic Hopf bifurcation of quasi-nonintegrable- Hamiltonian systems", *International Journal of Non-Linear Mechanics*, Vol. 34, No. 3, 437–447.

Zhu, W. Q. and Huang, Z. L., 2001. "Exact stationary solutions of stochastically excited and dissipated partially integrable Hamiltonian systems", *International Journal of Non-Linear Mechanics*, Vol. 36, 39–48.

Zhu, W. Q., Huang, Z. L., and Suzuki, Y., 2002. "Stochastic averaging and Lyapunov exponent of quasi-partially-integrable-Hamiltonian system", *International Journal of Nonlinear Mechanics*, Vol. 37, No. 3, 419–437.

Zhu, W. Q., Huang, Z. L., and Yang, Y. Q., 1997. "Stochastic averaging of quasi-integrable -Hamiltonian systems", *Journal of Applied Mechanics*, Vol. 64, No. 4, 975–984.

Zhu, W. Q. and Lei, Y., 1989. "First-passage time for state transition of randomly excited systems", in *Proceedings of the 47th International Statistical Institute Meeting*, Paris, 517–531.

Zhu, W. Q. and Lei, Y., 1991. "A stochastic theory of cumulative fatigue damage", *Probabilistic Engineering Mechanics*, Vol. 6, 222–227.

Zhu, W. Q., Lin, Y. K., and Lei, Y., 1992. "On fatigue crack growth under random loading", *Engineering Fracture Mechanics*, Vol. 43, No. 1, 1–12.

Zhu, W. Q., Soong, T. T, and Lei, Y., 1994. "Equivalent nonlinear system method for stochastically excited Hamiltonian systems", *Journal of Applied Mechanics*, Vol. 61, 618–623.

Zhu, W. Q. and Yang, Y. Q., 1996. "Exact stationary solutions of stochastically excited and dissipated integrable Hamiltonian systems", *Journal of Applied Mechanics*, Vol. 63, 493–500.

Zhu, W. Q. and Yang, Y. Q., 1997. "Stochastic averaging of quasi-nonintegrable-Hamiltonian systems", *Journal of Applied Mechanics*, Vol. 64, 157–164.

INDEX

Printed in the United States
By Bookmasters